CW00933184

Tensor Analysis

Fridtjov Irgens

Tensor Analysis

 Springer

Fridtjov Irgens
Norwegian University of Science
   and Technology
Bergen, Norway

ISBN 978-3-030-03411-5   ISBN 978-3-030-03412-2   (eBook)
https://doi.org/10.1007/978-3-030-03412-2

Library of Congress Control Number: 2018960417

This Springer imprint is published by the registered company Springer Nature Switzerland AG
The registered company address is: Gewerbestrasse 11, 6330 Cham, Switzerland

Preface

This book presents tensors and tensor analysis as primary mathematical tools for students and researchers of engineering and engineering science. The presentation is based on the concepts of vectors and vector analysis in three-dimensional Euclidean space. A vector is primarily defined as a quantity representing a magnitude and a direction in space. Vectors are in general independent of any particular coordinate system and are therefore coordinate-invariant quantity. However, in most applications, vectors will be represented by components related to a coordinate system defined in the space.

The book is intended as a textbook that takes the subject matter to a rather advanced level. However, as the presentation starts elementary with geometrical vector algebra, the book may serve as a first introduction to tensors and tensor analysis.

Vectors and vector analysis are import tools in classical Newtonian mechanics in which forces, displacements, velocities, and accelerations are vector quantities. However, these tools are essential in all branches of mathematical physics and in two- and three-dimensional geometry.

Tensors and tensor analysis were developed by Gregorio Ricci-Curbastro (1853–1925) and his student Tullio Levi-Civita (1873–1941) as an extension of vector and vector analysis. Tensors are used to formulize the manipulation of physical quantities and geometrical entities arising from the study of mathematical physics and geometry. Tensor analysis is concerned with relations or laws for physical quantities and geometrical entities that remain valid regardless of the coordinate systems used to specify the quantities. Tensor analysis was used by Albert Einstein (1879–1955) to develop the theory of general relativity.

In the mechanics of bodies of continuously distributed material, the analysis of mechanical stress and deformation leads to the definition of stress tensors and tensors related to the deformation of the bodies, e.g., the strain tensors.

The word tensor means stress, and the stress tensor is defined to be independent of the chosen coordinate system and therefore said to be a coordinate-invariant quantity. The analysis of the state of stress was developed into a mathematical theory that became the beginning of tensor analysis.

In the mechanics of continuous materials, the mechanical and thermomechanical behaviors of materials are described by coordinate-invariant equations, i.e., tensor equations, relating stress tensors, and deformation tensors. These equations are called constitutive equations.

My contact with tensors and tensor analysis relates to the study of continuum mechanics, constitutive modeling of materials, the theory of elastic shells, rheology, and non-Newtonian fluids.

During my years as lecturer and professor at the Norwegian University of Science and Technology, I have taught courses in Applied Mechanics, Mechanics of Materials, and Fluid Mechanics for undergraduate students. I have written textbooks in Norwegian on Statics (*Statikk*), Dynamics (*Dynamikk*), Strength of Materials (*Fasthetslære*), and Engineering Mechanics (*Ingeniørmekanikk*). For graduate students, I have designed courses in Continuum Mechanics, Rheology, and Tensor Analysis. The graduate courses were based on my compendia and my two books written in Norwegian: *Kontinuumsmekanikk* published in 1974 and *Tensor Analyse* published in 1982, both by Tapir Akademisk Forlag, Trondheim, Norway. The compendia were translated into English.

The major part of the material from these books and the compendia was presented and published as my Springer book *Continuum Mechanics* in 2008 and in my Springer book *Rheology and Non-Newtonian Fluids* in 2014, in which tensors were applied. However, I came to feel that there was enough material in the Norwegian books and compendia that had not been included in my book *Continuum Mechanics* to make it opportune or relevant to introduce this new book which I have called *Tensor Analysis*. Although some of the material from my book *Continuum Mechanics* is included in the new book, I have made an effort to improve the presentation in the present book.

The development of the materials in the book has been inspired by my former colleagues and graduate students at the Norwegian University of Science and Technology (NTNU). During the preparation of the present text, Professor Hugo Atle Jakobsen at NTNU has given valuable support. He has read through the manuscript in many stages and given me inspiration and good advice. I thank him and acknowledge his valuable comments and corrections. My friend, English teacher cand.philol Karin Hals, has read through the manuscript and sorted out some grammatical errors in addition to giving me good advice concerning the English language.

Finally, many thanks to my wife Eva, who has helped me with the index and otherwise supported and encouraged me during the writing and the preparation of the book.

Bergen, Norway   Fridtjov Irgens
September 2018

A Short Presentation of the Contents of This Book

Chapter 1 introduces some of the necessary mathematical foundation for the presentations in the following chapters. Vectors and the algebra of vectors in two- and three-dimensional Euclidean space are first given a geometrical presentation. After having introduced a Cartesian coordinate system in three-dimensional Euclidean space, I define: "A vector is a coordinate invariant quantity uniquely expressed by a magnitude and a direction in space and obeys the parallelogram law by addition of other vectors representing similar quantities." Examples of vector quantities from mechanics are forces, displacements, velocities, and accelerations. Vectors are in this book denoted by small boldface Latin letter, e.g., $\mathbf{a}$ and $\mathbf{b}$. The algebra of vectors is then given in terms of scalar components of the vectors related to Cartesian coordinates systems with axes x_1, x_2, and x_3, e.g., a_i and b_i for vector components. The basic law in classical mechanics is Newton's second law relating the force $\mathbf{f}$ on a body of mass m and the acceleration $\mathbf{a}$ of the body: $\mathbf{f} = m\mathbf{a} \Leftrightarrow f_i = ma_i$. In the algebra of vectors, matrices and some aspects of matrix algebra are needed. Matrices are described by Latin letters, e.g., A and b, and the matrix elements by, e.g., A_{ij} and b_i.

Chapter 2 introduces the equations of motions for bodies of continuous material. The concept of stress as force per unit area $\mathbf{t}$ on a material surface provides a convenient introduction to the mathematical definition of a tensor. The word "tensor" means stress, and tensor analysis was originally a stress analysis. The primary stress tensor is the Cauchy stress tensor $\mathbf{T}$.

Chapter 3 presents a general definition of tensors and tensor fields in three-dimensional Euclidean space. Based on the presentation of mechanical stress and the stress tensor in Chap. 2, tensors in three-dimensional space are then defined as multilinear scalar-valued functions of vectors, and the definition is independent of a coordinate system. For example, a tensor of second order, exemplified by the stress tensor $\mathbf{T}$, is a bilinear scalar-valued function of two vectors. This definition of tensors is inspired by the presentation of Jaunzemis [3], (the number in the brackets [ ] refers to the reference list on page ix). The definition implies that tensors are coordinate-invariant quantities, and directly leads to the definition of tensor components in Cartesian coordinate systems, as will be shown in Chap. 6, to the definition of tensor components in general coordinate systems. Tensors are in this book denoted by capital boldface Latin letter, e.g., **A** and **B**, with an exception for tensors of first order for which the vector notation is kept, e.g., **a** and **b**.

The definition of tensors preferred in the present book also connects tensor algebra and analysis to vector algebra and vector analysis as presented in Chap. 1 in a straightforward and natural manner. In the literature, tensors are defined in different ways. In some presentations of tensor analysis, e.g., McConnell [6], Spain [7], and Lawden [4], tensors are defined by their components in an n-dimensional space without the application of the geometrical vector concept. The components are functions of n variables or coordinates that obey certain transformation rules when the coordinates undergo a linear transformation. Green and Zerna [1] use the same approach to tensors in two- and three-dimensional space. In other presentations, the special concept of Cartesian tensors is introduced, e.g., Malvern [5] and Hunter [2], restricting tensor analysis to only Cartesian coordinate systems. Malvern [5] also defines tensors in general coordinate systems by their components obeying linear laws of transformations when changing coordinate systems, and he introduces tensors as polyadics. Section 3.2.2 briefly presents and discusses the alternative definitions of tensors.

Some aspects of rigid body dynamics are given as an example of application of the tensor analysis. The inertia tensor is introduced in this chapter.

In Chap. 4, a new family of tensors related to the deformation of continuous materials is introduced. Examples are the strain tensor that is applied to describe the behavior of elastic materials and the rate of deformation tensor related to the mechanics of viscous fluids.

Chapter 5 gives a brief presentation of constitutive equations for elastic materials and viscous fluids which are presented as tensor equations relating the tensor concept of stress to the tensors describing deformation, rate of deformation, and rotation.

Chapter 6 introduces general coordinate systems in three-dimensional Euclidean space. From the general definition of tensor presented in Chap. 3, different sets of components are defined. The algebra of tensors is now expressed through these components. General expressions for differential operators as the gradient, the divergence, and the rotation or curl of vectors and tensors are presented.

Chapter 7 shows how tensor equations discussed in Chaps. 4 and 5 are presented in general coordinates.

Chapter 8 presents surface geometry in three-dimensional Euclidean space. The surface represents a two-dimensional of non-Euclidean space called a two-dimensional Riemannian space. The material in this chapter is inspired by the presentation in the book by Green and Zerna [1]. The theory developed in this chapter is suitable for analysis in the theory of elastic shells.

Chapter 9 contains the most common integral theorems in two- and three-dimensional Euclidean space applied in continuum mechanics and mathematical physics.

Appendix: The book is supplied with an Appendix containing problems related to the text and with solutions to these problems. Most of the problems are referred to in the relevant chapters of the main text. The problems are meant to familiarize the reader with the formulas and the usage of them. To show the solutions to the problems in the main text would have made the text less readable.

References

1. Green AE, Zerna W (1968) Theoretical elasticity, 2nd edn. Oxford University Press, London
2. Hunter SC (1976) Mechanics of continuous media. Ellis Horwood, Chichester
3. Jaunzemis W (1967) Continuum mechanics. MacMillan, New York
4. Lawden DF (1967) An introduction to tensor calculus and relativity. Science Paperbacks and Methuen & Co LTD, London
5. Malvern LE (1969) Introduction to the mechanics of a continuous medium. Prentice-Hall Inc., Englewood Cliffs
6. McConnell AJ (1947) Application of tensor analysis. Dover Publications, Inc., New York
7. Spain B (1956) Tensor calculus. Oliver and Boyd, Edinburgh and London

Contents

Symbols

$A = (A_{ij}), A_{ij}$	Matrix, matrix elements
$A^T = (A_{i\alpha})$	Transposed matrix of the matrix $A = (A_{\alpha i})$
$AA^{-1} \Rightarrow AA^{-1} = 1$	Inverse matrix of the matrix A
$A_{ij}, A^{ij}, A^i_j, A^j_i$	Tensor components in general curvilinear coordinate system
$A_{\alpha\beta}, A^{\alpha\beta}, A^\alpha_\beta, A^\beta_\alpha$	Surface tensor components
$\mathbf{A}, A_{ij}..$	Tensor, tensor components in Cartesian coordinate system
$\mathbf{A(r}, t)$	Tensor field
$\mathbf{a}, \mathbf{a(r}, \text{t}), a_i, a^i$	Vector, vector field, and vector components
$\mathbf{a}(x, t) \equiv \mathbf{a}(x_1, x_2, x_3, \text{t})$	Vector field
$\mathbf{a} = [a_1, a_2, a_3]$	Vector and vector components in Cartesian coordinate systems
$\mathbf{a}_3 = \underset{\sim}{a}_i \mathbf{g}^i$	Unit normal vector to a surface in E_2 with components $\underset{\sim}{a}_i$ in E_3
$a = (a_i) = \begin{pmatrix} a_1 \\ a_2 \\ a_3 \end{pmatrix}$ $\equiv \{a_1 \; a_2 \; a_3\}$	Vector matrix, column matrix
$a^T = (a_1 \; a_2 \; a_3)$	Row matrix and transposed matrix of the column matrix $a = (a_i) = \{a_1 \; a_2 \; a_3\}$
$\mathbf{b}$	Body force, binormal vector of a space curve
$\mathbf{a}$	Acceleration
$\mathbf{a}_\alpha, \mathbf{a}^\alpha$	Base vectors, reciprocal base vectors
$a_{\alpha\beta}, a^{\alpha\beta}$	Fundamental parameters and reciprocal parameters of first order
$\mathbf{B}$	Left deformation tensor, curvature tensor
$B_{\alpha\beta}$	Fundamental parameter of second order

$B_{\alpha\beta}$	Components of the curvature tensor
C	Center of mass
$\mathbf{C}$	Green's deformation tensor, right deformation tensor
$C_{\alpha\beta}$	Fundamental parameters of third order
c	Wave velocity
$\mathbf{c}$	Vorticity vector
c_α, c^α	Covariant and contravariant components of a surface vector $\mathbf{c}$
$c_k, \underset{\sim}{c}^k$	Components of a surface vector $\mathbf{c}$ in general coordinates y in E_3
$\mathbf{D}, D_{ij}$	Rate of deformation tensor, rates of deformation
E_2, E_3	Two- and three-dimensional Euclidean space
$\mathbf{E}$	Green's strain tensor, tensor for small deformation
$\tilde{\mathbf{E}}$	Euler's and Almansi's strain tensor
$e_{ijk}, e_{\alpha\beta}$	Permutation symbol
$\mathbf{e}$	Unit vector
$\mathbf{e}_i$	Base vector in Cartesian coordinate system
$\mathbf{e}_i^y$	Unit vectors in the directions of the base vectors $\mathbf{g}_i$
$\mathbf{F}$	Deformation gradient tensor
$\mathbf{f}$	Force
g	Gravitational force per unit mass
g_{ij}, g^{ij}	Fundamental parameters of a general coordinate system in E_3
$g_K^i, g_{Ki}, g^{Ki}, g_i^K$	Euclidean shifters
$\mathbf{g}_i, \mathbf{g}^i$	Base vectors and reciprocal base vectors in a general curvilinear coordinate system
$\mathbf{H}, H_{ij}$	Displacement gradient tensor, displacement gradients
I, I_p	Second moment of area, polar moment of area
$\mathbf{I}, I_{ij}$	Inertia tensor, elements of the inertia tensor
$\mathbf{I} \equiv \mathbf{I}_2 = \alpha\mathbf{1}$	Isotropic tensor of second order
$\mathbf{I}_4 \Leftrightarrow I_{4ijkl} = \delta_{ik}\delta_{jl}$	Isotropic tensor of fourth order
J, J_y^x, J_x^y	Jacobians
K	Present configuration, kinetic energy
K_0	Reference configuration
$\mathbf{l}$	Angular momentum
m	Mass
$\mathbf{n}$	Principal curve normal, unit normal to a space curve, unit surface normal
Ox	Cartesian coordinate system with origin O
p	Pressure, thermodynamic pressure
$\mathbf{P}$	Permutation tensor
Q	Transformation matrix
$\mathbf{Q}$	Rotation tensor
R, R_2	Radius, two-dimensional Riemannian space

$\mathbf{R}$	Rotation tensor, Riemann–Christoffel tensor
$\tilde{\mathbf{R}}$	Rotation tensor for small deformations
$\mathbf{r}, \mathbf{r}_o$	Place vector, particle vector
$\mathbf{S}$	Symmetric tensor, second Piola–Kirchhoff stress tensor
s	Arc length
$\mathbf{T}$	Cauchy's stress tensor
$\mathbf{T}_o$	First Piola–Kirchhoff stress tensor
t	Time
$\mathbf{t}$	Stress vector, unit tangent vector of a space curve
$\mathbf{U}$	Right stretch tensor
$\mathbf{u}$	Displacement vector
u^1, u^2	Surface coordinates
$\mathbf{V}$	Left stretch tensor
$\mathbf{v}$	Velocity vector
$\mathbf{W}$	Rate of rotation tensor, spin tensor, vorticity tensor
$\mathbf{w}$	Angular velocity vector
x, x_i, X, X_i	Cartesian coordinates
y, y^i, Y^K	General space coordinate system in E_3, coordinates
$\mathbf{z}$	Rotation vector
α	Thermal expansion coefficient
$\alpha, \beta, \ldots$	Scalars
$\alpha(\mathbf{r}, t), \alpha(x, \mathrm{t})$	Scalar field $\alpha = \det\left(a_{\alpha\beta}\right)$
$\equiv \alpha(x_1, x_2, x_3, \mathrm{t})$	
$\Gamma^i_{jk}, \Gamma_{ijk}$	Christoffel symbols
γ	Shear strain, Gauss or total curvature
$\dot{\gamma}$	Rate of shear strain, magnitude of shear rate
δ_{ij}, δ^i_j	Kronecker delta
$\varepsilon, \dot{\varepsilon}$	Longitudinal strain, rate of longitudinal strain
$\varepsilon_v, \dot{\varepsilon}_v$	Volumetric strain, volumetric strain rate
$\varepsilon_{ijk}, \varepsilon^{ijk}, \varepsilon_{\alpha\beta}, \varepsilon^{\alpha\beta}$	Permutation symbols
η	Modulus of elasticity
θ	Temperature
κ	Bulk modulus, bulk viscosity, curvature of a space curve
$\tilde{\kappa}$	Principal curvature, curvature on a surface
λ, λ_i	Lamé constant, stretch, coordinate stretch
μ	Shear modulus of elasticity, viscosity, Lamé constant, mean curvature
ν	Poisson's ratio, kinematic viscosity
ρ	Density
σ	Normal stress, geodesic curvature
τ	Torsion of a space curve, shear stress, yield shear stress
R, θ, z	Cylindrical coordinates
r, θ, ϕ	Spherical coordinates

$\mathbf{1}$	Unit matrix
$\mathbf{1}_4 \equiv \mathbf{1} \bar{\otimes} \mathbf{1}$	Fourth-order unit tensor
$\Leftrightarrow 1_{4ijkl} = \delta_{ik}\delta_{jl}$	
$\mathbf{1} \otimes \mathbf{1} \Leftrightarrow \delta_{ij}\delta_{kl},$	Fourth-order isotropic tensors
$\underline{\mathbf{1}} = \mathbf{1} \underline{\otimes} \mathbf{1} \Leftrightarrow \underline{1}_{ijkl} = \delta_{il}\delta_{jk}$	
$\mathbf{I} \equiv \mathbf{I}_2 = \alpha\mathbf{1}$	Isotropic tensor of second order
$\mathbf{I}_4 \Leftrightarrow I_{4ijkl} = \delta_{ik}\delta_{jl}$	Isotropic tensor of fourth order
$\mathbf{1} \equiv \mathbf{1}_2 \Leftrightarrow 1_{ij} = \delta_{ij}$	Unit tensor of second order
$\mathbf{I}, I_{ij}$	Inertia tensor, elements of the inertia tensor
I, II, III	Principal invariants of second-order tensor
$\bar{I}, \bar{II}, \bar{III}, \quad \tilde{I}, \tilde{II}, \tilde{III}$	Moment invariant, trace invariants of second-order tensor
$\nabla( ) = \mathbf{e}_i \frac{\partial()}{\partial x_i}$	Del operator in Cartesian coordinates x
$\nabla( ) = \mathbf{g}_i \frac{\partial()}{\partial y^i}$	Del operator in general space coordinates y
$\overleftarrow{\nabla} \equiv \overleftarrow{\partial}_k \mathbf{e}_k, \quad \overleftarrow{\nabla} \equiv \overleftarrow{\partial}_k \mathbf{e}_k$	Right operator, left operator
$\text{grad } \alpha \equiv \frac{\partial \alpha}{\partial \mathbf{r}} \equiv \nabla\alpha$	Gradient of a scalar field $\alpha(\mathbf{r}, t)$
$\text{grad } \mathbf{A} \equiv \frac{\partial \mathbf{A}}{\partial \mathbf{r}}$	Gradient of a tensor field $\mathbf{A}(\mathbf{r}, t)$ of order n
$\Leftrightarrow A_{i \cdot \cdot j, k} \equiv \frac{\partial A_{i \cdot \cdot j}}{\partial x_k}$	Components of the gradient of a tensor field $\mathbf{A}(\mathbf{r}, t)$ in Cartesian coordinates x
$\Leftrightarrow A_{i \cdot \cdot j}\big|_k$	Components of the gradient of a tensor field $\mathbf{A}(\mathbf{r}, t)$ in general space coordinates y
$\text{div } \mathbf{a} \equiv \nabla \cdot \mathbf{a} = a_{i,i}$	Divergence of a vector field $\mathbf{a}(\mathbf{r}, t)$ in Cartesian coordinates x
$\text{div } \mathbf{a} \equiv \nabla \cdot \mathbf{a} = a^i\big|_i$	Divergence of a vector field $\mathbf{a}(\mathbf{r}, t)$ in general space coordinates y
$\text{rot } \mathbf{a} \equiv \text{curl } \mathbf{a}$ $= e_{ijk}a_{k,j}\mathbf{e}_i$	Rotation of a vector field $\mathbf{a}(\mathbf{r}, t)$ in Cartesian coordinates x
$\text{rot } \mathbf{a} \equiv \text{curl } \mathbf{a}$ $= e^{ijk}a_k\big|_j\mathbf{e}_i$	Rotation of a vector field $\mathbf{a}(\mathbf{r}, t)$ in general space coordinates y
$\text{rot } \mathbf{A} \equiv \text{curl } \mathbf{A}$	Rotation of a tensor field $\mathbf{A}(\mathbf{r}, t)$ of order n
$\Leftrightarrow \varepsilon_{ijk}A_{r \cdot \cdot k, j}\,\mathbf{e}_i \otimes \mathbf{e}_{r \cdot \cdot}$	In Cartesian coordinates x
$\Leftrightarrow \varepsilon^{ijk}A_{r \cdot \cdot k}\big|_j\mathbf{g}_i \otimes \mathbf{g}^r\cdot\cdot$	In general space coordinates y
$\det A, \text{ tr} A, \text{norm} A = ||A||$	Determinant, trace, norm of a matrix A
$\det \mathbf{A}, \text{ tr} \mathbf{A}, \text{norm} \mathbf{A} = ||\mathbf{A}||$	Determinant, trace, norm of a tensor of second order $\mathbf{A}$
$\mathbf{v} \cdot \nabla = v_i \frac{\partial}{\partial x_i}$	Scalar operator
$\frac{\partial f(x,t)}{\partial x_i} \equiv f_{,i}$	Comma notation
$\dot{f} = \partial_t f \equiv \frac{\partial f}{\partial t}$	Material derivative of an intensive field function $f(\mathbf{r}_0, t)$
$\dot{f} = \partial_t f + f_{,i} v_i$ $\equiv \partial_t f + (\mathbf{v} \cdot \nabla)f$	Material derivative of an intensive field function $f(\mathbf{r}, t)$

$\mathbf{a} \cdot \mathbf{b} = a_i b_i$	Scalar product of two vectors $\mathbf{a}$ and $\mathbf{b}$
$= |\mathbf{a}||\mathbf{b}| \cos(\mathbf{a}, \mathbf{b})$	
$\mathbf{a} \cdot \mathbf{b} = a_i b_i$	Scalar product in Cartesian coordinate systems
$\mathbf{a} \cdot \mathbf{b} = a^i b_i$	Scalar product in general coordinate systems
$\mathbf{a} \times \mathbf{b} = |\mathbf{a}||\mathbf{b}| \sin(\mathbf{a}, \mathbf{b}) \mathbf{e}$	Vector product of two vectors $\mathbf{a}$ and $\mathbf{b}$
$\mathbf{a} \times \mathbf{b} = e_{ijk} a_i b_j \mathbf{e}_k$	In Cartesian coordinate systems
$\mathbf{a} \times \mathbf{b} = \varepsilon^{ijk} a_i b_j \mathbf{g}_k$	In general curvilinear coordinate systems
$|\mathbf{a}| = \sqrt{\mathbf{a} \cdot \mathbf{a}} \equiv \sqrt{a^i a_i}$	Magnitude of a vector
$\mathbf{A} \otimes \mathbf{B} = \mathbf{C}$	Tensor product of $\mathbf{A}$ and $\mathbf{B}$
$\Leftrightarrow \mathbf{A}[\mathbf{a}, ..] \mathbf{B}[\mathbf{b}, .]$	
$= \mathbf{C}[\mathbf{a}, .., \mathbf{b}, .]$	
$\Leftrightarrow A^i {}_{..} B_j {}_{.} = C^i {}_{..j}{}_{.}$	
$\mathbf{c} \otimes \mathbf{d} = \mathbf{E}$	Dyadic product $\equiv$ dyad
$\Leftrightarrow c_i d_j = E_{ij}$	
$\mathbf{c} \otimes \mathbf{d} \otimes \mathbf{f}$	Triad, polyad
$= \mathbf{F} \Leftrightarrow c_i d_j f_k = F_{ijk}$	
$\mathbf{B} = B_{ij} \mathbf{e}_i \otimes \mathbf{e}_j$	In Cartesian coordinates
	$= B^{ij} \mathbf{g}_i \otimes \mathbf{g}_j$ in general curvilinear coordinates
$\left. \begin{array}{l} \mathbf{A} \cdot \mathbf{a} = \mathbf{A}\mathbf{a} = \mathbf{b} \Rightarrow A^{ij} a_j = b^i \\ \mathbf{a} \cdot \mathbf{A} = \mathbf{c} \Rightarrow a^i A_{ij} = c_j \end{array} \right\}$	Linear mapping of a vector $\mathbf{a}$ on other vectors $\mathbf{b}$ and $\mathbf{c}$
$\mathbf{A}\mathbf{B} \equiv \mathbf{A} \cdot \mathbf{B} = \mathbf{C}$	
$\mathbf{C} : \mathbf{A} = \mathbf{B} \Rightarrow C_{ijkl} A_{kl} = B_{ij}$	Linear mapping of a tensor $\mathbf{A}$ of second order on other
$\mathbf{A} : \mathbf{C} = \mathbf{D} \Rightarrow A_{ij} C_{ijkl} = D_{kl}$	second-order tensors $\mathbf{B}$ and $\mathbf{D}$
$\mathbf{C} \cdot \mathbf{b} = \mathbf{A} \Leftrightarrow C_{ijk} b_k = A_{ij}$	Inner products or dot products
$\mathbf{B} \cdot \mathbf{D} = \mathbf{B}\mathbf{D} \Leftrightarrow B_{ik} D_{kl}$	
$\mathbf{C}\mathbf{B} = \mathbf{a} \Leftrightarrow C_{ijk} B_{jk} = a_i$	
$\mathbf{A} : \mathbf{B}$	Scalar product of two tensors of second order$\}$
	$= \begin{cases} A_{ij} B_{ij} & \text{in Cartesian coordinates} \\ A^{ij} B_{ij} & \text{in general coordinates} \end{cases}$
$(\mathbf{u}, \mathbf{e}_i)$	Angle between the vectors $\mathbf{u}$ and $\mathbf{e}_i$

Chapter 1
Mathematical Foundation

1.1 Matrices and Determinants

A *two-dimensional matrix* is a table or rectangular array of elements arranged in rows and columns. The matrix A with 2 rows and 3 columns is alternatively presented as:

$$\begin{pmatrix} A_{11} & A_{12} & A_{13} \\ A_{21} & A_{22} & A_{23} \end{pmatrix} \equiv (A_{\alpha i}) \equiv A \qquad (1.1.1)$$

The elements may be numbers or functions. The symbol $A_{\alpha i}$ represents an arbitrary element for which the *row number* (α) may take the values 1 or 2, while the *column number* (i) may have any of the values 1, 2, or 3. If not otherwise specified, lower case Greek letter indices shall represent the numbers 1 and 2, while lower case Latin letter indices shall represent the numbers 1, 2 and 3.

In the present exposition matrices will in general be denoted by capital Latin letters. An exception to this notation applies to matrices with one row or one column and is presented below. Many textbooks on Matrix Analysis prefer to use capital boldface letters for matrices. But this book reserves boldface letters for the coordinate invariant notation of vectors and tensors.

It is convenient to let the symbol $A_{\alpha i}$ represent both an element in the matrix A and the complete matrix A itself. The table and the two other symbols for matrices given by the identities (1.1.1), and also the symbol $A_{\alpha i}$ alone, now represent four different ways of presenting one and the same matrix. The matrix $A_{\alpha i}$, which has 2 rows and 3 columns, is called a 2×3 ("two by three") matrix. The matrices $B_{\alpha\beta}$ (2×2) matrix and C_{ij} $(3 \times 3$ matrix) have as many columns as rows and are called *square matrices*, due to the forms of their tables.

A *one-dimensional matrix* is an array of elements arranged in one column or one row. A matrix with one column is called a *column matrix* or a *vector matrix* and will in general be denoted by lower case Latin letters. For practical reasons it may

© Springer Nature Switzerland AG 2019
F. Irgens, *Tensor Analysis*,
https://doi.org/10.1007/978-3-030-03412-2_1

be convenient to write the elements of a column matrix on a horizontal rather than a vertical line, e.g.:

$$a \equiv (a_i) \equiv \begin{pmatrix} a_1 \\ a_2 \\ a_3 \end{pmatrix} \equiv \{a_1 \, a_2 \, a_3\} \qquad (1.1.2)$$

From a two-dimensional matrix $A = (A_{\alpha i})$ we may construct a *transposed matrix* $A^T = (A_{i\alpha})$ by interchanging rows and columns. For a $(3 \times 3$ matrix$)$ C:

$$C = (C_{ij}) \Leftrightarrow C^T = \left(C_{ij}^T\right) \Rightarrow C_{ij}^T = C_{ji} \qquad (1.1.3)$$

The transposed matrix a^T of a column matrix a is a *row matrix*:

$$a^T \equiv (a_1 \, a_2 \, a_3) \qquad (1.1.4)$$

An *n-dimensional matrix* is represented by a set of elements with n indices. For example will the elements D_{ijk} represent a *three-dimensional matrix*. The matrix algebra, as presented below, is designed for one- and two-dimensional matrices.

Addition of matrices is defined only for matrices of the same size. The sum of two column matrices a and b is a column matrix c, and the sum of two 3×3 matrices A and B is a 3×3 matrix C:

$$a + b = c \Leftrightarrow a_i + b_i = c_i, \quad A + B = C \Leftrightarrow A_{ij} + B_{ij} = C_{ij} \qquad (1.1.5)$$

The element C_{ij} in the matrix C is obtained by adding corresponding elements A_{ij} and B_{ij} in the matrices A and B. We easily see that the operation addition is both associative and commutative. Addition also applies to *n*-dimensional matrices.

The product of a matrix $A = (A_{ij})$ by a term α is a matrix $\alpha A = (\alpha A_{ij})$. The *matrix product Ab* of square matrix A and a column matrix b is a column matrix c, and the *matrix product AB* of two matrices A and B is a new matrix C, such that:

$$Ab = c \Leftrightarrow \sum_{k=1}^{3} A_{ik} b_k = c_i, \quad AB = C \Leftrightarrow \sum_{k=1}^{3} A_{ik} B_{kj} = C_{ij} \qquad (1.1.6)$$

The element c_i in the column matrix c is obtained as the sum of the product pairs $A_{ik} b_k$ of the elements in row (i) of A and the elements b_k of b. The element C_{ij} in the matrix C is obtained as the sum of the all product pairs $A_{ik} B_{kj}$ of the elements in the row (i) of A and the elements in the column (j) of B.

We shall now introduce an important convention of great consequence when dealing with matrices in the index format.

The *Einstein's summation convention*: An index repeated once and only once in a term implies a summation over the number region of that index.

The convention is attributed to Albert Einstein [1879–1955]. By this convention we may write:

$$\sum_{k=1}^{3} A_{ik}B_{kj} \equiv A_{ik}B_{kj} \tag{1.1.7}$$

The summation convention does not apply if the *summation index* is repeated more than once. For example:

$$\sum_{k=1}^{3} A_{ik}B_{kj}a_k \neq A_{ik}B_{kj}a_k \tag{1.1.8}$$

An index that does not imply summation is called a *free index*. In the inequality (1.1.8) the index (k) on the left hand side is a summation index, while the index (k) on the right hand side is a free index. A summation index is also called a "*dummy*" *index* because it may be replaced by a different letter without changing the result of the summation. For example:

$$A_{ik}B_{kj} = A_{il}B_{lj} \tag{1.1.9}$$

Multiplication of matrices is not a commutative operation, which is shown as follows:

$$A_{ik}B_{kj} \neq B_{ik}A_{kj} \Leftrightarrow AB \neq BA \tag{1.1.10}$$

Multiplication of matrices is an associative operation. Let A, B, C, D and E be 3×3 matrices such that:

$$D = AB \Leftrightarrow D_{ij} = A_{ik}B_{kj}, \quad E = DC \Leftrightarrow E_{il} = D_{ij}C_{jl}$$

Then:

$$E = DC = (AB)C \Rightarrow E_{il} = D_{ij}C_{jl} = \left(A_{ik}B_{kj}\right)C_{jl} = A_{ik}\left(B_{kj}C_{jl}\right) = A_{ik}B_{kj}C_{jl}$$

These results are conveniently stated as: $E = (AB)C = A(BC) = ABC$:

$$(AB)C = A(BC) = ABC \tag{1.1.11}$$

From the definition (1.1.6) it follows that:

$$(AB)^T = B^T A^T \tag{1.1.12}$$

Note that in general $b_k A_{ki} \neq A_{ik}b_k = c_i \Rightarrow Ab = c$. The product $b_k A_{ki} = d_i$ may alternatively be expressed as:

$$b_k A_{ki} \equiv A_{ki} b_k = (A_{ik}^T) b_k = d_i \Leftrightarrow A^T b = d \tag{1.1.13}$$

Two *unit matrices* or *identity matrices* are now defined:

$$1 = (\delta_{\alpha\beta}) = \begin{pmatrix} 1 & 0 \\ 0 & 1 \end{pmatrix}, \quad 1 = (\delta_{ij}) = \begin{pmatrix} 1 & 0 & 0 \\ 0 & 1 & 0 \\ 0 & 0 & 1 \end{pmatrix} \tag{1.1.14}$$

The element symbols $\delta_{\alpha\beta}$ and δ_{ij} are called *Kronecker deltas*, named after Leopold Kronecker [1823–1891]. The elements have the following values:

$$\delta_{ij} = 1 \quad \text{when } i = j, \quad \delta_{ij} = 0 \quad \text{when } i \neq j \tag{1.1.15}$$

The effect of a unit matrix is shown in the following example.

$$A1 = 1A = A \Leftrightarrow A_{ik}\delta_{kj} = \delta_{ik}A_{kj} = A_{ij}$$

In the definition of the determinant of a matrix we need two special symbols and matrices:

The *permutation symbol* $e_{\alpha\beta} = 0, 1, \text{or} - 1$, such that: $(e_{\alpha\beta}) = \begin{pmatrix} 0 & 1 \\ -1 & 0 \end{pmatrix}$ (1.1.16)

The *permutation symbol* e_{ijk}:

$$= \begin{cases} 0 \text{ when two or three indices are equal} \\ 1 \text{ when the indices form a cyclic permutation of the numers 123} \\ -1 \text{ when the indices form a cyclic permutaton of the numbers 321} \end{cases}$$

e_{ijk} are elements of a three-dimensional matrix: the *permutation matrix* : (e_{ijk}).

$$\tag{1.1.17}$$

For example: $e_{122} = e_{333} = 0, e_{123} = e_{231} = e_{312} = 1, e_{321} = e_{213} = e_{132} = -1$. It follows from the definitions (1.1.16), (1.1.17) that: $e_{ijk} = e_{kij} = -e_{kji}, e_{\alpha\beta} = -e_{\beta\alpha} = e_{\alpha\beta 3}$. We also find that:

$$e_{ijk} = \frac{1}{2}(i-j)(j-k)(k-i) \tag{1.1.18}$$

The permutation symbol e_{ijk} and the Kronecker delta δ_{ij} are related through the identity:

$$e_{ijk}e_{rsk} = \delta_{ir}\delta_{js} - \delta_{is}\delta_{jr} \tag{1.1.19}$$

The validity of the identity (1.1.19) may be tested by selecting different sets of the free indices i, j, r, and s on both sides of the equation.

The *determinants*, det A, of 2×2 and 3×3 square matrices A are defined respectively by:

$$\det A \equiv \det\left(A_{\alpha\beta}\right) \equiv e_{\alpha\beta}A_{1\alpha}A_{2\beta} \equiv e_{\alpha\beta}A_{\alpha 1}A_{\beta 2} \tag{1.1.20}$$

$$\det A \equiv \det\left(A_{ij}\right) \equiv e_{ijk}A_{1i}A_{2j}A_{3k} \equiv e_{ijk}A_{i1}A_{j2}A_{k3} \tag{1.1.21}$$

From the definitions it follows that: $\det A^T = \det A$, and that the determinant of a matrix is zero when two rows or two columns are identical. For example:

$$A_{1j} = A_{2j} \Rightarrow \det A = 0$$

By inspection we find that:

$$e_{\alpha\beta}A_{\alpha\gamma}A_{\beta\lambda} = (\det A)e_{\gamma\lambda}, \quad e_{ijk}A_{ir}A_{js}A_{kt} = (\det A)e_{rst} \tag{1.1.22}$$

This result may be used to prove the *multiplication theorem for determinants*:

$$\det(AB) = (\det A)(\det B) \tag{1.1.23}$$

Proof of the theorem for 3×3 matrices: Using formula (1.1.21) for det B, then formula (1.1.22) for det A, and finally formula (1.1.21) for the determinant $\det(AB)$ of the matrix product AB, we obtain:

$$(\det A)(\det B) = (\det A)(e_{rst}B_{r1}B_{s2}B_{t3}) = (\det A\, e_{rst})(B_{r1}B_{s2}B_{t3}) = \left(e_{ijk}A_{ir}A_{js}A_{kt}\right)(B_{r1}B_{s2}B_{t3})$$
$$= e_{ijk}(A_{ir}B_{r1})\left(A_{js}B_{s2}\right)(A_{kt}B_{t3}) = \det(AB) \Rightarrow (\det A)(\det B) = \det(AB)$$

The determinant of a matrix A may be expressed as a linear function of the elements A_{ir} in any arbitrarily chosen row or column. The coefficients in the linear function are called the *cofactors* $\mathrm{Co}\,A_{ir}$, or the *algebraic complements*, to the corresponding elements A_{ir}. For instance:

$$\det A = e_{ijk}A_{i1}A_{j2}A_{k3} = A_{i1}\left(e_{ijk}A_{j2}A_{k3}\right) = A_{i1}\mathrm{Co}A_{i1} \Rightarrow \mathrm{Co}A_{i1} = e_{ijk}A_{j2}A_{k3}$$

By inspection we find:

$$\mathrm{Co}\,A_{ir} = \frac{1}{2}e_{ijk}e_{rst}A_{js}A_{kt} = \frac{\partial(\det A)}{\partial A_{ir}} \tag{1.1.24}$$

$$\mathrm{Co}\,A_{\alpha\gamma} = e_{\alpha\beta}e_{\gamma\lambda}A_{\beta\lambda} = \frac{\partial(\det A)}{\partial A_{\alpha\gamma}} \tag{1.1.25}$$

$$\det A = \sum_{i} A_{ir}\,\mathrm{Co}\,A_{ir} = \sum_{i} A_{ri}\,\mathrm{Co}\,A_{ri}, \quad r = 1, 2, \text{ or } 3 \tag{1.1.26}$$

The cofactors are elements in the matrix Co A, and we find that:

$$(\det A)1 = A^T \text{Co} A = A \text{Co} A^T = (\text{Co} A)A^T$$
$$\Leftrightarrow (\det A)\delta_{ij} = A_{ki}\text{Co} A_{kj} = A_{ik}\text{Co} A_{jk} = (\text{Co} A_{ik})A_{jk} \tag{1.1.27}$$

If the product of two square matrices A and B is equal to the unit matrix, i.e. $1 = (\delta_{ij})$, then we call one matrix the *inverse matrix* to the other. The inverse matrix to A is denoted A^{-1}:

$$AA^{-1} = A^{-1}A = 1 \tag{1.1.28}$$

The inverse matrix A^{-1} may be determined as follows. Using the formulas (1.2.27), (1.2.28) we get:

$$A^{-1}(\det A) = A^{-1}(\det A)1 = A^{-1}A \text{Co} A^T = 1\text{Co} A^T = \text{Co} A^T$$
$$\Rightarrow A^{-1} = \frac{1}{\det A}\text{Co} A^T \Leftrightarrow A^{-1}_{ij} = \frac{1}{\det A}\text{Co} A_{ji} \tag{1.1.29}$$

It follows that a condition for the inverse matrix A^{-1} to exist, is that the determinant det A is not equal to zero.

Natural powers of the matrices A and A^{-1}, where n is a natural number, are defined by:

$$A^n = AA\cdots A, \quad A^{-n} = A^{-1}A^{-1}\cdots A^{-1} \tag{1.1.30}$$

In addition to the determinant, det A, of a matrix A, we shall also need the *trace*, tr A, of A and the *norm* of A, denoted by "norm A "or by $\|A\|$. These quantities are defined respectively by:

$$\text{tr} A \equiv A_{kk} = A_{11} + A_{22} + A_{33} \tag{1.1.31}$$

$$\text{norm} A \equiv \|A\| \equiv \sqrt{\text{tr} AA^T} = \sqrt{A_{ij}A_{ij}} \tag{1.1.32}$$

The norm of the matrix A is also called the *magnitude* of the matrix A.

1.2   Cartesian Coordinate Systems. Scalars and Vectors

In order to localize physical objects in space and to define motion of bodies in space, we need a *reference body* or *reference frame*, here for short called a *reference* and denoted by *Rf*. The reference may be the earth, a space laboratory, or the Milky Way. A quantity that is not defined relative to a reference will be called a *reference invariant quantity*, an *objective quantity*, or a *reference invariant*. The distance between two space points and the temperature in a body are two examples of reference invariants. A quantity defined relative to a reference, will be called a *reference related quantity*. The velocity of a body with respect to a reference frame gives an example of a reference related quantity.

Figure 1.1 shows a reference *Rf* in three-dimensional space, a reference point *O* fixed in *Rf* and three orthogonal axes x_i that intersect in *O*. The point *O* and the axes represent an *orthogonal Cartesian right-handed coordinate system* denoted by *Ox*. The point *O* is called the *origin of the coordinate system*. The axes represent a *right-handed system* in the following sense: If the right hand is held about the x_3-axis such that the fingers point in the direction of a $90°$-rotation of the positive x_1-axis toward the positive x_2-axis, the thumb then points to the direction of the positive x_3-axis. If the direction of one of the axes is reversed, or two of the axes are interchanged, the new system becomes a *left-handed system*.

The origin *O* and a pair of any two coordinate axes define three orthogonal planes: $Ox_1x_2, Ox_2x_3, Ox_3x_1$. The distances x_1, x_2, and x_3 from these planes to a point *P* in space are positive or negative according to which side of the planes the point is. For instance, x_1 is positive when *P* is on the same side of the Ox_2x_3-plane as the positive x_1-axis. The three distances x_i are called the *coordinates* of the point *P* relative to an orthogonal right-handed Cartesian coordinate system *Ox*. The x_i-axes are called *coordinate axes. Coordinate planes* are defined by $x_1 =$ constant, $x_2 =$ constant, or $x_3 =$ constant. The lines of intersection between these planes are the *coordinate lines*. Three coordinate planes and three coordinate lines intersect in every point *P* in space.

The geometry presented in the three-dimensional physical space in which a Cartesian coordinates may be used, is called Euclidean and is based on the parallel postulate of Euclid [ca. 330 − 275BCE]. The space is called a *three-dimensional Euclidean space E_3*. A plane, in which we may use an *orthogonal right-handed Cartesian coordinate system Ox_1x_2*, Fig. 1.2, with two orthogonal axes represents a *two-dimensional Euclidean space E_2*. A curved surface embedded in the

Fig. 1.1 Cartesian coordinate system

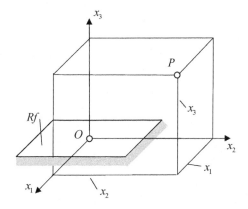

Fig. 1.2 Two-dimensional
Euclidean space E_2. Cartesian
coordinate system Ox_1x_2.
Resultant displacement vector
u. Component displacement
vectors $\mathbf{u}_1$ and $\mathbf{u}_2$

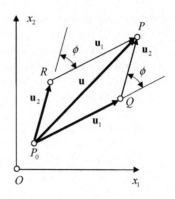

three-dimensional Euclidean space E_3, e.g. a spherical surface, in which we cannot introduce a Cartesian coordinate system, is called a *two-dimensional Riemannian space R_2*, after Georg Friedrich Bernhard Riemann [1826–1866]. Chapter 8 Surface Geometry presents vector and tensor analyses in two-dimensional Riemannian space R_2.

In the general exposition we shall use orthogonal Cartesian right-handed coordinate systems until Chap. 6 where general curvilinear coordinate analysis is introduced. The quantities we call scalars, vectors, and tensors are defined to be invariant with respect to our choice of coordinate system, and they are therefore called *coordinate invariants*. Reference invariant quantities are coordinate invariant, but in general the opposite does not apply.

The following definition of scalars is useful and important.

A *scalar invariant*, or for short a *scalar*, is a coordinate invariant quantity uniquely expressed by a magnitude.

Examples of scalars: the distance between two points in space, the temperature θ, and the pressure in a fluid p. Scalars will in the present exposition preferably be denoted by Greek letters: α, β, γ, the pressure p being an exception.

1.2.1 Displacement Vectors

Figure 1.2 presents Euclidean two-dimensional space E_2 and a Cartesian coordinate system Ox_1x_2. A small body marked as a point is at time t_0 at the position or *place P_0* on the plane and is displaced to position or place P on the plane at time t. The displacement is given by the arrow **u**, representing a length u and a direction in the plane. The quantity **u** is called a *displacement vector*. The figure shows the displacement vector **u** decomposed into two *component displacements vectors* $\mathbf{u}_1$ and $\mathbf{u}_2$ in either of two ways: either by the displacement vector $\mathbf{u}_1$ followed by the displacement vector $\mathbf{u}_2$, or by the displacement $\mathbf{u}_2$ followed by the displacement $\mathbf{u}_1$. The construction of the displacement **u** from the displacements $\mathbf{u}_1$ and $\mathbf{u}_2$ is also shown by the parallelogram P_0QPR. The displacement $\mathbf{u}$ is called the *resultant* of the components displacements $\mathbf{u}_1$ and $\mathbf{u}_2$. The resultant $\mathbf{u}$ is also called the *geometrical sum* of the two components $\mathbf{u}_1$ and $\mathbf{u}_2$.

The construction presented in Fig. 1.2 is called *vector addition* and may be presented as:

$$\mathbf{u}_1 + \mathbf{u}_2 = \mathbf{u}_2 + \mathbf{u}_1 = \mathbf{u} \tag{1.2.1}$$

When the resultant displacement $\mathbf{u}$ is constructed by the parallelogram on the components displacements $\mathbf{u}_1$ and $\mathbf{u}_2$, we say that displacements follow the *parallelogram law* by addition. Equation (1.2.1) may also be interpreted as a *composition* $\mathbf{u}$ of the two vectors $\mathbf{u}_1$ and $\mathbf{u}_2$, or as a *decomposition* of the vector $\mathbf{u}$ into the two vectors $\mathbf{u}_1$ and $\mathbf{u}_2$.

Each of the displacement vectors $\mathbf{u}, \mathbf{u}_1$, and $\mathbf{u}_2$ represents a *magnitude* given by the length of the displacement vector and a *direction*. The magnitudes are presented by:

$$|\mathbf{u}|, \quad |\mathbf{u}_1|, \quad \text{and} \quad |\mathbf{u}_2| \tag{1.2.2}$$

The magnitude $|\mathbf{u}|$ of the resultant displacement $\mathbf{u}$ may be calculated by the cosine-formula:

$$|\mathbf{u}|^2 = |\mathbf{u}_1|^2 + |\mathbf{u}_2|^2 + 2|\mathbf{u}_1||\mathbf{u}_2| \cos \phi \tag{1.2.3}$$

ϕ is the angle in Fig. 1.2 between the two component displacements $\mathbf{u}_1$ and $\mathbf{u}_2$.

It follows from Fig. 1.2 that if the two components $\mathbf{u}_1$ and $\mathbf{u}_2$ have the same magnitude but opposite directions, then the geometrical sum has neither a magnitude nor a direction and is called a *zero vector* $\mathbf{0}$. In this case we write:

$$\mathbf{u}_1 + \mathbf{u}_2 = \mathbf{0} \Rightarrow \mathbf{u}_2 = -\mathbf{u}_1 \tag{1.2.4}$$

The vector $-\mathbf{u}_1$ is interpreted as a vector of the same magnitude as $\mathbf{u}_1$ but in the opposite direction of $\mathbf{u}_1$.

General definitions: A *product of a scalar α and a vector* $\mathbf{a}$ is a new vector $\mathbf{b} = \alpha\mathbf{a}$ with magnitude $b = \alpha a$, and in the direction of $\mathbf{a}$ when α is positive and in the opposite direction of $\mathbf{a}$ if α is negative. A *unit vector* $\mathbf{e}$ in the direction of a vector $\mathbf{a}$ is defined by the expression:

$$\mathbf{e} = \frac{\mathbf{a}}{|\mathbf{a}|} \Rightarrow \mathbf{a} = |\mathbf{a}|\mathbf{e} \tag{1.2.5}$$

The unit vector $\mathbf{e}$ defined by formula (1.2.5) is called the *directional vector* related to the vector $\mathbf{a}$.

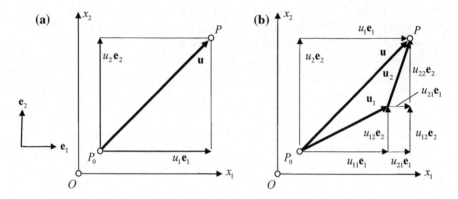

Fig. 1.3 **a** Base vectors e_1 and e_2. Scalar components u_1 and u_2. **b** Vector addition: $\mathbf{u} = \mathbf{u}_1 + \mathbf{u}_2 \Rightarrow$ $u_1 = u_{11} + u_{21}$, $u_2 = u_{12} + u_{22}$

In a plane we introduce an orthogonal Cartesian coordinate system Ox, Fig. 1.3, with two axes: x_1-axis and x_2-axis. We define *base vectors* as directional vectors in the positive directions of the coordinate axes:

$$\begin{array}{l} e_1 \text{ in the direction of the positive } x_1\text{-axis,} \\ e_2 \text{ in the direction of the positive } x_2\text{-axis} \end{array} \tag{1.2.6}$$

In Fig. 1.3a the displacement vector $\mathbf{u}$ is decomposed into two vectors: $u_1 e_1$ and $u_2 e_2$.

$$\mathbf{u} = u_1 e_1 + u_2 e_2 \tag{1.2.7}$$

The quantities u_1 and u_2 are called the *scalar components* or the *Cartesian component s* of the vector $\mathbf{u}$ in the coordinate system Ox. Note that in general the scalar components may be positive or negative scalars. The magnitude $|\mathbf{u}|$ of the vector $\mathbf{u}$ may now be expressed by the formula:

$$|\mathbf{u}| = \sqrt{(u_1)^2 + (u_2)^2} \tag{1.2.8}$$

Figure 1.3b presents the displacement vector $\mathbf{u}$ as the resultant of two displacement components $\mathbf{u}_1$ and $\mathbf{u}_2$. The three vectors $\mathbf{u}$, $\mathbf{u}_1$ and $\mathbf{u}_2$ are decomposed into components in the directions of the coordinate axes such that:

$$\mathbf{u} = u_1 e_1 + u_2 e_2, \quad \mathbf{u}_1 = u_{11} e_1 + u_{12} e_2, \quad \mathbf{u}_2 = u_{21} e_1 + u_{22} e_2 \tag{1.2.9}$$

Fig. 1.4 The resultant **u** of
three vectors $\mathbf{u}_1$, $\mathbf{u}_2$, and $\mathbf{u}_3$

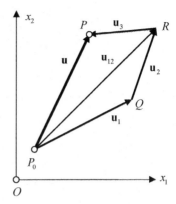

From Fig. 1.3b it follows that:

$$\mathbf{u} = \mathbf{u}_1 + \mathbf{u}_2 \Rightarrow u_1 = u_{11} + u_{21}, \quad u_2 = u_{12} + u_{22} \tag{1.2.10}$$

The results (1.2.9) are easily generalized to additions of more then two displacement components. As an example Fig. 1.4 demonstrates a geometrical summation of three displacement vectors $\mathbf{u}_1, \mathbf{u}_2$, and $\mathbf{u}_3$ to one resultant displacement vector **u**:

$$\mathbf{u}_{12} = \mathbf{u}_1 + \mathbf{u}_2, \quad \mathbf{u} = \mathbf{u}_{12} + \mathbf{u}_3 \Rightarrow$$
$$\mathbf{u} = \mathbf{u}_1 + \mathbf{u}_2 + \mathbf{u}_3 \Rightarrow \tag{1.2.11}$$
$$u_1 = u_{11} + u_{21} + u_{31}, \quad u_2 = u_{12} + u_{22} + u_{32}$$

As is easily seen the order of geometrical summation is arbitrary. The polygon $P_0 Q R P$ in Fig. 1.4 is called a *vector polygon*.

A generalization to a geometrical sum of n displacements vectors $\mathbf{u}_i \ i = 1, 2, \ldots, n$ to a resultant displacement **u** is presented as:

$$\mathbf{u} = \sum_i \mathbf{u}_i \Leftrightarrow u_1 = \sum_i u_{i1}, \quad u_2 = \sum_i u_{i2} \tag{1.2.12}$$

In a three-dimensional Euclidian space we introduce an *orthogonal Cartesian coordinate system Ox*, Fig. 1.5, with three *base vectors* $\mathbf{e}_i$ as directional vectors in the positive directions of the coordinate axes. A displacement vector **u** may be represented by its *vector components* $\mathbf{u}_i$, or by its *scalar components* or *Cartesian components* u_i with respect to the coordinate system *Ox*:

$$\mathbf{u} = \sum_i \mathbf{u}_i \equiv u_i \mathbf{e}_i \Leftrightarrow \mathbf{u}_i = u_i \mathbf{e}_i \quad \text{(no summation w.r. to } i) \tag{1.2.13}$$

The vector **u** may also alternatively be presented as follows:

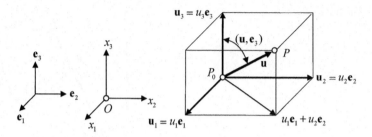

Fig. 1.5 Vector components $\mathbf{u}_i$ and Cartesian scalar components u_i of a vector $\mathbf{u}$ in an orthogonal Cartesian coordinate system Ox in three-dimensional space. Base vectors $\mathbf{e}_i$

$$\mathbf{u} = [u_1, u_2, u_3], \quad u = \begin{pmatrix} u_1 \\ u_2 \\ u_3 \end{pmatrix} = \{u_1 \; u_2 \; u_2\} \tag{1.2.14}$$

In the last formula u is the *vector matrix* related to the vector $\mathbf{u}$.

Let the symbols $(\mathbf{u}, \mathbf{e}_i)$ represent the angles between the vector $\mathbf{u}$ and the base vector $\mathbf{e}_i$. Then the Cartesian components u_i of the vector $\mathbf{u}$ in the coordinate system Ox are:

$$u_i = |\mathbf{u}| \cos(\mathbf{u}, \mathbf{e}_i) \tag{1.2.15}$$

Because the vector components $\mathbf{u}_i$ are orthogonal, the magnitude of the vector $\mathbf{u}$ may be determined as follows:

$$|\mathbf{u}_1 + \mathbf{u}_2| = \sqrt{u_1^2 + u_2^2}, \quad u = |\mathbf{u}| = \sqrt{\left(\sqrt{u_1^2 + u_2^2}\right)^2 + u_3^2} \Rightarrow$$
$$u = |\mathbf{u}| = \sqrt{u_1^2 + u_2^2 + u_3^2} \tag{1.2.16}$$

The resultant $\mathbf{u}$ of n component displacement vectors $\mathbf{u}_i \; i = 1, 2, \ldots, n$ may in principle be determined similarly to in a plane and is demonstrated in Fig. 1.6 for three component vectors $\mathbf{u}_1, \mathbf{u}_2$ and $\mathbf{u}_3$ and their resultant $\mathbf{u}$. In Fig. 1.6 the four vectors are projected onto an axis parallel to the x_2-axis. The components u_2, u_{12}, and u_{22} are positive, while the component u_{32} is negative. It follows from the figure that:

$$u_2 = u_{12} + u_{22} + u_{32}$$

Similarly we find:

$$u_1 = u_{11} + u_{21} + u_{31}, \quad u_3 = u_{13} + u_{23} + u_{33} \Rightarrow u_i = u_{1i} + u_{2i} + u_{3i}$$

Fig. 1.6 Resultant **u** of three component vectors **u**$_1$, **u**$_2$, and **u**$_3$. The figure shows the scalar components in the x_2-direction of the four vectors. Note that the component u_{32} of the vector **u**$_3$ is negative

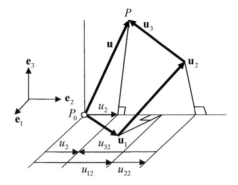

The figure consisting of the four vectors **u**, **u**$_1$, **u**$_2$ and **u**$_3$ is called a *space polygon*, and shows the vector **u** as a *geometrical sum* of the three component vectors **u**$_1$, **u**$_2$ and **u**$_3$.

A generalization to a geometrical sum of n displacements vectors **u**$_i$ $i = 1, 2, \ldots, n$ to a resultant displacement **u** is presented as:

$$\mathbf{u} = \sum_i \mathbf{u}_i \Leftrightarrow u_1 = \sum_i u_{i1}, \quad u_2 = \sum_i u_{i2}, \quad u_3 = \sum_i u_{i3} \qquad (1.2.17)$$

1.2.2 Vector Algebra

Definition of vectors:

> A *vector* is a coordinate invariant quantity uniquely expressed by a magnitude and a direction in space, and obeys the parallelogram law by addition of other vectors representing similar quantities.

Examples of vectors: displacements, velocities, accelerations, angular velocities, and mechanical forces. These vectors will be introduced in Chap. 2. The addition of vectors is exemplified by the addition of displacement components in Sect. 1.2.1.

Vectors will in the present exposition preferably be denoted by bold face lower case Latin letters: **a**, **b**, **c** . . . Bold face upper case Latin letters: **A**, **B**, **C** . . ., will primarily be reserved for quantities we define as tensors in the Chaps. 2 and 3.

In an orthogonal Cartesian coordinate system Ox a vector **a** is represented by its *Cartesian components* a_i, i.e. $\mathbf{a} = a_i \mathbf{e}_i$.

Scalar product of vectors. The *scalar product* of two vectors **a** and **b**, also called the *dot product*, is a scalar denoted by $\mathbf{a} \cdot \mathbf{b}$ ($''$**a** dot **b**$''$) and defined as the scalar:

$$\mathbf{a} \cdot \mathbf{b} = |\mathbf{a}||\mathbf{b}| \cos(\mathbf{a}, \mathbf{b}) \qquad (1.2.18)$$

Fig. 1.7 The scalar product
of the vectors **a** and **b**:
$\mathbf{a} \cdot \mathbf{b} = |\mathbf{a}||\mathbf{b}| \cos(\mathbf{a}, \mathbf{b})$

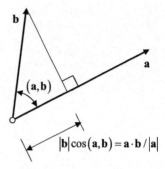

The symbol $(\mathbf{a}, \mathbf{b})$ represents the angle between the two vectors **a** and **b**. As demonstrated in Fig. 1.7 the scalar product $\mathbf{a} \cdot \mathbf{b} = |\mathbf{a}||\mathbf{b}| \cos(\mathbf{a}, \mathbf{b})$ is the product of the magnitude of the vector **a** and the normal projection of the vector **b** onto **a**. Because $\cos(\mathbf{a}, \mathbf{b})$ may be positive or negative the scalar product $\mathbf{a} \cdot \mathbf{b}$ may be positive or negative.

It follows from this definition (1.2.18) that the scalar product is commutative:

$$\mathbf{a} \cdot \mathbf{b} = \mathbf{b} \cdot \mathbf{a} \qquad (1.2.19)$$

Figure 1.8 shows that the scalar product follows the distributive rule:

$$\mathbf{a} \cdot (\mathbf{b} + \mathbf{c}) = \mathbf{a} \cdot \mathbf{b} + \mathbf{a} \cdot \mathbf{c} \qquad (1.2.20)$$

The scalar products of the base vectors $\mathbf{e}_i$ in a Cartesian coordinate system Ox are:

$$\mathbf{e}_i \cdot \mathbf{e}_j = \delta_{ij} \qquad (1.2.21)$$

According to the rule (1.2.20) and formula (1.2.21) the scalar product of the two vectors $\mathbf{a} = a_i \mathbf{e}_i$ and $\mathbf{b} = b_j \mathbf{e}_j$ becomes:

Fig. 1.8 The distribute rule
for the scalar product
$\mathbf{a} \cdot (\mathbf{b} + \mathbf{c}) = \mathbf{a} \cdot \mathbf{b} + \mathbf{a} \cdot \mathbf{c}$

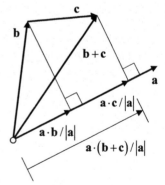

$$\mathbf{a} \cdot \mathbf{b} = (a_i \mathbf{e}_i) \cdot (b_j \mathbf{e}_j) = a_i b_j \, \mathbf{e}_i \cdot \mathbf{e}_j = a_i b_j \delta_{ij} = a_i b_i \Rightarrow$$
$$\mathbf{a} \cdot \mathbf{b} = a_i b_i \tag{1.2.22}$$

The *magnitude of a vector* **a**, i.e. the length of the vector arrow, may now be found from the formula:

$$|\mathbf{a}| = \sqrt{\mathbf{a} \cdot \mathbf{a}} = a_i a_i = a_1^2 + a_2^2 + a_3^2 \tag{1.2.23}$$

Vector product of vectors. Figure 1.9 shows two vectors **a** and **b** and a third vector $\mathbf{p} = \mathbf{a} \times \mathbf{b}$ ("**a** cross **b**") normal to **a** and **b**. The magnitude $|\mathbf{p}|$ of the vector $\mathbf{p} = \mathbf{a} \times \mathbf{b}$ is equal to the area A of the parallelogram spanned by the vectors **a** and **b**:

$$|\mathbf{p}| = |\mathbf{a} \times \mathbf{b}| = A = |\mathbf{a}||\mathbf{b}| \sin(\mathbf{a}, \mathbf{b}) \tag{1.2.24}$$

Here $(\mathbf{a}, \mathbf{b})$ represents the smallest angle between the vectors **a** and **b** (always less than $180°$), such that $\sin(\mathbf{a}, \mathbf{b})$ is always positive. The direction of the vector $\mathbf{p} = \mathbf{a} \times \mathbf{b}$ is determined by:
The right-hand rule: The right hand is held about the vector **p** with the fingers pointing in the direction of a rotation of **a** to **b** through the angle $(\mathbf{a}, \mathbf{b})$. The thumb then points in the direction of the vector $\mathbf{p} = \mathbf{a} \times \mathbf{b}$.
The vector $\mathbf{p} = \mathbf{a} \times \mathbf{b}$ is called the *vector product*, or the *cross product*, of the vectors **a** and **b**. Let **e** be the unit vector in the direction of $\mathbf{p} = \mathbf{a} \times \mathbf{b}$. Then we may write:

$$\mathbf{p} = \mathbf{a} \times \mathbf{b} = A\mathbf{e} = |\mathbf{a}||\mathbf{b}| \sin(\mathbf{a}, \mathbf{b})\mathbf{e} \tag{1.2.25}$$

From the definition (1.2.25) of the vector product $\mathbf{p} = \mathbf{a} \times \mathbf{b}$ it follows that the vector product is not commutative, i.e.:

$$\mathbf{a} \times \mathbf{b} = -\mathbf{b} \times \mathbf{a} \tag{1.2.26}$$

Let $\mathbf{e}_i$ be the base vectors in a right-handed orthogonal Cartesian coordinate system Ox. Then:

Fig. 1.9 Vector product **p** of the vectors **a** and **b**:
$\mathbf{p} = \mathbf{a} \times \mathbf{b} = |\mathbf{a}||\mathbf{b}| \sin(\mathbf{a}, \mathbf{b})\mathbf{e}$

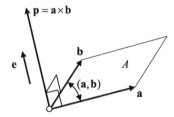

$$\left.\begin{array}{l} \mathbf{e}_1 \times \mathbf{e}_2 = \mathbf{e}_2 \times \mathbf{e}_3 = \mathbf{e}_3 \times \mathbf{e}_1 = 1 \\ \mathbf{e}_3 \times \mathbf{e}_2 = \mathbf{e}_2 \times \mathbf{e}_1 = \mathbf{e}_1 \times \mathbf{e}_3 = -1 \\ \mathbf{e}_1 \times \mathbf{e}_1 = \mathbf{e}_2 \times \mathbf{e}_2 = \mathbf{e}_3 \times \mathbf{e}_3 = 0 \end{array}\right\} \quad \mathbf{e}_i \times \mathbf{e}_j = e_{ijk}\mathbf{e}_k \qquad (1.2.27)$$

We shall now show, applying Figs. 1.10 and 1.11, that the vector product is distributive, which implies that for three vectors $\mathbf{a}, \mathbf{b}$ and $\mathbf{c}$ we may write:

$$\mathbf{a} \times (\mathbf{b}+\mathbf{c}) = \mathbf{a} \times \mathbf{b} + \mathbf{a} \times \mathbf{c} \qquad (1.2.28)$$

Because we will emphasize the geometrical nature of vector algebra, the following proof of the formula (1.2.28), i.e. the *distributive rule for the vector product*, is strictly geometrical.

Figure 1.10a shows three vectors $\mathbf{a}, \mathbf{b}$, and $\bar{\mathbf{b}}$, and a Cartesian coordinate system Ox with the x_3-axis oriented in the direction of the vector $\mathbf{a}$. The vector $\bar{\mathbf{b}}$ is the normal projection of the vector $\mathbf{b}$ onto the x_1x_2-plane. It is seen from the figure that areas of the parallelograms $OP_1P_2P_3O$ and $OP_1P_4P_5O$ are equal. This implies that

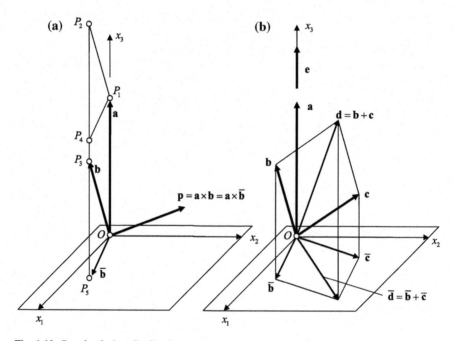

Fig. 1.10 Proof of the distributive rule for vector products: $\mathbf{a} \times (\mathbf{b} \times \mathbf{c}) = \mathbf{a} \times \mathbf{b} + \mathbf{a} \times \mathbf{c}$. (a) Proof of formula (1.2.29): Vector product $\mathbf{p} = \mathbf{a} \times \mathbf{b} = \mathbf{a} \times \bar{\mathbf{b}}$. The vector $\mathbf{a}$ is directed along the the x_3-axis. The vector $\bar{\mathbf{b}}$ is the normal projection of vector $\mathbf{b}$ onto the x_1 x_2-plane. (b) The vector $\mathbf{a}$ is directed along the the x_3-axis. Vector $\mathbf{d} = \mathbf{b}+\mathbf{c}$. Vectors $\bar{\mathbf{b}}, \bar{\mathbf{c}}$, and $\bar{\mathbf{d}}$ and are the normal projections of the vectors $\mathbf{b}$, $\mathbf{c}$ and $\mathbf{d}$ onto the x_1 x_2-plane

1.2 Cartesian Coordinate Systems. Scalars and Vectors

17

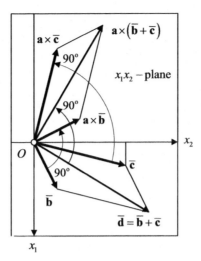

Fig. 1.11 The arrows of the vectors $\bar{\mathbf{b}}$, $\bar{\mathbf{c}}$, and $\bar{\mathbf{d}} = \bar{\mathbf{b}} + \bar{\mathbf{c}}$ in the x_1 x_2-plane are rotated $90°$ counter-clockwise and multiplied by the magnitude $|\mathbf{a}|$ of the vector $\mathbf{a}$. The result is a parallelogram of three vectors: $\mathbf{a} \times \bar{\mathbf{b}}$, $\mathbf{a} \times \bar{\mathbf{c}}$, and $\mathbf{a} \times \bar{\mathbf{d}} = \mathbf{a} \times (\bar{\mathbf{b}} + \bar{\mathbf{c}})$ From the figure it follows that: $\mathbf{a} \times \bar{\mathbf{d}} = \mathbf{a} \times (\bar{\mathbf{b}} + \bar{\mathbf{c}}) = \mathbf{a} \times \bar{\mathbf{b}} + \mathbf{a} \times \bar{\mathbf{c}}$ Using the formula (1.2.29) we may conclude that: $\mathbf{a} \times (\mathbf{b} + \mathbf{c}) = \mathbf{a} \times \mathbf{b}$, $\mathbf{a} \times \mathbf{c}$ This result is a proof of the distributive rule (1.2.28) for the vector product

the vector products $\mathbf{a} \times \mathbf{b}$ and $\mathbf{a} \times \bar{\mathbf{b}}$ are equal and are represented by a vector $\mathbf{p}$ perpendicular to the three vectors $\mathbf{a}, \mathbf{b}$, and $\bar{\mathbf{b}}$:

$$\mathbf{p} = \mathbf{a} \times \mathbf{b} = \mathbf{a} \times \bar{\mathbf{b}} \qquad (1.2.29)$$

Figure 1.10b shows the vectors $\mathbf{a}, \mathbf{b}, \mathbf{c}, \mathbf{d} = \mathbf{b} + \mathbf{c}$, the unit vector $\mathbf{e} = \mathbf{a}/|\mathbf{a}|$, and the normal projections $\bar{\mathbf{b}}$, $\bar{\mathbf{c}}$, and $\bar{\mathbf{d}} = \bar{\mathbf{b}} + \bar{\mathbf{c}}$ of $\mathbf{b}, \mathbf{c}$, and $\mathbf{d} = \mathbf{b} + \mathbf{c}$, onto the $x_1 x_2$-plane in a Cartesian coordinate system Ox. The x_3-axis is oriented in the direction of the vector $\mathbf{a}$.

Figure 1.11 shows the $x_1 x_2$-plane with the vectors $\bar{\mathbf{b}}, \bar{\mathbf{c}}$, and $\bar{\mathbf{d}} = \bar{\mathbf{b}} + \bar{\mathbf{c}}$. The three vectors are rotated $90°$ counter-clockwise and multiplied by the magnitude $|\mathbf{a}|$ to give the three vectors $\mathbf{a} \times \bar{\mathbf{b}}, \mathbf{a} \times \bar{\mathbf{c}}$, and $\mathbf{a} \times \bar{\mathbf{d}} = \mathbf{a} \times (\bar{\mathbf{b}} + \bar{\mathbf{c}})$. It then follows from the figure that:

$$\mathbf{a} \times (\bar{\mathbf{b}} + \bar{\mathbf{c}}) = \mathbf{a} \times \bar{\mathbf{b}} + \mathbf{a} \times \bar{\mathbf{c}} \qquad (1.2.30)$$

We apply the result (1.2.29) to each of the three vectors in this equation and obtain the distributive rule for vector products (1.2.28).

Let three vectors $\mathbf{a}, \mathbf{b}$, and $\mathbf{c}$ be represented by their components in a Cartesian coordinate system Ox:

$$\mathbf{a} = a_i \mathbf{e}_i, \quad \mathbf{b} = b_j \mathbf{e}_j, \quad \mathbf{c} = c_k \mathbf{e}_k \tag{1.2.31}$$

From the definition (1.2.25) of the vector product $\mathbf{p} = \mathbf{a} \times \mathbf{b}$, the formulas (1.2.27), and the distributive rule (1.2.28) it follows that:

$$\begin{aligned}
\mathbf{a} \times \mathbf{b} = \mathbf{c} &\Rightarrow (a_i \mathbf{e}_i) \times (b_j \mathbf{e}_j) = a_i b_j \mathbf{e}_i \times \mathbf{e}_j = a_i b_j e_{ijk} \mathbf{e}_k = c_k \mathbf{e}_k \Rightarrow \\
\mathbf{a} \times \mathbf{b} = \mathbf{c} &\Rightarrow a_i b_j e_{ijk} \mathbf{e}_k = c_k \mathbf{e}_k \Rightarrow c_k = a_i b_j e_{ijk}
\end{aligned} \tag{1.2.32}$$

$$\mathbf{a} \times \mathbf{b} = a_i b_j e_{ijk} \mathbf{e}_k \tag{1.2.33}$$

Using formula (1.1.21) for the determinant A of a matrix (A_{ij}), we rewrite formula (1.2.33):

$$\mathbf{a} \times \mathbf{b} = \det \begin{pmatrix} a_1 & a_2 & a_3 \\ b_1 & b_2 & b_3 \\ \mathbf{e}_1 & \mathbf{e}_2 & \mathbf{e}_3 \end{pmatrix} \tag{1.2.34}$$

Scalar triple product. The *scalar triple product* $[\mathbf{abc}]$ of the three vectors $\mathbf{a}$, $\mathbf{b}$, and $\mathbf{c}$, also called the *box product*, is defined by:

$$[\mathbf{abc}] \equiv (\mathbf{a} \times \mathbf{b}) \cdot \mathbf{c} = \mathbf{a} \cdot (\mathbf{b} \times \mathbf{c}) = \mathbf{b} \cdot (\mathbf{c} \times \mathbf{a}) \tag{1.2.35}$$

Using formula (1.2.33) for the vector product and formula (1.2.22) for the scalar product, we obtain the result:

$$\mathbf{a} \times \mathbf{b} = a_i b_j e_{ijk} \mathbf{e}_k, \quad (\mathbf{a} \times \mathbf{b}) \cdot \mathbf{c} = a_i b_j e_{ijk} c_k \Rightarrow$$
$$[\mathbf{abc}] = e_{ijk} a_i b_j c_k = \det \begin{pmatrix} a_1 & a_2 & a_3 \\ b_1 & b_2 & b_3 \\ c_1 & c_2 & c_3 \end{pmatrix} \tag{1.2.36}$$

The base vectors $\mathbf{e}_i$ are related according to:

$$[\mathbf{e}_1 \mathbf{e}_2 \mathbf{e}_3] = 1, \quad [\mathbf{e}_3 \mathbf{e}_2 \mathbf{e}_3] = 1 \quad \text{etc.} \Leftrightarrow [\mathbf{e}_i \mathbf{e}_j \mathbf{e}_k] = e_{ijk} \tag{1.2.37}$$

This relation may be taken to define an *orthogonal right-handed system of unit vectors*.

1.3 Cartesian Coordinate Transformations

Vectors and tensors, to be introduced in Chap. 2 and generally defined in Chap. 3, are coordinate invariant quantities, which in any coordinate system we have chosen, are represented by their components. It is important to determine the relations

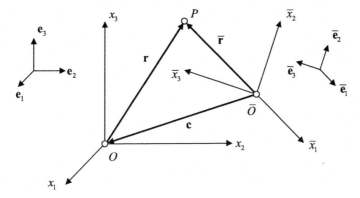

Fig. 1.12 Two right-handed orthogonal Cartesian coordinate systems: Ox with base vectors $\mathbf{e}_i$, and $\bar{O}\bar{x}$ with base vectors $\bar{\mathbf{e}}_i$. Place P defined by place vectors $\mathbf{r}$ and $\bar{\mathbf{r}}$

between sets of components relative to different coordinate systems. Let Ox and $\bar{O}\bar{x}$ denote two right-handed orthogonal Cartesian coordinate systems, see Fig. 1.12. When the Ox-system has been given, the system $\bar{O}\bar{x}$ is determined by the position vector $\mathbf{c}$ from the origin $\bar{O}$ of the $\bar{O}\bar{x}$-system to the origin O and the base vectors $\bar{\mathbf{e}}_i$.

$$\mathbf{c} = c_k \mathbf{e}_k = \bar{c}_i \bar{\mathbf{e}}_i \tag{1.3.1}$$

$$\bar{\mathbf{e}}_i = Q_{ik}\mathbf{e}_k, \quad \mathbf{e}_k = Q_{ik}\bar{\mathbf{e}}_i, \quad Q_{ik} = \cos(\bar{\mathbf{e}}_i, \mathbf{e}_k) = \bar{\mathbf{e}}_i \cdot \mathbf{e}_k \tag{1.3.2}$$

The elements Q_{ik} are called *direction cosines* and are elements in a matrix that is called the *transformation matrix* Q for the coordinate transformation from Ox to $\bar{O}\bar{x}$. The 9 elements Q_{ik} are, as will be shown below, connected through 7 relations.

Both Ox and $\bar{O}\bar{x}$ shall be right-handed orthogonal Cartesian coordinate systems and the base vectors $\mathbf{e}_i$ and $\bar{\mathbf{e}}_i$ shall be unit vectors. These requirements are satisfied for $\bar{\mathbf{e}}_i$ by the conditions:

$$\bar{\mathbf{e}}_i \cdot \bar{\mathbf{e}}_j = \delta_{ij}, \quad [\bar{\mathbf{e}}_1 \bar{\mathbf{e}}_2 \bar{\mathbf{e}}_3] = 1 \tag{1.3.3}$$

Confer the formulas (1.2.21) and (1.2.37). Now, also using formula (1.1.21), we obtain:

$$\bar{\mathbf{e}}_i \cdot \bar{\mathbf{e}}_j = (Q_{ik}\mathbf{e}_k) \cdot (Q_{jl}\mathbf{e}_l) = Q_{ik}Q_{jl}(\mathbf{e}_k \cdot \mathbf{e}_l) = Q_{ik}Q_{jl}\delta_{kl} = Q_{ik}Q_{jk}$$
$$[\bar{\mathbf{e}}_1 \bar{\mathbf{e}}_2 \bar{\mathbf{e}}_3] = [(Q_{1r}\mathbf{e}_r)(Q_{2s}\mathbf{e}_s)(Q_{3t}\mathbf{e}_t)] = Q_{1r}Q_{2s}Q_{3t}[\mathbf{e}_r \mathbf{e}_s \mathbf{e}_t] = Q_{1r}Q_{2s}Q_{3t}e_{rst} = \det Q$$

These results show, when compared with the conditions (1.3.3), that the transformation matrix Q has to satisfy the conditions:

$$Q_{ik}Q_{jk} = \delta_{ij} \Leftrightarrow QQ^T = 1 \Leftrightarrow Q^{-1} = Q^T, \quad \det Q = 1 \tag{1.3.4}$$

A matrix with these properties is called an *orthogonal matrix*. From the formulas (1.3.2) we see that the rows in the matrix Q represent the components in the Ox-system of the orthogonal vectors $\bar{\mathbf{e}}_i$, and that the columns in Q represent the components in the $\bar{O}\bar{x}$-system of the orthogonal vectors $\mathbf{e}_k$.

A place P in three-dimensional space may either be defined by the place vector $\mathbf{r}$ from the origin O to P, or by the place vector $\bar{\mathbf{r}}$ from the origin $\bar{O}$ to P. From Fig. 1.12 we obtain:

$$\bar{\mathbf{r}} = \mathbf{c} + \mathbf{r} \tag{1.3.5}$$

Let the coordinates of the place P be x_k in the Ox-system and $\bar{x}_i$ in the $\bar{O}\bar{x}$-system. Then:

$$\mathbf{r} = x_k\mathbf{e}_k = x_k(Q_{ik}\bar{\mathbf{e}}_i) = Q_{ik}x_k\bar{\mathbf{e}}_i, \quad \text{and} \quad \bar{\mathbf{r}} = \bar{x}_i\bar{\mathbf{e}}_i \tag{1.3.6}$$

The representation of the relation (1.3.5) in the coordinate system $\bar{O}\bar{x}$ provides the *coordinate transformation formula*:

$$\bar{x}_i = \bar{c}_i + Q_{ik}x_k \Leftrightarrow \bar{x} = \bar{c} + Qx \tag{1.3.7}$$

The inverse transformation is easily found to be:

$$x_k = -c_k + Q_{ik}\bar{x}_i \Leftrightarrow x = -c + Q^T\bar{x} \tag{1.3.8}$$

For a vector $\mathbf{a}$ with the components a_k in Ox and $\bar{a}_i$ in $\bar{O}\bar{x}$, we get:

$$\begin{aligned} \mathbf{a} = \bar{a}_i\bar{\mathbf{e}}_i = \bar{a}_iQ_{ik}\mathbf{e}_k = a_k\mathbf{e}_k = a_kQ_{ik}\bar{\mathbf{e}}_i = Q_{ik}a_k\bar{\mathbf{e}}_i \Leftrightarrow \\ \bar{a}_i = Q_{ik}a_k \Leftrightarrow \bar{a} = Qa, \quad a_k = Q_{ik}\bar{a}_i \Leftrightarrow a = Q^T\bar{a} \end{aligned} \tag{1.3.9}$$

In plane, two-dimensional analysis we introduce plane Cartesian coordinate systems Ox and $\bar{O}\bar{x}$ with the the x_α-axes and the $\bar{x}_\alpha$-axes in the plane, Fig. 1.13. The $\bar{x}_1$-axis makes the angle θ with respect to the x_1-axis. Then:

$$Q = (Q_{\alpha\beta}) = (\cos(\bar{\mathbf{e}}_\alpha, \mathbf{e}_\beta)) = \begin{pmatrix} \cos\theta & \sin\theta \\ -\sin\theta & \cos\theta \end{pmatrix} \quad \text{or}$$

$$Q = (Q_{ik}) = (\cos(\bar{\mathbf{e}}_i, \mathbf{e}_k)) = \begin{pmatrix} \cos\theta & \sin\theta & 0 \\ -\sin\theta & \cos\theta & 0 \\ 0 & 0 & 1 \end{pmatrix} \tag{1.3.10}$$

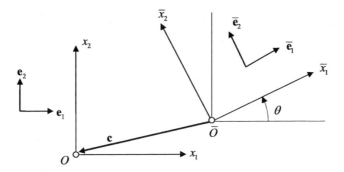

Fig. 1.13 Transformation of plane Cartesian coordinates: from the Ox-system to the $\bar{O}\bar{x}$-system

1.4 Curves in Space

Let the parameter p be a scalar variable and let the symbol $\mathbf{a}(p)$ represent a unique vector $\mathbf{a}$ for every value of the parameter p. The components of the vector $\mathbf{a}$ in a Cartesian coordinate system Ox are then also functions of the parameter p:

$$\mathbf{a}(p) = a_i(p)\mathbf{e}_i \qquad (1.4.1)$$

The derivative $d\mathbf{a}/dp$ is defined by the limiting process:

$$\frac{d\mathbf{a}}{dp} = \lim_{\Delta p \to 0} \frac{\mathbf{a}(p + \Delta p) - \mathbf{a}(p)}{\Delta p} \equiv \lim_{\Delta p \to 0} \frac{\Delta \mathbf{a}(p, \Delta p)}{\Delta p} \qquad (1.4.2)$$

The following differentiation rules are easily tested:

$$\frac{d\mathbf{a}}{dp} = \frac{da_i}{dp}\mathbf{e}_i, \quad \frac{d(\alpha\mathbf{a})}{dp} = \frac{d\alpha}{dp}\mathbf{a} + \alpha\frac{d\mathbf{a}}{dp}$$

$$\frac{d(\mathbf{a}\cdot\mathbf{b})}{dp} = \frac{d\mathbf{a}}{dp}\cdot\mathbf{b} + \mathbf{a}\cdot\frac{d\mathbf{b}}{dp}, \quad \frac{d(\mathbf{a}\times\mathbf{b})}{dp} = \frac{d\mathbf{a}}{dp}\times\mathbf{b} + \mathbf{a}\times\frac{d\mathbf{b}}{dp} \qquad (1.4.3)$$

The function $\mathbf{r} = \mathbf{r}(p) = x_i(p)\mathbf{e}_i$ describes a collection of places in space that we call a *space curve*. If the function $\mathbf{r}(p)$ is a continuous function of the parameter p and has continuous first derivatives $d\mathbf{r}/dp$, the curve is called a *smooth curve*. Figure 1.14 illustrates the space curve in a Cartesian coordinate system Ox. Three places are marked on the curve in Fig. 1.14: P_0 for parameter value p_0 and with place vector $\mathbf{r}(p_0)$, P for parameter value p and with place vector $\mathbf{r}(p)$, and a place given by the place vector $\mathbf{r}(p + \Delta p)$.

Fig. 1.14 Space curve $\mathbf{r}(p)$.
Tangent vector $\mathbf{t} = d\mathbf{r}/ds$

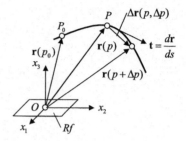

A tangent vector to the curve is defined by:

$$\frac{d\mathbf{r}}{dp} = \lim_{\Delta p \to 0} \frac{\mathbf{r}(p + \Delta p) - \mathbf{r}(p)}{\Delta p} \equiv \lim_{\Delta p \to 0} \frac{\Delta \mathbf{r}(p, \Delta p)}{\Delta p} = \frac{d\mathbf{r}}{dp} = \frac{dx_i}{dp}\mathbf{e}_i \qquad (1.4.4)$$

A curve consisting of a finite number of smooth curves is called a *piecewise smooth curve*. The length s of a smooth curve between two places P_0, for parameter p_0, and P for parameter p, is given by the *arc length formula*:

$$s(p) = \int_{p_0}^{p} \sqrt{\frac{d\mathbf{r}}{d\bar{p}} \cdot \frac{d\mathbf{r}}{d\bar{p}}} d\bar{p} = \int_{p_0}^{p} \sqrt{\frac{dx_i \, dx_i}{d\bar{p} \, d\bar{p}}} d\bar{p} \qquad (1.4.5)$$

We assume that there is a one-to-one correspondence between the parameter p and the arc length s. The formula (1.4.5) may then be used to substitute the parameter p by the *arc length* s as a new curve parameter. The space curve will now be defined by the place vector:

$$\mathbf{r} = \mathbf{r}(s) = x_i(s)\mathbf{e}_i \qquad (1.4.6)$$

From the formulas (1.4.5) and (1.4.6) we compute:

$$\frac{ds}{dp} = \sqrt{\frac{dx_i \, dx_i}{dp \, dp}} \Rightarrow ds = |d\mathbf{r}|, \quad d\mathbf{r} = dx_i \mathbf{e}_i \qquad (1.4.7)$$

We define the *tangent vector* $\mathbf{t}$ to the curve and at the place $\mathbf{r}(s)$, by the unit vector:

$$\mathbf{t} = \frac{d\mathbf{r}}{ds}, \quad t_i = \frac{dx_i}{ds} \qquad (1.4.8)$$

A *normal vector* to the space curve at a place $\mathbf{r}(s)$ is a vector perpendicular to the tangent vector at the place. The infinite number of such normal vectors defines the *plane of normals* at the place. Two normal vectors are of special interest: the *principal normal* vector $\mathbf{n}$ and the *binormal* vector $\mathbf{b}$. Before defining these two vectors we define the *curvature* κ and the *radius of curvature* ρ of the curve and at the place $\mathbf{r}(s)$:

$$\kappa = \left|\frac{d\mathbf{t}}{ds}\right| \equiv \left|\frac{d^2\mathbf{r}}{ds^2}\right|, \quad \rho = \frac{1}{\kappa} \tag{1.4.9}$$

The *principal normal vector* $\mathbf{n}$ to the curve is now defined by:

$$\mathbf{n} = \frac{1}{\kappa}\frac{d\mathbf{t}}{ds} = \frac{1}{\kappa}\frac{d^2\mathbf{r}}{ds^2} \Leftrightarrow n_i = \frac{1}{\kappa}\frac{d^2 x_i}{ds^2} \tag{1.4.10}$$

The plane defined by the tangent vector $\mathbf{t}$ and the principal normal $\mathbf{n}$ is called the *osculating plane*, Fig. 1.15. The unit vector $\mathbf{b} = \mathbf{t} \times \mathbf{n}$ is called the *binormal vector* to the space curve:

$$\mathbf{b} = \mathbf{t} \times \mathbf{n} \Rightarrow b_i = e_{ijk}t_j n_k \tag{1.4.11}$$

We can show that the vector $d\mathbf{b}/ds$ is parallel to the principal normal vector $\mathbf{n}$.

$$\left.\begin{array}{l} \mathbf{b}\cdot\mathbf{b} = 1 \Rightarrow \frac{d\mathbf{b}}{ds}\cdot\mathbf{b} = 0 \\ \mathbf{b}\cdot\mathbf{t} = 0 \Rightarrow \\ \frac{d\mathbf{b}}{ds}\cdot\mathbf{t} = -\mathbf{b}\cdot\frac{d\mathbf{t}}{ds} = -\mathbf{b}\cdot(\kappa\mathbf{n}) = 0 \end{array}\right\} \Rightarrow$$

$\dfrac{d\mathbf{b}}{ds}$ is perpendicular to both $\mathbf{b}$ and $\mathbf{t}$, and therefore parallel to $\mathbf{n}$

We use this result to define the *torsion* τ *of the curve* by the formula:

$$\frac{d\mathbf{b}}{ds} = -\tau\mathbf{n} \Rightarrow \tau = -\mathbf{n}\cdot\frac{d\mathbf{b}}{ds} \tag{1.4.12}$$

It follows from the formula $(1.4.12)_2$ that the torsion τ of a curve is positive if for increasing value of the parameter s the principal normal $\mathbf{b}$ rotates about the tangent vector $\mathbf{t}$ clockwise when observed in the direction of $\mathbf{t}$.

We shall now present the three *Frenet–Serret formulas* for a space curve, named after Jean Frederic Frenet [1816–1900] and Joseph Alfred Serret [1819–1885]:

Fig. 1.15 Space curve $\mathbf{r}$ (s) with tangent vector $\mathbf{t}$, principal normal $\mathbf{n}$, binormal $\mathbf{b}$, plane of normals, and osculating plane

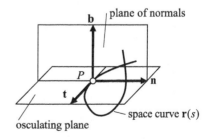

$$\frac{d\mathbf{t}}{ds} = \kappa\mathbf{n} \Rightarrow \frac{dt_i}{ds} = \kappa n_i$$

$$\frac{d\mathbf{n}}{ds} = \tau\mathbf{b} - \kappa\mathbf{t} \Rightarrow \frac{dn_i}{ds} = \tau b_i - \kappa t_i \qquad (1.4.13)$$

$$\frac{d\mathbf{b}}{ds} = -\tau\mathbf{n} \Rightarrow \frac{db_i}{ds} = -\tau n_i$$

The first and the last of these formulas are given above in the formulas (1.4.10) and (1.4.12). The second of the formulas (1.4.13) is derived as follows:

(a) $\mathbf{b} \cdot \mathbf{n} = 0 \Rightarrow \mathbf{b} \cdot \dfrac{d\mathbf{n}}{ds} = -\dfrac{d\mathbf{b}}{ds} \cdot \mathbf{n} = \tau$

(b) $\mathbf{n} \cdot \mathbf{n} = 1 \Rightarrow \dfrac{d\mathbf{n}}{ds} \cdot \mathbf{n} = 0 \Rightarrow \left[\dfrac{d\mathbf{n}}{ds} + \kappa\mathbf{t}\right] \cdot \mathbf{n} = 0$

(c) $\mathbf{n} \cdot \mathbf{t} = 0 \Rightarrow \dfrac{d\mathbf{n}}{ds} \cdot \mathbf{t} + \mathbf{n} \cdot \dfrac{d\mathbf{t}}{ds} = \dfrac{d\mathbf{n}}{ds} \cdot \mathbf{t} + \mathbf{n} \cdot (\kappa\mathbf{n}) = 0 \Rightarrow \left[\dfrac{d\mathbf{n}}{ds} + \kappa\mathbf{t}\right] \cdot \mathbf{t} = 0$

The results (b) and (c) show that the vector $[d\mathbf{n}/ds + \kappa\mathbf{t}]$ is parallel to the binormal $\mathbf{b}$, and the result (a) shows that the magnitude of this vector is the torsion τ. Hence:

$$\frac{d\mathbf{n}}{ds} + \kappa\mathbf{t} = \tau\mathbf{b} \Rightarrow \frac{d\mathbf{n}}{ds} = \tau\mathbf{b} - \kappa\mathbf{t} \Rightarrow (1.4.13)_2$$

From the result (a) we obtain the following formula for the torsion of the space curve:

$$\tau = \mathbf{b} \cdot \frac{d\mathbf{n}}{ds} = (\mathbf{t} \times \mathbf{n}) \cdot \frac{d\mathbf{n}}{ds} = \left[\mathbf{t}\mathbf{n}\frac{d\mathbf{n}}{ds}\right] = e_{ijk} t_i n_j \frac{dn_k}{ds} \qquad (1.4.14)$$

An alternative expression for the torsion τ of a space curve is obtained as follows:

Fig. 1.16 Helix $\mathbf{r} = R\mathbf{e}_R(\theta) + z\mathbf{e}_z$ Tangent $\mathbf{t}$. Principal normal $\mathbf{n}$. Binormal $\mathbf{b}$. Cylindrical coordinates (R, θ, z)

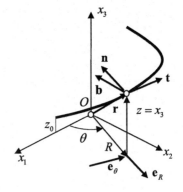

$$\frac{d\mathbf{n}}{ds} = \frac{d}{ds}\left(\frac{1}{\kappa}\frac{d\mathbf{t}}{ds}\right) = \frac{d}{ds}\left(\frac{1}{\kappa}\frac{d^2\mathbf{r}}{ds^2}\right) = \frac{-1}{\kappa^2}\frac{d\kappa}{ds}\frac{d^2\mathbf{r}}{ds^2} + \frac{1}{\kappa}\frac{d^3\mathbf{r}}{ds^3} = -\frac{1}{\kappa}\frac{d\kappa}{ds}\mathbf{n} + \frac{1}{\kappa}\frac{d^3\mathbf{r}}{ds^3} \Rightarrow$$

$$\tau = (\mathbf{t}\times\mathbf{n})\cdot\frac{d\mathbf{n}}{ds} = \left(\frac{d\mathbf{r}}{ds}\times\frac{d^2\mathbf{r}}{ds^2}\frac{1}{\kappa}\right)\cdot\left(-\frac{1}{\kappa}\frac{d\kappa}{ds}\mathbf{n} + \frac{1}{\kappa}\frac{d^3\mathbf{r}}{ds^3}\right) = \frac{1}{\kappa^2}\left[\frac{d\mathbf{r}}{ds}\frac{d^2\mathbf{r}}{ds^2}\frac{d^3\mathbf{r}}{ds^3}\right] \Rightarrow$$

$$\tau = \frac{1}{\kappa^2}\left[\frac{d\mathbf{r}}{ds}\frac{d^2\mathbf{r}}{ds^2}\frac{d^3\mathbf{r}}{ds^3}\right] \tag{1.4.15}$$

Example 1.1 Helix A helix is a space curve defined by the place vector $\mathbf{r}(R,\theta,z) = R\mathbf{e}_R(\theta) + z\mathbf{e}_z$ in Fig. 1.16, where we have introduced *cylindrical coordinates* (R,θ,z) as defined in detail in Fig. 1.20:

$$\mathbf{r} = R\mathbf{e}_R(\theta) + z\mathbf{e}_3 = x_i\mathbf{e}_i, \quad x_1 = R\cos\theta, \quad x_2 = R\sin\theta, \quad x_3 = z = b\theta + z_0$$

$R, b,$ and z_0 are constant parameters

The angle θ will be chosen as curve parameter $p = \theta$. The arc length formula (1.4.5) gives:

$$s(\theta) = \int_0^\theta \sqrt{\frac{d\mathbf{r}}{d\bar\theta}\cdot\frac{d\mathbf{r}}{d\bar\theta}}d\bar\theta = \int_0^\theta \sqrt{\frac{dx_i}{d\bar\theta}\frac{dx_i}{d\bar\theta}}d\bar\theta = \int_0^\theta \sqrt{(-R\sin\theta)^2 + (R\cos\theta)^2 + b^2}d\bar\theta = \sqrt{R^2 + b^2}\,\theta \Rightarrow$$

$$\frac{d\theta}{ds} = \frac{1}{\sqrt{R^2 + b^2}}$$

The tangent vector $\mathbf{t}$, the curvature κ, the radius of curvature ρ, the principal normal vector $\mathbf{n}$, the binormal vector $\mathbf{b}$, and the torsion τ are computed from the formulas (1.4.8), (1.4.9), (1.4.10), (1.4.11), (1.4.15):

$$\mathbf{t} = \frac{d\mathbf{r}}{ds} = \frac{d\mathbf{r}}{d\theta}\frac{d\theta}{ds} = \frac{1}{\sqrt{R^2 + b^2}}[-R\sin\theta, R\cos\theta, b]$$

$$\mathbf{t} = \frac{1}{\sqrt{R^2 + b^2}}(R\mathbf{e}_\theta + b\mathbf{e}_z)$$

$$\frac{d^2\mathbf{r}}{ds^2} = \frac{1}{R^2 + b^2}[-R\cos\theta, -R\sin\theta, 0] = \frac{-R}{R^2 + b^2}\mathbf{e}_R$$

$$\kappa = \left|\frac{d^2\mathbf{r}}{ds^2}\right| = \frac{R}{R^2 + b^2}, \quad \rho = \frac{1}{\kappa} = \frac{R^2 + b^2}{R}$$

$$\mathbf{n} = \frac{1}{\kappa}\frac{d^2\mathbf{r}}{ds^2} = -\mathbf{e}_R$$

$$\mathbf{b} = \mathbf{t} \times \mathbf{n} = \det \begin{pmatrix} t_1 & t_2 & t_3 \\ n_1 & n_2 & n_3 \\ \mathbf{e}_1 & \mathbf{e}_2 & \mathbf{e}_3 \end{pmatrix} \Rightarrow \mathbf{b} = \frac{1}{\sqrt{R^2+b^2}}[b\sin\theta, -b\cos\theta, R]$$

$$= \frac{1}{\sqrt{R^2+b^2}}(R\mathbf{e}_z - b\mathbf{e}_\theta)$$

$$\frac{d^3\mathbf{r}}{ds^3} = [R\sin\theta, -R\cos\theta, 0]\frac{1}{R^2+b^2}\frac{1}{\sqrt{R^2+b^2}}, \quad \tau = \frac{1}{\kappa^2}\left[\frac{d\mathbf{r}}{ds}\frac{d^2\mathbf{r}}{ds^2}\frac{d^3\mathbf{r}}{ds^3}\right]$$

$$= \frac{1}{\kappa^2}\det\begin{pmatrix} -R\sin\theta & R\cos\theta & b \\ -R\cos\theta & -R\sin\theta & 0 \\ R\sin\theta & -R\cos\theta & 0 \end{pmatrix}\frac{1}{\sqrt{R^2+b^2}}\frac{1}{R^2+b^2}\frac{1}{R^2+b^2}\frac{1}{\sqrt{R^2+b^2}} \Rightarrow \tau = \frac{b}{R^2+b^2}$$

1.5   Dynamics of a Mass Particle

A *mass particle* is a body of mass m and which is small enough for the extension of the body to be neglected, and the position P of the body may at any time t be localized by a *place* $\mathbf{r}(t)$ in space. Figure 1.17 illustrates a reference Rf, a Cartesian coordinate system Ox, and the path of the particle as it moves from the position P_0 at the time t_0 to the position P at the time t. The length of the particle path from P_0 to P is denoted by the *path length* $s(t)$.

The *velocity* $\mathbf{v}$ and the *acceleration* $\mathbf{a}$ of the particle are defined by the vectors:

$$\mathbf{v} = \mathbf{v}(t) = \frac{d\mathbf{r}}{dt} \equiv \dot{\mathbf{r}}, \quad \mathbf{a} = \mathbf{a}(t) = \frac{d\mathbf{v}}{dt} \equiv \dot{\mathbf{v}} = \ddot{\mathbf{r}} \tag{1.5.1}$$

Using the definition of the tangent vector $\mathbf{t}$ from the formula (1.4.8) and the principal normal vector $\mathbf{n}$ from formula (1.4.10), we obtain:

$$\mathbf{v} = \frac{d\mathbf{r}}{dt} = \frac{d\mathbf{r}}{ds}\frac{ds}{dt} = \frac{ds}{dt}\mathbf{t} \equiv \dot{s}\mathbf{t} \equiv v\mathbf{t} \tag{1.5.2}$$

Fig. 1.17 Motion of mass particle of mass m. Velocity $\mathbf{v}$. Acceleration $\mathbf{a}$. Path length s (t)

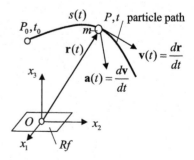

$$\mathbf{a} = \frac{d\mathbf{v}}{dt} = \frac{d^2s}{dt^2}\mathbf{t} + \frac{ds}{dt}\left(\frac{dt\,ds}{ds\,dt}\right) = \ddot{s}\mathbf{t} + \frac{\dot{s}^2}{\rho}\mathbf{n} \qquad (1.5.3)$$

The acceleration vector has two distinct components:

$$\text{tangential acceleration: } a_t = \ddot{s}.$$
$$\text{normal acceleration: } a_n = \frac{\dot{s}^2}{\rho} \equiv \frac{v^2}{\rho} \qquad (1.5.4)$$

The mass particle moves under the influence of a *force* given by the vector $\mathbf{f}$ according to *Newton's second law of motion*:

$$\mathbf{f} = m\mathbf{a} \qquad (1.5.5)$$

The equation of motion of a mass particle (1.5.5) may be decomposed into the three directions $\mathbf{t}, \mathbf{n}$, and $\mathbf{b}$:

$$f_t = ma_t, \quad f_n = ma_n, \quad f_b = 0 \qquad (1.5.6)$$

Example 1.2 Mass Particle Moving on a Helical Constraint
Figure 1.18 illustrates a particle with mass m on a fixed curve with the form of the helix presented in Example 1.1. The particle moves subjected to the vertical gravitational force mg and two constraining forces: N in the direction of the principal normal $\mathbf{n}$ and B in the direction of the binormal $\mathbf{b}$. The geometry of the helix is given in Example 1.1. We shall find expressions for the motion given by the angle $\theta(t)$ and the constraining forces N and B. The initial conditions are: $\theta(0) = \theta_0$, $\dot{\theta}(0) = 0$. The acceleration $\mathbf{a}$ consists of a tangential acceleration and a normal acceleration:

$$\mathbf{a} = a_t\mathbf{t} + a_n\mathbf{n}, \quad a_t = \ddot{s} = \sqrt{R^2 + b^2}\,\ddot{\theta}$$
$$a_n = \frac{\dot{s}^2}{\rho} = \frac{1}{\rho}\left(\frac{ds}{d\theta}\dot{\theta}\right)^2 = \frac{R}{R^2 + b^2}\left(\sqrt{R^2 + b^2}\,\dot{\theta}\right)^2 = R(\dot{\theta})^2$$

Fig. 1.18 Particle of mass m moving on the helix $\mathbf{r} = r\mathbf{e}_R(\theta) + x_3\,\mathbf{e}_3$. Vertical gravitational force $(-mg)\mathbf{e}_3$. Constraining forces $N\mathbf{n}$ and $B\mathbf{b}$

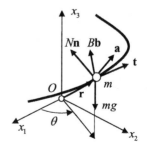

Newton's second law of motion (1.5.6) yields:

$$\mathbf{f} = m\mathbf{a} \Rightarrow$$

$$f_t = ma_t \Rightarrow (-mg\mathbf{e}_3)\cdot\mathbf{t} = m\sqrt{R^2+b^2}\,\ddot{\theta} \Rightarrow \ddot{\theta} = \frac{-bg}{R^2+b^2}, \quad \Rightarrow \dot{\theta} = \frac{-bg}{R^2+b^2}t \Rightarrow$$

$$\Rightarrow \theta = \frac{-bg}{2(R^2+b^2)}t^2 + \theta_0$$

$$f_n = ma_n \Rightarrow N + (-mg\mathbf{e}_3)\cdot\mathbf{n} = mR(\dot{\theta})^2 \Rightarrow N = mR(\dot{\theta})^2 = \frac{mRb^2g^2}{(R^2+b^2)^2}t^2$$

$$f_b = 0 \Rightarrow B + (-mg\mathbf{e}_3)\cdot\mathbf{b} = 0 \Rightarrow B = \frac{mgR}{\sqrt{R^2+b^2}}$$

1.6   Scalar Fields and Vector Fields

Scalars and vectors, representing physical or geometrical quantities related to places $\mathbf{r}$ in a defined region in space, are represented by fields. Here we define two kinds of fields:

$$\alpha = \alpha(\mathbf{r},t) \quad \text{a scalar field,} \quad \mathbf{a} = \mathbf{a}(\mathbf{r},t) \quad \text{a vector field} \qquad (1.6.1)$$

The symbol t is the *time*. The temperature θ in a body is an example of a scalar field, while the velocity $\mathbf{v}$ of a particle or a body is a vector field. If the time t does not appear as an independent variable in a field function, the field is said to be a *steady field*. If the time t is the only argument in the field function, the field is called a *uniform field*. The fields $\alpha = \alpha(\mathbf{r},t)$ and $\mathbf{a} = \mathbf{a}(\mathbf{r},t)$ are also denoted by:

$$\alpha = \alpha(\mathbf{r},t) = \alpha(x_1,x_2,x_3,t) \equiv \alpha(x,t), \quad \mathbf{a} = \mathbf{a}(\mathbf{r},t) = \mathbf{a}(x_1,x_2,x_3,t) \equiv \mathbf{a}(x,t)$$
$$(1.6.2)$$

The partial derivatives of a field function $f(\mathbf{r},t)$ with respect to the time t and the space coordinates x_i in the Ox-system and $\bar{x}_i$ in the $\bar{O}\bar{x}$-system, are expressed by the symbols:

$$\partial_t f \equiv \frac{\partial f}{\partial t}, \quad f_{,i} \equiv \frac{\partial f}{\partial x_i}, \quad \bar{f}_{,i} \equiv \frac{\partial f}{\partial \bar{x}_i}, \quad f_{,i\,,j} \equiv f_{,ij} \equiv f_{,ji} \equiv \frac{\partial^2 f}{\partial x_i x_j} \qquad (1.6.3)$$

Using a comma to indicate partial derivative with respect to a space coordinate is called the *comma notation*. For the partial derivatives of the vector components $a_i(x,t)$ of a vector $\mathbf{a}$ with respect to the space coordinates x_i in the Ox-system, and the partial derivatives of the vector components $\bar{a}_i(\bar{x},t)$ of the vector $\mathbf{a}$ with respect to the space coordinates $\bar{x}_i$ in the $\bar{O}\bar{x}$-system, we introduce the comma notations:

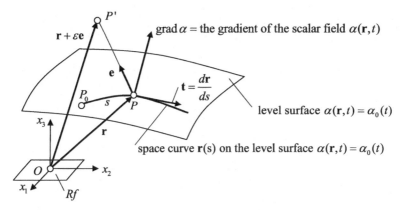

Fig. 1.19 Level surface $\alpha(\mathbf{r}, t) = \alpha_0(t)$ at the time t with a space curve $\mathbf{r}(s)$. s = the arc length of the space curve from the reference point P_0 to the point P at the place $\mathbf{r}$. The gradient of the scalar field $\alpha(\mathbf{r}, t)$ grad α is a normal vector to the level surface

$$a_{i,j} \equiv \frac{\partial a_i}{\partial x_j}, \quad \bar{a}_{i,j} \equiv \frac{\partial \bar{a}_i}{\partial \bar{x}_j} \tag{1.6.4}$$

The equation $\alpha(\mathbf{r}, t) = \alpha_0(t)$, where $\alpha_0(t)$ is a function of time only, represents at any time t a surface in space, see Fig. 1.19. The surface is called a *level surface* for the scalar field $\alpha(\mathbf{r}, t)$. From the place P on the surface given by the place vector $\mathbf{r}$, we introduce an axis in an arbitrary direction given by the unit vector $\mathbf{e} = e_i \mathbf{e}_i$. A coordinate ε along the axis is introduced representing the distance from the place P to a place P' on the axis. The place P' is now given by the place vector $\mathbf{r} + \varepsilon \mathbf{e}$. The value of the scalar field α at the place P' is equal to $\alpha(\mathbf{r} + \varepsilon \mathbf{e}, t)$. The rate of change of α at the place $\mathbf{r}$ for the direction $\mathbf{e}$ is defined by the *directional derivative* of α at the place $\mathbf{r}$ and in the direction $\mathbf{e}$, and is given by:

$$\left.\frac{\partial \alpha}{\partial \varepsilon}\right|_{\varepsilon=0} = \left[\frac{\partial \alpha}{\partial(x_i + \varepsilon e_i)} \cdot \frac{\partial(x_i + \varepsilon e_i)}{\partial \varepsilon}\right]_{\varepsilon=0} = \frac{\partial \alpha}{\partial x_i} e_i \equiv \alpha_{,i} e_i \tag{1.6.5}$$

For an arbitrarily given direction $\mathbf{e}$ the directional derivative defined in formula (1.6.5) is a scalar field, and it follows from formula (1.6.5) that the components $\partial \alpha / \partial x_i$ define a vector field: the *gradient of the scalar field* $\alpha(\mathbf{r}, t)$ denoted by gradα :

$$\text{grad}\alpha = \frac{\partial \alpha}{\partial x_i} \mathbf{e}_i \equiv \alpha_{,i} \mathbf{e}_i \tag{1.6.6}$$

The formula (1.6.5) for the *directional derivative* of the scalar field $\alpha(\mathbf{r}, t)$ in the direction $\mathbf{e} = e_i \mathbf{e}_i$ may now be presented as:

$$\left.\frac{\partial \alpha}{\partial \varepsilon}\right|_{\varepsilon=0} = (\mathrm{grad}\,\alpha) \cdot \mathbf{e} \tag{1.6.7}$$

Geometrically the gradient of the scalar field $\alpha(\mathbf{r}, t)$ represents a normal vector to the level surface $\alpha(\mathbf{r}, t) = \alpha_0(t)$. This fact may be seen as follows. Let the function $\mathbf{r}(s) = x_i(s)\mathbf{e}_i$ represent an arbitrary space curve at the time t on the surface $\alpha(\mathbf{r}, t) = \alpha_0(t)$, see Fig. 1.19, where the parameter s is the arc length of the curve measured from some reference point on the curve, and with as the tangent vector $\mathbf{t} = d\mathbf{r}/ds$ from the formula (1.4.8) to the curve. The equation $\alpha(\mathbf{r}(s), t) = \alpha_0$ is then satisfied for all values of s. Thus:

$$\frac{\partial \alpha}{\partial s} = \frac{\partial \alpha}{\partial x_i}\frac{dx_i}{ds} \equiv \alpha_{,i}\frac{dx_i}{ds} = (\mathrm{grad}\,\alpha)\cdot\frac{d\mathbf{r}}{ds} = 0 \tag{1.6.8}$$

The vector $d\mathbf{r}/ds$ is a unit tangent vector to the curve and therefore a tangent to the level surface. The result (1.6.8) shows that the vector $\mathrm{grad}\,\alpha$ is perpendicular to the tangent to an arbitrary curve on the surface. Hence:

> The gradient to a scalar field $\alpha(\mathbf{r}, t)$ represents a normal vector to the level surface $\alpha(\mathbf{r}, t) = \alpha_0$.

From formula (1.6.7) it now follows that:

> The maximum value of the directional derivative of a scalar field $\alpha(\mathbf{r}, t)$ on the level surface $\alpha(\mathbf{r}) = \alpha_0$ is given by the magnitude of the gradient of the scalar field $\alpha(\mathbf{r}, t)$. The vector $\mathrm{grad}\,\alpha$ points in the direction of increasing α–values.

The gradient of a scalar field $\alpha(\mathbf{r}, t)$ may alternatively be denoted by:

$$\mathrm{grad}\,\alpha \equiv \frac{\partial \alpha}{\partial \mathbf{r}} \equiv \nabla \alpha \tag{1.6.9}$$

The symbol ∇ is called "del" and represents an operator, the *del-operator*, which in Cartesian coordinates is given by:

$$\nabla \equiv \mathbf{e}_i \frac{\partial}{\partial x_i} \equiv \mathbf{e}_i\,\partial_i \tag{1.6.10}$$

Let $\alpha(\mathbf{r}, t)$ be a scalar field and $\mathbf{r}(s) = x_i(s)\mathbf{e}_i$ any space curve with the tangent vector $\mathbf{t} = d\mathbf{r}/ds$. Then:

$$\frac{\partial \alpha}{\partial s} = \frac{\partial \alpha}{\partial x_i}\frac{dx_i}{ds} \equiv \alpha_{,i}\frac{dx_i}{ds} = (\mathrm{grad}\,\alpha)\cdot\mathbf{t} \tag{1.6.11}$$

The result (1.6.8) is again obtained if the curve lies on the level surface $\alpha(\mathbf{r}, t) = \alpha_0(t)$.

The *divergence of a vector field* $\mathbf{a}(\mathbf{r}, t)$ is a scalar field $\mathrm{div}\,\mathbf{a}$ defined by:

$$\text{div}\,\mathbf{a} \equiv \nabla \cdot \mathbf{a} = (\mathbf{e}_i \partial_i) \cdot (a_k \mathbf{e}_k) = a_{k,i}\,\mathbf{e}_i \cdot \mathbf{e}_k = a_{k,i}\,\delta_{ik} = a_{i,i} \Rightarrow$$

$$\text{div}\,\mathbf{a} \equiv \nabla \cdot \mathbf{a} = a_{i,i} \equiv \frac{\partial a_i}{\partial x_i} = \frac{\partial a_1}{\partial x_1} + \frac{\partial a_2}{\partial x_2} + \frac{\partial a_3}{\partial x_3} \qquad (1.6.12)$$

The symbol $\nabla \cdot \mathbf{a}$ may be interpreted as "the scalar product" of the del-operator (1.6.10) and the vector $\mathbf{a}$.

We shall show that div $\mathbf{a}$ is a scalar field, i.e. a coordinate invariant quantity, by showing that the value of div $\mathbf{a}$ is the same in all coordinate systems. Using successively Eqs. (1.6.4), (1.3.9), the chain rule of differentiation, and finally the result $Q_{ik}Q_{ij} = \delta_{kj}$ obtained from Eq. (1.3.4), we obtain:

$$\bar{a}_{i,i} \equiv \frac{\partial \bar{a}_i}{\partial \bar{x}_i} = \frac{\partial (Q_{ik}a_k)}{\partial x_j}\frac{\partial x_j}{\partial \bar{x}_i} = Q_{ik}\frac{\partial a_k}{\partial x_j}Q_{ij} = Q_{ik}Q_{ij}a_{k,j} = \delta_{kj}a_{k,j} = a_{k,k} \qquad \text{Q.E.D}$$

The *divergence of the gradient of a scalar field* $\alpha(\mathbf{r},t)$ is a new scalar field of special interest in applications, and is therefore given a special symbol:

$$\nabla^2 \alpha \equiv \text{div}\,(\text{grad}\,\alpha) \equiv \nabla \cdot \nabla \alpha = \alpha_{,kk} \equiv \frac{\partial^2 \alpha}{\partial x_k \partial x_k} = \frac{\partial^2 \alpha}{\partial \bar{x}_k \partial \bar{x}_k} \qquad (1.6.13)$$

The symbol ∇^2 is called the *Laplace operator*, named after Pierre-Simon Laplace [1749–1827]. In a Cartesian coordinate system Ox the expression for the Laplace operator is:

$$\nabla^2 = \frac{\partial^2}{\partial x_k \partial x_k} \equiv \frac{\partial^2}{\partial x_1^2} + \frac{\partial^2}{\partial x_2^2} + \frac{\partial^2}{\partial x_3^2} \qquad (1.6.14)$$

The *rotation of a vector field* $\mathbf{a}(\mathbf{r},t)$ is the vector:

$$\text{rot}\,\mathbf{a} \equiv \text{curl}\,\mathbf{a} \equiv \nabla \times \mathbf{a} = e_{ijk}a_{k,j}\,\mathbf{e}_i \qquad (1.6.15)$$

The name and symbol *curl* is very often used in English literature. The symbol $\nabla \times \mathbf{a}$, read as "del cross $\mathbf{a}$", may be interpreted as a vector product of the del-operator and the vector $\mathbf{a}$. In a Cartesian coordinate system the rotation of a vector field may be computed from a determinant of a matrix:

$$\nabla \times \mathbf{a} = \det \begin{pmatrix} \partial/\partial x_1 & \partial/\partial x_2 & \partial/\partial x_3 \\ a_1 & a_2 & a_3 \\ \mathbf{e}_1 & \mathbf{e}_2 & \mathbf{e}_3 \end{pmatrix} \qquad (1.6.16)$$

The following identities may be derived by using the identity (1.1.19), see Problem 1.6:

Fig. 1.20 Cylindrical
coordinates (R, θ, z)

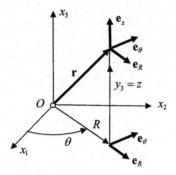

(a) $\mathbf{a} \times (\mathbf{b} \times \mathbf{c}) = (\mathbf{a} \cdot \mathbf{c})\mathbf{b} - (\mathbf{a} \cdot \mathbf{b}) \cdot \mathbf{c}$

(b) $\nabla^2 \mathbf{a} = \nabla(\nabla \cdot \mathbf{a}) - \nabla \times (\nabla \times \mathbf{a})$

(c) $(\mathbf{a} \cdot \nabla)\mathbf{a} = (\nabla \times \mathbf{a}) \times \mathbf{a} + \nabla(\mathbf{a} \cdot \mathbf{a}/2)$

(d) $\nabla \times (\alpha \mathbf{a}) = (\nabla \alpha) \times \mathbf{a} + \alpha(\nabla \times \mathbf{a})$

(1.6.17)

Cylindrical coordinates (R, θ, z) are defined and presented in Fig. 1.20. The unit vectors $\mathbf{e}_R$, $\mathbf{e}_\theta$, and $\mathbf{e}_z$ in direction of the coordinates may be expressed in terms of the base vectors $\mathbf{e}_i$ of the Cartesian coordinate system Ox:

$$\mathbf{e}_R = \cos\theta\,\mathbf{e}_1 + \sin\theta\,\mathbf{e}_2, \quad \mathbf{e}_\theta = -\sin\theta\,\mathbf{e}_1 + \cos\theta\,\mathbf{e}_2, \quad \mathbf{e}_z = \mathbf{e}_3 \quad (1.6.18)$$

The *del-operator* ∇, the *Laplace-operator* ∇^2, the *divergence* and the *rotation* of a vector field $\mathbf{a}(\mathbf{r}, t)$ in cylindrical coordinates are presented in Sect. 6.5.5 and given by:

$$\nabla = \mathbf{e}_R \frac{\partial}{\partial R} + \mathbf{e}_\theta \frac{1}{R}\frac{\partial}{\partial \theta} + \mathbf{e}_z \frac{\partial}{\partial z}$$

$$\nabla^2 = \frac{1}{R}\frac{\partial}{\partial R}\left(R\frac{\partial}{\partial R}\right) + \frac{1}{R^2}\frac{\partial^2}{\partial \theta^2} + \frac{\partial^2}{\partial z^2}$$

(1.6.19)

$$\text{div}\,\mathbf{a} \equiv \nabla \cdot \mathbf{a} = \frac{1}{R}\frac{\partial}{\partial R}(Ra_R) + \frac{1}{R}\frac{\partial a_\theta}{\partial \theta} + \frac{\partial a_z}{\partial z}$$

$$\text{rot}\,\mathbf{a} = \mathbf{e}_R\left[\frac{1}{R}\frac{\partial a_z}{\partial \theta} - \frac{\partial a_\theta}{\partial z}\right] + \mathbf{e}_\theta\left[\frac{\partial a_R}{\partial z} - \frac{\partial a_z}{\partial R}\right]$$

$$+ \mathbf{e}_z\left[\frac{1}{R}\frac{\partial}{\partial R}(Ra_\theta) - \frac{1}{R}\frac{\partial a_R}{\partial \theta}\right]$$

(1.6.20)

Spherical coordinates (r, θ, ϕ) are defined and presented in Fig. 1.21. Unit vectors $\mathbf{e}_r$, $\mathbf{e}_\theta$, and $\mathbf{e}_\phi$ in direction of the coordinates are expressed in terms of the base vectors of the Cartesian coordinate system Ox:

Fig. 1.21 Spherical
coordinates (r, θ, ϕ)

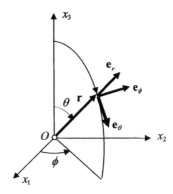

$$\mathbf{e}_r = \sin\theta\cos\phi\,\mathbf{e}_1 + \sin\theta\sin\phi\,\mathbf{e}_2 + \cos\theta\,\mathbf{e}_3$$
$$\mathbf{e}_\theta = \cos\theta\cos\phi\,\mathbf{e}_1 + \cos\theta\sin\phi\,\mathbf{e}_2 - \sin\theta\,\mathbf{e}_3 \qquad (1.6.21)$$
$$\mathbf{e}_\phi = -\sin\phi\,\mathbf{e}_1 + \cos\phi\,\mathbf{e}_2$$

The del-operator, the Laplace-operator, the diver-gence of a vector field $\mathbf{a}(\mathbf{r}, t)$ in spherical coor-dinates are presented in Sect. 6.5.5 and given by:

$$\nabla = \mathbf{e}_r\frac{\partial}{\partial r} + \mathbf{e}_\theta\frac{1}{r}\frac{\partial}{\partial\theta} + \mathbf{e}_\phi\frac{1}{r\sin\theta}\frac{\partial}{\partial\phi} \qquad (1.6.22)$$

$$\nabla^2 = \frac{1}{r^2}\frac{\partial}{\partial r}\left(r^2\frac{\partial}{\partial r}\right) + \frac{1}{r^2\sin\theta}\frac{\partial}{\partial\theta}\left(\sin\theta\frac{\partial}{\partial\theta}\right) + \frac{1}{r^2\sin^2\theta}\frac{\partial^2}{\partial\phi^2} \qquad (1.6.23)$$

$$\operatorname{div}\mathbf{a} \equiv \nabla\cdot\mathbf{a} = \frac{1}{r^2}\frac{\partial}{\partial r}\left(r^2 a_r\right) + \frac{1}{r\sin\theta}\frac{\partial}{\partial\theta}\left(\sin\theta a_\theta\right) + \frac{1}{r\sin\theta}\frac{\partial a_\phi}{\partial\phi}$$

$$\operatorname{rot}\mathbf{a} = \mathbf{e}_r\left[\frac{1}{r\sin\theta}\frac{\partial}{\partial\theta}\left(\sin\theta a_\phi\right) - \frac{1}{r\sin\theta}\frac{\partial a_\theta}{\partial\phi}\right] \qquad (1.6.24)$$
$$+ \mathbf{e}_\theta\left[\frac{1}{r\sin\theta}\frac{\partial a_r}{\partial\phi} - \frac{\partial}{\partial r}\left(ra_\phi\right)\right] + \mathbf{e}_\phi\left[\frac{1}{r}\frac{\partial}{\partial r}\left(ra_\theta\right) - \frac{1}{r}\frac{\partial a_r}{\partial\theta}\right]$$

Problems 1.1–1.6 with solutions see Appendix

Chapter 2
Dynamics. The Cauchy Stress Tensor

This chapter introduces the concept tensors by defining the original tensor, i.e. *Cauchy's stress tensor*. However, first some basic principles of dynamics related to continuous materials: solids, fluids, and gases, have to be presented. *Dynamics*, which is the science of motion of bodies with mass and the forces that cause this motion, is often subdivided into *Kinematics* and *Kinetics*. Kinematics is the geometry of motion with velocity and acceleration as the most important concepts. The kinematics of continuous materials is the subject matter in Sect. 2.1. Kinetics treats the interrelationship between forces and the motion they cause. *Continuum mechanics* is the science of motion and behaviour of continuous materials. Section 2.2 introduces the types of forces generally considered in Continuum Mechanics, and the equations of motion that apply to all continuous materials. Section 2.3 Stress Analysis discusses the internal forces in a continuum. *Material models* and *constitutive equations* are used to describe the response of continuous materials to motion, deformation, and forces. Such models and equations are presented in Chap. 5.

2.1 Kinematics

2.1.1 *Lagrangian Coordinates and Eulerian Coordinates*

A portion of the continuous material we are considering is called a *body*. Figure 2.1 illustrates a reference *Rf,* to which motion will be referred, an *orthogonal right—handed Cartesian coordinate system Ox*, with base vectors $\mathbf{e}_i$ and a body at two different times: a *reference time* t_0 and the *present time t*, i.e. the time at which the body is under investigation. A material point in the body is called a *particle*. At the reference time t_0 the particle is denoted P_0, and localized by the place vector $\mathbf{r}_0$, which we call the *particle vector*, or by the three coordinate X_i in the system *Ox*. At

© Springer Nature Switzerland AG 2019
F. Irgens, *Tensor Analysis*,
https://doi.org/10.1007/978-3-030-03412-2_2

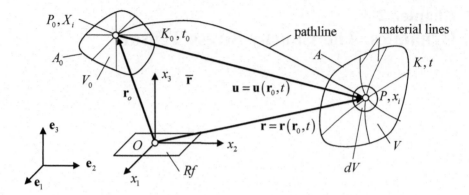

Fig. 2.1 Cartesian coordinate system Ox fixed in reference Rf. Body of continuous material. The body has volume V_0 and surface area A_0 in a reference configuration K_0, and volume V and surface area A in the present configuration K. Particle r_0 at the place $r = r(r_0,t)$ at the time t in K. Volume element dV. Displacement vector $u = u(r_0,t)$

the present time t the particle is denoted P and localized by the place vector r or by the three coordinate x_i. We choose the alternative notations for a particle and for the place of the particle at the present time t:

$$\text{particle: } r_0, \quad [X_1, X_2, X_3], \quad X_i, \quad X \equiv \begin{pmatrix} X_1 \\ X_2 \\ X_3 \end{pmatrix} \equiv \{X_1 X_2 X_3\} \qquad (2.1.1)$$

$$\text{place: } r, \quad [x_1, x_2, x_3], \quad x_i, \quad x \equiv \begin{pmatrix} x_1 \\ x_2 \\ x_3 \end{pmatrix} \equiv \{x_1 \ x_2 \ x_3\} \qquad (2.1.2)$$

The set of places that represents the body at any time is called a *configuration of the body* at that time. The configuration of the body at the present time t is called the *present configuration K*. The configuration of the body corresponding to the reference time t_0 is called the *reference configuration K_0*. The body has at any time t a volume V and a surface area A. It is assumed that the body consists of the same particles at all times and that the surface of the body is a closed material surface in the continuum and in general consists at all times of the same particles.

The motion of the continuum may be described by functional relationships between the place vector r and the place vector r_0, and between their components x_i and X_i:

$$r = r(r_0, t), \quad x_i = x_i(X_1, X_2, X_3, t) \equiv x_i(X, t) \qquad (2.1.3)$$

In Fig. 2.1 the motion of the particle is indicated by a *pathline*. The particle r_0 moves along this space curve. The relations (2.1.3) represent a one-to-one mapping between the places $\mathbf{r}_0$ in K_0 and the places $\mathbf{r}$ in K. The motion of the body may also be represented by the *displacement vector*:

$$\mathbf{u} = \mathbf{u}(\mathbf{r}_0, t) = \mathbf{r}(\mathbf{r}_0, t) - \mathbf{r}_0 \Leftrightarrow u_i = u_i(X, t) = x_i(X, t) - X_i \qquad (2.1.4)$$

The motion of the body from configuration K_0 to configuration K will generally lead to deformation of the body, i.e. geometrical figures will change their shapes and sizes during the motion. Deformation is illustrated in Fig. 2.1 by *material lines*, which in K_0 are chosen to coincide with the coordinate lines. In K the deformed material lines represent a curvilinear coordinate system, and an analysis of this system would give us information about the state of deformation of the body relative to the reference configuration K_0. A coordinate system imbedded in a body, and which deforms with it, is called a *convected coordinate system*. Such coordinate systems will be analyzed and utilized in Sect. 7.5.

The body consists at any time t of the same particles. The surface of the body is a closed material surface in the continuum and consists at any time of the same particles.

Physical quantities are separated into extensive quantities and intensive quantities. An *extensive quantity* is given for a body and may be a function of the volume V or the mass m of the body. Examples are the mass itself, the linear momentum of the body, and the kinetic energy of the body. The value of an extensive quantity is the sum of its values for the parts into which the body may be divided.

An *intensive quantity* is related to the particles in a body and independent of the volume or the mass of the body. Examples are pressure, temperature, mass density, and particle velocity. An intensive quantity may represent the intensity of an extensive quantity and gives quantities per unit volume or per unit mass. For instance, *mass density* is defined as mass per unit volume. In general an intensive quantity that is given per unit volume is characterized as a *density*. Kinetic energy per unit volume is thus a *kinetic energy density*. For short it is customary to use the word *density* also when we actually mean mass density. An intensive quantity defined per unit mass, is characterized as a *specific quantity*. For instance, the velocity of a particle is also linear momentum per unit mass, and thus represents the intensity related to the extensive quantity called the *linear momentum of the body*. Velocity is therefore *specific linear momentum*.

An intensive physical quantity may be expressed by a field as either a *particle function* $f(\mathbf{r}_0, t)$, which is a function of the particle vector $\mathbf{r}_0$ and the time t, or as a *place function* $f(\mathbf{r}, t)$, which is a function of the place vector $\mathbf{r}$ and the time t. Alternatively the particle function may be expressed as a function of the particle coordinates X and the time t, and the place function may be expressed as a function of the place coordinates x and the time t. Thus we have the alternative forms:

$$f(\mathbf{r}_0, t) = f(X, t) \equiv f(X_1, X_2, X_3, t) \quad particle \, function \qquad (2.1.5)$$

$$f(\mathbf{r}, t) = f(x, t) \equiv f(x_1, x_2, x_3, t) \quad place \, function \qquad (2.1.6)$$

The form $f(\mathbf{r}_0, t)$ for a particle function rather than the form $f(X, t)$ is used when we want to emphasize the coordinate invariant property of the quantity that the field function represents. To get from a place function for a particle related intensive quantity to a particle function we introduce the motion $\mathbf{r}(\mathbf{r}_0, t)$ of the particle and write alternatively:

$$f(\mathbf{r}, t) = f(\mathbf{r}(\mathbf{r}_0, t), t) \Leftrightarrow f(x, t) = f(x(X, t), t) \qquad (2.1.7)$$

The coordinates $(X, t) \equiv (X_1, X_2, X_3, t)$ are alternatively called *Lagrangian coordinates*, named after Joseph Louis Lagrange [1736–1813], *material coordinates*, *particle coordinates*, or *reference coordinates*. The application of these coordinates is called *Lagrangian description* or *reference description*.

The coordinates $(x, t) \equiv (x_1, x_2, x_3, t))$ are alternatively called *Eulerian coordinates*, named after Leonhard Euler [1736–1813], or *space coordinates*, and their application for *Eulerian description* or *spacial description*.

According to Truesdell and Toupin [1], the Lagrangian coordinates were introduced by Euler in 1762, while Jean le Rond D'Alembert [1717–1783] was the first to use the Eulerian coordinates in 1752. In general Continuum Mechanics Lagrangian coordinates and the reference description are the most common. The same holds true in Solid Mechanics. However, in Fluid Mechanics, due to the large displacements and the complex deformations resulting from the fluid motion, it is usually necessary and most practical to use Eulerian coordinates and spacial description.

2.1.2 Material Derivative of Intensive Quantities

Let the particle function $f(\mathbf{r}_0, t)$, which may be a scalar or a vector, represent an arbitrary *intensive physical quantity*. For a particular choice of the particle vector $\mathbf{r}_0$ or the particle coordinates X the function $f(\mathbf{r}_0, t)$ is connected to that particle $\mathbf{r}_0$ at all times t. The change per unit time, i.e. the time rate of change, of f when attached to the particle $\mathbf{r}_0$ is called the *material derivative* of the particle function f and is denoted by f with a superimposed dot.

$$\dot{f} \equiv \frac{df}{dt}\bigg|_{\mathbf{r}_0 = \text{constant}} = \frac{\partial f(\mathbf{r}_0, t)}{\partial t} \equiv \partial_t f(\mathbf{r}_0, t) \Leftrightarrow$$

$$\dot{f} \equiv \frac{df}{dt}\bigg|_{X = \text{constant}} = \frac{\partial f(X, t)}{\partial t} \equiv \partial_t f(X, t) \qquad (2.1.8)$$

Other names for this quantity are the *substantial derivative*, the *particle derivative*, and the *individual derivative*.

The *velocity* **v** and the *acceleration* **a** of the particle X are defined by:

$$\mathbf{v}(\mathbf{r}_0,t) = \dot{\mathbf{r}} = \partial_t \mathbf{r}(\mathbf{r}_0,t) \Leftrightarrow \mathbf{v}(X,t) = \dot{\mathbf{r}} = \partial_t \mathbf{r}(X,t) \qquad (2.1.9)$$

$$\mathbf{a}(\mathbf{r}_0,t) = \dot{\mathbf{v}} = \ddot{\mathbf{r}} = \partial_t^2 \mathbf{r}(\mathbf{r}_0,t) \Leftrightarrow \mathbf{a}(X,t) = \dot{\mathbf{v}} = \ddot{\mathbf{r}} = \partial_t^2 \mathbf{r}(X,t) \qquad (2.1.10)$$

Using the definition (2.1.4) of the displacement vector **u**, we have as alternative expressions:

$$\mathbf{v} = \dot{\mathbf{u}} = \partial_t \mathbf{u}(\mathbf{r}_0,t) = \partial_t \mathbf{u}(X,t), \quad \mathbf{a} = \dot{\mathbf{v}} = \ddot{\mathbf{u}} = \partial_t^2 \mathbf{u}(\mathbf{r}_0,t) = \partial_t^2 \mathbf{u}(X,t) \quad (2.1.11)$$

The components of the velocity and the acceleration in the coordinate system Ox are:

$$v_i = \dot{u}_i = \partial_t u_i(X,t) = \dot{x}_i = \partial_t x_i(X,t) \qquad (2.1.12)$$

$$a_i = \dot{v}_i = \ddot{x} = \partial_t^2 x_i(X,t) = \ddot{u}_i \qquad (2.1.13)$$

Let the place function $f(\mathbf{r},t) = f(x,t)$ represent an intensive quantity, which may be a scalar or a vector. For any particular choice of **r** or coordinates x the function is attached to the particular place **r** or x in space. The local change of $f(\mathbf{r},t) = f(x,t)$ per unit time is:

$$\partial_t f(\mathbf{r},t) \equiv \frac{\partial f(\mathbf{r},t)}{\partial t} \Leftrightarrow \partial_t f(x,t) \equiv \frac{\partial f(x,t)}{\partial t} \qquad (2.1.14)$$

In order to find the *material derivative of the place function* $f(\mathbf{r},t) = f(x,t)$, we attach the function to the particle $\mathbf{r}_0$ that takes the position x at time t and use the form (2.1.7). The definition (2.1.8) of the material derivative of the field function $f(x(X,t),t)$ representing an intensive quantity leads us to the result:

$$\dot{f} = \frac{\partial f(x,t)}{\partial t} + \frac{\partial f(x,t)}{\partial x_i} \frac{\partial x_i(X,t)}{\partial t} \equiv \partial_t f + f_{,i} \partial_t x_i$$

The last factor in the last term represents the components v_i of the particle velocity. We then have an expression for the material derivative of a place function $f(x,t)$ representing an intensive quantity for a particle X:

$$\dot{f} = \partial_t f + v_i f_{,i} \qquad (2.1.15)$$

We use the *del-operator* ∇ from the formula (1.6.10) and introduce the operator:

$$\mathbf{v} \cdot \nabla = v_i \frac{\partial}{\partial x_i} \qquad (2.1.16)$$

This operator represents a scalar product of the velocity vector and the del-operator. The formula (2.1.15) may now alternatively be written as:

$$\dot{f} = \partial_t f + (\mathbf{v} \cdot \nabla) f \tag{2.1.17}$$

The material derivative $\dot{f}$ in the formula (2.1.17) contains two parts: The *local part* $\partial_t f$ and the *convective part* $(\mathbf{v} \cdot \nabla) f$ *of the material derivative.* The local part $\partial_t f$ registers the change of f at the place $\mathbf{r}$. In Fluid Mechanics, for instance, recording of intensive quantities like pressure, temperatures and velocities will usually be performed with stationary instruments. A stationary instrument can register f at a definite place $\mathbf{r}$ and provides values of f for the particles passing through the place $\mathbf{r}$. The recording of such an instrument can therefore only give the local part $\partial_t f$ of the material derivative $\dot{f}$. The convective part, $(\mathbf{v} \cdot \nabla) f$ of the material derivative $\dot{f}$ represents the time rate of change of f due to the fact that the particle, which at time t is at place $\mathbf{r}$, moves to a new position in space.

The *particle acceleration* $\mathbf{a}$ may be computed from the velocity field $\mathbf{v}(\mathbf{r}, t)$ by application of the formula (2.1.17). We write:

$$\mathbf{a} = \dot{\mathbf{v}} = \partial_t \mathbf{v} + (\mathbf{v} \cdot \nabla) \mathbf{v} \Leftrightarrow a_i = \dot{v}_i = \partial_t v_i + v_k v_{i,k} \tag{2.1.18}$$

From the result (2.1.18) we see that the particle acceleration $\mathbf{a}$ consists of the *local acceleration* $\partial_t \mathbf{v}$ and the *convective acceleration* $(\mathbf{v} \cdot \nabla) \mathbf{v}$.

2.1.3   Material Derivative of Extensive Quantities

A body of continuous matter and with volume $V(t)$ has constant mass m. This statement is called the *principle of conservation of mass.* Mass per unit volume is called the mass density, or for short, the *density.* Let the scalar field $\rho = \rho(\mathbf{r}, t)$ represent the density at the place $\mathbf{r}$, and let dV be *differential element of volume.* In a Cartesian coordinate system Ox the differential element of volume is presented as a rectangular parallelepiped as shown in Fig. 2.2 with volume $dV = dx_1 dx_2 dx_3$. In other coordinate systems and in general curvilinear coordinate system the volume element is defined and presented in Sect. 9.3. The mass m of a body of volume $V(t)$ at the present time t, may then be expressed by a volume integral:

$$m = \int_{V(t)} \rho \, dV \tag{2.1.19}$$

Let $f(\mathbf{r}, t)$ be a place function that represents an intensive quantity expressed per unit mass, i.e. a specific quantity, and let $F(t)$ be the corresponding extensive quantity for a body with volume $V(t)$:

Fig. 2.2 Differential volume
element in a Cartesian
coordinate system
$Ox : dV = dx_1 dx_2 dx_3$

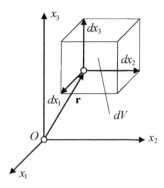

$$F(t) = \int_{V(t)} f\rho dV \qquad (2.1.20)$$

For instance, $f(\mathbf{r}, t)$ may be the kinetic energy per unit mass, $\mathbf{v} \cdot \mathbf{v}/2 = v_i v_i/2$ and $F(t)$ the total kinetic energy for the body.

The time rate of change of $F(t)$ attached to the body is called the *material derivative* $\dot{F}$ of *the extensive quantity F(t)*. We will find that:

$$\dot{F} = \dot{F}(t) = \int_{V(t)} \dot{f}\rho dV \qquad (2.1.21)$$

This result may be derived as follows. The body is considered to consist of many small elements with volumes and mass symbolized by dV and ρdV respectively. The contribution to $F(t)$ from the volume element dV is $f\rho dV$. The integral in (2.1.20) represents the sum of contributions from all volume elements. The volume elements may change both in form and size with time, but the mass $\rho \, dV$ of any element is constant by the principle of mass conservation. The time rate of change of the quantity $f\rho dV$ is therefore $\dot{f}\rho dV$. The time rate of change of $F(t)$ is then the sum of the element contributions $\dot{f}\rho dV$, and thus equal to the integral in formula (2.1.21).

2.2 Equations of Motion

2.2.1 Euler's Axioms

Newton's laws for the motion of a mass particle cannot in the strict sense logically be transferred to apply to a body of continuously distributed matter. It has been customary in some texts to claim that this transformation is possible: From

Newton's second law for motion of a body, originally by Newton for a *mass particle*, and *Newton's third law* for the interaction between mass particles, the equations of motion for a system of mass particles are derived as two vector equations: firstly an equation equating the resultant of all the external forces on the system to the change in the sum of the linear momentums of the particles, and secondly an equation equating the sum of all the moments of the external forces about a point O to the sum of angular momentums of the particles about the same point O. The first vector equation may be referred to as *Newton's second law for a system of mass particles*. The second vector equation may be referred to as the *law of moments for a system of mass particles*. These two vector equations are then taken to apply to a continuum. The consideration is either that the continuum is a system of a large number of elementary particles, or as a system of infinitely many infinitely small particles. The first method of consideration fails on the ground that elementary particles do not obey Newtonian mechanics. Quantum mechanical issues are involved here, and the transfer from the micro world to the macro world must include statistical mechanics. The other method contains logical difficulties of mathematical nature.

In Continuum Mechanics two axioms are postulated: *Euler's first axiom*, which corresponds to Newton's second law for a system a system of particles, and *Euler's second axiom*, which has its parallel in the law of moments for a system of particles. The two axioms are presented below as Eqs. (2.2.6) and (2.2.7). The law of action and reaction of forces, i.e. Newton's third law, and of couples and counter couples for interaction between two bodies now follows as a consequence of the Eulerian axioms. This will be demonstrated in Sect. 2.2.2.

From a continuum at time t we now consider in Fig. 2.3 the present configuration K of a body having mass m, volume V, and surface A. The body is assumed to be acted upon by two types of forces: *body force* $\mathbf{b}$ per unit mass, and *contact force* $\mathbf{t}$ per unit area over the surface of the body. The Cartesian coordinate system Ox is fixed in the reference Rf.

The body forces are external forces originating from sources outside of the body, and they represent actions at a distance from the surroundings. It is for practical reasons that all body forces are considered to be given per unit mass. Gravitation, centrifugal forces, electrostatic and magnetic influences are examples of body forces. The most typical body force is the *gravitational force g* in the parallel constant field of gravity, with its standard value $g = 9.81$ N/kg. Normally it is assumed that the body forces are independent of the state of the body, but in general the body forces may depend upon the position in space of the particle on which they act. At the place $\mathbf{r}$ we express the body force by:

$$\mathbf{b} = \mathbf{b}(\mathbf{r}, t) = \mathbf{b}(x, t) \qquad (2.2.1)$$

If a body force is given as force per unit volume, it is called a *volume force*.

In a solid material the contact force represents the internal forces between the physical particles, atoms and molecules, on both sides of the surface A separating the body from its surroundings. *Surface force* and *traction* are other names of the

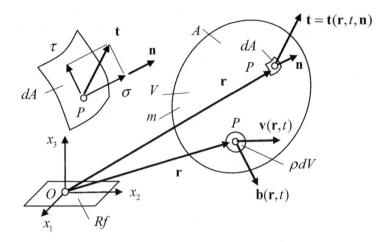

Fig. 2.3 Present configuration K at the time t of a body of continuous material with volume V, mass m, and surface area A. Density ρ = mass per unit volume. Reference Rf. Body force $\mathbf{b}(\mathbf{r}, t)$ = force per unit mass. Surface force, stress vector, $\mathbf{t}(\mathbf{r},t,\mathbf{n})$ = force per unit area. Velocity vector $\mathbf{v}(\mathbf{r},t)$. Element of area dA. Element of volume dV. Decomposition of the stress vector $\mathbf{t}$ into a normal stress σ and a shear stress τ

contact force. The vector $\mathbf{t}$ will in the present exposition be called the *stress vector*. The stress vector varies with the place $\mathbf{r}$, or the particle X, on the surface, with the time t, and with the normal $\mathbf{n}$ to the surface at the place $\mathbf{r}$. The vector $\mathbf{n}$ is defined to be a unit vector pointing out from the surface A at the place $\mathbf{r}$. Alternatively we use the expressions:

$$\mathbf{t} = \mathbf{t}(\mathbf{r}, t, \mathbf{n}) = \mathbf{t}(x, t, n) = \mathbf{t}(\mathbf{r}_o, t, \mathbf{n}) = \mathbf{t}(X, t, n) \qquad (2.2.2)$$

The components of the stress vector $\mathbf{t}$ in the direction $\mathbf{n}$ and in the tangent plane to the surface A at the place $\mathbf{r}$ are the *normal stress* σ and the *shear stress* τ respectively, see Fig. 2.2.

When a gas is considered as a continuous medium, the surface of a gaseous body is a material surface. However, if we take into consideration that the gas consists of molecules, moving about at great speeds, the surface of gaseous body is a mathematical surface through which molecules may pass, although the macromechanical flow through a surface element is assumed to be zero. The motion, or rather the momentum, of the molecules that pass through the boundary surface is in Continuum Mechanics represented by the contact forces. Intermolecular forces in a gas are negligible. When the boundary surface is an interface between a gas and a liquid or a solid material, the molecular motion is represented by a pressure and tangential forces, i.e. shear stresses, on the liquid surface or the surface of the solid material respectively. On an interface between two liquid bodies, the contact forces represent both intermolecular forces and molecular motion.

In special situations, it becomes necessary to add to the two types of forces introduced above, *body couples*, given as couple per unit mass, and surface couples or *couple stresses* on the surface of the body, given as couple (moment) per unit area. Body couples are for instance present as a result of an elastic strain wave passing through a body exposed to an electromagnetic field. Another example of the presence of body couples is given by the effect of the magnetic field of the earth on the needle of a compass. Couple stresses may appear when the molecular structure of a material is taken into consideration, and when dislocations in metals are considered. In this book these types of mechanical actions are not considered.

The body shown in Fig. 2.3 is subjected to a *resultant force* $\mathbf{f}$ given by:

$$\mathbf{f} = \int_A \mathbf{t}\, dA + \int_V \mathbf{b}\, dV \qquad (2.2.3)$$

The symbol dA represents a *differential element of area* of the surface A of the body, and the symbol dV is the differential element of volume of the body. When a surface is a part of a x_1x_2-plane in a Cartesian coordinate system Ox, the element of area may be represented by the rectangle $dA = dx_1 dx_2$. The expression for the element of area in general coordinates is defined and presented in Sect. 9.3.

The *resultant moment* $\mathbf{m}_O$ about the point O of all forces acting of the body is:

$$\mathbf{m}_O = \int_A \mathbf{r} \times \mathbf{t}\, dA + \int_V \mathbf{r} \times \mathbf{b}\, dV \qquad (2.2.4)$$

The velocity of a particle $\mathbf{r}_0$ at the place $\mathbf{r}$ in the body is given by the vector fields $\mathbf{v}(\mathbf{r}_0, t)$ or $\mathbf{v}(\mathbf{r}, t)$. The *linear momentum* $\mathbf{p}$ of the body and the *angular momentum* $\mathbf{l}_O$ of the body about the origin O are defined by the integrals:

$$\mathbf{p} = \int_V \mathbf{v}\rho\, dV, \quad \mathbf{l}_O = \int_V \mathbf{r} \times \mathbf{v}\rho\, dV \qquad (2.2.5)$$

Note that the velocity $\mathbf{v}$ may be interpreted as the specific linear momentum, i.e. momentum per unit mass. Likewise, $\mathbf{r} \times \mathbf{v}$ is the specific angular momentum about O.

The general laws of motion for a body of mass m, which at time t has the volume V and the surface A, are postulated as *Euler's two axioms* or *laws* and given by:

$$\mathbf{f} = \dot{\mathbf{p}} \equiv \int_V \dot{\mathbf{v}}\rho\, dV \equiv \int_V \mathbf{a}\rho\, dV \qquad \text{Euler's first axiom} \qquad (2.2.6)$$

$$\mathbf{m}_O = \dot{\mathbf{l}}_O = \int_V \mathbf{r} \times \dot{\mathbf{v}}\rho\, dV \equiv \int_V \mathbf{r} \times \mathbf{a}\rho\, dV \qquad \text{Euler's second axiom} \qquad (2.2.7)$$

$\mathbf{a} = \dot{\mathbf{v}}$ is the particle acceleration. When computing the material derivative of the extensive quantities $\mathbf{p}$ and $\mathbf{l}_O$, we have used the formula (2.1.21). Furthermore, we have used the development:

$$\frac{d}{dt}(\mathbf{r} \times \mathbf{v}) = \dot{\mathbf{r}} \times \mathbf{v} + \mathbf{r} \times \dot{\mathbf{v}} = \mathbf{v} \times \mathbf{v} + \mathbf{r} \times \mathbf{a} = \mathbf{r} \times \mathbf{a}$$

Euler's first axiom is also called the *law of balance of linear momentum of a body*. Euler's second axiom is also called the *law of balance of angular momentum of a body*.

The *center of mass* of a body is defined by the point C that at time t is at the place given by the place vector:

$$\mathbf{r}_C = \frac{1}{m} \int_V \mathbf{r} \rho \, dV \tag{2.2.8}$$

The center of mass is a reference invariant point that moves with the body. The weight of a body, which represents the resultant of the constant and parallel gravitational field near the surface of the earth, always has its line of action through the mass center C. The point C is therefore also called the *center of gravity* of the body. The velocity $\mathbf{v}_C$ and the acceleration $\mathbf{a}_C$ of the center of mass are found from the expressions:

$$\mathbf{v}_C = \dot{\mathbf{r}}_C = \frac{1}{m} \int_V \dot{\mathbf{r}} \rho \, dV = \frac{1}{m} \int_V \mathbf{v} \rho \, dV \tag{2.2.9}$$

$$\mathbf{a}_C = \dot{\mathbf{v}}_C = \ddot{\mathbf{r}}_C = \frac{1}{m} \int_V \dot{\mathbf{v}} \rho \, dV = \frac{1}{m} \int_V \mathbf{a} \rho \, dV \tag{2.2.10}$$

From the definition (2.2.5) (2.2.5)$_1$ and the formula (2.2.9), it follows that the linear momentum $\mathbf{p}$ of the body may be expressed by:

$$\mathbf{p} = m\mathbf{v}_C \tag{2.2.11}$$

Then, according to Euler's first axiom (2.2.6) the motion of the center of mass is governed by the equation:

$$\mathbf{f} = m\mathbf{a}_C \tag{2.2.12}$$

Thus, the center of mass of a body moves as a physical particle with mass equal to the mass m of the body, and subjected to the resultant force $\mathbf{f}$ on the body.

2.2.2 Newton's Third Law of Action and Reaction

Newton formulated his third law of action and reaction when two bodies interacted, without taking into consideration the extent of the two bodies. Using the two fundamental axioms of Euler, we shall derive and extend the third law to two bodies of arbitrary shapes and extensions.

Figure 2.4 shows a body, considered to consist of two parts I and II separated by the interface A'. The axioms (2.2.6) and (2.2.7) are now formulated for the body and for each of the parts I and II, using the point O' on the interface A' as a moment point. Subtractions of corresponding equations for the parts from the equations for the total body give as results:

$$-\int_{A',I} \mathbf{t}(\mathbf{r},t,\mathbf{n})\,dA - \int_{A',II} \mathbf{t}(\mathbf{r},t,-\mathbf{n})\,dA = \mathbf{0} \Leftrightarrow -\mathbf{f}_{12} - \mathbf{f}_{21} = \mathbf{0}$$

$$-\int_{A',I} \mathbf{r}' \times \mathbf{t}(\mathbf{r},t,\mathbf{n})\,dA - \int_{A',II} \mathbf{r}' \times \mathbf{t}(\mathbf{r},t,-\mathbf{n})\,dA = \mathbf{0} \Leftrightarrow -\mathbf{m}_{12} - \mathbf{m}_{21} = \mathbf{0}$$

$$(2.2.13)$$

The result: $-\mathbf{f}_{12} - \mathbf{f}_{21} = \mathbf{0}$ shows that the resultant force $\mathbf{f}_{12}$ of the contact forces on part I from part II, is equal to the resultant force $\mathbf{f}_{21}$ of the contact forces acting on part II from part I, but with opposite signs. The result: $-\mathbf{m}_{12} - \mathbf{m}_{21} = \mathbf{0}$ shows that the resultant moment $\mathbf{m}_{12}$ of the contact forces on part I from part II, is equal to the resultant moment $\mathbf{m}_{21}$ of the contact forces acting on part II from part I, but with opposite signs. Hence:

$$\mathbf{f}_{12} = -\mathbf{f}_{21}, \quad \mathbf{m}_{12} = -\mathbf{m}_{21} \qquad (2.2.14)$$

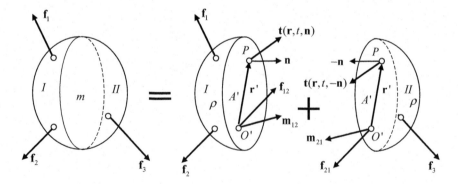

Fig. 2.4 Body of mass m and density ρ subjected to external forces $\mathbf{f}_1$, $\mathbf{f}_2$, and $\mathbf{f}_3$. The body consists of two parts I and II separated by an interface A'. On the interface A' on the parts I and II: resultant forces $\mathbf{f}_{12}$, $\mathbf{f}_{21}$, and resultant moments $\mathbf{m}_{12}$, $\mathbf{m}_{21}$ about the point O'

Fig. 2.5 Cauchy's lemma

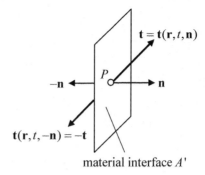

material interface A'

The resultant forces act through the moment point O, and the resultant moments are couples. The result (2.2.14) may be interpreted as a generalization of *Newton's third law of action and reaction*: To each action ($\mathbf{f}_{12}$ or $\mathbf{m}_{12}$) there is a reaction ($\mathbf{f}_{21}$ or $\mathbf{m}_{21}$).

Figure 2.5 illustrates a material interface A' with a particle P located by place vector $\mathbf{r}$, not shown in the figure. Let the area A' shrink to zero. Then it follows from the results (2.2.13) that:

$$\mathbf{t}(\mathbf{r}, t, \mathbf{n}) = -\mathbf{t}(\mathbf{r}, t, -\mathbf{n}) \tag{2.2.15}$$

This result is called *Cauchy's lemma*, named after Augustin Louis Cauchy [1789–1857]. The lemma shows that the stress vectors on the two sides of a material surface are equal but in opposite directions, as shown in Fig. 2.5.

2.2.3 Coordinate Stresses

Figure 2.6 shows the stress vectors $\mathbf{t}_k$ on three material surfaces through the particle P at the place $\mathbf{r}$ and perpendicular to the coordinate axes. The unit normals to these surfaces are the base vectors $\mathbf{e}_k$. The components of the stress vectors $\mathbf{t}_k$ are denoted T_{ik}. Hence:

$$\mathbf{t}_k = T_{ik}\mathbf{e}_i \Leftrightarrow \mathbf{e}_i \cdot \mathbf{t}_k = T_{ik} \tag{2.2.16}$$

The components T_{ik} will be called the *coordinate stresses* in the particle P or at the place $\mathbf{r}$. The coordinate stresses are elements of the *stress matrix* $T = (T_{ik})$ in P, or at $\mathbf{r}$, with respect to the coordinate system Ox, or with respect to the base vectors $\mathbf{e}_i$. The coordinate stresses T_{11}, T_{22}, and T_{33} are *normal stresses*, while the coordinate stresses $T_{ik}, i \neq k$, are *shear stresses*. The first index (i) of T_{ik} refers to the direction of the stress and the second index (k) refers to the normal vector $\mathbf{e}_k$ to the surface on which the stress acts. In the literature the meaning of the indices is often reversed.

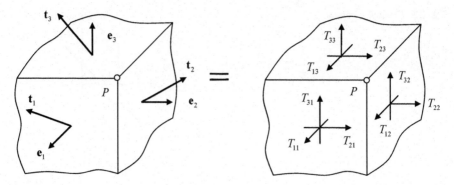

Fig. 2.6 Stress vectors $\mathbf{t}_i$ on material surfaces through a particle P and with unit normal vectors equal to the base vectors $\mathbf{e}_i$ in a Cartesian coordinate system Ox. Coordinate stresses T_{ik} at the particle P

Figure 2.7 shows two sides of a material surface in an x_1x_3-coordinate plane through a particle P. On the surface side with unit normal equal to the base vector $\mathbf{e}_2$ the stress vector is $\mathbf{t}_2$ with the three Cartesian components T_{2k}, where T_{22} is a normal stress, and T_{12} and T_{23} are shear stresses. On the side of the surface with unit normal $-\mathbf{e}_2$ the stress vector is, according to Cauchy's lemma, equal to $-\mathbf{t}_2$ with the three Cartesian components $-T_{2k}$, where T_{22} is a normal stress, and T_{12} and T_{23} are shear stresses.

Because the normal stress on both sides of the material surface in Fig. 2.7 points out from the surface this normal stress is called a *tensile stress*. A normal stress representing a pressure on the surface is called a *compressive stress*.

Based on the situation in Fig. 2.7 we may state the following sign rule for *coordinate stresses*:

A positive coordinate stress acts in direction of a positive coordinate axis on that side of a material coordinate surface facing the positive direction of a coordinate axis. On the side of the material surface facing the negative direction of a coordinate axis, the positive coordinate stress acts in the direction of a negative coordinate axis.

Fig. 2.7 Positive coordinate stresses

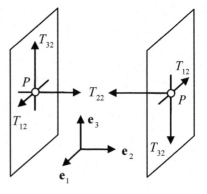

As will be shown in Sect. 2.2.5, the coordinate shear stresses T_{ik} and T_{ki} are equal, i.e. $T_{ik} = T_{ki}$, as long as we only consider the two types of forces: body forces **b** and contact forces **t**. Normally therefore the order of the indices is not really so important. An exception to the equality of shear stress pairs is provided when *body couples* and when *couple stresses* are present. However, in some derivations and applications it is advantageous to distinguish between T_{ik} and T_{ki}.

The literature uses a variety of symbols for coordinate stresses. When number indices are used, the following symbols may be found.

$$T_{ik} = S_{ik} = \sigma_{ik} = \tau_{ik} = -p_{ik} \tag{2.2.17}$$

In texts using *xyz*-coordinates, normal stresses are denoted as σ_x or σ_{xx} etc., and shear stresses as τ_{xy} or σ_{xy} etc.. Thus:

$$T = \begin{pmatrix} T_{11} & T_{12} & T_{13} \\ T_{21} & T_{22} & T_{23} \\ T_{31} & T_{32} & T_{33} \end{pmatrix} \equiv \begin{pmatrix} \sigma_x & \tau_{xy} & \tau_{xz} \\ \tau_{yx} & \sigma_y & \tau_{yz} \\ \tau_{zx} & \tau_{zy} & \sigma_z \end{pmatrix} \equiv \begin{pmatrix} \sigma_{xx} & \sigma_{xy} & \sigma_{xz} \\ \sigma_{yx} & \sigma_{yy} & \sigma_{yz} \\ \sigma_{zx} & \sigma_{zy} & \sigma_{zz} \end{pmatrix} \tag{2.2.18}$$

Example 2.1 Uniaxial State of Stress

A rod with axis along the x_1-axis in a Cartesian coordinate system Ox is subjected to an axial force N, Fig. 2.8.

Any cross-section of area A of the rod is subjected to a normal stress $\sigma = N/A$. The cross-section is free of shear stresses: $T_{21} = T_{31} = 0$. The state of stress is given by the stress matrix:

$$T = \begin{pmatrix} \sigma & 0 & 0 \\ 0 & 0 & 0 \\ 0 & 0 & 0 \end{pmatrix}, \quad \sigma = \frac{N}{A}$$

This state of stress is called uniaxial, a term that will be explained in Sect. 2.3.1.

Example 2.2 State of Pure Shear Stress

A thin-walled tube with a mean radius r and a wall thickness of h $(h \ll r)$ is subjected to a torsion moment, or *torque*, m_t, Fig. 2.9. The resultant of the stresses over the cross-section of the tube must be equal to the torsion moment. Due to symmetry the cross-section will only carry a constant shear stress, which we find to be $\tau = m_t/(2\pi r^2 h)$. Figure 2.9 shows the state of stress on a small, approximately plane, element of the tube wall. Related to the local Cartesian coordinate system Ox,

Fig. 2.8 Rod subjected to an axial force N. Normal stress σ on a cross-section. Uniaxial state of stress $\sigma = N/A$

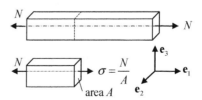

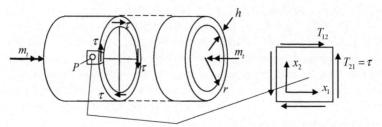

small, approximately plane, element of the tube wall

Fig. 2.9 Torsion of a thin—walled tube with mean radius r and wall thickness h. The tube is subjected to a torque m_t

shown on the element in Fig. 2.9, the non-zero coordinate stresses are the shear stresses T_{12} and T_{21}. Moment equilibrium of the element requires that: $T_{12} = T_{21} = \tau$.

The state of stress in a particle P in the wall of the tube is now expressed by the stress matrix:

$$T = \begin{pmatrix} 0 & \tau & 0 \\ \tau & 0 & 0 \\ 0 & 0 & 0 \end{pmatrix}, \quad \tau = \frac{m_t}{2\pi r^2 h}$$

The small, approximately plane, element of the tube wall shown in Fig. 2.9 is in a *state of pure shear stress*.

2.2.4 Cauchy's Stress Theorem and Cauchy's Stress Tensor

Figure 2.10 shows the *stress vector* $\mathbf{t}$ on a surface element dA with unit normal $\mathbf{n}$ through a particle P. The element may be a part of the boundary surface of a body or a material surface in a body. Let T_{ik} be the coordinate stresses in the particle related to a Cartesian coordinate system Ox with base vectors $\mathbf{e}_i$, and let:

$$\mathbf{t} = t_i\,\mathbf{e}_i, \quad \mathbf{n} = n_k\,\mathbf{e}_k \qquad (2.2.19)$$

Fig. 2.10 Stress vector $\mathbf{t}$ on a surface element dA through a particle P on the surface of a body. Unit normal vector $\mathbf{n}$

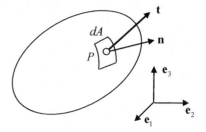

Then the *Cauchy's stress theorem* states that:

$$t_i = T_{ik} n_k \Leftrightarrow t = T n \qquad (2.2.20)$$

The proof of the theorem follows below.

Figure 2.11 shows a small body in the form of a tetrahedron, called a *Cauchy tetrahedron*, of volume V and with a surface consisting of four triangles. Three of the triangles are parallel to coordinate planes and have the areas A_k and are connected at the P. The fourth triangular plane has the area A and a unit normal vector **n** and has a distance h from particle P.

The body is subjected to a body force **b**, the stress vectors $-\mathbf{t}_k$ on the areas A_k and the stress vector **t** on the fourth triangle of area A. Euler's first axiom applied to the body results in the equations of motion:

$$\sum_{k=1}^{3} \int_{A_k} (-\mathbf{t}_k)\, dA + \int_A \mathbf{t}\, dA + \int_V \mathbf{b}\, dV = \int_V \mathbf{a}\, \rho\, dV \qquad (2.2.21)$$

If we let the vectors: $-\mathbf{t}_k$, **t**, $\mathbf{b}\rho$, and $\mathbf{a}\rho$ represent mean values on the respective surfaces and in the volume, the equation of motion (2.2.21) may be presented as:

$$-\mathbf{t}_k A_k + \mathbf{t} A + \mathbf{b}\, \rho\, V = \mathbf{a}\, \rho\, V \qquad (2.2.22)$$

The edges of the tetrahedron in Fig. 2.11 parallel to the base vectors $\mathbf{e}_k$ are denoted by h_k, and since **n** is a unit vector, we may write:

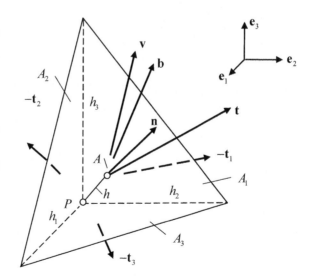

Fig. 2.11 Cauchy thetrahedron. Body of volume V and mass m. Body force **b**. Acceleration **a**. Surface consisting of four triangles: Three orthogonal triangles parallel to coordinate planes are connected at the particle P and have and have areas A_i, unit normals $\mathbf{e}_i$, and stress vectors $-\mathbf{t}_i$. A fourth triangle has area A, unit normal **n**, and stress vector **t**

$$n_k = \cos{(\mathbf{n}, \mathbf{e}_k)} = h/h_k \qquad (2.2.23)$$

The volume V of the tetrahedron may be expressed in four different ways as:

$$V = A \cdot h/3 = A_1 \cdot h_1/3 = A_2 \cdot h_2/3 = A_3 \cdot h_3/3$$

Using the result (2.2.23), we obtain the formulas:

$$V = A \cdot h/3, \qquad A_k = A n_k \qquad (2.2.24)$$

The results (2.2.24) are substituted into the equation of motion (2.2.22), and after division by A, we get:

$$-\mathbf{t}_k n_k + \mathbf{t} + \mathbf{b} \rho h/3 = \mathbf{a} \rho h/3$$

Now we let h approach zero. Then we are left with the following relation between the four stress vectors $\mathbf{t}$ and $-\mathbf{t}_k$ on planes through the particle P:

$$\mathbf{t} = \mathbf{t}_k n_k \qquad (2.2.25)$$

Using the relation (2.2.16) for the components of the vectors $\mathbf{t}_k$, we obtain from the formula (2.2.25) the result:

$$\mathbf{t} = t_i \mathbf{e}_i = T_{ik} \mathbf{e}_i n_k \Rightarrow t_i = T_{ik} n_k$$

This completes the proof for the Cauchy's stress theorem (2.2.20). The theorem shows how the coordinate stresses T_{ik}, or the stress matrix $T = (T_{ik})$, related to a Cartesian coordinate system Ox, completely determines the *state of stress* in a particle.

The unit normal $\mathbf{n}$ and the corresponding stress vector $\mathbf{t}$ are coordinate invariant properties. This implies that the relation (2.2.20) between the components of the two vectors $\mathbf{t}$ and $\mathbf{n}$ has a coordinate invariant character. The stress matrix T works as a vector operator, and we may state: The vector $\mathbf{t}$ is determined by the vector $\mathbf{n}$ through an operator $\mathbf{T}$, which in a Cartesian coordinate system Ox is represented by the matrix T. If a new coordinate system $\bar{O}\bar{x}$ with base vectors $\bar{\mathbf{e}}_i = Q_{ik} \mathbf{e}_k$ is introduced, we shall find the following relation between the components $\bar{t}_i$ of the stress vector $\mathbf{t}$, the components $\bar{n}_k$ of the vector $\mathbf{n}$, and the coordinate stresses $\bar{T}_{ik}$ in the coordinate system $\bar{O}\bar{x}$:

$$\bar{t}_i = \bar{T}_{ik} \bar{n}_k \Leftrightarrow \bar{t} = \bar{T} \bar{n} \qquad (2.2.26)$$

Because the two relations (2.2.20) and (2.2.26) express the same connection between the vectors $\mathbf{n}$ and $\mathbf{t}$, it is appropriate to introduce a coordinate invariant quantity, called the stress tensor $\mathbf{T}$, that let us express the connection between the stress vector **t** and the surface normal **n** in an invariant form. We say that the matrices T and $\bar{T}$ define the *stress tensor* **T** in the following sense:

The *stress tensor* **T** is a coordinate invariant intensive quantity that in any Cartesian coordinate system Ox is represented by the stress matrix T in that coordinate system. The coordinate stresses T_{ik} are called the *components of the stress tensor* in the Ox-system.

The word tensor derives from the Latin word "tensio" meaning tension and was originally the name of the stress matrix. Chapter 3 presents the general definition of tensors and their algebra.

The relations (2.2.20) and (2.2.26) between the components in any two Cartesian coordinate systems Ox and $\bar{O}\bar{x}$ of the stress vector **t**, the stress tensor **T** and the surface normal **n** represent the Cauchy's stress theorem:

Cauchy's stress theorem: The stress vector **t** on a surface through a particle P is uniquely determined by the stress tensor **T** in the particle and the unit normal **n** to the surface through the relation:

$$\mathbf{t} = \mathbf{T}\,\mathbf{n} = \mathbf{T} \cdot \mathbf{n} \Leftrightarrow t_i = T_{ik}n_k \Leftrightarrow t = T\,n \qquad (2.2.27)$$

The two coordinate invariant forms, **Tn** and **T** · **n**, in the relation (2.2.27) are equivalent. In this book the latter form is preferred of reasons that will be given later. The first form is sometimes chosen because it relates to the form of its matrix representation.

The normal stress σ on a surface with unit normal vector **n**, Fig. 2.12, is given by the scalar product of **n** and the stress vector **t**:

$$\sigma = \mathbf{n} \cdot \mathbf{t} = \mathbf{n} \cdot \mathbf{T} \cdot \mathbf{n} = n_i\,T_{ik}\,n_k = n^T\,T\,n \qquad (2.2.28)$$

The shear stress τ on the surface may be computed from:

$$\tau = |\mathbf{n} \times \mathbf{t}| = \sqrt{\mathbf{t} \cdot \mathbf{t} - \sigma^2} \qquad (2.2.29)$$

The formula $\tau = |\mathbf{n} \times \mathbf{t}|$ is obtained as follows:

$$|\mathbf{n} \times \mathbf{t}| = 1 \cdot |\mathbf{t}| \cdot sin(\mathbf{n}, \mathbf{t}) = |\mathbf{t}| \cdot (\tau/|\mathbf{t}|) = \tau$$

Fig. 2.12 Stress vector **t**, normal stress σ shear stresses τ and $\bar{\tau}$ on a surface with unit normal **n** through a particle P

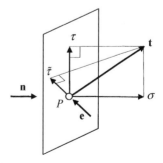

The projection $\tilde{\tau}$ of the shear stress τ in the direction given by a unit vector $\mathbf{e}$ in the surface may be expressed as the scalar product of the vectors $\mathbf{e}$ and $\mathbf{t}$:

$$\tilde{\tau} = \mathbf{e} \cdot \mathbf{t} = \mathbf{e} \cdot \mathbf{T} \cdot \mathbf{n} = e_i \, T_{ik} \, n_k = e^T \, T \, n \tag{2.2.30}$$

It follows from Eqs. (2.2.28) and (2.2.30), that:

$$T_{ik} = \mathbf{e}_i \cdot \mathbf{T} \cdot \mathbf{e}_k \tag{2.2.31}$$

The result (2.2.31) may be interpreted as a general relation between the stress tensor $\mathbf{T}$ and its components T_{ik} in a general Cartesian coordinate system Ox. The relation will now be used to find the relation between the stress matrices T and $\bar{T}$ in two coordinate systems Ox and $O\bar{x}$. The base vectors of the two systems are related through Eq. (1.3.2): $\bar{\mathbf{e}}_i = Q_{ik} \mathbf{e}_k$, where $Q = (Q_{ik})$ is the transformation matrix for the transformation from Ox to $O\bar{x}$. Using the general relation (2.2.31), we obtain:

$$\bar{T}_{ij} = \bar{\mathbf{e}}_i \cdot \mathbf{T} \cdot \bar{\mathbf{e}}_j = (Q_{ik} \mathbf{e}_k) \cdot \mathbf{T} \cdot (Q_{jl} \mathbf{e}_l) = Q_{ik} Q_{jl} \mathbf{e}_k \cdot \mathbf{T} \cdot \mathbf{e}_l = Q_{ik} Q_{jl} T_{kl} \Rightarrow$$

$$\bar{T}_{ij} = Q_{ik} Q_{jl} T_{kl} = Q_{ik} T_{kl} Q_{jl} \Leftrightarrow \bar{T} = QTQ^T \tag{2.2.32}$$

The two stress matrices, T and $\bar{T}$ represent the same state of stress, i.e. the same stress tensor $\mathbf{T}$. Each matrix is the representation of the tensor $\mathbf{T}$ in the respective coordinate system.

Example 2.3 Fluid at Rest. Isotropic State of Stress

In a fluid at rest all material surfaces through a fluid particle transmit the same normal stress, which is the pressure p, and the shear stress on the surfaces is zero. This is called an *isotropic state of stress*. The stress matrix related to any Cartesian coordinate system Ox for a fluid at rest is therefore:

$$T = \begin{pmatrix} -p & 0 & 0 \\ 0 & -p & 0 \\ 0 & 0 & -p \end{pmatrix} = -p \, 1 \Leftrightarrow T_{ik} = -p \, \delta_{ik}, \quad \text{isotropic state of stress}$$

$$\tag{2.2.33}$$

2.2.5   Cauchy's Equations of Motion

From the two axioms of Euler, formulas 2.2.6 and 2.2.7, field equations will be derived that represent the laws of balance of linear and angular momentum of particles in a continuum. Figure 2.13 shows a differential element of mass with volume $dV = dx_1 dx_2 dx_3$ about the particle P.

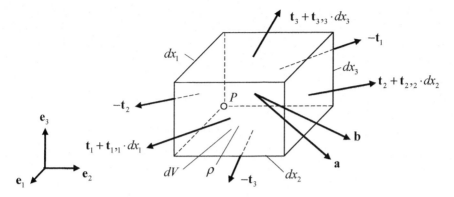

Fig. 2.13 Differential volume element $dV = dx_1 dx_2 dx_3$ about the particle P subjected to body force **b**, and surface forces $\mathbf{t}_i$. Density ρ and acceleration **a**. Base vectors $\mathbf{e}_i$ in a Cartesian coordinate system

The body in Fig. 2.13 is subjected to body forces and contact forces. The law of balance of linear momentum for the element, i.e. Euler's first axiom (2.2.6), yields:

$$(\mathbf{t}_{1,1} dx_1) \cdot (dx_2 dx_3) + (\mathbf{t}_{2,2} dx_2) \cdot (dx_3 dx_1)$$
$$+ (\mathbf{t}_{3,3} dx_3) \cdot (dx_1 dx_2) + \mathbf{b}\rho dV = \mathbf{a}\rho dV \Rightarrow$$

$$\mathbf{t}_{k,k} + \rho\,\mathbf{b} = \rho\,\mathbf{a} \qquad (2.2.34)$$

The terms in this equation represents mean values over the respective surfaces and in the volume, related to the particle P. When the expression (2.2.16) for the stress vectors $\mathbf{t}_k$ is substituted into Eq. (2.2.34), the following component form for the balance of linear momentum is obtained.

$$T_{ik,k} + \rho\,b_i = \rho\,a_i \quad \text{Cauchy's equations of motion in Cartesian coordinates}$$
$$(2.2.35)$$

If we now let the volume of the element shrink to zero, Eq. (2.2.35) become field equations related to particle P. Equation (2.2.35) is called *Cauchy's equations of motion, Cauchy's first law of motion*, or for short the *Cauchy equations*.

We now introduce the concept of the *divergence of a tensor of second order*. The divergence of the stress tensor is a vector, div **T**, with components $T_{ik,k}$ in the Ox-system:

$$\operatorname{div} \mathbf{T} = T_{ik,k} \mathbf{e}_i \qquad (2.2.36)$$

The definition (2.2.36) of div **T** is coordinate invariant, a fact that will be demonstrated in Sect. 3.4 on Tensor Fields. Indirectly the coordinate invariance follows from the Cauchy equations (2.2.35): Since b_i and a_i are components of vectors, the terms $T_{ik,k}$ must also be components of a vector, i.e. a coordinate invariant quantity. The Cauchy equations may now be written in an index free form:

$$\operatorname{div}\mathbf{T} + \rho\mathbf{b} = \rho\mathbf{a} = \rho\ddot{\mathbf{u}} \quad \text{Cauchy's equations of motion} \tag{2.2.37}$$

In a *xyz*-notation the Cauchy equations are:

$$
\begin{aligned}
\frac{\partial\sigma_x}{\partial x} + \frac{\partial\tau_{xy}}{\partial y} + \frac{\partial\tau_{xz}}{\partial z} + \rho\,b_x &= \rho\,a_x = \rho\ddot{u}_x \\[4pt]
\frac{\partial\sigma_y}{\partial y} + \frac{\partial\tau_{yz}}{\partial z} + \frac{\partial\tau_{yx}}{\partial x} + \rho\,b_y &= \rho\,a_y = \rho\ddot{u}_y \\[4pt]
\frac{\partial\sigma_z}{\partial z} + \frac{\partial\tau_{zx}}{\partial x} + \frac{\partial\tau_{zy}}{\partial y} + \rho\,b_z &= \rho\,a_z = \rho\ddot{u}_z
\end{aligned}
\tag{2.2.38}
$$

The Cauchy equations of motion in cylindrical coordinates and in spherical coordinates are presented in Sect. 7.7.

Under the assumption that the forces on a body of continuous material only are contact forces and body forces, the law of balance of angular momentum, i.e. Euler's second axiom (2.2.7), implies that the stress matrix is symmetric:

$$T^T = T \Leftrightarrow T_{ki} = T_{ik} \tag{2.2.39}$$

This result may be proved as follows. Let the state of stress in the neighbourhood of the particle P be given by a homogeneous state of stress given by the stress matrix T in P plus an additional state of stress given by the stress matrix ΔT. Since the matrix T now represents a homogeneous stress field and thus satisfies the Cauchy equations: $T_{ik,k} = 0$, the additional stress matrix ΔT must satisfy the Cauchy equations (2.2.35) and thus balance the body forces $\mathbf{b}$ and the acceleration $\mathbf{a}$. Figure 2.14 shows an element of volume $dV = dx_1\,dx_2\,dx_3$ that contains the particle P. The element is subjected to the homogeneous stress field T. The law of balance of angular momentum applied to the element provides three component equations. The x_3-component equation is:

$$(T_{21}\cdot dx_2\,dx_3)\cdot dx_1 - (T_{12}\cdot dx_1\,dx_3)\cdot dx_2 = 0 \Rightarrow \quad T_{21} = T_{12}$$

Similar results are obtained for the other component equations. The results prove the statement (2.2.39). The symmetry of the stress matrix is a coordinate invariant property, and we therefore say that *the stress tensor is symmetric*. The result (2.2.39) may be interpreted as the law of balance of angular momentum for a particle and is also called *Cauchy's second law of motion*. Normally the symmetry of the stress matrix is assumed a priori such that the matrix is considered to contain only six independent elements rather then nine. Thus the laws of motion for a particle are the three Cauchy equations of motion (2.2.35) or the vector Eq. (2.2.37).

From the symmetry (2.2.39) of the coordinate stresses we may extract the following statement, see Fig. 2.15:

Fig. 2.14 Volume element $dV = dx_1 dx_2 dx_3$ about the particle P subjected to a homogeneous state of stress T. The shear stresses T_{31} and T_{32} are not shown on the element

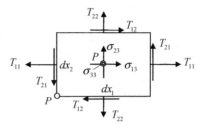

Fig. 2.15 Equal shear stresses τ on two orthogonal surfaces

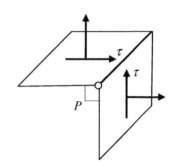

On two orthogonal surfaces through a particle the shear stress components normal to the line intersecting the surfaces are equal.

The truth of the statement follows from the fact that according to the result (2.2.39), the statement is true for the shear stress components normal to the line intersecting any two coordinate planes through the particle.

Example 2.4 Pressure in a Fluid at Rest

Figure 2.16 shows a vessel containing a homogeneous liquid of constant density ρ. The fluid is at rest and is subjected to the constant gravitational force g in the negative x_3-direction. The state of stress in the fluid is represented by the stress matrix (2.2.33) in Example 2.3, i.e. $T_{ik} = -p\delta_{ik}$. The pressure p is generally a function of the position coordinates x_i. For a fluid at rest the Cauchy equations (2.2.35) are reduced to:

$$(-p\,\delta_{ik})_{,k} + \rho\, b_i = 0 \Rightarrow$$

$$-p_{,i} + \rho\, b_i = 0 \Leftrightarrow -\nabla p + \rho \mathbf{b} = \mathbf{0} \qquad (2.2.40)$$

Equation (2.2.40) is called the *equilibrium equation for a fluid*. In this example:

$$b_3 = -g, \quad b_1 = b_2 = 0,$$

and Eq. (2.2.40) yield:

$$\frac{\partial p}{\partial x_1} = 0, \frac{\partial p}{\partial x_2} = 0, \frac{\partial p}{\partial x_3} = -\rho g$$

Fig. 2.16 Fluid of density ρ at rest in a vessel. Body force $\mathbf{b} = -g e_3$. Atmospheric pressure pa

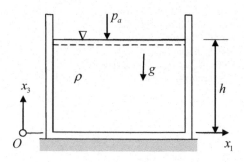

Partial integrations of these equations and use of the boundary condition: $p = p_a$, the atmospheric pressure, for $z = h$ on the free liquid surface, provide the following expression for the pressure in the liquid:

$$p(x_3) = p_a + \rho g (h - x_3)$$

2.3 Stress Analysis

2.3.1 Principal Stresses and Principal Stress Directions

Let us assume that the stress tensor is known in a particle, and let the unit vector $\mathbf{n}$ be a normal vector to a plane through the particle. In Sect. 2.2.4 we have seen that using the Cauchy stress theorem we may compute the stress vector, the normal stress, and the shear stress on the plane. We shall now show that in general there are three orthogonal planes through the particle that are free of shear stress. In Sect. 2.3.3 we show that the normal stresses on these planes include the maximum and the minimum normal stress on planes through the particle. The result of the investigation may be formulated in the following theorem.

The principal stress theorem: For any state of stress there exists, through a particle, three orthogonal planes free of shear stresses. The planes are called the *principal stress planes*, the three unit normals $\mathbf{n}_i$ to the planes are the *principal stress directions*, and the normal stresses σ_i on the planes are the *principal stresses* in the particle. The principal stresses include the maximum normal stress and minimum normal stress on planes through the particle.

First we will show that planes without shear stress through the particle do exist. We search for a plane, defined by its unit normal vector $\mathbf{n}$, on which the stress vector $\mathbf{t}$ is parallel to $\mathbf{n}$. This unit normal vector has to satisfy the relation:

$$\mathbf{t} = \mathbf{T} \cdot \mathbf{n} = \sigma \mathbf{n} \tag{2.3.1}$$

The normal stress σ is the principal stress on the plane. Equation (2.3.1) may be rewritten in the matrix format:

$$(\sigma 1 - T)\, n = 0 \Leftrightarrow (\sigma\, \delta_{ik} - T_{ik})\, n_k = 0 \tag{2.3.2}$$

This is a set of three linear, homogeneous equations for the three unknown components n_k.

The condition that the set of Eq. (2.3.2) has a solution for n_k is that the determinant of the coefficient matrix for n_k is equal to zero.

$$\det(\sigma 1 - T) = 0 \Rightarrow \quad \det \begin{pmatrix} \sigma - T_{11} & -T_{12} & -T_{13} \\ -T_{21} & \sigma - T_{22} & -T_{23} \\ -T_{31} & -T_{32} & \sigma - T_{33} \end{pmatrix} = 0 \Rightarrow$$

$$\sigma^3 - I\sigma^2 + II\sigma - III = 0 \tag{2.3.3}$$

The three coefficients I, II, and III are called the *principal stress invariants* and are expressed by:

$$I = T_{kk} = \operatorname{tr} T$$
$$II = \frac{1}{2} [T_{ii}\, T_{kk} - T_{ik}\, T_{ik}] = \frac{1}{2} \left[(\operatorname{tr} T)^2 - (\operatorname{norm} T)^2 \right] \tag{2.3.4}$$
$$III = \det T$$

The fact that the three coefficients I, II, and III are coordinate invariant quantities, will formally be shown later. But from a physical point of view we may conclude that if the cubic Eq. (2.3.3) has a solution, this solution cannot depend upon the coordinate system applied to find it. Thus, if the solution of Eq. (2.3.3) is coordinate invariant, the coefficients in the equation from which the solution is found, must also be coordinate invariant. It will be shown that the cubic Eq. (2.3.3), called the *characteristic equation of the stress tensor*, has three real roots: σ_1, σ_2, and σ_3.

The mathematical problem related to Equations (2.3.1–2.3.4) is known as an *eigenvalue problem*: σ_i are the *eigenvalues* and the corresponding $\mathbf{n}_i$ are the *eigenvectors* of the tensor $\mathbf{T}$.

Any cubic equation has at least one real root, which we shall denote by σ_3. The corresponding principal direction is denoted by $\mathbf{n}_3$. If we now choose a coordinate system Ox for which the base vector $\mathbf{e}_3$ is equal to $\mathbf{n}_3$, the stress vector $\mathbf{t}_3$ will be equal to $\sigma\, \mathbf{e}_3$ and the shear stress components $T_{\alpha 3} = 0$. The stress matrix in this coordinate system is then:

$$T = \begin{pmatrix} T_{11} & T_{12} & 0 \\ T_{21} & T_{22} & 0 \\ 0 & 0 & \sigma_3 \end{pmatrix} \tag{2.3.5}$$

To find the two other principal stresses σ_1 and σ_2, and the corresponding principal directions $\mathbf{n}_1$ and $\mathbf{n}_2$ the three Eq. (2.3.2) must be solved. These equations are now reduced to:

$$\left(\sigma\,\delta_{\alpha\beta} - T_{\alpha\beta}\right)n_\beta = 0, \quad \left(\sigma - \sigma_3\right)n_3 = 0 \tag{2.3.6}$$

The principal stress σ is either σ_1 and σ_2. Let us first assume that $\sigma \neq \sigma_3$. Then it follows from the last of Eq. (2.3.6) that $n_3 = 0$. Thus the corresponding principal stress direction $\mathbf{n}$ is parallel to the $x_1\,x_2$-plane and therefore normal to the principal direction $\mathbf{n}_3$. Figure 2.17 shows a particle P and a material surface normal to the principal direction $\mathbf{n}$. Let ϕ be the angle between $\mathbf{n}$ and the x_1-axis. Then:

$$\mathbf{n} = [\cos\phi,\ \sin\phi,\ 0]$$

The first two of Eq. (2.3.6) become:

$$\begin{aligned}(\sigma - T_{11})\cos\phi - T_{12}\sin\phi = 0 \\ - T_{21}\cos\phi + (\sigma - T_{22})\sin\phi = 0\end{aligned} \tag{2.3.7}$$

For this set of linear equations to have a solution for the angle ϕ the determinant of the coefficient matrix has to be zero. Thus:

$$(\sigma - T_{11})(\sigma - T_{22}) - T_{12}\,T_{21} = 0 \Rightarrow$$

$$\sigma^2 - (T_{11} + T_{22})\,\sigma + T_{11}\,T_{22} - T_{12}\,T_{21} = 0 \tag{2.3.8}$$

The solution of this quadratic equation is, when the symmetry property of the matrix T is utilized:

$$\begin{matrix}\sigma_1 \\ \sigma_2\end{matrix} = \frac{T_{11} + T_{22}}{2} \pm \sqrt{\left(\frac{T_{11} - T_{22}}{2}\right)^2 + (T_{12})^2} \tag{2.3.9}$$

Fig. 2.17 Principal stress surface with principal stress σ and principal direction $\mathbf{n}$

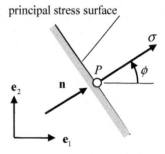

principal stress surface

The radicand can never be negative, which means that the two roots σ_1 and σ_2 are real. The angle ϕ which determines the principal stress directions are determined from Eq. (2.3.7), from which we find the results:

$$\tan \phi_1 = \frac{\sigma_1 - T_{11}}{T_{12}} \quad \text{(for } \sigma = \sigma_1\text{)}, \qquad \tan \phi_2 = \frac{\sigma_2 - T_{11}}{T_{12}} \quad \text{(for } \sigma = \sigma_2\text{)}$$

$$(2.3.10)$$

If any two principal stresses are unequal, for instance $\sigma_1 \neq \sigma_2$, the corresponding principal directions $\mathbf{n}_1$ and $\mathbf{n}_2$ are orthogonal. This result may be demonstrated using Eqs. (2.3.10) and (2.3.9). Another way of proving that the principal directions $\mathbf{n}_1 = n_{1\alpha}\,\mathbf{e}_\alpha$ and $\mathbf{n}_2 = n_{2\alpha}\,\mathbf{e}_\alpha$ are orthogonal if $\sigma_1 \neq \sigma_2$ is as follows. From Eq. (2.3.6) we obtain:

$$\left(\sigma_1 \delta_{\alpha\beta} - T_{\alpha\beta}\right) n_{1\beta}\, n_{2\alpha} = 0, \qquad \left(\sigma_2 \delta_{\alpha\beta} - T_{\alpha\beta}\right) n_{2\beta}\, n_{1\alpha} = 0$$

The two sets of equations are subtracted and due to the symmetry of the stress matrix T, we get:

$$\left(\sigma_1 - \sigma_2\right) n_{1\alpha}\, n_{2\alpha} = \left(\sigma_1 - \sigma_2\right) \mathbf{n}_1 \cdot \mathbf{n}_2 = 0$$

Since $\sigma_1 \neq \sigma_2$, it follows that $\mathbf{n}_1 \cdot \mathbf{n}_2 = 0$. This result proves that the principal stress directions $\mathbf{n}_1$ and $\mathbf{n}_2$ are orthogonal if $\sigma_1 \neq \sigma_2$.

We have now in fact proved that the three roots in the cubic Eq. (2.3.3) are real, and that if the three roots are all different, the principal stress directions are orthogonal. Of physical reasons it is clear that the principal stresses and the principal stress directions are coordinate invariant properties of the stress tensor.

The stress vectors on the principal stress planes are given by:

$$\mathbf{t}_i = \sigma_i\, \mathbf{n}_i \qquad\qquad (2.3.11)$$

Because the principal stress planes, by definition, are free of shear stress, the stress matrix with respect to a Cartesian coordinate system Ox with base vectors $\mathbf{e}_i$ coinciding with the principal stress directions $\mathbf{n}_i$, is a diagonal matrix:

$$T = \begin{pmatrix} \sigma_1 & 0 & 0 \\ 0 & \sigma_2 & 0 \\ 0 & 0 & \sigma_3 \end{pmatrix} \qquad\qquad (2.3.12)$$

The principal stress invariants (2.3.4) now take the simple forms:

$$I = \sigma_1 + \sigma_2 + \sigma_3, \quad II = \sigma_1 \sigma_2 + \sigma_2 \sigma_3 + \sigma_3 \sigma_1, \quad III = \sigma_1 \sigma_2 \sigma_3 \qquad (2.3.13)$$

The result shows that *I, II,* and *III* are coordinate invariant properties, i.e. they are scalars. The principal directions of stress $\mathbf{n}_i$ are also called the *principal axes of stress* in the particle.

The two roots in Eq. (2.3.9) coincide only if $T_{12} = 0$ and $T_{11} = T_{22}$. In that case:

$$\sigma_1 = \sigma_2 = T_{11} = T_{22}$$

From Eq. (2.3.7) it follows that the angle ϕ in that case becomes indeterminate. This means that any direction in a plane parallel to the x_1x_2-plane is a principal direction, and that the stress on any plane parallel to the x_3-axis, is a principal stress equal to $\sigma_1 = \sigma_2$. The situation is called a *plane-isotropic state of stress*, see Example 2.5 below.

We now return to Eq. (2.3.6) and consider the possibility that the principal stress σ_1 is equal to σ_3. We may still choose $n_3 = 0$ and obtain the solution given by Eqs. (2.3.9–2.3.10), but if $\sigma = \sigma_1 = \sigma_3$, we obviously shall find that any direction $\mathbf{n}$ in the plane parallel to the x_1x_3-plane is a principal direction of stress with principal stress $\sigma = \sigma_1 = \sigma_3$.

If all three roots are equal: $\sigma_1 = \sigma_2 = \sigma_3$, all directions are principal stress directions. The stress matrix will be the same in all coordinate systems and equal to the scalar σ multiplied by the unit matrix:

$$T_{ik} = \sigma\,\delta_{ik} \Leftrightarrow T = \sigma\,\mathbf{1} \tag{2.3.14}$$

In this case the stress tensor $\mathbf{T}$ is called an *isotropic tensor*, and we have an *isotropic state of stress*. Because the state of stress in a fluid at rest is isotropic, with the pressure p as the principal stress, see Example 2.3 and the stress matrix (2.2.33), and because water is the most typical fluid, and is called "hydro" in Latin (=hudor in Greek), the state of stress (2.3.14) is also called a *hydrostatic state of stress*.

Equation (2.3.2), that determine the principal directions $\mathbf{n}_i$, do not determine the direction of the arrow of the vectors $\mathbf{n}_i$. If $\mathbf{n}_i$ is a principal direction so is the vector $-\mathbf{n}_i$.

The general case when all the principal stresses are different from zero is called a *triaxial state of stress*. If only two of the principal stresses in a particle are different from zero, the particle is in a *biaxial state of stress*. This is also called *plane state of stress*; see Example 2.5 below. A *uniaxial state of stress* has only one non-zero principal stress, as in Example 2.1.

Example 2.5 Biaxial State of Stress

The container shown in Fig. 2.18 has a thin-walled circular cylindrical main part. The mean radius of the cylinder is r and the wall thickness is $h \ll r$. The container is subjected to an internal pressure p. We shall present expressions for the stresses in the cylindrical wall.

With respect to the local Cartesian coordinate system shown on the small element of the container wall in Fig. 2.17, the non-zero coordinate stresses are the normal stresses T_{11}, T_{22} and T_{33}. The normal stress T_{33} on the element surface representing the outside of the container is zero, while the normal stress T_{33} on the element

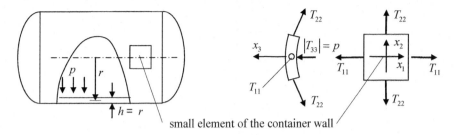

small element of the container wall

Fig. 2.18 Circular thin—walled cylindrical container with internal pressure p. Mean radius r and wall—thickness h

surface representing the inside if the container is equal to $T_{33} = -p$. The normal stress T_{11} on cross-sections of the container and the normal stress T_{22} on planes through the axis of the container may be found by equilibrium consideration to be:

$$T_{11} = \frac{r}{2h}p, \quad T_{22} = \frac{r}{h}p \tag{2.3.15}$$

Because $t \ll r$, the absolute value of the stress T_{33} is much smaller than T_{11} and T_{22} from the formula (2.3.15). The state of stress in the cylindrical wall may therefore be represented by the stress matrix:

$$T = \begin{pmatrix} T_{11} & 0 & 0 \\ 0 & T_{22} & 0 \\ 0 & 0 & T_{33} \end{pmatrix} \approx \begin{pmatrix} 1 & 0 & 0 \\ 0 & 2 & 0 \\ 0 & 0 & 0 \end{pmatrix} \frac{r}{2h}p \tag{2.3.16}$$

The state of stress represented by the stress matrix (2.3.16), with $T_{33} = 0$, is an example of a *plane state of stress* and a *biaxial state of stress*.

When a thin-walled spherical shell with middle radius r and wall thickness $h(\ll r)$ is subjected to an internal pressure p, the normal stress on any meridian plane through the shell is:

$$\sigma = \frac{r}{2h}p \tag{2.3.17}$$

On a small element of the wall of the shell analogous to the element in Fig. 2.18 for the cylindrical container, the state of stress may approximately be given by the stress matrix:

$$T = \begin{pmatrix} T_{11} & 0 & 0 \\ 0 & T_{22} & 0 \\ 0 & 0 & T_{33} \end{pmatrix} \approx \begin{pmatrix} 1 & 0 & 0 \\ 0 & 1 & 0 \\ 0 & 0 & 0 \end{pmatrix} \frac{r}{2h}p \tag{2.3.18}$$

This special kind of plane state of stress is called *plane-isotropic state of stress*.

Fig. 2.19 Principal stresses
σ_1 and σ_2, and principal
directions in a thin—walled
tube subjected to a torque M

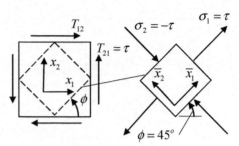

small plane elements of the tube wall

Example 2.6 Principal Stresses in a State of Pure Shear Stress
The state of stress in a thin-walled tube subjected to a torsion moment M has been
presented in Example 2.2. The shear stresses on a small, approximately plane,
element of the tube wall are again presented here in Fig. 2.19.

Based on the stress matrix presented in Example 2.2 and the formulas (2.3.9) and
(2.3.10) we obtain the results:

$$\begin{matrix} \sigma_1 \\ \sigma_2 \end{matrix} = \frac{T_{11} + T_{22}}{2} \pm \sqrt{\left(\frac{T_{11} - T_{22}}{2}\right)^2 + (T_{12})^2} = \begin{matrix} \tau \\ -\tau \end{matrix}$$

$$\tan \phi_1 = \frac{\sigma_1 - T_{11}}{T_{12}} = \frac{\tau}{\tau} = 1 \quad \Rightarrow \quad \phi_1 = 45°$$

Figure 2.19 illustrates the results by the principal stresses on a small plane element
of the tube wall. With respect to the local Cartesian coordinate system with the
$\bar{x}_\alpha$-axes parallel to the principal stress directions, stress matrix is:

$$\bar{T} = \begin{pmatrix} 1 & 0 & 0 \\ 0 & -1 & 0 \\ 0 & 0 & 0 \end{pmatrix} \tau$$

Both matrices T and $\bar{T}$ in this example represent the same state of stress: a *state of
pure shear stress.*

2.3.2 Stress Deviator and Stress Isotrop

The stress matrix T may uniquely be decomposed into a trace-free matrix T' and an
isotropic matrix T^o:

$$T = T' + T^o, \quad T^o = \left(\frac{1}{3} \operatorname{tr} T\right) \mathbf{1}, \quad T' = T - T^o, \quad \operatorname{tr} T' = 0 \qquad (2.3.19)$$

The matrix T' represents a stress tensor $\mathbf{T}'$ which we shall call the *stress deviator*, while the matrix T^o represents a stress tensor $\mathbf{T}^o$ which we shall call the *stress isotrop*:

$$\mathbf{T} = \mathbf{T}' + \mathbf{T}^o, \quad \mathbf{T}^o = \left(\frac{1}{3}\operatorname{tr} T\right)\mathbf{1}, \quad \mathbf{T}' = \mathbf{T} - \mathbf{T}^o, \quad \operatorname{tr}\mathbf{T}' = 0 \qquad (2.3.20)$$

Another name for the tensor $\mathbf{T}^o$ is the *hydrostatic stress tensor*. Related to the principal axes of stress, all the matrices T, T', and T^o have diagonal form. The stress deviator is represented by the matrix:

$$T' = \begin{pmatrix} \sigma_1' & 0 & 0 \\ 0 & \sigma_2' & 0 \\ 0 & 0 & \sigma_3' \end{pmatrix}, \qquad \sigma_i' = \sigma_i - \frac{1}{3}(\sigma_1 + \sigma_2 + \sigma_3)$$

Because the tensors $\mathbf{T}$ and $\mathbf{T}'$ have coinciding principle directions, we say that the tensors $\mathbf{T}$ and $\mathbf{T}'$ are *coaxial tensors*.

The principal deviator stresses are determined from the characteristic Eq. (2.3.3) for the stress deviator $\mathbf{T}'$. Since the first principal invariant of $\mathbf{T}'$ by its definition in the formulas (2.3.19) is equal to zero, i.e. $I' = \operatorname{tr} T' = 0$, the characteristic equation is reduced to:

$$(\sigma')^3 + II'\sigma' - III' = 0 \qquad (2.3.21)$$

II' and III' are the second and third principal invariants of the stress deviator $\mathbf{T}'$:

$$II' = -\frac{1}{2}T_{ik}' T_{ik}' = -\frac{1}{2}(\operatorname{norm} T')^2 \leq 0, \quad III' = \det T' \qquad (2.3.22)$$

Decomposition of the Stress Deviator T′

The matrix T' of the stress deviator $\mathbf{T}'$ in a general Cartesian coordinate system Ox may be decomposed into five matrices each representing a state of pure shear stress. First we write:

$$\operatorname{tr}\mathbf{T}' = T_{11}' + T_{22}' + T_{33}' = 0 \Rightarrow \quad T_{22}' = -T_{11}' - T_{33}' \qquad (2.3.23)$$

The matrix T' may then be presented by the decomposition:

$$T' = \begin{pmatrix} T'_{11} & T'_{12} & T'_{13} \\ T'_{21} & T'_{22} & T'_{23} \\ T'_{31} & T'_{32} & T'_{33} \end{pmatrix} = T'_{11} \begin{pmatrix} 1 & 0 & 0 \\ 0 & -1 & 0 \\ 0 & 0 & 0 \end{pmatrix} + T'_{33} \begin{pmatrix} 0 & 0 & 0 \\ 0 & -1 & 0 \\ 0 & 0 & 1 \end{pmatrix}$$

$$+ T'_{12} \begin{pmatrix} 0 & 1 & 0 \\ 1 & 0 & 0 \\ 0 & 0 & 0 \end{pmatrix} + T'_{23} \begin{pmatrix} 0 & 0 & 0 \\ 0 & 0 & 1 \\ 0 & 1 & 0 \end{pmatrix} + T'_{31} \begin{pmatrix} 0 & 0 & 1 \\ 0 & 0 & 0 \\ 1 & 0 & 0 \end{pmatrix}$$

$$(2.3.24)$$

When the five matrices on the right-hand side are compared with the stress matrices presented in Example 2.2 and Example 2.6, we see that they all five represent states of pure shear stress.

Applications of the decomposition (2.3.20) will be demonstrated in Sect. 5.2.1 for isotropic, linearly elastic materials and in Sect. 5.3.3 for linearly viscous fluids.

2.3.3 Extreme Values of Normal Stress

The three principal stresses in a particle represent the extreme values for normal stress on planes through the particle. To show this we first choose a coordinate system Ox-system with base vectors e_i parallel to the principal stress directions n_i. The stress matrix in this coordinate system is the diagonal matrix (2.3.12) having the elements:

$$T_{ik} = \sigma_i \delta_{ik} \qquad (2.3.25)$$

For convenience the principal stresses σ_i are now ordered such that:

$$\sigma_3 \leq \sigma_2 \leq \sigma_1 \qquad (2.3.26)$$

The normal stress σ on a plane with unit normal n is given by formula (2.2.28):

$$\sigma = n \cdot T \cdot n = \sum_{i,k} n_i \sigma_i \delta_{ik} n_k = \sigma_1 n_1^2 + \sigma_2 n_2^2 + \sigma_3 n_3^2 \qquad (2.3.27)$$

Due to the arrangement (2.3.26) and because n is a unit vector, i.e.: $n \cdot n = n_1^2 + n_2^2 + n_3^2 = 1$, we find from the result (2.3.27) that:

$$\sigma_3 \left(n_1^2 + n_2^2 + n_3^2 \right) \leq \sigma \leq \sigma_1 \left(n_1^2 + n_2^2 + n_3^2 \right) \Rightarrow$$

$$\sigma_3 \leq \sigma \leq \sigma_1 \qquad (2.3.28)$$

From this result it follows that when the principal stresses are arranged as in (2.3.26), then:

$$\sigma_{max} = \sigma_1, \quad \sigma_{min} = \sigma_3 \qquad (2.3.29)$$

The largest principal stress is therefore the maximum normal stress in the particle on planes through the particle, and the smallest principal stress is the minimum normal stress in the particle on planes through the particle.

2.3.4 Maximum Shear Stress

The shear stress τ on a material plane with unit normal $\mathbf{n}$ is given by formula (2.2.29). The normal projection $\tilde{\tau}$ of the stress vector $\mathbf{t}$ onto the direction $\mathbf{e}$ in the plane is determined by formula (2.2.30): $\tilde{\tau} = \mathbf{e} \cdot \mathbf{t} = \mathbf{e} \cdot \mathbf{T} \cdot \mathbf{n} = e_i T_{ik} n_k$. From Fig. 2.20 it follows that $\tilde{\tau} \leq \tau$. The equality sign applies when the unit vector $\mathbf{e}$ lies in the plane through $\mathbf{t}$ and $\mathbf{n}$.

We shall determine the unit vectors $\mathbf{n}$ and $\mathbf{e}$ such that $\tilde{\tau}$ becomes a maximum. Once again, we choose the representation (2.3.25) for the elements of the stress matrix and the principal stresses are arranged according to (2.3.26). Then:

$$\tilde{\tau} = \mathbf{e} \cdot \mathbf{T} \cdot \mathbf{n} = \sum_{i,k} e_i \, \sigma_i \, \delta_{ik} \, n_k = \sigma_1 \, e_1 \, n_1 + \sigma_2 \, e_2 \, n_2 + \sigma_3 \, e_3 \, n_3$$

Now, since $\mathbf{e}$ and $\mathbf{n}$ are orthogonal vectors:

$$\mathbf{e} \cdot \mathbf{n} = e_1 n_1 + e_2 n_2 + e_3 n_3 = 0 \Rightarrow \quad e_2 n_2 = -e_1 n_1 - e_3 n_3.$$

Hence we may write:

$$\tilde{\tau} = (\sigma_1 - \sigma_2)e_1 n_1 + (\sigma_2 - \sigma_3)(-e_3 \, n_3) \qquad (2.3.30)$$

The terms $(\sigma_1 - \sigma_2)$ and $(\sigma_2 - \sigma_3)$ are both non-negative. In order to make $\tilde{\tau}$ as large as possible we must make the terms $e_1 \, n_1$ and $-e_3 n_3$ as large as possible.

Fig. 2.20 Stress vector t, normal stress σ, and shear stresses τ and $\tilde{\tau}$ on a material plane through a particle P

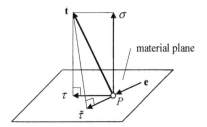

Fig. 2.21 Stress vector **t**, normal stress σ and shear stress $\tilde{\tau} = \tau$ on a material plane through a particle P. The unit vectors **e** and **n** are parallel to the principal directions $\mathbf{n}_1$ and $\mathbf{n}_3$

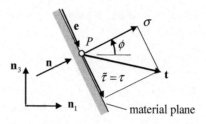

The vectors **e** and **n** are unit vectors, and the absolute values of e_1, e_3, n_1, and n_3 become largest if we set: $e_2 = n_2 = 0$. This means that we should choose **e** and **n** in a plane parallel to the principal directions $\mathbf{n}_1$ and $\mathbf{n}_3$, as shown in Fig. 2.21. Since the figure plane is a principal stress plane, the stress vector **t** has no component in the $\mathbf{n}_2$-direction. This implies that $\tilde{\tau} = \tau$, i.e. the shear stress on the plane according to formula (2.2.29). From Fig. 2.21 we find that: $n_1 = -e_3 = \cos\phi$, $n_3 = e_1 = \sin\phi$, and Eq. (2.3.29) gives:

$$\tilde{\tau} = \tau = (\sigma_1 - \sigma_2)\sin\phi\,\cos\phi + (\sigma_2 - \sigma_3)(\cos\phi)\sin\phi \Rightarrow$$
$$\tau = \frac{1}{2}(\sigma_1 - \sigma_3)\sin 2\phi$$

From this result we conclude that the maximum shear stress is:

$$\tau_{max} = \frac{\sigma_1 - \sigma_3}{2} \quad \text{for } 2\phi = \frac{\pi}{2} \quad \Rightarrow \quad \phi = 45°.$$

Thus we have found that:

$$\tau_{max} = \frac{1}{2}(\sigma_{max} - \sigma_{min}) \tag{2.3.31}$$

> The maximum shear stress acts on planes that are inclined 45° with respect to the principal directions of the largest and the smallest principal stresses.

There are four such planes. Figure 2.22 shows the orientation of one of these planes and how the maximum shear stress τ_{max} acts on that plane.

Figure 2.22a shows a small volume element about a particle P, with sides parallel to the initial coordinate surfaces, and with coordinate stresses T_{ik}. Figure 2.22b shows an element about P with sides parallel to the principal stress planes of the stress tensor **T**. Figure 2.22c shows a triangular prism about P with the maximum shear stress τ_{max} on a surface that is inclined 45° with respect to the principal axes corresponding to σ_{max} and σ_{min}.

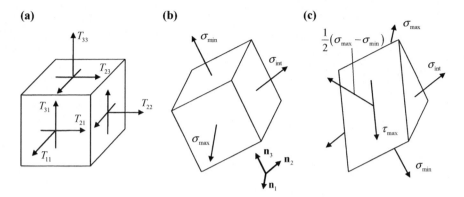

Fig. 2.22 a Coordinate stresses T_{ik} . **b** Principal stresses: σ_{max}, σ_{min}, σ_{int}. **c** Plane with maximum shear stress τ_{max}

2.3.5   State of Plane Stress

If a surface free of stress exists through a particle, the particle is in a *state of plane stress*, or a *state of biaxial stress*. This is often the case in engineering problems and especially where the stresses are at their extreme values. For instance, the surface of machine parts and structural elements may be free of loads, but the stresses in the surface may be very high.

Let the plane state of stress in a particle P be defined by the condition: $T_{i3} = 0$, which implies that the plane normal to the x_3-axis is stress free at the particle. The state of stress in the particle P is illustrated in Fig. 2.23a and is determined by the three coordinate stresses T_{11}, T_{22}, and T_{12}. The stress free plane through P normal to the x_3-direction is a *principal stress plane* with the principal stress $\sigma_3 = 0$. When discussing plane stress we do not use the ordering $\sigma_3 \leq \sigma_2 \leq \sigma_1$ for the three principal stresses.

The two other principal stresses, σ_1 and σ_2, are given by formula (2.3.9), repeated here:

$$\begin{matrix} \sigma_1 \\ \sigma_2 \end{matrix} = \frac{T_{11} + T_{22}}{2} \pm \sqrt{\left(\frac{T_{11} - T_{22}}{2}\right)^2 + (T_{12})^2} \qquad (2.3.32)$$

The principal directions are represented by the angles ϕ_1 and ϕ_2, see Fig. 2.23b. From Eq. (2.3.10) we obtain:

$$\tan \phi_1 = \frac{\sigma_1 - T_{11}}{T_{12}}, \quad \tan \phi_2 = \frac{\sigma_2 - T_{11}}{T_{12}} \qquad (2.3.33)$$

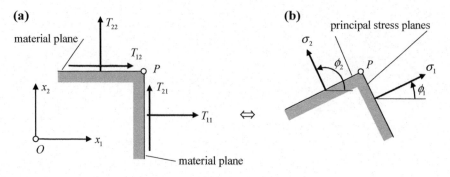

Fig. 2.23 State of plane stress. **a** Coordinate stresses. **b** Principal stresses

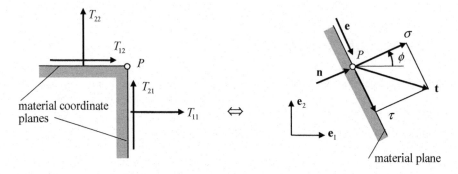

Fig. 2.24 State of plane stress. Normal stress σ and shear stress τ on a materialplane through a particle P and with unit normal **n**

Since σ_1 and σ_2 may both be positive, both be negative, or have different signs, we must remember that the third principal stress, $\sigma_3 = 0$, may represent either σ_{max} or σ_{min}.

Formulas for stresses on planes perpendicular to the stress free plane, i.e. planes parallel to the x_3-direction, will now be developed. Figure 2.24 presents the situation. The unit normal **n** to the plane and the unit vector **e** in the plane are given by:

$$\mathbf{n} = [\cos\phi,\ \sin\phi, 0], \quad \mathbf{e} = [\sin\phi, -\cos\phi, 0] \tag{2.3.34}$$

ϕ is the angle between the direction of the normal **n** and the x_1-direction.

The components of the stress vector **t** on the plane are obtained from Cauchy's stress theorem (2.2.20) with $T_{21} = T_{12}$.

$$t_\alpha = T_{\alpha\beta}\, n_\beta = \begin{cases} t_1 = T_{11}\cos\phi + T_{12}\sin\phi \\ t_2 = T_{12}\cos\phi + T_{22}\sin\phi \end{cases}$$

The normal stress σ and the shear stress τ on the plane are then:

$$\sigma = \mathbf{n} \cdot \mathbf{t} = n_\alpha t_\alpha = T_{11} \cos^2 \phi + T_{22} \sin^2 \phi + 2T_{12} \sin \phi \cos \phi$$

$$\tau = \mathbf{e} \cdot \mathbf{t} = e_\alpha t_\alpha = (T_{11} - T_{22}) \sin \phi \cos \phi - T_{12}(\cos^2 \phi - \sin^2 \phi)$$

Using the trigonometric formulas:

$$\sin 2\phi = 2 \sin \phi \cos \phi, \quad \cos 2\phi = \cos^2 \phi - \sin^2 \phi \qquad (2.3.35)$$

we may transform the expressions for σ and τ to:

$$\sigma(\phi) = \frac{1}{2}(T_{11} + T_{22}) + \frac{1}{2}(T_{11} - T_{22}) \cos 2\phi + T_{12} \sin 2\phi \qquad (2.3.36)$$

$$\tau(\phi) = \frac{1}{2}(T_{11} - T_{22}) \sin 2\phi - T_{12} \cos 2\phi \qquad (2.3.37)$$

The extreme values of σ and τ in the formulas (2.3.36) and (2.3.37) may be determined as follows. First we develop an expression for $[\sigma - (T_{11} + T_{22})/2]$ from formula (2.3.36). We then add the square of this expression and the square of τ obtained from formula (2.3.37). The result is:

$$\left[\sigma(\phi) - \frac{T_{11} + T_{22}}{2}\right]^2 + [\tau(\phi)]^2 = \left[\frac{T_{11} - T_{22}}{2}\right]^2 + [T_{12}]^2 \qquad (2.3.38)$$

From this equation we see that the normal stress σ obtains its extreme values when the shear stress $\tau = 0$, and that the extreme values are given by the formulas (2.3.32). The extreme values of the shear stress τ occur when $\sigma = (T_{11} + T_{22})/2$ and are:

$$\tau = \pm\sqrt{\left[\frac{T_{11} - T_{22}}{2}\right]^2 + [T_{12}]^2} = \pm(\sigma_1 - \sigma_2) \qquad (2.3.39)$$

It is clear from the discussion above that the maximum shear stress in the particle P is in general not given by this result, but only when the two principal stresses σ_1 and σ_2 have opposite signs.

2.3.6 Mohr-Diagram for State of Plane Stress

The stress analysis of plane stress may be illustrated graphically in a diagram. This graphic method has many important applications and gives a concentrated presentation of all aspects of the state of plane stress. The graphic method is also applicable in the analysis of any symmetric tensor of second order in two dimensions, e.g. the strain tensor for small deformation in a surface in Sect. 4.3.7.

Let us first assume that the principal stresses and principal stress directions are known. In a coordinate system with axes parallel to the principal directions, Eqs. (2.3.36–2.3.38) become:

$$\sigma(\phi) = \frac{1}{2}(\sigma_1 + \sigma_2) + \frac{1}{2}(\sigma_1 - \sigma_2)\cos 2\phi, \quad \tau(\phi) = \frac{1}{2}(\sigma_1 - \sigma_2)\sin 2\phi \quad (2.3.40)$$

$$\left[\sigma(\phi) - \frac{\sigma_1 + \sigma_2}{2}\right]^2 + [\tau(\phi)]^2 = \left[\frac{\sigma_1 - \sigma_2}{2}\right]^2 \quad (2.3.41)$$

In a plane Cartesian coordinate system with σ and τ as coordinates, see Fig. 2.25, Eq. (2.3.41) describes a circle of radius $(\sigma_1 - \sigma_2)/2$ and with center C on the σ-axis at a distance $(\sigma_1 + \sigma_2)/2$ from the origin O. This circle is called *Mohr's stress circle* after Otto Mohr [1835–1918]. Figure 2.25 will be called a *Mohr-diagram*. The points on the circle will be called stress points.

The stress point S having coordinates $(\sigma, \tau) \equiv (\sigma(\phi), \tau(\phi))$ represents the stresses on the physical plane that makes the angle ϕ with the principal direction for

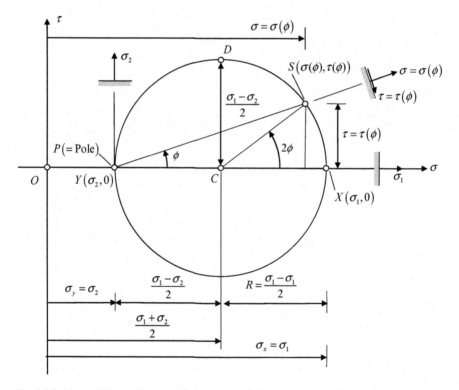

Fig. 2.25 Mohr—diagram for state of plane stress. Pole P of normals. Principal stresses σ_1 and σ_2. Stresses $\sigma(\phi)$ and $\tau(\phi)$ on planes perpendicular to the stress free plane with unit normal $\mathbf{n} = [\cos\phi, \sin\phi, 0]$

σ_1. The central angle between the σ-axis and the radius CS is equal to $2\,\phi$. This may be seen as follows. From the Mohr-diagram we derive the formulas:

$$\sin 2\phi = \frac{2\,\tau}{\sigma_1 - \sigma_2}, \quad \cos 2\phi = \frac{2\sigma - (\sigma_1 + \sigma_2)}{\sigma_1 - \sigma_2} \tag{2.3.42}$$

and these formulas are confirmed by the formulas (2.3.40).

The stresses, σ' and τ', on a plane with a normal making the angle $\phi + \pi/2$ with the σ_1-direction, i.e. a plane perpendicular to the plane defined by the angle ϕ, are by formula (2.3.40) given as:

$$\sigma' = \frac{1}{2}(\sigma_1 + \sigma_2) + \frac{1}{2}(\sigma_1 - \sigma_2) \cos\left[2\left(\phi + \frac{\pi}{2}\right)\right] = \frac{1}{2}(\sigma_1 + \sigma_2) - \frac{1}{2}(\sigma_1 - \sigma_2)\cos 2\phi$$

$$\tau' = \frac{1}{2}(\sigma_1 - \sigma_2)\sin\left[2\left(\phi + \frac{\pi}{2}\right)\right] = -\frac{1}{2}(\sigma_1 - \sigma_2)\sin 2\phi = -\tau$$

The point $Y(\sigma_2, 0)$ in the Mohr-diagram Fig. 2.25 also represents a *pole of normals* to the planes parallel to the x_3-direction. From the figure we see that the line from the pole $P = Y(\sigma_2, 0)$ to the stress point S makes an angle ϕ with the x_1-direction: The periphery angle XYS is half of the central angle XCS. The line YS is therefore parallel to the normal to the plane with the stresses $\sigma = \sigma(\phi)$ and $\tau = \tau(\phi)$. Using this property of the pole we can include all information about the stresses on planes parallel to the x_3-direction in the Mohr-diagram. Figure 2.25 shows how this may be presented.

The Mohr-diagram may also be constructed in the more general case of plane stress when the state of stress is given by the coordinate stresses T_{11}, T_{22}, and T_{12}. In a plane Cartesian coordinate system with $\sigma = \sigma(\phi)$ and $\tau = \tau(\phi)$ as coordinates, Eq. (2.3.38) describes a circle of radius r and center C:

$$r = \sqrt{\left[\frac{T_{11} - T_{22}}{2}\right]^2 + [T_{12}]^2}, \quad C = \left(\frac{T_{11} - T_{22}}{2}, 0\right) \tag{2.3.43}$$

The coordinates for each point on the circle represent the normal stress and shear stress on a plane through the particle and parallel to the x_3-axis. Two points on the circle are known from the coordinate stresses:

$$Y = Y(T_{22}, T_{12}), \quad X = X(T_{11}, -T_{21})$$

Figure 2.26 shows this Mohr-diagram. The stress point $X = (T_{11}, T_{12})$ represents the stresses on the plane normal to the x_1-direction, i.e. the plane for which $\phi = 0$. The stress point $Y = (T_{22}, T_{12})$ represents the stresses on the plane normal to the x_1-direction, that is the plane for which $\phi = \pi/2$. The two points X and Y lie on the a diameter of the stress circle that makes the angle 2ϕ with the σ-axis, measured in the clockwise direction from the σ-axis.

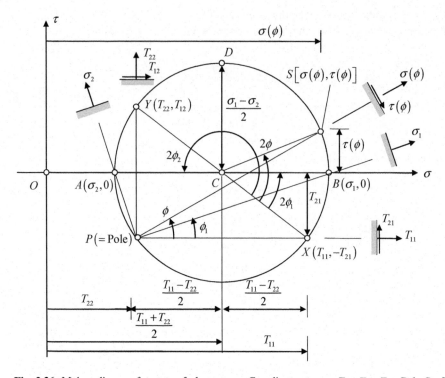

Fig. 2.26 Mohr—diagram for state of plane stress. Coordinate stresses T_{11}, T_{22}, T_{12}. Pole P of normals. Principal stresses σ_1 and σ_2. Principal stress directions by the angles ϕ_1 and ϕ_2. Stresses $\sigma(\phi)$ and $\tau(\phi)$ on planes perpendicular to the stress free plane with unit normal $\mathbf{n} = [\cos\phi, \sin\phi, 0]$

From the Mohr-diagram in Fig. 2.26 we may derive the formula 2.3.33 and the results:

$$\phi_1 = \frac{1}{2}\arctan\frac{2T_{12}}{T_{11} - T_{22}}, \quad \phi_2 = \phi_1 + \frac{\pi}{2} \qquad (2.3.44)$$

Problem 2.1–2.4 with solutions see Appendix.

Reference

1. Truesdell C, Toupin R (1960) The classical field theories. In: Handbuch der Physik, vol 3/1. Springer, Berlin

Chapter 3
Tensors

3.1 Definition of Tensors

In Sect. 2.2.4 the *Cauchy stress tensor* **T** was defined. The stress tensor is the "original" tensor as the word *tensor* means stress. We shall use the definition of the stress tensor as an introduction to the general concept of tensors.

We consider a body of continuous material and a material surface A in the body. At a place **r** a positive side of the surface is defined by a unit vector **n** as a normal pointing out from the surface. In a Cartesian coordinate system Ox with base vectors $\mathbf{e}_k$ the normal vector **n** has the components: n_k, i.e. $\mathbf{n} = n_k \mathbf{e}_k$. The contact force on the positive side of the surface is represented by the stress vector **t** with Cartesian components: t_i, i.e. $\mathbf{t} = t_i \mathbf{e}_i$. The contact forces on positive coordinate surfaces through the place **r** are the stress vectors $\mathbf{t}_k$ with Cartesian components T_{ik}, i.e. $\mathbf{t}_k = T_{ik} \mathbf{e}_i$. The components T_{ik} are called the *coordinate stresses*. The Cauchy stress theorem by Eq. (2.2.27) states that:

$$\mathbf{t} = \mathbf{T} \cdot \mathbf{n} \tag{3.1.1}$$

$$t_i = T_{ik} n_k, \quad t = Tn \tag{3.1.2}$$

We may interpret the symbol **T** in the relation (3.1.1) as a function with the vector **n** as an argument and the vector **t** as the value of the function: For any value of the argument **n** the relations (3.1.1), or (3.1.2), produce a vector **t**. We may express this fact by stating that the tensor **T** represents a *vector-valued function of a vector*. The component version (3.1.2) of the relation (3.1.1) shows how the function **T** operates.

It follows that if we substitute the argument vector **n** in the relation (3.1.1) by an arbitrary vector **a**, we will again get a vector $\mathbf{c} = \mathbf{T} \cdot \mathbf{a}$ as the value of the function. If we let the vector **a** be equal to the sum of two vectors **d** and **h**, i.e. $\mathbf{a} = \mathbf{d} + \mathbf{g}$, we find that:

© Springer Nature Switzerland AG 2019
F. Irgens, *Tensor Analysis*,
https://doi.org/10.1007/978-3-030-03412-2_3

$$\mathbf{T} \cdot \mathbf{a} = \mathbf{T} \cdot (\mathbf{d} + \mathbf{g}) = \mathbf{T} \cdot \mathbf{d} + \mathbf{T} \cdot \mathbf{g} \Leftrightarrow T_{ik} a_k = T_{ik}(d_k + g_k) = T_{ik} d_k + T_{ik} g_k$$
$$(3.1.3)$$

The symbol $\mathbf{T}$ therefore represents a linear function of the vector argument $\mathbf{a}$. The relationship $\mathbf{c} = \mathbf{T} \cdot \mathbf{a}$ is called a *linear mapping of vectors* or a *linear transformation of vectors*. Both names are used in the literature. The tensor $\mathbf{T}$ represents in Eq. (3.1.1) a *linear vector-valued function of a vector*.

The normal stress σ on the positive side of the surface A and the shear stress $\tilde{\tau}$ in an arbitrary direction $\mathbf{e}$ on the positive side of the surface A are given by; see Fig. 2.11 and Eqs. (2.2.28) and (2.2.30):

$$\sigma = \mathbf{n} \cdot \mathbf{t} = \mathbf{n} \cdot \mathbf{T} \cdot \mathbf{n}, \quad \tilde{\tau} = \mathbf{e} \cdot \mathbf{t} = \mathbf{e} \cdot \mathbf{T} \cdot \mathbf{n} \qquad (3.1.4)$$

In the *Ox*-system Eq. (3.1.4) have the representations:

$$\sigma = n_i T_{ik} n_k = n^T T n, \quad \tilde{\tau} = e_i T_{ik} n_k = e^T T n \qquad (3.1.5)$$

In the relations (3.1.4) $\mathbf{T}$ represents a function of two vector arguments $\mathbf{n}$ and $\mathbf{n}$, or $\mathbf{e}$ and $\mathbf{n}$, and with the scalars σ and $\tilde{\tau}$ as function values. The function $\mathbf{T}$ is linear with respect to both argument vectors. We may state that the tensor $\mathbf{T}$ represents a *bilinear scalar-valued function of two vectors*. The relations (3.1.5) show how the function operates. The coordinate stresses T_{ik} are called the *components of the stress tensor* $\mathbf{T}$ in the coordinate system *Ox*.

In Continuum Mechanics and in field theory in general, the properties we call tensors appear primarily in relations between scalars and vectors, as shown in Eqs. (3.1.1) and (3.1.4), and secondarily, as we shall see, in relations between already established tensors. Among the many possible definitions of the tensor concept, this book chooses the following general and completely coordinate invariant definition:

A tensor $\mathbf{A}$ of order n is a multilinear scalar-valued function of n argument vectors.

The word multilinear means that the function is linear in every argument vector. The value of the function is a scalar. Based on this definition we see from Eq. (3.1.4) that the stress tensor $\mathbf{T}$ is a bilinear scalar-valued function of two argument vectors.

To see the implications of the general definition of tensors, we start to investigate the properties of a *tensor of first order*. Let $\alpha = \mathbf{a}[\mathbf{b}]$ be a linear scalar-valued function of a vector $\mathbf{b}$. Then $\mathbf{a}$ is a tensor of first order. In a Cartesian coordinate system *Ox* with base vectors $\mathbf{e}_i$ we may compute the function values:

$$a_i = \mathbf{a}[\mathbf{e}_i] \qquad (3.1.6)$$

Due to the linear property of the tensor $\mathbf{a}$, the scalar value α for an arbitrarily chosen argument vector $\mathbf{b} = b_i \mathbf{e}_i$ may be computed thus:

$$\alpha = \mathbf{a}[\mathbf{b}] = \mathbf{a}[b_i \mathbf{e}_i] = b_i \mathbf{a}[\mathbf{e}_i] = b_i a_i \tag{3.1.7}$$

We see that the three function values a_i represent the tensor $\mathbf{a}$ in the Ox-system. For this reason the function values a_i are called the *components of the tensor* $\mathbf{a}$ *in the Ox-system*. The matrix $a = \{a_1\ a_2\ a_3\}$ is the *tensor matrix* of the tensor $\mathbf{a}$ in the Ox-system.

We shall now develop a relation between the components of the tensor $\mathbf{a}$ in two different Cartesian coordinate systems Ox and $\bar{O}\bar{x}$, for which the base vectors $\mathbf{e}_j$ and $\bar{\mathbf{e}}_i$ are connected through the transformation matrix:

$$Q = (Q_{ij}), \quad Q_{ij} = \cos(\bar{\mathbf{e}}_i \cdot \mathbf{e}_j), \quad \bar{\mathbf{e}}_i = Q_{ij}\mathbf{e}_j, \quad \mathbf{e}_j = Q_{ij}\bar{\mathbf{e}}_i \tag{3.1.8}$$

In the two coordinate systems the tensor of first order $\mathbf{a}$ is represented by the component sets:

$$a_j = \mathbf{a}[\mathbf{e}_j], \quad \bar{a}_i = \mathbf{a}[\bar{\mathbf{e}}_i] \tag{3.1.9}$$

We find that:

$$\mathbf{a}[\bar{\mathbf{e}}_i] = \mathbf{a}[Q_{ij}\mathbf{e}_j] = Q_{ij}\mathbf{a}[\mathbf{e}_j] \Rightarrow$$

$$\bar{a}_i = Q_{ij}a_j \Leftrightarrow \bar{a} = Qa \Leftrightarrow a = Q^T\bar{a} \tag{3.1.10}$$

This is also the transformation formula for the components of a vector $\mathbf{a}$. We may therefore make the statement that *vectors are tensors of first order*. From the expressions (3.1.7) it follows that the scalar-valued function $\mathbf{a}[\mathbf{b}]$ is equal to the scalar product of the two vectors $\mathbf{a}$ and $\mathbf{b}$.

$$\alpha = \mathbf{a}[\mathbf{b}] = \mathbf{a}[b_i \mathbf{e}_i] = b_i \mathbf{a}[\mathbf{e}_i] = b_i a_i = \mathbf{b} \cdot \mathbf{a} \tag{3.1.11}$$

In accordance with the general definition of tensors given above, it is convenient to consider *scalars* as *tensors of zero order*.

Let $\mathbf{A}[\mathbf{b}, \mathbf{c}]$ be *a bilinear scalar-valued function of two vectors* $\mathbf{b}$ and $\mathbf{c}$. The value of the function is a scalar α, and the function is linear in each vector argument vector.

$$\alpha = \mathbf{A}[\mathbf{b}, \mathbf{c}], \quad \mathbf{A}[\beta\,\mathbf{b}, \gamma\mathbf{c}] = \beta\gamma\mathbf{A}[\mathbf{b}, \mathbf{c}] \tag{3.1.12}$$

We say that $\mathbf{A}$ is *a tensor of second order*. The symbol for tensors will in this book in general be denoted by capital bold face Latin letters. Exceptions are for tensors of zero order, i.e. scalars, for which we prefer small Greek letters, and for tensors of first order, for which we prefer lower case bold letters in accordance with what has been decided previously for vectors.

The *components of the second order tensor* **A** *in a Cartesian coordinate system* Ox with base vectors $\mathbf{e}_i$ are defined as the following values of the tensor **A**:

$$A_{ij} = \mathbf{A}[\mathbf{e}_i, \mathbf{e}_j] \tag{3.1.13}$$

The scalar value α for two arbitrarily chosen argument vectors:

$$\mathbf{b} = b_i \mathbf{e}_i, \quad \mathbf{c} = c_j \mathbf{e}_j \tag{3.1.14}$$

may now be computed as follows. The second order tensor **A** is a bilinear function of two vectors **b** and **c**. Thus, according to the formulas $(3.1.12)_2$:

$$\mathbf{A}[\mathbf{b}, \mathbf{c}] = \mathbf{A}[b_i \mathbf{e}_i, c_j \mathbf{e}_j] = b_i c_j \mathbf{A}[\mathbf{e}_i, \mathbf{e}_j] \Rightarrow$$

$$\alpha = \mathbf{A}[\mathbf{b}, \mathbf{c}] = b_i c_j A_{ij} = b^T A c \tag{3.1.15}$$

The $3^2 = 9$ tensor components A_{ij}, or the *tensor matrix* $A = (A_{ij})$, represent the tensor **A** in the Ox-system.

The expressions (3.1.4) for the normal stress σ on a surface with unit normal **n** and the shear stress $\tilde{\tau}$ in a direction **e** on the surface may now be written as:

$$\sigma = \mathbf{T}[\mathbf{n}, \mathbf{n}] = n_i n_j T_{ij} = n^T T n, \quad \tilde{\tau} = \mathbf{T}[\mathbf{e}, \mathbf{n}] = e_i n_j T_{ij} = e^T T n \tag{3.1.16}$$

The coordinate stresses T_{ij} are components of the stress tensor **T** in the Ox-system.

The relation between the tensor components A_{ij} and $\bar{A}_{ij}$ in two Cartesian coordinate systems Ox and $\bar{O}\bar{x}$ with base vectors related through the formulas (1.3.2) by the transformation matrix Q, are found as follows. By definition:

$$\bar{A}_{ij} = \mathbf{A}[\bar{\mathbf{e}}_i, \bar{\mathbf{e}}_j], \quad A_{kl} = \mathbf{A}[\mathbf{e}_k, \mathbf{e}_l] \tag{3.1.17}$$

Now:

$$\mathbf{A}[\bar{\mathbf{e}}_i, \bar{\mathbf{e}}_j] = \mathbf{A}[Q_{ik}\mathbf{e}_k, Q_{jl}\mathbf{e}_l] = Q_{ik}Q_{jl}\mathbf{A}[\mathbf{e}_k, \mathbf{e}_l] \Rightarrow$$

$$\bar{A}_{ij} = Q_{ik}Q_{jl}A_{kl} \Leftrightarrow \bar{A} = QAQ^T \tag{3.1.18}$$

The inverse transformation is:

$$A_{kl} = Q_{ik}Q_{jl}\bar{A}_{ij} \Leftrightarrow A = Q^T \bar{A} Q \tag{3.1.19}$$

In some presentations in the literature the relations (3.1.18) and (3.1.19) are used to define a tensor of second order:

A tensor of second order is an invariant quantity which in every Cartesian coordinate system Ox is represented by a two-dimensional matrix $A = (A_{ij})$, such that the tensor matrices in any two coordinate systems Ox and $\bar{O}\bar{x}$ are related by the formulas (3.1.18) and (3.1.19).

This definition is the most practical one in n-dimensional spaces in which the concept of vectors is abstract since it is not possible to use geometrical figures in the same fashion as in the three-dimensional *Euclidian space E_3*. The definition chosen in the present exposition, that a tensor is a multilinear scalar-valued function of vectors, is preferred because the definition is obviously coordinate invariant, and because it is the most convenient definition when we want to introduce tensor components in general curvilinear coordinate systems E_3, as we shall see in Chap. 6

.

For convenience we shall let all the symbols $\mathbf{A}$, A, and A_{ij} represent one and the same tensor of second order. The bold face notation $\mathbf{A}$ is preferred when it is important to emphasize the coordinate invariance of the property described by the tensor. For a tensor of first $\mathbf{a}$, i.e. a vector, we use alternatively the symbols $\mathbf{a}$, a, and a_i.

An *isotropic tensor* is a tensor represented by the same matrix in all Cartesian coordinate systems. Isotropic tensors of second, third, and fourth order will be presented below.

The *unit tensor of second order* is denoted by the tensor symbol $\mathbf{1}$ and is defined by the scalar product of the two argument vectors $\mathbf{b}$ and $\mathbf{c}$:

$$\alpha = \mathbf{1}[\mathbf{b}, \mathbf{c}] = \mathbf{b} \cdot \mathbf{c} \tag{3.1.20}$$

The *components of the unit tensor $\mathbf{1}$* in a Cartesian coordinate system Ox are given by a *Kronecker delta*:

$$\delta_{ij} = \mathbf{1}[\mathbf{e}_i, \mathbf{e}_j] = \mathbf{e}_i \cdot \mathbf{e}_j \Leftrightarrow \mathbf{1} = (\delta_{ij}) \tag{3.1.21}$$

It follows from this result, and also from the transformation formula (3.1.18) for components of second order tensors, that $\bar{1} = Q1Q^T = 1$. The unit tensor is represented by the unit matrix in all Cartesian coordinate systems and is thus an *isotropic tensor of second order*.

A tensor of third order $\mathbf{C}$ is in a Cartesian coordinate system Ox represented by $3^3 = 27$ components:

$$C_{ijk} = \mathbf{C}[\mathbf{e}_i, \mathbf{e}_j, \mathbf{e}_k] \tag{3.1.22}$$

which may be considered to be the elements in a three-dimensional matrix C. The relation between the tensor components C_{ijk} in the Ox-system and the tensor components $\bar{C}_{rst}$ in another Cartesian coordinate system $O\bar{x}$ is:

$$\bar{C}_{rst} = Q_{ri} Q_{sj} Q_{tk} C_{ijk} \tag{3.1.23}$$

The derivation of this result follows the development of the relation (3.1.18).

The scalar triple product of three vectors **a**, **b**, and **c** defined by the formulas (1.2.35) and (1.2.36):

$$\alpha = [\mathbf{abc}] = (\mathbf{a} \times \mathbf{b}) \cdot \mathbf{c} = e_{ijk} a_i b_j c_k \tag{3.1.24}$$

may be represented by a third order isotropic tensor **P** called the *permutation tensor*, such that:

$$\alpha = [\mathbf{abc}] = \mathbf{P}[\mathbf{a}, \mathbf{b}, \mathbf{c}] \tag{3.1.25}$$

From the formulas (1.2.37) it follows that the components of **P** in a Cartesian coordinate system Ox are the permutation symbols e_{ijk}:

$$\mathbf{P}[\mathbf{e}_i, \mathbf{e}_j, \mathbf{e}_k] = [\mathbf{e}_i \mathbf{e}_j \mathbf{e}_k] = e_{ijk} \tag{3.1.26}$$

Defining the permutation tensor P by its Cartesian components (3.1.25), we obtain:

$$\alpha = \mathbf{P}[\mathbf{a}, \mathbf{b}, \mathbf{c}] = a_i\, b_j\, c_k \mathbf{P}[\mathbf{e}_i, \mathbf{e}_j, \mathbf{e}_k] = a_i\, b_j\, c_k e_{ijk} = [\mathbf{abc}]$$

Because the components of the tensor **P** will be the same in any Cartesian coordinate system, the permutation tensor **P** is an isotropic tensor of third order.

A tensor **A** of order n is in a Cartesian coordinate system Ox represented by 3^n components:

$$A_{ij\cdots} = \mathbf{A}[\mathbf{e}_i, \mathbf{e}_j, \cdots] \tag{3.1.27}$$

The $A_{ij\cdots}$ components may symbolically be presented by an n-dimensional matrix A. The relations between the components $A_{ij\cdots}$ and the components $\bar{A}_{rs\cdots}$ in an $O\bar{x}$-system are:

$$\bar{A}_{rs\cdots} = Q_{ri} Q_{sj} \cdots A_{ij\cdots} \Leftrightarrow A_{ij\cdots} = Q_{ri} Q_{sj} \cdots \bar{A}_{rs\cdots} \tag{3.1.28}$$

These relations may be used in an alternative definition of a tensor of order n:

A tensor **A** of order n is a coordinate invariant quantity, which in every Cartesian coordinate system is represented by an n-dimensional matrix, such that the components in two Cartesian coordinate systems Ox and $O\bar{x}$ are related by formula (3.1.28).

A tensor **A** of order n (>1) is symmetric/antisymmetric with respect to two argument vectors, or two component indices, if:

$$\begin{aligned}
\mathbf{A}[\cdot, \cdots, \mathbf{a}, \cdot, \mathbf{b}, \cdot] &= \pm \mathbf{A}[\cdot, \cdots, \mathbf{b}, \cdot, \mathbf{a}, \cdot] \Leftrightarrow \\
\mathbf{A}[\cdot, \cdots, \mathbf{e}_i, \cdot, \mathbf{e}_j, \cdot] &= \pm \mathbf{A}[\cdot, \cdots, \mathbf{e}_j, \cdot, \mathbf{e}_i, \cdot] \Leftrightarrow A_{\cdots i \cdots j \cdot} = \pm A_{\cdots j \cdot i \cdot}
\end{aligned} \tag{3.1.29}$$

The signs ($\pm$) imply symmetry/antisymmetry respectively.

The tensor **A** is *completely symmetric/antisymmetric* if the symmetry/antisymmetry property applies to any two argument vectors, or any two component indices. The stress tensor **T** is an example of a symmetric tensor of second order. With reference to Eq. (3.1.17):

$$\mathbf{T}[\mathbf{e}, \mathbf{n}] = \mathbf{T}[\mathbf{n}, \mathbf{e}] \Leftrightarrow \mathbf{T}[\mathbf{e}_i, \mathbf{e}_j] = \mathbf{T}[\mathbf{e}_j, \mathbf{e}_i], \quad T_{ij} = T_{ji} \Leftrightarrow T = T^T$$

The unit tensor **1** is a symmetric tensor of second order: $\delta_{ij} = \delta_{ji}$. A completely symmetric tensor **S** of third order has components that satisfy the conditions:

$$S_{ijk} = S_{ikj} = S_{jki} = S_{jik} = S_{kij} = S_{kji} \tag{3.1.30}$$

The number of distinct components different from zero is reduced from 27 for a general third order tensor **S**, to 10 for a completely symmetric third order tensor **S**. A completely antisymmetric tensor of third order has only one distinct component α different from zero, such that the tensor is a product of a scalar α and the permutation tensor **P**. The proof of this statement is given as Problem 3.1.

3.2 Tensor Algebra

Tensors of the same order n may be added or subtracted, and the results are new tensors of order n. The *sum* of two second order tensors **A** and **B** is defined by:

$$\mathbf{A} + \mathbf{B} = \mathbf{C} \Leftrightarrow \mathbf{A}[\mathbf{a}, \mathbf{b}] + \mathbf{B}[\mathbf{a}, \mathbf{b}] = \mathbf{C}[\mathbf{a}, \mathbf{b}] \tag{3.2.1}$$

This means that the scalar value of **C** for some argument vectors **a** and **b**, is obtained by adding the scalar values of **A** and **B** for the same argument vectors. It follows that:

$$\mathbf{A} + \mathbf{B} = \mathbf{C} \Leftrightarrow A_{ij} + B_{ij} = C_{ij} \Leftrightarrow A + B = C \tag{3.2.2}$$

The *difference* **A** − **B** = **D** is defined similarly. Addition and subtraction of tensor of other orders are defined analogously.

If the scalar value of tensor **A** is equal to the negative scalar value of a tensor **B** of the same order for all sets of the same argument vectors, we write:

$$\mathbf{A} = -\mathbf{B} \Leftrightarrow A = -B \tag{3.2.3}$$

The sum **A** + **B** is then a *zero tensor* **O** with all components equal to zero in any Cartesian coordinate system.

Tensor Product
The *tensor product* of a tensor **A** of order m and a tensor **B** of order n is a tensor **C** of order $(m + n)$, and defined by:

$$\mathbf{A} \otimes \mathbf{B} = \mathbf{C} \Leftrightarrow \mathbf{A}[\mathbf{a}, \cdots]\mathbf{B}[\mathbf{b}, \cdot] = \mathbf{C}[\mathbf{a}, \cdots, \mathbf{b}, \cdot] \Leftrightarrow A_{i \cdots}B_{j \cdots} = C_{i \cdots j \cdots} \qquad (3.2.4)$$

The tensor product is a multilinear scalar-valued function of all the argument vectors of the two factor tensors such that the scalar value of the tensor product is the product of the scalar values of the factor tensors. In general $\mathbf{A} \otimes \mathbf{B} \neq \mathbf{B} \otimes \mathbf{A}$, i.e. the tensor product is not commutative. However, tensor products are distributive. For example, for a tensor $\mathbf{A}$ of order m and two tensors $\mathbf{B}$ and $\mathbf{C}$ of order n we find:

$$\mathbf{A} \otimes (\mathbf{B} + \mathbf{C}) = \mathbf{A} \otimes \mathbf{B} + \mathbf{A} \otimes \mathbf{C} \qquad (3.2.5)$$

The tensor products $\mathbf{c} \otimes \mathbf{d}$ and $\mathbf{d} \otimes \mathbf{c}$ of two vectors $\mathbf{c}$ and $\mathbf{d}$ are tensors of second order and are called *dyadic products*, or for short *dyads*, of the two vectors:

$$\mathbf{c} \otimes \mathbf{d} = \mathbf{E} \Leftrightarrow \mathbf{c}[\mathbf{a}]\mathbf{d}[\mathbf{b}] = \mathbf{E}[\mathbf{a}, \mathbf{b}] \Leftrightarrow c_i d_j = E_{ij} \Leftrightarrow E = (c_i d_j)$$
$$\mathbf{d} \otimes \mathbf{c} = \mathbf{F} \Leftrightarrow \mathbf{d}[\mathbf{a}]\mathbf{c}[\mathbf{b}] = \mathbf{F}[\mathbf{a}, \mathbf{b}] \Leftrightarrow d_i c_j = F_{ij} \Leftrightarrow F = (d_i c_j) = E^T \qquad (3.2.6)$$

We see that $E = F^T$. Tensor products of many vectors are called *polyads*, for example the *triad*:

$$\mathbf{c} \otimes \mathbf{d} \otimes \mathbf{f} = (\mathbf{c} \otimes \mathbf{d}) \otimes \mathbf{f} = \mathbf{c} \otimes (\mathbf{d} \otimes \mathbf{f})$$

In some presentations, see for instance Malvern [1], the multiplication symbol $\otimes$ in the tensor product is omitted. The tensor product in (3.2.4) is then denoted: $\mathbf{AB} = \mathbf{C}$. In the present book the product $\mathbf{AB}$ ($\neq \mathbf{A} \otimes \mathbf{B}$) is defined only for second order tensors and is called the *composition of the two second order tensors*, and defined by the relation (3.2.23) below.

The tensor product of a scalar α, i.e. tensor of order zero, and a tensor $\mathbf{B}$ of order n is a tensor $\mathbf{C}$ of order n:

$$\alpha \mathbf{B} = \mathbf{C} \Leftrightarrow \alpha \mathbf{B}[\mathbf{a}, \cdots] = \mathbf{C}[\mathbf{a}, \cdots] \Leftrightarrow \alpha B_{i \cdots} = C_{i \cdots} \Leftrightarrow \alpha B = C \qquad (3.2.7)$$

It may be shown that an *isotropic tensor of second order*, $\mathbf{I} \equiv \mathbf{I}_2$, always is the product of a scalar α and the unit tensor $\mathbf{1}$.

$$\mathbf{I} \equiv \mathbf{I}_2 = \alpha \mathbf{1} \qquad (3.2.8)$$

The proof of this is given as Problem 3.2. It may also be shown that the general isotropic tensor of third order, denoted by $\mathbf{I}_3$, is a product of a scalar α and the permutation tensor $\mathbf{P}$.

$$\mathbf{I}_3 = \alpha \mathbf{P} \qquad (3.2.9)$$

Contraction

A *contraction* of a tensor of order n is an operation on the matrix of the tensor leading to a new tensor of order $(n - 2)$. As an example, let $\mathbf{C}$ be a tensor of order 3 for which the components in two Cartesian coordinate systems Ox and $\bar{O}\bar{x}$ are related according to the formulas (3.1.23):

$$\bar{C}_{rst} = Q_{ri}\, Q_{sj}\, Q_{tk}\, C_{ijk}$$

Then the component sets: C_{irr}, C_{rjr}, and C_{rrk} represent three different tensors of order $(3 - 2 = 1)$, in this case three different vectors. These operations are called *contractions*: two indices in the tensor matrix are set equal and a summation is implied over the region for this index. That the result of a contraction represents components of a new tensor of order two less then the original tensor will now be shown for the contraction C_{rrk} of the tensor **C**. In the component relation above we set $s = r$ and perform the summation with respect to the index r:

$$\bar{C}_{rrt} = (Q_{ri}\, Q_{rj})\, Q_{tk}\, C_{ijk} = \delta_{ij}\, Q_{tk}\, C_{ijk} \Rightarrow \bar{C}_{rrt} = Q_{tk}\, C_{iik} \quad \text{Q.E.D}$$

Let **B** be a tensor of second order with components B_{ij} and $\bar{B}_{kl}$ in two Cartesian coordinate systems. The contraction $B_{ii} = \operatorname{tr} B$ results in a new tensor of order $(2 - 2) = 0$, i.e. a scalar. This scalar is called the *trace of* B and is denoted $\operatorname{tr} \mathbf{B}$. The value of the scalar is equal to the trace of the tensor matrix in any Cartesian coordinate system.

$$\operatorname{tr} \mathbf{B} = \operatorname{tr} B = B_{ii} = \operatorname{tr} \bar{B} = \bar{B}_{ii} \tag{3.2.10}$$

As an example: The trace of the stress tensor **T** is equal to the sum of the normal coordinate stresses on any set of three orthogonal planes:

$$\operatorname{tr} \mathbf{T} = \operatorname{tr} T = T_{ii} = T_{11} + T_{22} + T_{33} = \sigma_1 + \sigma_2 + \sigma_3 \tag{3.2.11}$$

Thus: the sum of normal stresses on three orthogonal planes is independent of the coordinate system. This fact is already shown in Sect. 2.3.1 by the formulas (2.3.4) and (2.3.13).

Inner Products or Dot Products

The *scalar product* of two tensors **A** and **B** of second order is defined as the scalar:

$$\alpha = \mathbf{A} : \mathbf{B} = A_{ij}\, B_{ij} \tag{3.2.12}$$

The following reasoning proves that the expression $\mathbf{A} : \mathbf{B}$ is coordinate invariant. The tensor product of **A** and **B** is a tensor of order $(2 + 2) = 4$ with components $A_{ij}B_{kl}$. Two contractions leading to $A_{ij}B_{ij}$ reduce the order of the tensor to $(4 - 2 - 2) = 0$, i.e. a tensor of order zero, or a scalar, and thus a coordinate invariant quantity.

The following results are easily proved for tensors of second order:

$$\mathbf{A} : \mathbf{B} = \mathbf{B} : \mathbf{A}, \quad \mathbf{A} : (\mathbf{B} + \mathbf{C}) = \mathbf{A} : \mathbf{B} + \mathbf{A} : \mathbf{C} \tag{3.2.13}$$

In the literature, e.g. Malvern [1], two different scalar products are introduced for tensors of second order:

$$\mathbf{A} : \mathbf{B} = A_{ij}B_{ij}, \quad \mathbf{A} \cdot\!\cdot \mathbf{B} = A_{ij}B_{ji} \tag{3.2.14}$$

The latter scalar product is given an alternative presentation by formula (3.3.2) below.

Linear mappings of a vector **a** onto another vector, **b** or **c**, are given by a second order tensor **A**:

$$\mathbf{A} \cdot \mathbf{a} = \mathbf{b} \Rightarrow A_{ij}\, a_j = b_i \Leftrightarrow A\, a = b \tag{3.2.15}$$

$$\mathbf{a} \cdot \mathbf{A} = \mathbf{c} \Rightarrow a_i A_{ij} = c_j \Leftrightarrow a^T A = c^T \tag{3.2.16}$$

To see that the components b_i and c_j represent proper vectors, we argue as follows: The tensor product $\mathbf{A} \otimes \mathbf{a}$ is a tensor order 3 with components $A_{ij}\, a_k$. The contraction: $A_{ij}\, a_j$ leads to a tensor of order $(3 - 2) = 1$, i.e. to the vectors $A_{ij}\, a_j = b_i$. Similar arguments prove that $a_i A_{ij}$ represents the components of a proper vector.

The linear mapping in (3.2.15) is alternatively written as:

$$\mathbf{A}\mathbf{a} = \mathbf{b} \tag{3.2.17}$$

This notation is attractive because it is analogous to notation for the related matrix product: $Aa = b$ of the matrices representing the tensors. In the present exposition the notation $\mathbf{A} \cdot \mathbf{a}$ is preferred because it fits in with the general definition of the dot product given in the formulas (3.2.22) below, but also because of the symmetry it provides in the following expression (3.2.18). Using the formulas (3.1.13–3.1.15) we find that:

$$\alpha = \mathbf{A}[\mathbf{b}, \mathbf{c}] = b_i c_j A_{ij} = b_i A_{ij} c_j = b^T A\, c = \mathbf{b} \cdot \mathbf{A} \cdot \mathbf{c} \tag{3.2.18}$$

This notation has already been used to express the normal stress and the shear stress on a surface by the stress tensor. See the formulas (3.1.4) and (3.1.5). A special application of the notation expressed in (3.2.18) is:

$$A_{ij} = \mathbf{A}[\mathbf{e}_i, \mathbf{e}_j] = \mathbf{e}_i \cdot \mathbf{A} \cdot \mathbf{e}_j \tag{3.2.19}$$

Let **C** be a fourth order tensor. This tensor provides *linear mappings* of a tensor of second order **A** onto another tensor of second order, **B** or **D**, by:

$$\mathbf{C} : \mathbf{A} = \mathbf{B} \Leftrightarrow C_{ijkl} A_{kl} = B_{ij}, \quad \mathbf{A} : \mathbf{C} = \mathbf{D} \Leftrightarrow A_{ij} C_{ijkl} = D_{kl} \tag{3.2.20}$$

The scalar product of **B**, or **D**, obtained from the formulas (3.2.20), and a second order tensor **E** are the scalars:

$$\alpha = \mathbf{B} : \mathbf{E} = \mathbf{E} : \mathbf{B} = \mathbf{E} : \mathbf{C} : \mathbf{A} = E_{ij}\, C_{ijkl}\, A_{kl}$$
$$\beta = \mathbf{D} : \mathbf{E} = \mathbf{E} : \mathbf{D} = \mathbf{A} : \mathbf{C} : \mathbf{E} = A_{ij}\, C_{ijkl}\, E_{kl}$$
(3.2.21)

The scalar product of two vectors, the scalar product of two tensors of second order, and the products in the linear mapping in the formulas (3.2.15), (3.2.16), and (3.2.20) are all called *inner products* or *dot products*.

We now generalize the concept of dot products. Let **A** be a tensor of order (n) and **B** a tensor of order (m). We define the dot product **C** of **A** and **B** and the *double dot product* **D** of **A** and **B** by the operations:

dot product: $\mathbf{A} \cdot \mathbf{B} = \mathbf{C} \Leftrightarrow A_{i..k}B_{k...j} = C_{i.....j}$ tensor of order $(n+m-2)$
double dot product: $\mathbf{A} : \mathbf{B} = \mathbf{D} \Leftrightarrow A_{i..kl}B_{kl...j} = D_{i.....j}$ tensor of order $(n+m-4)$
(3.2.22)

The dot product of two tensors **A** and **B** of second order is a tensor **C** of second order and is called a *composition of the two second order tensors*.

$$\mathbf{A} \cdot \mathbf{B} = \mathbf{C} \Leftrightarrow A_{ik}B_{kj} = C_{ij} \Leftrightarrow AB = C \qquad (3.2.23)$$

Due to the matrix form of this type of product the composition is alternatively presented as:

$$\mathbf{A} \cdot \mathbf{B} \equiv \mathbf{AB} = \mathbf{C} \Leftrightarrow AB = C \quad \text{composition of second order tensors} \qquad (3.2.24)$$

In general: $\mathbf{A}\,\mathbf{B} \neq \mathbf{B}\,\mathbf{A}$. It follows that:

$$(\mathbf{A}\,\mathbf{B})\mathbf{C} = \mathbf{A}(\mathbf{B}\,\mathbf{C}), \quad \text{presented as } \mathbf{ABC}$$
$$\mathbf{A}(\mathbf{B}+\mathbf{C}) = \mathbf{A}\,\mathbf{B} + \mathbf{A}\,\mathbf{C}, \quad (\mathbf{A}+\mathbf{B})\mathbf{C} = \mathbf{AC} + \mathbf{BC}$$
(3.2.25)

The following operations for the tensor product, the scalar product, and linear mapping of vectors are easily verified by their component versions.

$$(\mathbf{a} \otimes \mathbf{b}) \cdot \mathbf{c} = \mathbf{a}(\mathbf{b} \cdot \mathbf{c}), \quad \mathbf{c} \cdot (\mathbf{a} \otimes \mathbf{b}) = (\mathbf{c} \cdot \mathbf{a})\mathbf{b}$$
$$(\mathbf{a} \otimes \mathbf{b}){:}(\mathbf{c} \otimes \mathbf{d}) = (\mathbf{a} \cdot \mathbf{c})(\mathbf{b} \cdot \mathbf{d}) = a_i\, b_j\, c_i\, d_j$$
$$\mathbf{a} \cdot (\mathbf{b} \otimes \mathbf{c}) \cdot \mathbf{d} = (\mathbf{a} \cdot \mathbf{b})(\mathbf{c} \cdot \mathbf{d}) = a_i b_i c_j d_j$$
(3.2.26)

In Sect. 3.1 we have defined the stress tensor alternatively as a linear vector-valued function of a vector, by Eq. (3.1.3), and as a bilinear scalar-valued function of two vectors, by Eqs. (3.1.4) and (3.1.5). In the present exposition the general definition of a *tensor of order n* is as a *multilinear scalar-valued functions of n vectors*. The formulas (3.2.20) express the fourth order tensor $\mathbf{C}$ as two linear tensor-valued functions of a second order tensor $\mathbf{A}$:

$$\mathbf{B} = \mathbf{C} : \mathbf{A} \quad \text{and} \quad \mathbf{D} = \mathbf{A} : \mathbf{C},$$

while the formulas (3.2.21) may be interpreted as expressing the fourth order tensors $\mathbf{C}$ as a bilinear scalar-valued function of two second order tensors $\mathbf{E}$ and $\mathbf{A}$.

The Quotient Theorem

In the presentation of the Cauchy stress theorem (2.2.27) we have derived a set of linear equations between the components in a Cartesian coordinate system Ox of the stress vector $\mathbf{t}$ on a material surface in a material particle and the unit normal vector $\mathbf{n}$ on the surface:

$$t_i = T_{ik}\, n_k \tag{3.2.27}$$

The coefficients T_{ik} in these equations are the coordinate stresses. The component relation (3.2.27) is valid in any Cartesian coordinate system. By the formulas (2.2.32) we found the relations between the coordinate stresses in two different Cartesian coordinate systems:

$$\bar{T}_{ij} = Q_{ik}Q_{jl}T_{kl} \Leftrightarrow \bar{T} = QTQ^T \tag{3.2.28}$$

According to the general definition (3.1.19) of tensors of second order the result (3.2.28) shows that the matrices T and $\bar{T}$ represent a tensor of second order.

Let us generalize to the following situation. Suppose that we have developed linear relations between the components of two tensors of first order $\mathbf{a}$ and $\mathbf{b}$ in two Cartesian coordinate systems Ox and $\bar{O}\bar{x}$:

$$a_i = C_{ij}b_j \Leftrightarrow a = Cb, \quad \bar{a}_r = \bar{C}_{rs}\bar{b}_s \Leftrightarrow \bar{a} = \bar{C}\bar{b} \tag{3.2.29}$$

The two coordinate systems are related through the transformation matrix Q, such that:

$$\bar{a}_r = Q_{ri}\, a_i, \quad b_j = Q_{sj}\,\bar{b}_s \tag{3.2.30}$$

We assume that for any choice of tensor $\mathbf{b}$ the outcome of the relations (3.2.29) is the components of a tensor of first order $\mathbf{a}$. We can then prove that the "quotients" C and $\bar{C}$ in the relations (3.2.29) represent the components of a second order tensor $\mathbf{C}$ in the respective coordinate systems, such that:

$$\bar{C}_{rs} = Q_{ri}\, Q_{sj}\, C_{ij} \tag{3.2.31}$$

Proof We start with the relations $(3.2.29)_2$, then apply the relations $(3.2.30)_1$, $(3.2.29)_1$, $(3.2.30)_2$, and finally the relations $(3.2.29)_2$:

$$\bar{C}_{rs}\bar{b}_s = \bar{a}_r = Q_{ri}\,a_i = Q_{ri}\left(C_{ij}\,b_j\right) = Q_{ri}\,C_{ij}\left(Q_{sj}\,\bar{b}_s\right) \Rightarrow \left(\bar{C}_{rs} - Q_{ri}\,Q_{sj}C_{ij}\right)\bar{b}_s = 0$$

$$(3.2.32)$$

This result is to be valid for all choices of the argument tensor **b**, which means that the components $\bar{b}_s$ may be chosen freely. That implies that the term $(\bar{C}_{rs} - Q_{ri}\,Q_{sj}C_{ij})$ must be zero, and the relations (3.2.31) are proved, i.e. the matrices C and $\bar{C}$ represent a second order tensor **C**.

The example above is generalized to:

The *quotient theorem*: Given a linear tensor-valued function of a tensor **B** with the value **A** such that in any Cartesian coordinate system Ox the component matrices B of **B** and A of **A** are related through the coefficient matrix C, and such that A represents a tensor for any choice of the tensor **B**, then the matrix C represents the components in Ox of a tensor **C**.

Another example of application of the quotient theorem may be as follows. Let **B** be a tensor of second order and **a** a tensor of first order. Suppose that we have developed the following linear relation between the matrices of the tensors **B** and **a**:

$$a_i = C_{ijk}\,B_{jk} \qquad (3.2.33)$$

The relation is to be valid in any Ox-system and for any argument tensor **B**. Then the "quotient" elements C_{ijk} in the relations (3.2.33) are the components of a tensor of third order **C**. The component relation (3.2.33) may be generalized to the coordinate invariant form:

$$\mathbf{a} = \mathbf{C} : \mathbf{B} \qquad (3.2.34)$$

Tensor Equations

A *tensor equation* is a coordinate invariant equation of tensors. All terms in the equation have to be tensors of the same order. In the component format all terms must contain the same free indices. An example of a tensor equation is:

$$A_{ij} + C_{ijk}\,b_k = B_{ij} + a_i\,c_j \Leftrightarrow \mathbf{A} + \mathbf{C}\cdot\mathbf{b} = \mathbf{B} + \mathbf{a}\otimes\mathbf{c} \qquad (3.2.35)$$

It often is convenient to develop physical or geometrical equations in a special Cartesian coordinate system. If such an equation can be identified as a tensor equation, it automatically is valid in any other coordinate system. In Chap. 6 we shall see how the components of tensors are defined in general curvilinear coordinates, and how a tensor equation is written in any general coordinate system. We should always try to formulate equations between physical or geometrical quantities that are represented by scalars, vectors, and tensors in a coordinate invariant format, which in fact means that the equations should be tensor equations.

3.2.1   Isotropic Tensors of Fourth Order

In this section we present four special isotropic tensors of fourth order and a general fourth order isotropic tensor. First, we see that because δ_{ij} represents the components of an isotropic second order tensor, i.e. the second order unit tensor $\mathbf{1} \equiv \mathbf{1}_2$, the components $\delta_{ij}\,\delta_{kl}$ represent a fourth order isotrop tensor which is the tensor product of $\mathbf{1}$ by itself: $\mathbf{1} \otimes \mathbf{1}$. Two other isotropic tensors of fourth order are defined by their components: (1) the *fourth order unit tensor* denoted $\mathbf{1}_4 \equiv \mathbf{1}\overline{\otimes}\mathbf{1}$ with components in Cartesian coordinate systems: $1_{ijkl} = \delta_{ik}\,\delta_{jl}$, and (2) a tensor denoted $\underline{\mathbf{1}} = \mathbf{1}\underline{\otimes}\mathbf{1}$ with components in Cartesian coordinate systems: $\underline{1}_{ijkl} = \delta_{il}\,\delta_{jk}$. The three fourth order isotropic tensors are then:

$$\mathbf{1} \otimes \mathbf{1} \Leftrightarrow \delta_{ij}\,\delta_{kl}, \quad \mathbf{1}_4 \equiv \mathbf{1}\overline{\otimes}\mathbf{1} \Leftrightarrow 1_{ijkl} = \delta_{ik}\,\delta_{jl}, \quad \underline{\mathbf{1}} \equiv \mathbf{1}\underline{\otimes}\mathbf{1} \Leftrightarrow \underline{1}_{ijkl} = \delta_{il}\,\delta_{jk}$$

$$(3.2.36)$$

The fourth order unit tensor $\mathbf{1}_4$ has the symmetry properties:

$$1_{ijkl} = 1_{kjil} = 1_{ilkj} = 1_{klij} \tag{3.2.37}$$

The reason for the name unit tensor for the tensor $\mathbf{1}_4 \equiv \mathbf{1}\overline{\otimes}\mathbf{1}$ becomes apparent in the dot product of the tensor $\mathbf{1}_4$ and any second order tensor $\mathbf{B}$:

$$\mathbf{1}_4{:}\mathbf{B} = \mathbf{B} \Leftrightarrow 1_{ijkl}\,B_{kl} = \delta_{ik}\,\delta_{jl}\,B_{kl} = B_{ij}$$
$$\mathbf{B} : \mathbf{1}_4 = \mathbf{B} \Leftrightarrow B_{ij}\,1_{ijkl} = B_{ij}\,\delta_{ik}\,\delta_{jl} = B_{kl}$$

$$(3.2.38)$$

The fourth order unit tensor $\mathbf{1}_4$ may be decomposed into a symmetric part $\mathbf{1}_4^s$ and an antisymmetric part $\mathbf{1}_4^a$:

$$\mathbf{1}_4 \equiv \mathbf{1}\overline{\otimes}\mathbf{1} = \mathbf{1}_4^s + \mathbf{1}_4^a \tag{3.2.39}$$

$$\mathbf{1}_4^s = \frac{1}{2}(\mathbf{1}\overline{\otimes}\mathbf{1} + \mathbf{1}\underline{\otimes}\mathbf{1}) \Leftrightarrow 1_{ijkl}^s = \frac{1}{2}\left(\delta_{ik}\,\delta_{jl} + \delta_{il}\,\delta_{jk}\right) = 1_{jikl}^s = 1_{ijlk}^s = 1_{klij}^s \quad (3.2.40)$$

$$\mathbf{1}_4^a = \frac{1}{2}(\mathbf{1}\overline{\otimes}\mathbf{1} - \mathbf{1}\underline{\otimes}\mathbf{1}) \Leftrightarrow 1_{ijkl}^a = \frac{1}{2}\left(\delta_{ik}\,\delta_{jl} - \delta_{il}\,\delta_{jk}\right) = -1_{jikl}^a = -1_{ijlk}^a = 1_{klij}^a$$

$$(3.2.41)$$

The tensor product $\mathbf{1} \otimes \mathbf{1}$ has the following property in a dot product with a second order tensor $\mathbf{B}$:

$$(\mathbf{1} \otimes \mathbf{1}){:}\mathbf{B} = (\mathrm{tr}\,\mathbf{B})\,\mathbf{1} \Leftrightarrow \delta_{ij}\,\delta_{kl}\,B_{kl} = B_{kk}\,\delta_{ij} \tag{3.2.42}$$

It may be shown that the *general isotropic tensor of fourth order* is given by:

$$\mathbf{I}_4 = 2\mu\,\mathbf{1}_4^s + 2\theta\,\mathbf{1}_4^a + \lambda\,(\mathbf{1} \otimes \mathbf{1}) \Leftrightarrow$$
$$I_{4ijkl} = \mu\big(\delta_{ik}\,\delta_{jl} + \delta_{il}\,\delta_{jk}\big) + \theta\big(\delta_{ik}\,\delta_{jl} - \delta_{il}\,\delta_{jk}\big) + \lambda\,\delta_{ij}\,\delta_{kl} \qquad (3.2.43)$$

The parameters μ, θ, and λ are three scalars. The dot product of the general isotropic tensor of fourth order $\mathbf{I}_4$ and a symmetric second order tensor $\mathbf{E}$ is a symmetric second order tensor:

$$\mathbf{T} = \mathbf{I}_4\!:\!\mathbf{E} \Leftrightarrow T_{ij} = I_{4ijkl}\,E_{kl} \qquad (3.2.44)$$

Because the tensor $\mathbf{E}$ is symmetric, the antisymmetric part $\mathbf{1}_4^a$ of the isotropic tensor $\mathbf{I}_4$ does not contribute in the dot product. Thus:

$$\mathbf{T} = \mathbf{I}_4\!:\!\mathbf{E} = \mathbf{I}_4^s\!:\!\mathbf{E} = \big(2\mu\,\mathbf{1}_4^s + \lambda\,\mathbf{1} \otimes \mathbf{1}\big)\!:\!\mathbf{E}, \quad \mathbf{I}_4^s = 2\mu\,\mathbf{1}_4^s + \lambda\,\mathbf{1} \otimes \mathbf{1} \qquad (3.2.45)$$

The tensor $\mathbf{I}_4^s$ is the symmetric part of the isotropic tensor $\mathbf{I}_4$ and has the symmetry property:

$$I_{4ijkl}^s = I_{4jikl}^s = I_{4ijlk}^s = I_{4klij}^s \qquad (3.2.46)$$

From Eqs. (3.2.43) and (3.2.45) it follows that:

$$\mathbf{T} = 2\mu\,\mathbf{E} + \lambda(\mathrm{tr}\,\mathbf{E})\,\mathbf{1} \Leftrightarrow T_{ij} = 2\mu\,E_{ij} + \lambda\,E_{kk}\,\delta_{ij} \qquad (3.2.47)$$

In Sect. 5.2 we identify $\mathbf{T}$ as the stress tensor, $\mathbf{E}$ as the *strain tensor for small deformations*, defined in Sect. 4.3, and furthermore the formula (3.2.47) as the *generalized Hooke's law* for an *isotropic, linearly elastic material*, confer the formula (5.2.17). The scalars μ and λ are then called the *Lamé constants*, after Gabriel Lamé [1795–1870], and are related to the modulus of elasticity η and Poisson's ratio ν for the material. The parameter μ is the *shear modulus*, and the parameter λ is equal to $\kappa - 2\mu/3$, where κ is the *bulk modulus of elasticity*, confer the formulas (5.2.9), (5.2.11), and (5.2.19).

In Sect. 5.3 the formula (3.2.47) represents the stress contribution $\mathbf{T}$ due to the viscosity in a Newtonian fluid if $\mathbf{E}$ is replaced by the *rate of deformation tensor* $\mathbf{D}$, also called the *rate of strain tensor*, confer the formulas (5.3.11). The tensor $\mathbf{D}$ is defined in Sect. 4.4. The parameter μ is now the *dynamic viscosity* also called the *shear viscosity* and $\lambda = \kappa - 2\mu/3$, where κ is the *bulk viscosity*.

The tensor equation (3.2.47) has also other applications when we want to describe the state of stress in isotropic viscoelastic materials.

3.2.2 Tensors as Polyadics

The polyads $\mathbf{e}_i \otimes \mathbf{e}_j$ of the base vectors $\mathbf{e}_i$ in a coordinate system Ox and the polyads $\bar{\mathbf{e}}_k \otimes \bar{\mathbf{e}}_l$ of the base vectors $\bar{\mathbf{e}}_k$ in a coordinate system $\bar{O}\bar{x}$ may be interpreted as tensors of second order. The relations between the two sets of base vectors $\mathbf{e}_i$ and $\bar{\mathbf{e}}_k$ are:

$$Q_{ki} = \cos(\bar{\mathbf{e}}_k, \mathbf{e}_i) \Leftrightarrow \bar{\mathbf{e}}_k = Q_{ki}\,\mathbf{e}_i \Leftrightarrow \mathbf{e}_i = Q_{ki}\,\bar{\mathbf{e}}_k$$

Then the components of the tensors $\mathbf{e}_i \otimes \mathbf{e}_j$ are:

$$\delta_{ki}\,\delta_{lj} \text{ in } Ox \quad \text{and} \quad Q_{ki}\,Q_{lj} \text{ in } O\bar{x}$$

The components of the tensor $\bar{\mathbf{e}}_k \otimes \bar{\mathbf{e}}_l$ are:

$$Q_{ki}\,Q_{lj} \text{ in } Ox \quad \text{and} \quad \delta_{ki}\,\delta_{lj} \text{ in } O\bar{x}$$

Let $\mathbf{B}$ be any tensor of second order with components:

$$B_{ij} \text{ in } Ox \quad \text{and} \quad \bar{B}_{kl} \text{ in } O\bar{x} \Leftrightarrow \bar{B}_{kl} = Q_{ki}\,Q_{lj}\,B_{ij}, \quad B_{ij} = Q_{ki}\,Q_{lj}\,\bar{B}_{kl}$$

Then the tensors $B_{ij}\mathbf{e}_i \otimes \mathbf{e}_j$ and $\bar{B}_{kl}\,\bar{\mathbf{e}}_k \otimes \bar{\mathbf{e}}_l$ are both identical to the tensor $\mathbf{B}$. In order to see this, we evaluate the components of the two tensors in the two coordinate systems.

$$B_{ij}\,\mathbf{e}_i \otimes \mathbf{e}_j \Rightarrow B_{ij}\,\delta_{ki}\,\delta_{lj} = B_{kl} \text{ in } Ox \quad \text{and} \quad B_{ij}\,Q_{ki}\,Q_{lj} = \bar{B}_{kl} \text{ in } O\bar{x}$$
$$\bar{B}_{kl}\,\bar{\mathbf{e}}_k \otimes \bar{\mathbf{e}}_l \Rightarrow \bar{B}_{kl}\,Q_{ki}\,Q_{lj} = B_{ij} \text{ in } Ox \quad \text{and} \quad \bar{B}_{kl}\,\delta_{ki}\,\delta_{lj} = \bar{B}_{ij} \text{ in } O\bar{x}$$

Thus we may write:

$$\mathbf{B} \equiv B_{ij}\,\mathbf{e}_i \otimes \mathbf{e}_j \equiv \bar{B}_{kl}\,\bar{\mathbf{e}}_k \otimes \bar{\mathbf{e}}_l \tag{3.2.48}$$

A linear combination of dyads of vectors is called a *dyadic*. The formula (3.2.48) shows how a second order tensor may be expressed as a dyadic.

It is easy to see how this tensor representation may be extended to tensors of order n by using *polyadics*, i.e. linear combinations of polyads. For tensors of first, second and third order the polyadic representations are:

$$\mathbf{a} \equiv a_i\,\mathbf{e}_i, \quad \mathbf{B} \equiv B_{ij}\,\mathbf{e}_i \otimes \mathbf{e}_j, \quad \mathbf{C} \equiv C_{ijk}\,\mathbf{e}_i \otimes \mathbf{e}_j \otimes \mathbf{e}_k \tag{3.2.49}$$

A contraction in a tensor may now be performed by replacing a tensor multiplication by a dot multiplication, which is indicated by replacing the sign ($\otimes$) by a dot ($\cdot$) in the polyadic representation of the tensor. For example:

$$C_{ijk}\,\mathbf{e}_i \otimes \mathbf{e}_j \otimes \mathbf{e}_k = C_{ijk}(\mathbf{e}_i \otimes \mathbf{e}_j) \otimes \mathbf{e}_k \Rightarrow C_{ijk}(\mathbf{e}_i \cdot \mathbf{e}_j) \otimes \mathbf{e}_k = C_{ijk}\,\delta_{ij}\,\mathbf{e}_k = C_{iik}\,\mathbf{e}_k$$
$$\tag{3.2.50}$$

We now have three *alternative definitions of tensors*: (1) as scalar-valued multilinear functions of vectors, (2) as coordinate invariant quantities defined in

Cartesian coordinate systems by components that are related by the formulas (3.1.28), and (3) as polyadics. In general *curvilinear coordinate systems* vectors and tensors are represented by more than one set of components. This will be demonstrated in Sect. 6.3 for vectors and in Sect. 6.4 for tensors.

3.3 Tensors of Second Order

Most of the important tensors relevant in continuum mechanics are of second order, the prominent example being the stress tensor $\mathbf{T}$. It is therefore natural to give second order tensors special attention and to investigate their properties thoroughly. Related to a tensor $\mathbf{A}$ of second order we define the *transposed tensor* $\mathbf{A}^T$ by:

$$\mathbf{A}^T \Leftrightarrow \mathbf{A}^T[\mathbf{b}, \mathbf{c}] = \mathbf{A}[\mathbf{c}, \mathbf{b}] \Leftrightarrow (A^T)_{ij} = A_{ji} \qquad (3.3.1)$$

The matrix of $\mathbf{A}^T$ is the matrix A^T. The second scalar product in (3.2.14) may now be presented as:

$$\mathbf{A} \cdot \cdot \mathbf{B} = \mathbf{A} : \mathbf{B}^T = A_{ij} B_{ji} \qquad (3.3.2)$$

A tensor $\mathbf{B}$ of second order is *symmetric/antisymmetric* if it is symmetric/antisymmetric with respect to the argument vectors:

Symmetric $\mathbf{B}$: $\mathbf{B}[\mathbf{b}, \mathbf{c}] = \mathbf{B}[\mathbf{c}, \mathbf{b}] \Leftrightarrow B_{ij} = B_{ji} \Leftrightarrow B = B^T \Leftrightarrow \mathbf{B} = \mathbf{B}^T$

Antisymmetric $\mathbf{B}$: $\mathbf{B}[\mathbf{b}, \mathbf{c}] = -\mathbf{B}[\mathbf{c}, \mathbf{b}] \Leftrightarrow B_{ij} = -B_{ji} \Leftrightarrow B = -B^T \Leftrightarrow \mathbf{B} = -\mathbf{B}^T$

$$(3.3.3)$$

Any second order tensor $\mathbf{B}$ may uniquely be linearly decomposed into a symmetric tensor $\mathbf{S}$ and an antisymmetric tensor $\mathbf{A}$:

$$\mathbf{B} = \mathbf{S} + \mathbf{A}, \quad \mathbf{S} = \mathbf{S}^T = \frac{1}{2}\left(\mathbf{B} + \mathbf{B}^T\right), \quad \mathbf{A} = -\mathbf{A}^T = \frac{1}{2}\left(\mathbf{B} - \mathbf{B}^T\right) \Leftrightarrow \qquad (3.3.4)$$

$$S = S^T = \frac{1}{2}\left(B + B^T\right), \quad S_{ij} = S_{ji} = \frac{1}{2}\left(B_{ij} + B_{ji}\right) \equiv B_{(ij)}$$

$$A = -A^T = \frac{1}{2}\left(B - B^T\right), \quad A_{ij} = -A_{ji} = \frac{1}{2}\left(B_{ij} - B_{ji}\right) \equiv B_{[ij]}$$

$$(3.3.5)$$

The tensor $\mathbf{B}$ is specified when values for the 9 components B_{ij} in an Ox-system are given. The symmetric part S of B contains 6 distinct components $S_{ij}(=S_{ji})$, while the antisymmetric part A of B contains 3 distinct components A_{ij}. Together the two tensors $\mathbf{S}$ and $\mathbf{A}$ contain the same information as the original second order tensor $\mathbf{B}$.

To any vector $\mathbf{a}$ there is an antisymmetric tensor $\mathbf{A}$ of second order that contains the same information as the vector. The vector $\mathbf{a}$ and the tensor $\mathbf{A}$ are called *dual quantities*, and with the permutation tensor $\mathbf{P}$, the dual tensor $\mathbf{A}$ to a vector $\mathbf{a}$ is defined by:

$$\mathbf{A} = -\mathbf{P} \cdot \mathbf{a} = -\mathbf{a} \cdot \mathbf{P} \Leftrightarrow A_{ij} = -e_{ijk}a_k = -a_k e_{kij} \Leftrightarrow A = \begin{pmatrix} 0 & -a_3 & a_2 \\ a_3 & 0 & -a_1 \\ -a_2 & a_1 & 0 \end{pmatrix}$$

$$(3.3.6)$$

It follows that:

$$\mathbf{a} = -\frac{1}{2}\mathbf{P} : \mathbf{A} = -\frac{1}{2}\mathbf{A} : \mathbf{P} \Leftrightarrow$$
$$a_i = -\frac{1}{2}e_{ijk}A_{jk} = -\frac{1}{2}A_{jk}e_{jki} \Leftrightarrow a = -\{A_{23}\,A_{31}\,A_{12}\} \qquad (3.3.7)$$

In some presentations in the literature the vector $\mathbf{a}$ and the transposed tensor $\mathbf{A}^T$ are defined as dual quantities.

When the vector $\mathbf{a}$ and the antisymmetric second order tensor $\mathbf{A}$ are dual quantities, we may express the vector product $\mathbf{a}$ with any other vector $\mathbf{b}$ by:

$$\mathbf{a} \times \mathbf{b} = \mathbf{A} \cdot \mathbf{b} \Leftrightarrow \mathbf{b} \times \mathbf{a} = \mathbf{b} \cdot \mathbf{A} \qquad (3.3.8)$$

These relations are easily checked by writing out their component versions. These alternative expressions for vector products are convenient in tensor equations and in their matrix representations.

Three important *scalar invariants* related to a tensor $\mathbf{A}$ of second order are defined by:

$$\text{the } trace \text{ of } \mathbf{A} : \text{tr}\,\mathbf{A} = \text{tr}\,A = A_{kk} \qquad (3.3.9)$$

$$\text{the } determinant \text{ of } \mathbf{A} : \det \mathbf{A} = \det A \qquad (3.3.10)$$

$$\text{the } norm \text{ of } \mathbf{A} : \text{norm}\,\mathbf{A} \equiv \|\mathbf{A}\| = \sqrt{\mathbf{A} : \mathbf{A}} = \sqrt{\text{tr}\,(\mathbf{A}\,\mathbf{A}^T)} = \sqrt{\text{tr}\,(A\,A^T)} = \sqrt{A_{ij}\,A_{ij}}$$

$$(3.3.11)$$

It follows from their definitions that tr $\mathbf{A}$ and norm $\mathbf{A}$ are coordinate invariant quantities. Now we shall demonstrate that the determinant of $\mathbf{A}$, det $\mathbf{A}$, also is an invariant and thus a scalar. Let Q be the transformation matrix in a coordinate transformation from a system Ox to a system $\bar{O}\bar{x}$. Then since det $Q = \det Q^T = 1$, we obtain from the multiplication theorem for determinants in formula (1.1.23):

$$\det \bar{A} = \det(QAQ^T) = (\det Q)(\det A)(\det Q^T) = \det A$$

If det $\mathbf{A} \neq 0$, we may determine the *inverse tensor* $\mathbf{A}^{-1}$ of $\mathbf{A}$ from the tensor equation:

$$A^{-1}A = 1 = A\,A^{-1}, \quad A^{-1}A = 1 = A\,A^{-1} \tag{3.3.12}$$

A tensor **A** is called an *orthogonal tensor of second order* if:

$$A^T = A^{-1} \quad \text{and} \quad \det A = 1 \Rightarrow A^T A = AA^T = 1 \Leftrightarrow$$
$$A^T A = AA^T = 1 \Leftrightarrow A_{ik}A_{jk} = \delta_{ij}, \quad \text{and} \quad \det A = 1 \tag{3.3.13}$$

It follows from the definition (3.3.13) that the matrix A of an orthogonal tensor **A** is an orthogonal matrix. This implies that the columns, or rows, of the matrix of an orthogonal tensor represent the components of an orthogonal set of unit vectors, which is called an *orthonormal set of vectors*. In the same order as they appear in the matrix they form a *right-handed system of three vectors* in the same sense as the base vectors in a Cartesian right-handed coordinate system form a right-handed system. In the linear vector mapping $\mathbf{a} = \mathbf{A} \cdot \mathbf{b}$, where **A** is an orthogonal tensor, the tensor **A** represents a rotation of the argument vector **b**, such that |**a**| = |**b**|. This property will be further discussed in Sect. 3.7 Q-Rotation of Vectors and Tensors of Second Order.

The following rules may be directly transferred from matrix algebra. For any two second order tensors **A** and **B**:

$$(\mathbf{AB})^T = \mathbf{B}^T\mathbf{A}^T, \quad (\mathbf{AB})^{-1} = \mathbf{B}^{-1}\mathbf{A}^{-1} \tag{3.3.14}$$

The cofactor Co A of the matrix A of a second order tensor **A** represents a tensor Co **A**, the *cofactor tensor*. From the formulas (1.1.29), noting that $\mathbf{A}^{-T} = (\mathbf{A}^{-1})^T$, we obtain:

$$\text{Co } \mathbf{A} = \mathbf{A}^{-T}\det \mathbf{A} \tag{3.3.15}$$

The formulas (1.1.24) imply that:

$$\text{Co } \mathbf{A} = \frac{\partial(\det \mathbf{A})}{\partial \mathbf{A}} \Leftrightarrow \text{CoA}_{ij} = \frac{\partial(\det A)}{\partial A_{ij}} \tag{3.3.16}$$

3.3.1 Symmetric Tensors of Second Order

Let **S** be a symmetric tensor of second order and **a** and **b** two orthogonal unit vectors. Referring to Fig. 3.1, we define the *vector* **s** *of the tensor* **S** *for the direction* **a**, the projection σ of this vector on the direction **a**, and the projection τ of the vector **s** on **b**:

$$\mathbf{s} = \mathbf{S} \cdot \mathbf{a} \Leftrightarrow s_i = S_{ik} a_k \Leftrightarrow s = S a \tag{3.3.17}$$

$$\sigma = \mathbf{a} \cdot \mathbf{s} = \mathbf{a} \cdot \mathbf{S} \cdot \mathbf{a} = a_i S_{ik} a_k = a^T S a \tag{3.3.18}$$

$$\tau = \mathbf{b} \cdot \mathbf{s} = \mathbf{b} \cdot \mathbf{S} \cdot \mathbf{a} = b_i S_{ik} a_k = b^T S a \tag{3.3.19}$$

σ and τ are respectively called the *normal component of the tensor* $\mathbf{S}$ *for the direction* $\mathbf{a}$, and the *orthogonal shear component of the tensor* $\mathbf{S}$ *for the orthogonal directions* $\mathbf{a}$ *and* $\mathbf{b}$. The names "normal component" and "shear component" are taken from the corresponding quantities related to the stress tensor $\mathbf{T}$, confer the formulas (2.2.27), (2.2.28), and (2.2.30). Note that the three vectors $\mathbf{a}$, $\mathbf{b}$ and $\mathbf{s}$ do not necessarily lie in one and the same plane.

We may also write:

$$\sigma = \mathbf{S}[\mathbf{a}, \mathbf{a}]$$
$$\tau = \mathbf{S}[\mathbf{b}, \mathbf{a}] = \mathbf{S}[\mathbf{b}, \mathbf{a}] \tag{3.3.20}$$

If we choose $\mathbf{a} = \mathbf{e}_1$ and $\mathbf{b} = \mathbf{e}_2$, we get:

$$\sigma = \mathbf{S}[\mathbf{e}_1, \mathbf{e}_1] = S_{11}, \quad \tau = \mathbf{S}[\mathbf{e}_2, \mathbf{e}_1] = S_{21} \tag{3.3.21}$$

Hence in the matrix S the elements on the diagonal represent normal components, while the off-diagonal elements are shear components.

Any symmetric second order tensor has mathematically the same properties as the stress tensor $\mathbf{T}$. In two dimensions we may analyze the properties in a Mohr–diagram and otherwise use the formulas developed in Sects. 2.3.5 and 2.3.6. In the general three-dimensional case the three *principal values* $\sigma = \sigma_i$ and the three *principal directions* $\mathbf{a} = \mathbf{a}_i$ of the tensor $\mathbf{S}$ are determined from the condition:

$$\mathbf{s} = \mathbf{S} \cdot \mathbf{a} = \sigma \mathbf{a} \tag{3.3.22}$$

This equation is organized into the algebraic equations:

Fig. 3.1 Vector s of the tensor S for the direction a. Normal component σ of the tensor S for the direction a. Shear component τ of the tensor S for the directions a and b

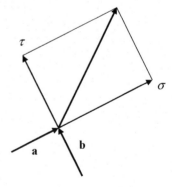

$$(\sigma\mathbf{1} - \mathbf{S}) \cdot \mathbf{a} = \mathbf{0} \Leftrightarrow (\sigma\mathbf{1} - S)a = 0 \tag{3.3.23}$$

A solution of these three equations requires that the determinant of the coefficient matrix $(\sigma\mathbf{1} - S)$ is zero:

$$\det(\sigma\mathbf{1} - S) = 0 \Rightarrow \sigma^3 - I\sigma^2 + II\sigma - III = 0 \tag{3.3.24}$$

This equation is called the *characteristic equation of the tensor second order* **S**. The coefficients *I*, *II*, and *III* are the *principal invariants* of the tensor **S**:

$$I = \operatorname{tr}\mathbf{S} = \sigma_1 + \sigma_2 + \sigma_3$$

$$II = \frac{1}{2}\left[(\operatorname{tr}\mathbf{S})^2 - \operatorname{tr}\mathbf{S}^2\right] \equiv \frac{1}{2}\left[(\operatorname{tr}\mathbf{S})^2 - (\operatorname{norm}\mathbf{S})^2\right] = \sigma_1\sigma_2 + \sigma_2\sigma_3 + \sigma_3\sigma_1 \tag{3.3.25}$$

$$III = \det\mathbf{S} = \frac{1}{6}\left[(\operatorname{tr}\mathbf{S})^3 - 3\operatorname{tr}\mathbf{S}\operatorname{tr}\mathbf{S}^2 + 2\operatorname{tr}\mathbf{S}^3\right] = \sigma_1\sigma_2\sigma_3$$

These formulas are obtained directly from the formulas (2.3.4), except for the second variation in the formula for the invariant III, where an application of the formula (3.3.36) has been utilized. The three principal values σ_i are determined from Eq. (3.3.24), and the three principal directions $\mathbf{a}_i$ are then determined from Eq. (3.3.23). The following properties and results, also demonstrated for the stress tensor **T** in Sect. 2.3.1, apply to all symmetric second order tensors **S**. The principal values are real, and if they are all different, the principal directions are orthogonal. If two principal values are equal, but different from the third principal value, then all directions normal to the principal direction related to the third principal value, are principal directions. If all three principal values are equal, the tensor is isotropic, and any direction is a principal direction. The three principal directions $\mathbf{a}_i$ are said to represent the *principal axes* of the tensor **S**. The principal values σ_i are also called the *eigenvalues* and the principal direction $\mathbf{a}_i$ the *eigenvectors* to the symmetric tensor **S**. The mathematical problem related to Eqs. (3.3.23–3.3.25) is an *eigenvalue problem*.

In a Cartesian coordinate system Ox with base vectors $\mathbf{e}_i \equiv \mathbf{a}_i$ the symmetric second order tensor **S** has the matrix representation:

$$(\mathbf{S}[\mathbf{a}_i, \mathbf{a}_j]) = \begin{pmatrix} \sigma_1 & 0 & 0 \\ 0 & \sigma_2 & 0 \\ 0 & 0 & \sigma_3 \end{pmatrix} = (\sigma_i\delta_{ij}) \tag{3.3.26}$$

The principal values of the tensor represent extreme values for normal components. If the principal values are ordered such that: $\sigma_3 \le \sigma_2 \le \sigma_1$, then:

$$\sigma_{\max} = \sigma_1, \quad \sigma_{\min} = \sigma_3 \tag{3.3.27}$$

The maximum value of the orthogonal shear component is:

$$\tau_{max} = \frac{1}{2}(\sigma_{max} - \sigma_{min}) \quad \text{for } \mathbf{a} = \pm(\mathbf{a}_1 + \mathbf{a}_3)/\sqrt{2}, \quad \mathbf{b} = \pm(\mathbf{a}_1 - \mathbf{a}_3)/\sqrt{2}$$

$$(3.3.28)$$

Let the elements a_{ki} be the components in the Cartesian coordinate system Ox of the principal directions $\mathbf{a}_k$:

$$\mathbf{a}_k = a_{ki}\mathbf{e}_i \Leftrightarrow a_{ki} = \mathbf{a}_k \cdot \mathbf{e}_i \Rightarrow \mathbf{e}_i = a_{ki}\mathbf{a}_k \qquad (3.3.29)$$

Then we may write:

$$S_{ij} = \mathbf{S}[\mathbf{e}_i, \mathbf{e}_j] = \mathbf{S}[a_{ki}\mathbf{a}_k, a_{lj}\mathbf{a}_l] = \mathbf{S}[\mathbf{a}_k, \mathbf{a}_l]a_{ki}a_{lj} = \sum_{k,l} \sigma_k \delta_{kl} a_{ki} a_{lj} \Rightarrow$$

$$S_{ij} = \sum_k \sigma_k a_{ki} a_{kj} \Leftrightarrow \mathbf{S} = \sum_k \sigma_k \mathbf{a}_k \otimes \mathbf{a}_k \qquad (3.3.30)$$

This formula also follows directly from formula (3.2.48) when we choose $\mathbf{a}_k$ as base vectors.

Powers of a tensor $\mathbf{A}$ of second order is defined by:

$$\mathbf{A}^n = \mathbf{A}\mathbf{A}\cdots\mathbf{A} \quad (n \text{ compositions}) \qquad (3.3.31)$$

It may be shown that for a symmetric tensor of second order $\mathbf{S}$ the result (3.3.30) implies that, see Problem 3.3:

$$\mathbf{S}^n = \sum_k (\sigma_k)^n \mathbf{a}_k \otimes \mathbf{a}_k, \quad n = 1, 2, 3, \cdots \qquad (3.3.32)$$

The tensors $\mathbf{S}^n$ and $\mathbf{S}$ have the same principal directions and are therefore called *coaxial tensors*. Note that the exponent n is a natural number. A symmetric second order tensor is called a *positive definite tensor* if:

$$\mathbf{c} \cdot \mathbf{S} \cdot \mathbf{c} > 0 \quad \text{for all vectors } \mathbf{c} \neq \mathbf{0} \qquad (3.3.33)$$

From the definition (3.3.33) it follows that the principal values and the principal invariants of a positive definite symmetric tensor all are positive, see Problem 3.4. It also follows that the trace, the determinant, and the norm of a positive definite tensor $\mathbf{S}$ are positive.

For any real number α we define real powers of a positive definite tensor $\mathbf{S}$ by:

$$\mathbf{S}^\alpha = \sum_k (\sigma_k)^\alpha \mathbf{a}_k \otimes \mathbf{a}_k, \quad \alpha \text{ is a real number} \qquad (3.3.34)$$

The tensors $\mathbf{S}^\alpha$ are positive definite tensors, and $\mathbf{S}^\alpha$ and $\mathbf{S}$ are *coaxial tensors*. The inverse tensor $\mathbf{S}^{-1}$ is obtained from formula (3.3.34) when $\alpha = -1$. For the special case $\alpha = 1/2$, we write:

$$\mathbf{S}^{1/2} = \sqrt{\mathbf{S}} \tag{3.3.35}$$

All powers with exponents $n = 1, 2, 3, \ldots$ of a symmetric tensor $\mathbf{S}$ of second order may be expressed by $\mathbf{S}$, $\mathbf{S}^2$, and the principal invariants I, II, and III of the tensor. In order to see this, we first prove the following theorem.

Cayley–Hamilton Theorem A symmetric tensor $\mathbf{S}$ of second order satisfies its own characteristic equation (3.3.24), such that:

$$\mathbf{S}^3 - I\,\mathbf{S}^2 + II\,\mathbf{S} - III\mathbf{1} = \mathbf{0} \tag{3.3.36}$$

The theorem is named after Arthur Cayley [1821–1895] and William Rowan Hamilton [1805–1865]. Proof of the theorem: The characteristic equation (3.3.24) of the tensor $\mathbf{S}$ is satisfied by the principal values σ_k of the tensor. If we multiply Eq. (3.3.24) by the tensor product $\mathbf{a}_k \otimes \mathbf{a}_k$ and sum with respect to the index k, we obtain:

$$\sum_k \left[(\sigma_k)^3 \mathbf{a}_k \otimes \mathbf{a}_k - I(\sigma_k)^2 \mathbf{a}_k \otimes \mathbf{a}_k + II\sigma_k\, \mathbf{a}_k \otimes \mathbf{a}_k - III\, \mathbf{a}_k \otimes \mathbf{a}_k \right] = 0$$

This result is directly transferred to Eq. (3.3.36) through application of formula (3.3.32), and the Cayley–Hamilton theorem is proved. By Eq. (3.3.36) all powers of $\mathbf{S}$ with natural number exponents may be expressed by the tensors $\mathbf{S}$ and $\mathbf{S}^2$, and the principal invariants I, II, and III of $\mathbf{S}$.

3.3.2 Alternative Invariants of Second Order Tensors

The following sets of alternative invariants of a symmetric tensor of second order sometimes appear in the literature. The *moment invariants*:

$$\bar{I} = \text{tr}\,\mathbf{S} = I, \quad \overline{II} = \frac{1}{2}\text{tr}\,\mathbf{S}^2 = \frac{1}{2}I^2 - II, \quad \overline{III} = \frac{1}{3}\text{tr}\,\mathbf{S}^3 \tag{3.3.37}$$

The *trace invariants*:

$$\begin{aligned}
\tilde{I} &= \text{tr}\,\mathbf{S} = \mathbf{1} : \mathbf{S} = I, \\
\tilde{II} &= \text{tr}\,\mathbf{S}^2 = \mathbf{S} : \mathbf{S} = I^2 - 2II \\
\tilde{III} &= \text{tr}\,\mathbf{S}^3 = \mathbf{S} : (\mathbf{SS}) = 3III + I^3 - 3I \cdot II
\end{aligned} \tag{3.3.38}$$

3.3.3   Deviator and Isotrop of Second Order Tensors

A symmetric second order tensor $\mathbf{S}$ may be decomposed uniquely in a trace free
deviator $\mathbf{S}'$ and an *isotrop* $\mathbf{S}^o$:

$$\mathbf{S} = \mathbf{S}' + \mathbf{S}^o \tag{3.3.39}$$

$$\mathbf{S}^o = \frac{1}{3}(\operatorname{tr}\mathbf{S})\mathbf{1} \Leftrightarrow S_{ij}^o = \frac{1}{3}S_{kk}\,\delta_{ij} \quad \text{isotrop} \tag{3.3.40}$$

$$\mathbf{S}' = \mathbf{S} - \mathbf{S}^o, \quad \operatorname{tr}\mathbf{S}' = 0 \quad \text{deviator} \tag{3.3.41}$$

$\mathbf{S}'$ and $\mathbf{S}$ are *coaxial tensors*, i.e. the two tensors have the same principal directions.
The principal invariants and principal values of $\mathbf{S}'$ are:

$$I' = 0,$$
$$II' = -\frac{1}{2}\operatorname{tr}(\mathbf{S}')^2 = II - \frac{1}{3}I^2, \tag{3.3.42}$$
$$III' = \det \mathbf{S}'$$

$$\sigma_i' = \sigma_i - \frac{1}{3}I \Leftrightarrow \sigma_1' = \frac{1}{3}(2\sigma_1 - \sigma_2 - \sigma_3) \quad \text{etc.} \tag{3.3.43}$$

It has been shown in Sect. 2.3.2 that the stress deviator $\mathbf{T}'$ may be decomposed into
five states of shear. This type of decomposition is possible for any deviator $\mathbf{S}'$ of a
second order tensor $\mathbf{S}$.

3.4   Tensor Fields

Tensors in continuum mechanics represent often intensive quantities or properties
related to particles or places in space at a given time t. This means that these tensors
really are *tensor fields*. A tensor $\mathbf{A}$ which is a tensor field is denoted by either
$\mathbf{A}(\mathbf{r}_0, t)$ where $\mathbf{r}_0$ represents a particle, or by $\mathbf{A}(\mathbf{r}, t)$, where $\mathbf{r}$ represent a place in
space. At each place $\mathbf{r}$ and at the time t the symbol $\mathbf{A}(\mathbf{r}, t)$ represents a tensor. The
tensor fields $\mathbf{A}(\mathbf{r})$ and $\mathbf{A}(\mathbf{r}_0)$ are called *steady tensor fields*, and the tensor fields
$\mathbf{A}(\mathbf{r})$ and $\mathbf{A}(\mathbf{r}_0)$ are called *uniform tensor fields*. Time and space derivatives of
tensor fields are defined below as new tensor fields. In this section we shall
introduce some of the most important of these derived tensors fields.
    The local time derivative of a tensor field $\mathbf{A}(\mathbf{r}, t)$ is a tensor field $\partial_t \mathbf{A}$ with
components $\partial_t A_{i..}$ in an Ox-system:

$$\partial_t \mathbf{A} \Leftrightarrow \partial_t A_{i\cdot\cdot} = \frac{\partial A_{i\cdot\cdot}}{\partial t} \tag{3.4.1}$$

The definitions of the gradient, the divergence, and the rotation of a tensor field may vary somewhat in the literature, and this may lead to confusion. In the present exposition these quantities will first be defined in an analogous manner to how the corresponding quantities were presented for scalars and vectors in Sect. 1.6. Then the del-operator will be introduced, which together with the dyadic representation will indicate two possible definitions of the gradient, the divergence, and the rotation of a tensor field.

3.4.1   Gradient, Divergence, and Rotation of Tensor Fields

The gradient of a scalar field $\alpha(\mathbf{r}, t)$ has been defined by formula (1.6.6). The divergence and rotation of a vector field $\mathbf{a}(\mathbf{r}, t)$ are defined by the formulas (1.6.12) and (1.6.15) respectively. The divergence of a second order tensor is defined by the formula (2.2.36). These results will now be generalized.

Let $\mathbf{A}(\mathbf{r}, t)$ be a tensor field of order n with components $A_{i\cdot\cdot j}$ in a Cartesian coordinate systems Ox with base vectors $\mathbf{e}_i$. The components of the tensor field in another Cartesian coordinate system $\bar{O}\bar{x}$ with base vectors $\bar{\mathbf{e}}_r$ are denoted $\bar{A}_{r\cdot\cdot s}$ such that:

$$\bar{A}_{r\cdot\cdot s} = Q_{ri}\cdot\cdot Q_{sj} A_{i\cdot\cdot j}, \quad Q_{ri} = \frac{\partial \bar{x}_r}{\partial x_i} = \frac{\partial x_i}{\partial \bar{x}_r} = \bar{\mathbf{e}}_r \cdot \mathbf{e}_i \tag{3.4.2}$$

It follows that:

$$\begin{aligned} \frac{\partial \bar{A}_{r\cdot\cdot s}}{\partial \bar{x}_t} &= Q_{ri}\cdot\cdot Q_{sj} \frac{\partial A_{i\cdot\cdot j}}{\partial x_k} \frac{\partial x_k}{\partial \bar{x}_t} = Q_{ri}\cdot\cdot Q_{sj} \frac{\partial A_{i\cdot\cdot j}}{\partial x_k} Q_{tk} \Rightarrow \\ \frac{\partial \bar{A}_{r\cdot\cdot s}}{\partial \bar{x}_t} &= Q_{ri}\cdot\cdot Q_{sj} Q_{tk} \frac{\partial A_{i\cdot\cdot j}}{\partial x_k} \Leftrightarrow \bar{A}_{r\cdot\cdot s,t} = Q_{ri}\cdot\cdot Q_{sj} Q_{tk} A_{i\cdot\cdot j,k} \end{aligned} \tag{3.4.3}$$

We use this result to define the *gradient of a tensor field* $\mathbf{A}(\mathbf{r}, t)$ *of order* n as a tensor field of order $(n+1)$ with the components:

$$\frac{\partial A_{i\cdot\cdot j}}{\partial x_k} \equiv A_{i\cdot\cdot j,k} \text{ in the } Ox\text{-system}, \quad \frac{\partial \bar{A}_{r\cdot\cdot s}}{\partial \bar{x}_t} \equiv \bar{A}_{r\cdot\cdot s,t} \text{ in the } \bar{O}\bar{x}\text{-system} \tag{3.4.4}$$

This tensor is denoted alternatively by grad $\mathbf{A}$ and $\partial \mathbf{A}/\partial \mathbf{r}$. If the tensor field $\mathbf{A}$ is presented as a polyadic, the gradient of $\mathbf{A}$ may be presented as follows:

$$\mathbf{A} = A_{i \cdot \cdot j} \mathbf{e}_i \otimes \cdots \otimes \mathbf{e}_j \Rightarrow \text{grad } \mathbf{A} \equiv \frac{\partial \mathbf{A}}{\partial \mathbf{r}} = A_{i \cdot \cdot j, k} \mathbf{e}_i \otimes \cdots \otimes \mathbf{e}_j \otimes \mathbf{e}_k \tag{3.4.5}$$

For a tensor field of order 0, i.e. a scalar field $\alpha(\mathbf{r}, t)$, this formula agrees with the formula (1.6.6): grad $\alpha = \alpha_{,i} \mathbf{e}_i$. The *gradient of a vector field* $\mathbf{a}(\mathbf{r}, t) = a_i(x, t) \mathbf{e}_i$ is a second order tensor with components $a_{i,k}$ in the Cartesian coordinate system Ox:

$$\text{grad } \mathbf{a} = a_{i,k} \mathbf{e}_i \otimes \mathbf{e}_k \tag{3.4.6}$$

Let $\mathbf{r}(s) = x_i(s) \, \mathbf{e}_i$ represent a space curve, where s is the arc length parameter along the curve. The unit tangent vector to the curve is defined by the formula (1.4.8):

$$\mathbf{t} = \frac{d\mathbf{r}}{ds} = \frac{dx_i}{ds} \mathbf{e}_i = t_i \mathbf{e}_i \tag{3.4.7}$$

A tensor field $\mathbf{A}(\mathbf{r}, t)$ of order n becomes a function of the arc length parameter s along the curve $\mathbf{r}(s) = x_i(s) \, \mathbf{e}_i$, and we may compute the tensor field $\partial \mathbf{A} / \partial s$ of order $(n-1)$:

$$\frac{\partial \mathbf{A}}{\partial s} = A_{i \cdot \cdot j, k} \frac{dx_k}{ds} \mathbf{e}_i \otimes \cdots \otimes \mathbf{e}_j \Rightarrow \frac{\partial \mathbf{A}}{\partial s} = (\text{grad } \mathbf{A}) \cdot \mathbf{t} \Leftrightarrow \frac{\partial A_{i \cdot \cdot j}}{\partial s} = A_{i \cdot \cdot j, k} t_k \tag{3.4.8}$$

The *divergence of a tensor field* $\mathbf{A}(\mathbf{r}, t)$ *of order* n, with the components $A_{i \cdot \cdot rj}$, is defined as the tensor field "div $\mathbf{A}$" of order $(n-1)$ with components $A_{i \cdot \cdot rk, k}$.

$$\mathbf{A} = A_{i \cdot \cdot rj} \mathbf{e}_i \otimes \cdots \otimes \mathbf{e}_r \otimes \mathbf{e}_j \Rightarrow \text{div } \mathbf{A} = A_{i \cdot \cdot rk, k} \mathbf{e}_i \otimes \cdots \otimes \mathbf{e}_r \tag{3.4.9}$$

From this general definition it follows that the *divergence of a vector field* $\mathbf{a}(\mathbf{r}, t)$ is the scalar $a_{k,k}$, in agreement with formula (1.6.12), and that the *divergence of a second order tensor* $\mathbf{A}$ is a vector $\mathbf{a} = \text{div } \mathbf{A}$:

$$\begin{aligned} \mathbf{a} = a_k \mathbf{e}_k &\Rightarrow \text{div } \mathbf{a} \equiv \nabla \cdot \mathbf{a} = a_{k,k} \\ \mathbf{A} = A_{ij} \mathbf{e}_i \otimes \mathbf{e}_j &\Rightarrow \text{div } \mathbf{A} = A_{ik,k} \mathbf{e}_i = \mathbf{a} \Rightarrow a_i = A_{ik,k} \end{aligned} \tag{3.4.10}$$

The del-operator ∇ has been introduced by the formula (1.6.10), and the divergence of a second order tensor has already been introduced in formula (2.2.36) in connection with the development of the Cauchy equations of motion in Sect. 2.2.5.

The *divergence of the gradient of a tensor field* $\mathbf{A} = A_{i \cdot \cdot j} \mathbf{e}_i \otimes \cdots \otimes \mathbf{e}_j$ *of order* n, i.e. div grad $\mathbf{A}$, becomes:

$$\begin{aligned} \text{div grad } \mathbf{A} = A_{i \cdot \cdot j, k, k} \mathbf{e}_i \otimes \cdots \otimes \mathbf{e}_j &= A_{i \cdot \cdot j, kk} \mathbf{e}_i \otimes \cdots \otimes \mathbf{e}_j = \left(A_{i \cdot \cdot j} \mathbf{e}_i \otimes \cdots \otimes \mathbf{e}_j \right)_{,kk} \\ &= \nabla^2 \mathbf{A} \Rightarrow \\ \text{div grad } \mathbf{A} = A_{i \cdot \cdot j, kk} \mathbf{e}_i &\otimes \cdots \otimes \mathbf{e}_j = \nabla^2 \mathbf{A} \end{aligned} \tag{3.4.11}$$

The symbol ∇^2 is the *Laplace-operator* defined by the formula (1.6.14).

The *gradient of the divergence of a tensor field A of order n*, i.e. grad div A, is:

$$\text{grad div } \mathbf{A} = A_{i\cdot\cdot k,k,l}\mathbf{e}_i \otimes \cdot\cdot \otimes \mathbf{e}_l = A_{i\cdot\cdot k,kl}\mathbf{e}_i \otimes \cdot\cdot \otimes \mathbf{e}_l \tag{3.4.12}$$

The *rotation of a tensor field* $\mathbf{A}(\mathbf{r},t)$ *of order* n with components $A_{i\cdot\cdot rk}$ in a Cartesian coordinate system Ox, is a tensor "rot A" of order n, also denoted "curl A". The tensor is defined by its components $e_{ijk}A_{i\cdot\cdot rk,j}$:

$$\mathbf{A} = A_{i\cdot\cdot rk}\mathbf{e}_i \otimes \cdot\cdot \otimes \mathbf{e}_r \otimes \mathbf{e}_k \Rightarrow \text{rot } \mathbf{A} = \text{curl } \mathbf{A} = e_{ijk}A_{i\cdot\cdot rk,j}\mathbf{e}_i \otimes \cdot\cdot \otimes \mathbf{e}_r \tag{3.4.13}$$

The rotation of a vector $\mathbf{a} = a_i\mathbf{e}_i$ is the vector $e_{ijk}a_{k,j}\mathbf{e}_i$, in agreement with the formula (1.6.15):

$$\mathbf{a} = a_i\mathbf{e}_i \Rightarrow \text{rot } \mathbf{a} \equiv \text{curl } \mathbf{a} \equiv \nabla \times \mathbf{a} = e_{ijk}a_{k,j}\mathbf{e}_i \tag{3.4.14}$$

The gradient, the divergence, and the rotation of a tensor field $\mathbf{A}(\mathbf{r}_0,t)$ related to the place $\mathbf{r}_0$ in the reference configuration K_0 of a continuum, are defined in similar manners as above but are denoted respectively as Grad A, Div A, and Rot A.

3.4.2 Del-Operators

The definitions of gradient, divergence, and rotation of tensors fields are not universal. The literature defines for any tensor field the following tensors fields: *right-gradient, left-gradient, right-divergence, left-gradient, right-divergence, left-divergence, right-rotation*, and *left-rotation*. For instance, in the books by Malvern [1] and Jaunzemis [2], the following vector operators are introduced:

$$\overleftarrow{\nabla} \equiv \partial_k\,\mathbf{e}_k \quad right-operator, \quad \overrightarrow{\nabla} \equiv \nabla \equiv \mathbf{e}_i\partial_i \quad left-operator \tag{3.4.15}$$

The first operator is called a *right-operator* because it operates from the right. Note that the *left-operator* is identical to the del-operator defined by formula (1.6.10). Applying the operators in (3.4.15) to a tensor field $\mathbf{A}$ of order n, we obtain:

$$\mathbf{A} \otimes \overleftarrow{\nabla} = (A_{i\cdot\cdot j}\mathbf{e}_i \otimes \cdot\cdot \otimes \mathbf{e}_j) \otimes \overleftarrow{\partial}_k\mathbf{e}_k = A_{i\cdot\cdot j,k}\mathbf{e}_i \otimes \cdot\cdot \otimes \mathbf{e}_j \otimes \mathbf{e}_k \quad right-gradient$$

$$\overrightarrow{\nabla} \otimes \mathbf{A} = \mathbf{e}_i\,\overrightarrow{\partial}_i \otimes (A_{j\cdot\cdot k}\mathbf{e}_j \otimes \cdot\cdot \otimes \mathbf{e}_k) = A_{j\cdot\cdot k,i}\mathbf{e}_i \otimes \mathbf{e}_j \otimes \cdot\cdot \otimes \mathbf{e}_k \quad left-gradient$$

$$\tag{3.4.16}$$

The two tensors defined by the expressions (3.4.16) are represented by the same components, but the components are organized differently. The tensor defined in Sect. 3.4.1 as grad A, i.e. the gradient of the tensor field A by the formula (3.4.5), is now seen to be the right-gradient of the tensor field A:

$$\text{grad } \mathbf{A} \equiv \mathbf{A} \otimes \overleftarrow{\nabla} = A_{ij,k}\mathbf{e}_i \otimes \mathbf{e}_j \otimes \mathbf{e}_k \qquad (3.4.17)$$

The right- and left-gradient of a vector **a** are second order tensors where one tensor is the transposed of the other:

$$\text{right-gradient:} \quad \text{grad } \mathbf{a} = \mathbf{a} \otimes \overleftarrow{\nabla} = a_{i,k}\mathbf{e}_i \otimes \mathbf{e}_k \Rightarrow (\mathbf{a}\overleftarrow{\nabla})_{ik} = \partial_k a_i \equiv a_{i,k} \cdots$$

$$(3.4.18)$$

$$\text{grad} \quad \overrightarrow{\nabla} \otimes \mathbf{a} = (\text{grad } \mathbf{a})^T = a_{k,i}\mathbf{e}_i \otimes \mathbf{e}_k \Rightarrow (\overrightarrow{\nabla}\mathbf{a})_{ik} = \partial_i a_k \equiv a_{k,i} \qquad (3.4.19)$$

The right-divergence and left–divergence of a tensor field **A** are defined respectively by:

$$\mathbf{A} \cdot \overleftarrow{\nabla} = (A_{i \cdot \cdot j}\mathbf{e}_i \otimes \cdots \otimes \mathbf{e}_j) \cdot (\overleftarrow{\partial_k}\mathbf{e}_k) = A_{i \cdot \cdot j,k}\mathbf{e}_i \otimes \cdots(\mathbf{e}_j \cdot \mathbf{e}_k) \Rightarrow$$
$$\mathbf{A} \cdot \overleftarrow{\nabla} = A_{i \cdot \cdot k,k}\mathbf{e}_i \otimes \cdots \quad right-divergence \qquad (3.4.20)$$

$$\overrightarrow{\nabla} \cdot \mathbf{A} = (\mathbf{e}_k\overrightarrow{\partial_k}) \cdot (A_{j \cdot \cdot i}\mathbf{e}_j \otimes \cdots \otimes \mathbf{e}_i) = A_{j \cdot \cdot i,k}(\mathbf{e}_k \cdot \mathbf{e}_j) \cdots \otimes \mathbf{e}_i \Rightarrow$$
$$\overrightarrow{\nabla} \cdot \mathbf{A} = A_{k \cdot \cdot i,k} \cdots \otimes \mathbf{e}_i \quad left-divergence \qquad (3.4.21)$$

We see that the definition (3.4.9) of the divergence of a second order tensor field **A** is a right-divergence:

$$\text{div } \mathbf{A} = \mathbf{A} \cdot \overleftarrow{\nabla} \qquad (3.4.22)$$

For a vector **a** the right-divergence and the left-divergence are identical, and we may skip the arrow over the "del" symbol ∇.

$$\text{div } \mathbf{a} = \mathbf{a} \cdot \overleftarrow{\nabla} = \overrightarrow{\nabla} \cdot \mathbf{a} = \nabla \cdot \mathbf{a} = a_{i,i} \qquad (3.4.23)$$

The divergence of the gradient of a tensor field **A** of second order may be given by:

$$\text{div grad } \mathbf{A} \equiv \left(\mathbf{A} \otimes \overleftarrow{\nabla}\right) \cdot \overleftarrow{\nabla} = (A_{ij,k}\mathbf{e}_i \otimes \mathbf{e}_j \otimes \mathbf{e}_k) \cdot \left(\overleftarrow{\partial_r}\mathbf{e}_r\right) = A_{ij,kr}\mathbf{e}_i \otimes \mathbf{e}_j \otimes \mathbf{e}_k \cdot \mathbf{e}_r \Rightarrow$$
$$\text{div grad } \mathbf{A} = A_{ij,kk}\mathbf{e}_i \otimes \mathbf{e}_j = (A_{ij}\mathbf{e}_i \otimes \mathbf{e}_j)_{,kk} = \nabla^2 \mathbf{A}$$

$$(3.4.24)$$

The result agrees with the formula (3.4.11).

3.4.3 Directional Derivative of Tensor Fields

Let the unit vector $\mathbf{e}$ define an axis originating from a place $\mathbf{r}$. The distance from the place $\mathbf{r}$ to a position on the axis is given by a local coordinate s. We consider the tensor field $\mathbf{A}(\mathbf{r} + s\mathbf{e}, t)$ and define the *directional derivative of the components of the tensor field* $\mathbf{A}(\mathbf{r}, t)$ *at the place* $\mathbf{r}$ *and in the direction* $\mathbf{e}$ by:

$$\frac{\partial A_{i\cdot\cdot j}(x_k + se_k, t)}{\partial s}\bigg|_{s=0} = \left[\frac{\partial A_{i\cdot\cdot j}}{\partial(x_k + se_k)}\frac{\partial(x_k + se_k)}{\partial s}\right]_{s=0} = A_{i\cdot\cdot j,k}e_k \qquad (3.4.25)$$

The *directional derivative of the tensor field* $\mathbf{A}(\mathbf{r}, t)$ *at the place* $\mathbf{r}$ *and in the direction* $\mathbf{e}$ is defined as the tensor:

$$\frac{\partial \mathbf{A}(\mathbf{r} + s\mathbf{e}, t)}{\partial s}\bigg|_{s=0} = (\text{grad } \mathbf{A}) \cdot \mathbf{e} \Leftrightarrow [(\text{grad } \mathbf{A}) \cdot \mathbf{e}]_{i\cdot\cdot j} \quad A_{i\cdot\cdot j,k}e_k \qquad (3.4.26)$$

3.4.4 Material Derivative of Tensor Fields

A particle, i.e. material point, in a continuum is identified by the place vector $\mathbf{r}_0$ in the reference configuration K_0 of the continuum. The *motion of the continuum* is then given by the place vector $\mathbf{r}(\mathbf{r}_0, t)$ for the particle $\mathbf{r}_0$ in the present configuration K of the continuum.

The *material derivative* of a tensor field of order n is a new tensor field of order n. For the tensor field $\mathbf{A}(\mathbf{r}_0, t)$ the material-derivative is:

$$\dot{\mathbf{A}} = \partial_t\mathbf{A} = \partial_t\mathbf{A}(\mathbf{r}_0, t) \Leftrightarrow \dot{A}_{i\cdot\cdot j} = \frac{\partial A_{i\cdot\cdot j}}{\partial t} \equiv \partial_t A_{i\cdot\cdot j} \qquad (3.4.27)$$

For the tensor field $\mathbf{A}(\mathbf{r}, t)$ we introduce the motion $\mathbf{r} = \mathbf{r}(\mathbf{r}_0, t)$, such that $\mathbf{A} = \mathbf{A}(\mathbf{r}(\mathbf{r}_0, t), t)$. Then:

$$\dot{\mathbf{A}} = \partial_t\mathbf{A} + \frac{\partial \mathbf{A}}{\partial \mathbf{r}} \cdot \frac{\partial \mathbf{r}}{\partial t} = \partial_t\mathbf{A} + (\text{grad } \mathbf{A}) \cdot \frac{\partial \mathbf{r}}{\partial t} \Leftrightarrow \dot{A}_{i\cdot\cdot j} = \partial_t A_{i\cdot\cdot j} + A_{i\cdot\cdot j,k}v_k \quad (3.4.28)$$

Alternatively we may write:

$$\dot{\mathbf{A}} = \partial_t\mathbf{A} + \mathbf{A} \otimes \overleftarrow{\nabla} \cdot \mathbf{v} \qquad (3.4.29)$$

We introduce the operator:

$$(\mathbf{v} \cdot \nabla) = (v_k \mathbf{e}_k) \cdot \left(\mathbf{e}_i \frac{\partial}{\partial x_i} \right) = v_k \frac{\partial}{\partial x_i} \mathbf{e}_k \cdot \mathbf{e}_i = v_k \frac{\partial}{\partial x_i} \delta_{ki} = v_k \frac{\partial}{\partial x_k} \Rightarrow$$

$$(\mathbf{v} \cdot \nabla) = v_k \frac{\partial}{\partial x_k} \tag{3.4.30}$$

The formula (3.4.29) may now be expressed by:

$$\dot{\mathbf{A}} = \partial_t \mathbf{A} + (\mathbf{v} \cdot \nabla)\mathbf{A} \tag{3.4.31}$$

3.5 Rigid-Body Dynamics. Kinematics

All material bodies are deformable and will in general be deformed when subjected to forces. However, when the deformations are small enough we may treat the body as rigid. Then subsequently, when the dynamics has been determined, the deformations of the body are taken into consideration. In the present section and the one to follow we shall include some fundamental aspects of rigid-body dynamics.

Rigid-body kinematics plays an important part in deformation analysis in Chap. 4, especially when large deformations are considered in Sect. 4.5. Rigid-body kinetics in the next section provides us with an example of an important symmetric second order tensor, the *inertia tensor*.

3.5.1 Pure Rotation About a Fixed Axis

Figure 3.2 shows a rigid body in two configurations: a reference configuration K_0 at time t_0 and the present configuration K at time $t > t_0$. The Cartesian coordinate system Ox is fixed in a reference Rf. The x_3-axis is perpendicular the plane of the figure. The rigid body rotates about the x_3-axis. The Cartesian coordinate system $O\bar{x}$ with base vectors $\bar{\mathbf{e}}_i$ and with the $\bar{x}_3$-axis coinciding with the x_3-axis is fixed in the body and coincides with the Ox-system at time t_0.

The rotation of the body and of the body fixed coordinate system $O\bar{x}$ with respect to the reference Rf is given by an *angle of rotation* $\theta(t)$. A particle $\mathbf{r}_0$ in the body has the place P_0 at the reference time t_0 and the place P at the present time t. These places are also given by the place vectors $\mathbf{r}_0$ and $\mathbf{r}$:

$$\mathbf{r}_0 = \bar{x}_i \bar{\mathbf{e}}_i \equiv X_i \bar{\mathbf{e}}_i, \quad \mathbf{r} = x_i \mathbf{e}_i \tag{3.5.1}$$

The coordinates in the $O\bar{x}$-system are also representing reference coordinates for particles in the body, i.e. $\bar{x}_i \equiv X_i$.

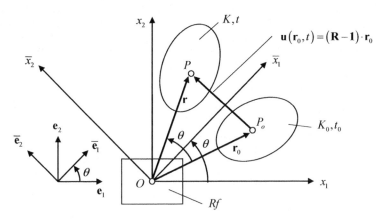

Fig. 3.2 Rigid body rotation about a fixed x_3-axis perpendicular to the plane of the figure. Present configuration of K of the rigid body. Reference configuration K_0 of the body. The Cartesian coordinate system Ox with base vectors $\mathbf{e}_i$ is attached to the reference RF. The Cartesian coordinate system $O\bar{x}$ with base vectors $\bar{\mathbf{e}}_i$ is fixed in the body. Displacement vector $\mathbf{u}(\mathbf{r}_0, t) = \mathbf{r} - \mathbf{r}_0(\mathbf{R} - 1) \cdot \mathbf{r}_0$

The coordinate systems Ox and $O\bar{x}$ are related through the transformation matrix Q:

$$x = Q^T\bar{x} \equiv Q^T X \Leftrightarrow x_i = Q_{ki}X_k \qquad (3.5.2)$$

$$Q_{ik}(\theta) = \cos(\bar{\mathbf{e}}_i, \mathbf{e}_j) \Rightarrow Q(\theta) = \begin{pmatrix} \cos\theta & \sin\theta & 0 \\ -\sin\theta & \cos\theta & 0 \\ 0 & 0 & 1 \end{pmatrix}, \quad \theta = \theta(t) \qquad (3.5.3)$$

The motion of the body is a *rigid-body rotation about a fixed axis*, i.e. the x_3-axis, and will be described by the place vector function $\mathbf{r}(\mathbf{r}_0, t)$:

$$\mathbf{r}(\mathbf{r}_0, t) = \mathbf{R}(t) \cdot \mathbf{r}_0 \Rightarrow x(X, t) = R(t)X \qquad (3.5.4)$$

$\mathbf{R}(t)$ is a second order tensor called the *rotation tensor* of rigid-body rotation about a fixed axis. By comparing the component relation $(3.5.4)_2$ with Eq. $(3.5.2)$, we see that:

$$R(t) = Q^T(t) \Leftrightarrow R_{ik}(t) = Q_{ki}(t) \qquad (3.5.5)$$

Because the transformation matrix Q is an orthogonal matrix it follows that the rotation tensor is an orthogonal tensor:

$$\mathbf{R}\mathbf{R}^T = 1, \quad \det \mathbf{R} = 1 \qquad (3.5.6)$$

The displacement of the particle $\mathbf{r}_0$ is given by the *displacement vector*:

$$\mathbf{u} = \mathbf{u}(\mathbf{r}_0, t) = \mathbf{r} - \mathbf{r}_0 = (\mathbf{R}(t) - \mathbf{1}) \cdot \mathbf{r}_0 \Rightarrow u(X, t) = x(X, t) - X = (R(t) - 1)X$$
(3.5.7)

3.5.2   Pure Rotation About a Fixed Point

Figure 3.3 shall illustrate a general rotation of a rigid body about a point O fixed in a reference *Rf*. The coordinate system Ox is fixed in the reference *Rf*. The rigid body is shown in its configuration K at the present time t and in a reference configuration K_0 at a reference time t_0. The coordinate system $O\bar{x}$ is fixed in the body and moves with it, and at the reference time t_0 the two coordinate systems Ox and $O\bar{x}$ coincide.

A particle $\mathbf{r}_0$ in the body moves from its position P_0 at the reference time t_o to the present position P, given by the place vector $\mathbf{r}$, at the present time t. Let x_i be the coordinates of the position P in the Ox-system, and let $\bar{x}_i$ be the coordinates of the position P_0 in the $O\bar{x}$-system. The coordinates of the particle $\mathbf{r}_0$ are then:

$$x_i = x_i(t) \text{ at the present time } t \text{ in the } Ox\text{-system}$$
$$X_i \equiv \bar{x}_i \text{ at the reference time } t_o \text{ in the } O\bar{x}\text{-system}$$
(3.5.8)

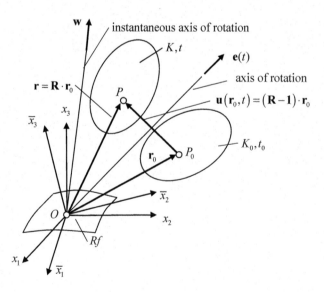

Fig. 3.3 Rigid body rotation about a fixed point O. Present configuration of K of the rigid body. Reference configuration K_0 of the rigid body. The Cartesian coordinate system Ox is attached to the reference *Rf*. The Cartesian coordinate system $O\bar{x}$ is fixed in the body. Displacement vector $\mathbf{u}(\mathbf{r}_0, t) = (\mathbf{R} - 1) \cdot \mathbf{r}_0$

The two coordinate systems $O x$ and $O \bar{x}$ are related through the relations:

$$x = Q^T \bar{x} \equiv Q^T X \tag{3.5.9}$$

$Q(t)$ is the transformation matrix relating the base vectors of the two systems:

$$Q(t) = (\bar{\mathbf{e}}_i \cdot \mathbf{e}_k) = (\cos(\bar{\mathbf{e}}_i, \mathbf{e}_k)) \tag{3.5.10}$$

The motion of the body is a *rigid-body rotation* about a fixed point O, and is described by the place function $\mathbf{r}(\mathbf{r}_0, t)$, where:

$$\mathbf{r}(\mathbf{r}_0, t) = \mathbf{R}(t) \cdot \mathbf{r}_0 \Rightarrow x(X, t) = R(t)X \tag{3.5.11}$$

The function $\mathbf{R}(t)$ is a second order tensor called the *rotation tensor* of the rigid-body rotation about the fixed point O. By comparing Eqs. (3.5.9) and (3.5.11), we see that:

$$R(t) = Q^T(t) \Leftrightarrow R_{ik}(t) = Q_{ki}(t) \tag{3.5.12}$$

As was the case for rigid body rotation about a fixed axis, we understand that the rotation tensor $\mathbf{R}(t)$ is an orthogonal tensor. It also follows that:

$$\mathbf{R}(t_0) = \mathbf{1} \quad \text{and} \quad x(X, t_0) = X \tag{3.5.13}$$

The motion (3.5.11) may also be described by the *displacement vector* $\mathbf{u}(\mathbf{r}_0, t)$ as shown in Fig. 3.3:

$$\mathbf{u}(\mathbf{r}_0, t) = \mathbf{r} - \mathbf{r}_0 = (\mathbf{R} - \mathbf{1}) \cdot \mathbf{r}_0 \Leftrightarrow u(X, t) = x(X, t) - X \tag{3.5.14}$$

Let A and B be two particles in the rigid body given respectively by the place vectors $\mathbf{r}_A = \mathbf{r}_A(t)$ and $\mathbf{r}_B = \mathbf{r}_B(t)$ at the present time t, and the place vectors $\mathbf{r}_{A0} = \mathbf{r}_{A0}(t_0)$ and $\mathbf{r}_{B0} = \mathbf{r}_{B0}(t_0)$ at the reference time t_0. For a vector $\mathbf{c} = \mathbf{c}(t)$ moving with the body and connecting the two particles A and B we write:

$$\mathbf{c} = \mathbf{c}(t) = \mathbf{r}_A(t) - \mathbf{r}_B(t), \quad \mathbf{c}_0 = \mathbf{c}(t_0) = \mathbf{r}_A(t_0) - \mathbf{r}_B(t_0) \tag{3.5.15}$$

It now follows from Eq. (3.5.11) that:

$$\begin{aligned}
\mathbf{c} = \mathbf{c}(t) &= \mathbf{R}(t) \cdot \mathbf{r}_{A0} - \mathbf{R}(t) \cdot \mathbf{r}_{B0} = \mathbf{R}(t) \cdot (\mathbf{r}_{A0} - \mathbf{r}_{B0}) \Rightarrow \\
\mathbf{c} = \mathbf{c}(t) &= \mathbf{R}(t) \cdot \mathbf{c}_0, \quad \mathbf{c}_0 = \mathbf{c}(t_0)
\end{aligned} \tag{3.5.16}$$

We shall use the expression: the vector $\mathbf{c}$ is the $\mathbf{R}-$*rotation* of the vector $\mathbf{c}_0$. It follows that the place vector $\mathbf{r}$ is the $\mathbf{R}-$*rotation* of the place vector $\mathbf{r}_0$.

It will now be demonstrated that at any time t there exists a material straight line in the body through the fixed point O which has the same position referred to Rf at the time t as it has at the reference time t_0. The line is called the *axis of rotation* related to the rotation $\mathbf{R}(t)$. Material particles on the line may have moved during the time interval from the time t_0 to the time t but with no resulting displacement. A unit vector $\mathbf{e}(t)$ on the axis of rotation must satisfy Eq. (3.5.14) for zero displacement:

$$\mathbf{R} \cdot \mathbf{e} = \mathbf{e} \Leftrightarrow (\mathbf{R} - \mathbf{1}) \cdot \mathbf{e} = \mathbf{0} \qquad (3.5.17)$$

The condition for this equation to have a solution for $\mathbf{e}(t)$ is:

$$\det(\mathbf{R} - \mathbf{1}) = 0 \qquad (3.5.18)$$

To prove that this condition is satisfied we use the fact that the rotation tensor $\mathbf{R}(t)$ is orthogonal: $\det \mathbf{R} = 1$ and $\mathbf{R}^T \mathbf{R} = \mathbf{1}$. We write: $\mathbf{R} - \mathbf{1} = \mathbf{R} - \mathbf{R}^T \mathbf{R} = (\mathbf{1} - \mathbf{R}^T)\mathbf{R} = -(\mathbf{R}^T - \mathbf{1})\mathbf{R}$. Now we use the multiplication theorem (1.1.23) for determinants and obtain:

$$\det(\mathbf{R} - \mathbf{1}) = -\det(\mathbf{R}^T - \mathbf{1}) \det \mathbf{R} = -\det(\mathbf{R} - \mathbf{1}) \Rightarrow \det(\mathbf{R} - \mathbf{1}) = 0 \Leftrightarrow$$
$$(3.5.18)$$

Thus we have shown that the unit vector $\mathbf{e}(t)$, which may be determined from Eq. (3.5.17), defines the *axis of rotation related to the rotation* $\mathbf{R}(t)$ of the body about the point O.

Note that the motion $\mathbf{r}(\mathbf{r}_0, t) = \mathbf{R}(t) \cdot \mathbf{r}_0$ of the rigid body when it moves from the reference configuration K_0 to the present configuration K, is not in general a rotation about a reference fixed axis of rotation defined at the present time t by the vector $\mathbf{e}(t)$. However, what has been demonstrated is that it is possible to get the body from K_0 to K by a pure rotation about this axis as defined by the vector $\mathbf{e}(t)$. The axis of rotation will in general change its position both relative to the rigid body and to the reference Rf.

The *angle of rotation* $\theta(t)$ that the body must rotate about the axis of rotation may be determined as follows. We choose an Ox-system such that the base vector $\mathbf{e}_3$ is parallel to the axis of rotation, i.e. $\mathbf{e}_3 = \mathbf{e}(t)$. This means that the transformation matrix $Q(t)$ is given by the formula (3.5.3) and the rotation matrix $R(\theta) = Q^T(\theta)$ is given by:

$$R(\theta) = \begin{pmatrix} \cos\theta & -\sin\theta & 0 \\ \sin\theta & \cos\theta & 0 \\ 0 & 0 & 1 \end{pmatrix}, \quad \theta = \theta(t) \qquad (3.5.19)$$

From this result we obtain: $\operatorname{tr} R = 2\cos\theta + 1$. Because $\mathbf{R}$ is a tensor the trace is an invariant and we have arrived at the coordinate invariant formula:

$$\cos\theta(t) = \frac{1}{2}\left(\text{tr }\mathbf{R}(\theta) - 1\right), \quad \theta = \theta(t) \tag{3.5.20}$$

This formula gives the angle of rotation in any chosen coordinate system.

We now continue to discuss the rotation of a rigid body about a fixed point O. The velocity $\mathbf{v}$ of the particle $\mathbf{r}_0$ when the particle is in the place $\mathbf{r}(\mathbf{r}_0, t)$, is found to be:

$$\mathbf{v} = \dot{\mathbf{r}} = \dot{\mathbf{R}} \cdot \mathbf{r}_o = \dot{\mathbf{R}} \cdot \left(\mathbf{R}^T \cdot \mathbf{r}\right) = \left(\dot{\mathbf{R}}\mathbf{R}^T\right) \cdot \mathbf{r} \Rightarrow$$
$$\mathbf{v} = \dot{\mathbf{r}} = \left(\dot{\mathbf{R}}\mathbf{R}^T\right) \cdot \mathbf{r} = \mathbf{W} \cdot \mathbf{r} \tag{3.5.21}$$

Here we have introduced the *rate of rotation tensor* $\mathbf{W}(t)$ defined by:

$$\mathbf{W} = \dot{\mathbf{R}}\mathbf{R}^T \tag{3.5.22}$$

The rate of rotation tensor is an antisymmetric tensor, as may be seen from the following development, applying the formula (1.1.12):

$$\mathbf{R}\mathbf{R}^T = \mathbf{1} \Rightarrow \dot{\mathbf{R}}\mathbf{R}^T + \mathbf{R}\dot{\mathbf{R}}^T = \mathbf{0} \Rightarrow \dot{\mathbf{R}}\mathbf{R}^T = -\mathbf{R}\dot{\mathbf{R}}^T = -(\dot{\mathbf{R}}\mathbf{R}^T)^T \Rightarrow \mathbf{W} = -\mathbf{W}^T.$$

From the formula (3.5.21) we obtain:

$$v_i = W_{ij}x_j \Rightarrow v_{i,k} = W_{ij}x_{j,k} = W_{ij}\delta_{jk} = W_{ik} \Rightarrow$$
$$W_{ik} = v_{i,k} \tag{3.5.23}$$

The dual vector $\mathbf{w}$ of the antisymmetric tensor $\mathbf{W}$ is by formula (3.3.7):

$$\mathbf{w} = -\frac{1}{2}\mathbf{P}:\mathbf{W} = -\frac{1}{2}\mathbf{W}:\mathbf{P} \Rightarrow w_i = \frac{1}{2}e_{ijk}W_{kj} = \frac{1}{2}e_{ijk}v_{k,j}, \quad \mathbf{w} = \frac{1}{2}\text{rot }\mathbf{v} \tag{3.5.24}$$

$\mathbf{P}$ is the permutation tensor. The vector $\mathbf{w}$ represents physically the same as the tensor $\mathbf{W}$ and is called the *angular velocity vector* or the *rotational velocity vector*. The first name is most common, although it is only when the body rotates about a fixed axis that is possible to associate the rotation of the body with a real angle of rotation. The inverse of the relation (3.5.24) is:

$$\mathbf{W} = -\mathbf{P}\cdot\mathbf{w} \Leftrightarrow W_{ij} = -W_{ji} = -e_{ijk}w_k \Leftrightarrow W = \begin{pmatrix} 0 & -w_3 & w_2 \\ w_3 & 0 & -w_1 \\ -w_2 & w_1 & 0 \end{pmatrix} \tag{3.5.25}$$

The formula (3.3.7), relating the dual vector and tensor quantities, can be used to develop the following alternative velocity distribution formulas from the result (3.5.21):

$$v = \dot{\mathbf{r}} = \mathbf{W} \cdot \mathbf{r} \Leftrightarrow v_i = \dot{x}_i = W_{ik} x_k, \quad v = \dot{\mathbf{r}} = \mathbf{w} \times \mathbf{r} \Leftrightarrow v_i = \dot{x}_i = e_{ijk} w_j x_k$$
$$(3.5.26)$$

The acceleration $\mathbf{a}$ of a particle $\mathbf{r}_0$ at the place $\mathbf{r}$ becomes:

$$\mathbf{a} = \dot{\mathbf{v}} = \ddot{\mathbf{r}} = \dot{\mathbf{W}} \cdot \mathbf{r} + \mathbf{W} \cdot \dot{\mathbf{r}} = \dot{\mathbf{w}} \times \mathbf{r} + \mathbf{w} \times \dot{\mathbf{r}} \Rightarrow$$
$$\mathbf{a} = \left(\dot{\mathbf{W}} + \mathbf{W}^2 \right) \cdot \mathbf{r} = \dot{\mathbf{w}} \times \mathbf{r} + \mathbf{w} \times (\mathbf{w} \times \mathbf{r}) \Leftrightarrow a_i = e_{ijk} \dot{w}_j x_k + e_{ijk} e_{kst} w_j w_s x_t$$
$$(3.5.27)$$

The tensor $\dot{\mathbf{W}}(t)$ is called the *angular acceleration tensor* and the vector $\dot{\mathbf{w}}(t)$ is called the *angular acceleration vector*. In general the angular velocity and the angular acceleration are non-parallel vectors.

The velocity distribution formula (3.5.26) shows that particles on a straight line through the point of rotation O and parallel to the angular velocity vector $\mathbf{w}$ are at the present time t instantaneously at rest. The straight line is called the *instantaneous axis of rotation* and is shown in Fig. 3.3. In the general case the instantaneous axis of rotation and the axis of rotation, defined by Eq. (3.5.17), do not coincide. The instantaneous axis of rotation will in general change its position both relative to the rigid body and to the reference *Rf*. Only when the rigid body is constrained to rotate about an axis fixed with respect to the reference *Rf*, as discussed in Sect. 3.5.1, will the instantaneous axis of rotation and the axis of rotation be one and the same axis at any time t.

We shall now introduce alternative expressions for the velocity and acceleration of a particle in body in rigid-body rotation about a fixed point. First we define the dual antisymmetric tensor $\mathbf{Z}$ to the position vector $\mathbf{r} = [x_1, x_2, x_3]$:

$$\mathbf{Z} = -\mathbf{P} \cdot \mathbf{r} \Leftrightarrow Z_{ij} = -Z_{ji} = -e_{ijk} x_k \Leftrightarrow Z = \begin{pmatrix} 0 & -x_3 & x_2 \\ x_3 & 0 & -x_1 \\ -x_2 & x_1 & 0 \end{pmatrix} \quad (3.5.28)$$

We use the formula $(3.3.8)_1$ to write:

$$\mathbf{w} \times \mathbf{r} = -\mathbf{r} \times \mathbf{w} = -\mathbf{Z} \cdot \mathbf{w} \quad (3.5.29)$$

From the formulas (3.5.25) and (3.5.28) we obtain:

$$\mathbf{w} \times (\mathbf{w} \times \mathbf{r}) = \mathbf{W} \cdot (\mathbf{w} \times \mathbf{r}) = \mathbf{W} \cdot (-\mathbf{Z} \cdot \mathbf{w}) = -(\mathbf{WZ}) \cdot \mathbf{w} \Rightarrow$$
$$\mathbf{w} \times (\mathbf{w} \times \mathbf{r}) = -(\mathbf{WZ}) \cdot \mathbf{w} \quad (3.5.30)$$

The particle velocity $\mathbf{v}$ from the formula (3.5.26) and the particle acceleration $\mathbf{a}$ from the formulas (3.5.27) may now be expressed by the alternative formulas:

$$\mathbf{v} = \dot{\mathbf{r}} = -\mathbf{Z} \cdot \mathbf{w} \Leftrightarrow v_i = \dot{x}_i = -Z_{ij}w_j = e_{ijk}w_j x_k \qquad (3.5.31)$$

$$\mathbf{a} = \dot{\mathbf{v}} = \ddot{\mathbf{r}} = -\mathbf{Z} \cdot \dot{\mathbf{w}} - (\mathbf{WZ}) \cdot \mathbf{w} \Leftrightarrow a_i = e_{ijk}\dot{w}_j x_k + e_{ijk}e_{kst}w_j w_s x_t \qquad (3.5.32)$$

The formulas (3.5.31) and (3.5.32) are convenient when rigid-body kinematics is presented or analyzed in a matrix format.

3.5.3   Kinematics of General Rigid-Body Motion

Figure 3.4 illustrates a rigid body in a general motion from a reference configuration K_0 at the time t_0 to the present configuration K at the present time t. A particle in the body moves from its position P_0 at the time t_0 to the present position P at the present time t. The coordinate system Ox is fixed in the reference Rf, while the coordinate system $\bar{O}\bar{x}$ moves rigidly with the body and coincides with the Ox-system at the reference time t_0. The particle is denoted by the place vector $\mathbf{r}_0$ or by the reference coordinates X in the coordinate system Ox. Let $\mathbf{u}(\mathbf{r}_0, t)$ be the displacement of the particle $\mathbf{r}_0$. The motion of the particle is given by:

$$\mathbf{r}(\mathbf{r}_0, t) = \mathbf{r}_0 + \mathbf{u}(\mathbf{r}_0, t) \Rightarrow x_i(X, t) = X_i + u_i(X, t) \qquad (3.5.33)$$

The origin $\bar{O}$ of the body–fixed coordinate system $\bar{O}\bar{x}$ moves according to the displacement vector $\mathbf{u}_0(t)$. Figure 3.4 illustrates that the motion (3.5.33) may be considered to be a combination of *translation* $\mathbf{u}_0(t)$ by which all particle of the body are given the same motion, and a pure rotation $\bar{\mathbf{r}}_0 = \mathbf{R}(t) \cdot \mathbf{r}_o$ about the point $\bar{O}$.

$$\mathbf{r}(\mathbf{r}_0, t) = \mathbf{u}_0(t) + \mathbf{R}(t) \cdot \mathbf{r}_0 \qquad (3.5.34)$$

The displacement $\mathbf{u}(\mathbf{r}_0, t)$ of the particle $\mathbf{r}_0$ may now be expressed by:

$$\mathbf{u}(\mathbf{r}_0, t) = \mathbf{u}_0(t) + (\mathbf{R}(t) - \mathbf{1}) \cdot \mathbf{r}_0 \qquad (3.5.35)$$

To the general rigid-body motion (3.5.34) the translation $\mathbf{u}_{\bar{O}}(t)$ contributes with a velocity and acceleration, representing the motion of the reference point $\bar{O}$:

$$\mathbf{v}_{\bar{O}} = \dot{\mathbf{u}}_{\bar{O}}(t), \quad \mathbf{a}_{\bar{O}} = \dot{\mathbf{v}}_{\bar{O}} = \ddot{\mathbf{u}}_{\bar{O}}(t) \qquad (3.5.36)$$

To find velocity and acceleration contribution from the pure rotation $\bar{\mathbf{r}}_0 = \mathbf{R}(t) \cdot \mathbf{r}_o$ about the point $\bar{O}$ we follow the developments from the formulas (3.5.21) to the resulting formulas (3.5.26) and (3.5.27). With the rate of rotation tensor $\mathbf{W}$ from formula (3.5.22) and the angular velocity vector (3.5.24) we obtain:

$$\mathbf{v} = \dot{\mathbf{r}} = \dot{\mathbf{u}}(t) = \mathbf{v}_{\bar{O}} + \mathbf{W} \cdot \bar{\mathbf{r}}_0 = \mathbf{v}_{\bar{O}} + \mathbf{w} \times \bar{\mathbf{r}}_0 \qquad (3.5.37)$$

$$\mathbf{a} = \mathbf{a}_{\bar{O}} + \left(\dot{\mathbf{W}} + \mathbf{W}^2\right) \cdot \bar{\mathbf{r}}_0 = \mathbf{a}_{\bar{O}} + \dot{\mathbf{w}} \times \bar{\mathbf{r}}_0 + \mathbf{w} \times (\mathbf{w} \times \bar{\mathbf{r}}_0) \qquad (3.5.38)$$

The tensor $\dot{\mathbf{W}}(t)$ is called the *angular acceleration tensor* and $\dot{\mathbf{W}}(t)$ and the vector $\underline{\mathbf{w}}(t)$ is called the *angular acceleration vector* of the general rigid-body motion.

Let P and Q be two points fixed in a rigid body, and let $\mathbf{r}_{Q/P}$ be the position vector from point P to point Q. Then the formulas (3.5.37) and (3.5.38) may be used to relate the velocities and accelerations of the two points. The results are:

$$\mathbf{v}_Q = \mathbf{v}_P + \mathbf{W} \cdot \mathbf{r}_{Q/P} = \mathbf{v}_P + \mathbf{w} \times \mathbf{r}_{Q/P} \qquad (3.5.39)$$

$$\mathbf{a}_Q = \mathbf{a}_P + \left(\dot{\mathbf{W}} + \mathbf{W}^2\right) \cdot \mathbf{r}_{Q/P} = \mathbf{a}_P + \dot{\mathbf{w}} \times \mathbf{r}_{Q/P} + \mathbf{w} \times \left(\mathbf{w} \times \mathbf{r}_{Q/P}\right) \qquad (3.5.40)$$

These formulas are respectively called the *velocity distribution formula* and the *acceleration distribution formula* for the motion of a rigid body.

The material derivative of a vector $\mathbf{c}(t)$ that rigidly follows the motion of the body is:

$$\dot{\mathbf{c}} = \mathbf{W} \cdot \mathbf{c} = \mathbf{w} \times \mathbf{c} \qquad (3.5.41)$$

In order to see this let $\mathbf{c} = \mathbf{r}_{Q/P} = \mathbf{r}_Q - \mathbf{r}_p$. Then:

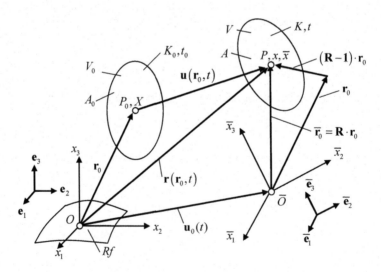

Fig. 3.4 Rigid-body motion from the reference configuration K_0 at the time t_0 to the present configuration K at the time t. The Cartesian coordinate system Ox is attached to the reference Rf. The Cartesian coordinate system $\bar{O}\bar{x}$ moves with the body. The motion consists of a sum of a translation $\mathbf{u}_0(t)$ and a pure rotation $\mathbf{r}_0 = \mathbf{R}(t) \cdot \mathbf{r}_0$ about the point $\bar{O}$

$$\dot{c} = \dot{r}_{Q/P} = \dot{r}_Q - \dot{r}_p = v_Q - v_p = W \cdot r_Q - W \cdot r_P = W \cdot (r_Q - r_p)$$
$$= W \cdot r_{Q/P} = w \times r_{Q/P} \Rightarrow$$
$$\dot{c} = W \cdot c = w \times c$$

Example 3.1 Rotating Circular Plate on a Rotating Arm

Figure 3.5 illustrates a system consisting of a thin circular plate with radius r, mass m, and mass center C that rotates about its horizontal axis with a constant angular velocity ω_3 with respect to an arm OC. The arm rotates with constant angular velocity ω_2 about a vertical axis. We shall determine the angular velocity vector w and the angular acceleration vector $\dot{w}$ for the plate, and the velocities and accelerations of the two points A and B on the plate.

To describe the motion of the ensemble we select the coordinate system Cx fixed to the rotating arm and shown in Fig. 3.5. The x_3-axis coincides with the symmetry axis of the plate, and the x_2-axis is vertical. The angular velocities of the arm and the plate are now:

$$w_a = \omega_2 e_2 \quad \text{for the arm,} \quad w = \omega_2 e_2 + \omega_3 e_3 \quad \text{for the plate}$$

The angular acceleration for the arm $\dot{w}_a$ and for the plate $\dot{w}$ becomes:

$$\dot{w}_a = \omega_2 \dot{e}_2 = \omega_2 w_a \times e_2 = \omega_2 \omega_2 e_2 \times e_2 = 0$$
$$\dot{w} = \omega_2 \dot{e}_2 + \omega_3 \dot{e}_3 = \omega_2 w_a \times e_2 + \omega_3 w_a \times e_3$$
$$= \omega_2 \omega_2 e_2 \times e_2 + \omega_3 \omega_2 e_2 \times e_3 \Rightarrow \dot{w} = \omega_3 \omega_2 e_1$$

The mass center of the plate C moves with the corresponding point on the arm and the velocity of C is:

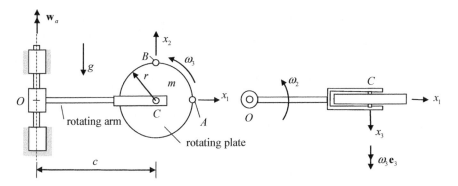

Fig. 3.5 Thin plate with radius r, mass m, and mass center C rotates about its horizontal axis, i.e. the x_3-axis, with constant angular velocity ω_3 with respect to an arm OC. The arm rotates with constant angular velocity ω_2 about a vertical axis

$$\mathbf{v}_C = \mathbf{w}_a \times \mathbf{r}_{C/O} = (\omega_2\mathbf{e}_2) \times (c\mathbf{e}_1) = -\omega_2 c\mathbf{e}_3$$

For the velocities of the points A and B on the plate we find, using formula (3.5.37):

$$\left.\begin{array}{l} \mathbf{v}_A = \mathbf{v}_C + \mathbf{w} \times \mathbf{r}_{A/C} = -\omega_2 c\mathbf{e}_3 + (\omega_2\mathbf{e}_2 + \omega_3\mathbf{e}_3) \times (r\,\mathbf{e}_1) \\ \mathbf{v}_B = \mathbf{v}_C + \mathbf{w} \times \mathbf{r}_{B/C} = -\omega_2 c\mathbf{e}_3 + (\omega_2\mathbf{e}_2 + \omega_3\mathbf{e}_3) \times (r\,\mathbf{e}_2) \end{array}\right\} \Rightarrow$$

$$\mathbf{v}_A = -\omega_2(c+r)\mathbf{e}_3 + \omega_3 r\mathbf{e}_2, \quad \mathbf{v}_B = -\omega_2 c\mathbf{e}_3 - \omega_3 r\mathbf{e}_1$$

The acceleration of the point C as a point on the arm is, using formula (3.5.38):

$$\mathbf{a}_C = \dot{\mathbf{w}}_a \times \mathbf{r}_{C/O} + \mathbf{w}_a \times \left(\mathbf{w}_a \times \mathbf{r}_{C/O}\right) = \mathbf{0} + (\omega_2\mathbf{e}_2) \times (\omega_2\mathbf{e}_2 \times c\mathbf{e}_1) \Rightarrow \mathbf{a}_C$$
$$= -\omega_2^2 c\mathbf{e}_1$$

For the acceleration of the point A on the plate we use formula (3.5.38):

$$\mathbf{a}_A = \mathbf{a}_C + \dot{\mathbf{w}} \times \mathbf{r}_{A/C} + \mathbf{w} \times \left(\mathbf{w} \times \mathbf{r}_{A/C}\right)$$

For the last two acceleration terms we obtain:

$$\dot{\mathbf{w}} \times \mathbf{r}_{A/C} = (\omega_3\omega_2\mathbf{e}_1) \times (r\mathbf{e}_1) = \mathbf{0}$$
$$\mathbf{w} \times \mathbf{r}_{A/C} = (\omega_2\mathbf{e}_2 + \omega_3\mathbf{e}_3) \times (r\mathbf{e}_1) = -\omega_2 r\mathbf{e}_3 + \omega_3 r\mathbf{e}_2 \Rightarrow$$
$$\mathbf{w} \times \left(\mathbf{w} \times \mathbf{r}_{A/C}\right) = (\omega_2\mathbf{e}_2 + \omega_3\mathbf{e}_3) \times (-\omega_2 r\mathbf{e}_3 + \omega_3 r\mathbf{e}_2) = -\left(\omega_2^2 + \omega_3^2\right)r\mathbf{e}_1$$

Thus the acceleration of the point A on the plate is:

$$\mathbf{a}_A = -\left[\omega_2^2(c+r) + \omega_3^2 r\right]\mathbf{e}_1$$

For the acceleration of the point B on the plate we use formula (3.5.38):

$$\mathbf{a}_B = \mathbf{a}_C + \dot{\mathbf{w}} \times \mathbf{r}_{B/C} + \mathbf{w} \times \left(\mathbf{w} \times \mathbf{r}_{B/C}\right)$$

For the last two acceleration terms we obtain:

$$\dot{\mathbf{w}} \times \mathbf{r}_{B/C} = (\omega_3\omega_2\mathbf{e}_1) \times (r\mathbf{e}_2) = \omega_3\omega_2 r\mathbf{e}_3$$
$$\mathbf{w} \times \mathbf{r}_{B/C} = (\omega_2\mathbf{e}_2 + \omega_3\mathbf{e}_3) \times (r\mathbf{e}_2) = -\omega_3 r\mathbf{e}_1 \Rightarrow$$
$$\mathbf{w} \times \left(\mathbf{w} \times \mathbf{r}_{A/C}\right) = (\omega_2\mathbf{e}_2 + \omega_3\mathbf{e}_3) \times (-\omega_3 r\mathbf{e}_1) = \omega_2\omega_3 r\mathbf{e}_3 - \omega_3^2 r\mathbf{e}_2$$

Thus the acceleration of the point B on the plate is:

$$\mathbf{a}_B = -\omega_2^2 c\,\mathbf{e}_1 - \omega_3^2 r\,\mathbf{e}_2 + 2\omega_2\omega_3 r\,\mathbf{e}_3$$

Note that the expressions for the velocities and accelerations for the points A and B on the plate are only valid when the two points are in the positions shown in Fig. 3.5. The kinetics of the ensemble in Fig. 3.1 will be discussed in Example 3.2 below.

3.6 Rigid-Body Dynamics. Kinetics

3.6.1 Rotation About a Fixed Point. The Inertia Tensor

The center of mass C of a body of mass m is defined by formula (2.2.8). For a rigid body the mass center is a point fixed with respect to the body, but does not necessarily coincide with a material point or particle in the body. The motion of the center of mass is governed by the equation of motion (2.2.12):

$$\mathbf{f} = m\mathbf{a}_C \qquad (3.6.1)$$

The vector $\mathbf{f}$ represents the resultant force on the body and $\mathbf{a}_C$ is the acceleration of the center of mass. The vector equation (3.6.1) represents three component equations.

A rigid body in free motion has six degrees of freedom. The motion of the center of mass is governed by the three coordinate functions of time. The rotation of the body may be described by three angles of rotations as three time functions. In Fig. 3.6 the coordinate system $Cx_1x_2x_3$ is fixed relative to a reference that translates with the center of mass C. The coordinate system $CX_1X_2X_3$ is fixed in the rigid body. The rotation of the body may be described t by the following procedures. The body is in the position with $OX_1X_2X_3 = Ox_1x_2x_3$. Then the body is rotated by an angle $\phi(t)$, the precession angle, about the x_3-axis. The body fixed coordinate system is now in the position $OX_1X_2X_3 = O\tilde{x}_1\tilde{x}_2x_3$. Next the body is rotated by an angle $\theta(t)$, the nutation angle, about the $\tilde{x}_1$-axis. The body fixed coordinate system is now in the position $OX_1X_2X_3 = O\tilde{x}_1\bar{x}_2X_3$. Finally the body is rotated by an angle

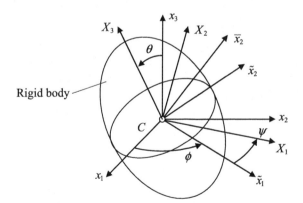

Fig. 3.6 Rotation of a rigid body about the center of mass C. Four coordinate systems: $Cx_1x_2x_3$ is fixed relative to a reference that translates with the mass center, C, $CX_1X_2X_3$ is fixed in the rigid body, $C\tilde{x}_1\tilde{x}_2x_3$ is obtained from $Cx_1x_2x_3$ by a rotation angle $\phi(t)$, the precession angle, about the x_3-axis, $C\tilde{x}_1\bar{x}_2x_3$ is obtained from $C\tilde{x}_1\tilde{x}_2X_3$ by a rotation angle $\theta(t)$ from $C\tilde{x}_1\tilde{x}_2X_3$ by a rotation angle $\theta(t)$, the nutation angle, about the $\tilde{x}_1$-axis, $CX_1X_2X_3$ is obtained from $C\tilde{x}_1\tilde{x}_2X_3$ by a rotation angle $\psi(t)$, the spin angle, about the X_3-axis

$\psi(t)$, the spin angle, about the X_3-axis. The body fixed coordinate system is now in its final position. The three angles $\phi(t)$, $\theta(t)$, $\psi(t)$ are called the *Eulerian angles*.

The additional three equations of motion are provided by the law of balance of angular momentum, i.e. Euler's second axiom (2.2.7). These additional equations of motion suitable for a body in rigid-body motion will now be developed.

We shall start by considering a rigid body that rotates about a fixed point O as shown in Fig. 3.7. First we compute the angular momentum l_O of the body about the point O from the definition formula $(2.2.5)_2$:

$$l_O = \int_V \mathbf{r} \times \mathbf{v} \rho dV \qquad (3.6.2)$$

We use the formula (3.5.28) to represent the particle velocity by: $\mathbf{v} = -\mathbf{Z} \cdot \mathbf{w}$, and the formula $(3.3.8)_1$ for the dual quantities $\mathbf{r}$ and $\mathbf{Z}$ to write: $\mathbf{r} \times \mathbf{v} = \mathbf{Z} \cdot \mathbf{v}$. Then we obtain:

$$\mathbf{r} \times \mathbf{v} = \mathbf{Z} \cdot \mathbf{v} = \mathbf{Z} \cdot (-\mathbf{Z} \cdot \mathbf{w}) = -\mathbf{Z}^2 \cdot \mathbf{w} \qquad (3.6.3)$$

By inspection we find that:

$$\mathbf{Z}^2 = \mathbf{r} \otimes \mathbf{r} - (\mathbf{r} \cdot \mathbf{r})\mathbf{1} \Leftrightarrow (\mathbf{Z}^2)_{ij} = x_i x_j - x_k x_k \delta_{ij} \qquad (3.6.4)$$

The angular momentum of the body about O, Eq. (3.6.2), then becomes:

$$l_O = \int_V \mathbf{r} \times \mathbf{v} \rho dV = \int_V \mathbf{r} \times (\mathbf{w} \times \mathbf{r}) \rho dV = \left[\int_V (-\mathbf{Z}^2) \rho dV \right] \cdot \mathbf{w} \Rightarrow$$

$$l_O = \int_V \mathbf{r} \times (\mathbf{w} \times \mathbf{r}) \rho dV = \mathbf{I} \cdot \mathbf{w} = I_{ij} w_j \mathbf{e}_i \qquad (3.6.5)$$

The quantity $\mathbf{I}$ is a symmetric tensor of second order defined by the formula:

Fig. 3.7 Rotation of a rigid body about a fixed point O. Volume V. Mass m. Center of mass C. Volume element dV. Element of mass ρdV. Particle velocity $\mathbf{v}$. Angular velocity $\mathbf{w}$. The Cartesian coordinate system Ox is attached to the reference Rf

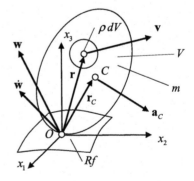

$$\mathbf{I} = \int_V \left(-\mathbf{Z}^2\right)\rho dV = \int_V \left[-\mathbf{r}\otimes\mathbf{r}+(\mathbf{r}\cdot\mathbf{r})\mathbf{1}\right]\rho dV \Leftrightarrow$$

$$I_{ij} = \int_V \left[-x_i x_j + x_k x_k \delta_{ij}\right]\rho dV \tag{3.6.6}$$

The tensor $\mathbf{I}$ is called the *inertia tensor* of the body with respect to the point O. The inertia tensor is symmetric and the six distinct components of the tensor in the Ox-system are three *moments of inertia with respect to the three coordinate axes:*

$$I_{11} = \int_V \left[x_2 x_2 + x_3 x_3\right]\rho\, dV, \quad I_{22} = \int_V \left[x_3 x_3 + x_1 x_1\right]\rho\, dV,$$

$$I_{33} = \int_V \left[x_1 x_1 + x_2 x_2\right]\rho\, dV \tag{3.6.7}$$

and three *products of inertia* with respect to the axes x_i- and x_j:

$$I_{ij} = I_{ji} = -\int_V x_i x_j \rho\, dV, \quad i \neq j \tag{3.6.8}$$

Because the inertia tensor is a symmetric tensor of second order, there exist three orthogonal axes through the point O with respect to which the products of inertia are zero. These axes are the *principal axes of inertia* of the body with respect to the point O. The corresponding moments of inertia are the *principal moments of inertia* I_i.

The following rules apply:

1. An axis of symmetry through the point O is a principal axis of inertia.
2. An axis normal to a plane of symmetry through O is a principal axis of inertia.
3. If two orthogonal axes through the point O are principal axes of inertia then the axis through O normal to the plane defined by the two orthogonal axes is a third principal axis of inertia.

Table 3.1 presents the moments of inertia of some characteristic homogeneous bodies of mass m with respect to principal axes of inertia. In general, an axis of symmetry and an axis normal to a plane of symmetry are principal axes of inertia.

If the coordinate axes coincide with the principal axes of inertia, the angular momentum vector (3.6.5) takes the form:

$$\mathbf{l}_O = \mathbf{I}\cdot\mathbf{w} = I_1 w_1\,\mathbf{e}_1 + I_2 w_2\,\mathbf{e}_2 + I_3 w_3\,\mathbf{e}_3 \tag{3.6.9}$$

Table 3.1 Moment of inertia of some homogeneous bodies

Homogeneous slender bar of mass m and length l
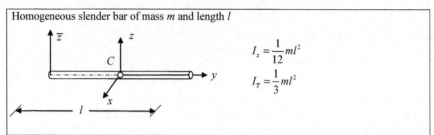 $$I_z = \frac{1}{12}ml^2$$ $$I_{\bar{z}} = \frac{1}{3}ml^2$$
Thin rectangular plate of mass m, length b, and width a
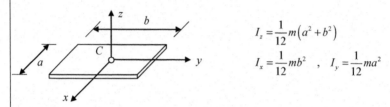 $$I_z = \frac{1}{12}m\left(a^2 + b^2\right)$$ $$I_x = \frac{1}{12}mb^2 \quad , \quad I_y = \frac{1}{12}ma^2$$
Thin circular plate of mass m and radius r
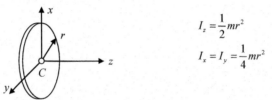 $$I_z = \frac{1}{2}mr^2$$ $$I_x = I_y = \frac{1}{4}mr^2$$
Homogeneous circular cylinder of mass m, length l, and radius r
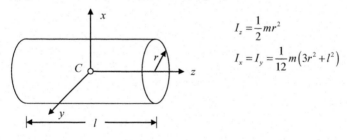 $$I_z = \frac{1}{2}mr^2$$ $$I_x = I_y = \frac{1}{12}m\left(3r^2 + l^2\right)$$

The Euler's second axiom, (2.2.7), equates the resultant moment $\mathbf{m}_O$ about the point O of the forces on the body and the material derivative of the angular momentum of the body about the point O:

$$\mathbf{m}_O = \dot{\mathbf{l}}_O \tag{3.6.10}$$

The material derivative of the angular momentum $\mathbf{l}_O$ about O, from Eq. (3.6.5), becomes:

$$\dot{\mathbf{l}}_O = \mathbf{I} \cdot \dot{\mathbf{w}} + \dot{\mathbf{I}} \cdot \mathbf{w} \tag{3.6.11}$$

The inertia tensor is constant when referred to the rigid body, i.e. all the elements I_{ij} in the formulas (3.6.7) and (3.6.8) are time-independent when referred to a coordinate system that follows the motion of the body. However, in a coordinate system fixed to the reference Rf to which the motion of the rigid body is referred, the components I_{ij} of the inertia tensor $\mathbf{I}$ are time–dependent. That means that the inertia tensor $\mathbf{I}$ is a time–dependent tensor when referred to the reference Rf. In order to find a suitable expression for the material derivative $\dot{\mathbf{I}}$ of the inertia tensor $\mathbf{I}$, we may argue as follows. Let $\mathbf{a}$ and $\mathbf{b}$ be two vectors that move rigidly with the rigid body. By the formula (3.5.41):

$$\dot{\mathbf{a}} = \mathbf{W} \cdot \mathbf{a}, \quad \dot{\mathbf{b}} = \mathbf{W} \cdot \mathbf{b} \tag{3.6.12}$$

Because $\mathbf{I}$ is a second order tensor, the product $\mathbf{a} \cdot \mathbf{I} \cdot \mathbf{b}$ is a time-independent scalar. We now compute:

$$\begin{aligned} \alpha = \mathbf{a} \cdot \mathbf{I} \cdot \mathbf{b} \Rightarrow \dot{\alpha} = \dot{\mathbf{a}} \cdot \mathbf{I} \cdot \mathbf{b} + \mathbf{a} \cdot \dot{\mathbf{I}} \cdot \mathbf{b} + \mathbf{a} \cdot \mathbf{I} \cdot \dot{\mathbf{b}} = 0 \Rightarrow \\ (\mathbf{W}\mathbf{a}) \cdot \mathbf{I} \cdot \mathbf{b} + \mathbf{a} \cdot \dot{\mathbf{I}} \cdot \mathbf{b} + \mathbf{a} \cdot \mathbf{I} \cdot (\mathbf{W}\mathbf{b}) = 0 \end{aligned} \tag{3.6.13}$$

$\mathbf{W}$ is an antisymmetric tensor, i.e. $\mathbf{W}^T = -\mathbf{W} \Rightarrow W_{ji} = -W_{ij}$, and therefore it follows that for the first and the last terms on the left–hand side of Eq. (3.6.13) we may write:

$$(\mathbf{W}\mathbf{a}) \cdot \mathbf{I} \cdot \mathbf{b} = \left(W_{ij}a_j\right)I_{ik}b_k = -W_{ji}a_jI_{ik}b_k = -a_jW_{ji}I_{ik}b_k = -\mathbf{a} \cdot (\mathbf{W}\mathbf{I}) \cdot \mathbf{b}$$
$$\mathbf{a} \cdot \mathbf{I} \cdot (\mathbf{W}\mathbf{b}) = a_iI_{ik}\left(W_{kj}b_j\right) = a_i\left(I_{ik}W_{kj}\right)b_j = \mathbf{a} \cdot (\mathbf{I}\mathbf{W}) \cdot \mathbf{b}$$

Equation (3.6.13) may now be rewritten to:

$$-\mathbf{a} \cdot (\mathbf{W}\mathbf{I}) \cdot \mathbf{b} + \mathbf{a} \cdot \dot{\mathbf{I}} \cdot \mathbf{b} + \mathbf{a} \cdot (\mathbf{I}\mathbf{W}) \cdot \mathbf{b} = \mathbf{a} \cdot \left(-\mathbf{W}\mathbf{I} + \dot{\mathbf{I}} + \mathbf{I}\mathbf{W}\right) \cdot \mathbf{b} = 0$$

Because this result is valid for any choice of the vectors $\mathbf{a}$ and $\mathbf{b}$, we conclude that the term in the parenthesis is a zero tensor of second order, i.e. $-\mathbf{W}\mathbf{I} + \dot{\mathbf{I}} + \mathbf{I}\mathbf{W} = \mathbf{0}$. Thus we have found the material derivative of the inertia tensor to be:

$$\dot{\mathbf{I}} = \mathbf{W}\mathbf{I} - \mathbf{I}\mathbf{W} \Leftrightarrow \dot{I}_{ij} = W_{ik}I_{kj} - I_{ik}W_{kj} \tag{3.6.14}$$

Note that the result (3.6.14) is valid for the material derivative of any tensor of second order that is constant in the body, not necessarily symmetric, i.e. for any *body–invariant tensor of second order*.

We now return to Euler's second axiom, (3.6.10), which by Eqs. (3.6.11) and (3.6.14), yields the *balance of angular momentum equation*:

$$\mathbf{m}_O = \dot{\mathbf{l}}_O = \mathbf{I} \cdot \dot{\mathbf{w}} + \dot{\mathbf{I}} \cdot \mathbf{w} = \mathbf{I} \cdot \dot{\mathbf{w}} + \mathbf{W}\mathbf{I} \cdot \mathbf{w} - \mathbf{I}\mathbf{W} \cdot \mathbf{w}$$

We shall now show that the vector $\mathbf{I}\mathbf{W} \cdot \mathbf{w}$ is zero. The formulas (3.5.25) and the antisymmetric property of the permutation symbol e_{kjl}, i.e. $e_{kjl} = -e_{ljk}$, is used to write:

$$(\mathbf{I}\mathbf{W}) \cdot \mathbf{w} = I_{ij}W_{jk}w_k\mathbf{e}_i = I_{ij}(e_{kjl}w_l)w_k\mathbf{e}_i = I_{ij}(e_{kjl}w_lw_k)\mathbf{e}_i = \mathbf{0}$$

The Euler's second axiom, i.e. the balance of angular momentum equation, therefore takes the form:

$$\mathbf{m}_O = \mathbf{I} \cdot \dot{\mathbf{w}} + \mathbf{W}\mathbf{I} \cdot \mathbf{w} \Leftrightarrow m_{Oi} = I_{ij}\dot{w}_j + W_{ik}I_{kl}w_l = I_{ij}\dot{w}_j + e_{ijk}w_jI_{kl}w_l \quad (3.6.15)$$

The three component equations are called *Euler's equations of motion for a rigid body*.

If the coordinate axes x_i are chosen to be parallel to the principal axes of inertia of the body with respect to the point O, the Euler equations become:

$$\begin{aligned} m_{O1} &= I_1\dot{w}_1 + (I_3 - I_2)w_3w_2 \\ m_{O2} &= I_2\dot{w}_2 + (I_1 - I_3)w_1w_3 \\ m_{O3} &= I_3\dot{w}_3 + (I_2 - I_1)w_2w_1 \end{aligned} \quad (3.6.16)$$

I_1, I_2, and I_3 are the *principal moments of inertia*.

3.6.2   General Rigid-Body Motion

For a general motion of a rigid body the expression of the angular momentum about a fixed point O is developed as follows. Figure 3.8 shows a rigid body of mass m and with the center of mass C. The position vector from the fixed point O to a particle P in the body is expressed by the vector sum $\mathbf{r}_C + \mathbf{r}$, where $\mathbf{r}_C$ is the place vector from O to the mass center C, and $\mathbf{r}$ is the place vector from C to the particle. The velocity of the particle is given by the formula (3.5.37), with reference to Fig. 3.8, and by use of the alternative form (3.5.29) for $\mathbf{w} \times \mathbf{r}$:

$$\mathbf{v} = \mathbf{v}_C + \mathbf{w} \times \mathbf{r} = \mathbf{v}_C - \mathbf{Z} \cdot \mathbf{w} \qquad (3.6.17)$$

The angular momentum of the rigid body about the fixed point O is according to Fig. 3.8, the formula (3.6.2), and the formula (3.6.17) for the velocity field:

$$\mathbf{l}_O = \int_V \mathbf{r} \times \mathbf{v}\rho dV = \int_V (\mathbf{r}_C + \mathbf{r}) \times \mathbf{v}\rho dV = \mathbf{r}_C \times \int_V \mathbf{v}\rho dV + \left[\int_V \mathbf{r}\rho dV \right] \times \mathbf{v}_C$$

$$+ \int_V \mathbf{r} \times (\mathbf{w} \times \mathbf{r})\,\rho dV \Rightarrow$$

$$\mathbf{l}_O = \mathbf{r}_C \times \int_V \mathbf{v}\rho dV + \left[\int_V \mathbf{r}\rho dV \right] \times \mathbf{v}_C + \int_V \mathbf{r} \times (\mathbf{w} \times \mathbf{r})\rho dV$$

$$(3.6.18)$$

The three integrals on the right-hand side of Eq. (3.6.18) are as follows: The first integral is equal to the *linear momentum* $m\mathbf{v}_C$ of the rigid body, confer the formulas (2.2.9) and (2.2.10). The second integral vanishes due to the definition (2.2.8) of the center of mass C. The third integral is denoted $\mathbf{l}_C$ and is called the *central angular momentum*. The angular momentum of the rigid body about the fixed point O is then:

$$\mathbf{l}_O = \mathbf{r}_C \times m\,\mathbf{v}_C + \mathbf{l}_C \qquad (3.6.19)$$

Note that the central angular momentum also represents the angular momentum of the body about a fixed point that at time t coincides with the center of mass, i.e. $O = C$ at time t.

The central angular momentum may be developed similarly to the derivation of the momentum in formula (3.6.5). Thus we may write:

Fig. 3.8 General rigid body motion. Volume V and mass m. Center of mass C. Volume element dV. Element of mass ρdV. Particle velocity $\mathbf{v}$. Velocity $\mathbf{v}_C$ and acceleration $\mathbf{a}_C$ of the mass center of mass. Angular velocity $\mathbf{w}$ and angular acceleration $\dot{\mathbf{w}}$ of the body

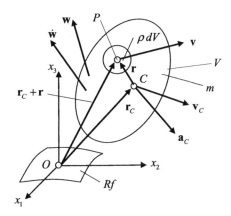

$$\mathbf{l}_C = \int_V \mathbf{r} \times (\mathbf{w} \times \mathbf{r})\rho dV = \mathbf{I}_C \cdot \mathbf{w} \qquad (3.6.20)$$

The quantity $\mathbf{I}_C$ is the inertia tensor of the body with respect to the mass center C. With reference to the formulas (3.6.5) and (3.6.6), and a Cartesian coordinate system Cx fixed in the body and with origin in C, not shown in Fig. 3.7, we write:

$$\mathbf{I}_C = \int_V \left[-\mathbf{r} \otimes \mathbf{r} + (\mathbf{r} \cdot \mathbf{r})\mathbf{1} \right]\rho dV \Leftrightarrow I_{Cij} = \int_V \left[-x_i x_j + x_k x_k \delta_{ij} \right]\rho dV \qquad (3.6.21)$$

The law of balance of angular momentum (3.6.10) now takes the form:

$$\begin{aligned} \mathbf{m}_O = \dot{\mathbf{l}}_O = \dot{\mathbf{r}}_C \times m\mathbf{v}_C + \mathbf{r}_C \times m\dot{\mathbf{v}}_C + \dot{\mathbf{l}}_C \Rightarrow \\ \mathbf{m}_O = \dot{\mathbf{l}}_O = \mathbf{r}_C \times m\mathbf{a}_C + \dot{\mathbf{l}}_C \end{aligned} \qquad (3.6.22)$$

The last term on the right-hand side may be transformed using the result (3.6.15).

$$\dot{\mathbf{l}}_C = \mathbf{I} \cdot \dot{\mathbf{w}} + \mathbf{W}\mathbf{I} \cdot \mathbf{w} \Leftrightarrow \dot{l}_C = \left(I_{ij}\dot{w}_j + W_{ik}I_{kl}w_l \right)\mathbf{e}_i = \left(I_{ij}\dot{w}_j + e_{ijk}w_j I_{kl}w_l \right)\mathbf{e}_i \qquad (3.6.23)$$

Example 3.2 Rotating Circular Plate on a Rotating Arm
For the ensemble presented in Example 3.1 we shall find the force $\mathbf{f}_a$ and the torque $\mathbf{m}_t$, see Fig. 3.9, which must be supplied to the plate in order to obtain the prescribed motion. When the ensemble is at rest the plate is only subjected to its weight mg.

The acceleration of the mass center C of the plate is found in Example 3.1: $\mathbf{a}_C = -\omega_3^2 c\mathbf{e}_1$. From symmetry it follows that the coordinate axes are principal axes of inertia and from the table of moments of inertia we get:

Fig. 3.9 Free-body diagram of the rotating plate

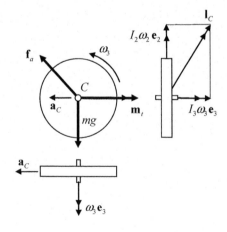

$$I_{11} = I_1 = I_{22} = I_2 = \frac{1}{4}mr^2, \quad I_{33} = I_3 = \frac{1}{2}mr^2$$

The central angular momentum of the plate becomes:

$$\mathbf{l}_C = I_2\omega_2\mathbf{e}_2 + I_3\omega_3\mathbf{e}_3 = \frac{1}{4}mr^2\omega_2\mathbf{e}_2 + \frac{1}{2}mr^2\omega_3\mathbf{e}_3$$

This vector is constant relative to rotating arm, and the angular velocity or the arm is $\mathbf{w}_a = \omega_2\mathbf{e}_2$. The fundamental laws of motion applied to the plate are now:

$$\mathbf{f} = m\mathbf{a}_C \Rightarrow$$

$$\mathbf{f}_a - mg\,\mathbf{e}_2 = m\left(-\omega_3^2 c\mathbf{e}_1\right)$$
$$= m\left(g\,\mathbf{e}_2 - \omega_3^2 c\mathbf{e}_1\right)$$

$$\mathbf{m}_C = \dot{\mathbf{l}}_C = \mathbf{w}_a \times \mathbf{l}_C \Rightarrow$$
$$\mathbf{m}_t = (\omega_2\mathbf{e}_2) \times (I_2\omega_2\mathbf{e}_2 + I_3\omega_3\mathbf{e}_3) \Rightarrow$$
$$\mathbf{m}_t = I_3\omega_3\omega_2\mathbf{e}_1 = \frac{1}{2}mr^2\omega_3\omega_2\mathbf{e}_1$$

The Euler equations (3.6.16) in this case yield:

$$\left.\begin{array}{l}
m_{C1} = I_1\dot{w}_1 + (I_3 - I_2)w_3w_2 = I_1w_3w_2 + (I_3 - I_2)w_3w_2 = I_3w_3w_2 \\
m_{C2} = I_2\dot{w}_2 + (I_1 - I_3)w_1w_3 = 0 + 0 = 0 \\
m_{C3} = I_3\dot{w}_3 + (I_2 - I_1)w_2w_1 = 0 + 0 = 0 \\
\Rightarrow \mathbf{m}_t = I_3\omega_3\omega_2\mathbf{e}_1 = \frac{1}{2}mr^2\omega_3\omega_2\mathbf{e}_1
\end{array}\right\} \Rightarrow$$

3.7 Q-Rotation of Vectors and Tensors of Second Order

Let $\mathbf{Q}$ be an orthogonal tensor of second order and $\mathbf{a}$ a vector. It follows from Eq. (3.5.16) that the vector $\mathbf{b} = \mathbf{Q} \cdot \mathbf{a}$ represents a rotation of the vector $\mathbf{a}$, such that the two vectors $\mathbf{a}$ and $\mathbf{b}$ have the same magnitude:

$$\mathbf{b} = \mathbf{Q} \cdot \mathbf{a} \Leftrightarrow |\mathbf{b}| = |\mathbf{a}| \qquad (3.7.1)$$

We call the vector $\mathbf{b}$ the Q-*rotation* of the vector $\mathbf{a}$. It follows from the discussion in Sect. 3.5.2 that there exists a direction given by the unit vector $\mathbf{e}$ such that:

$$\mathbf{Q} \cdot \mathbf{e} = \mathbf{e} \qquad (3.7.2)$$

The unit vector $\mathbf{e}$ represents the *axis of rotation* related to the tensor $\mathbf{Q}$. The vector mapping (3.7.1) represents a rotation of all vectors $\mathbf{a}$, as if they were fixed in a rigid body rotating an angle θ about an axis of rotation parallel to the unit vector **e**. We use the formula (3.5.20) to express the *rotation angle* θ:

$$\cos \theta = \frac{1}{2}(\mathrm{tr}\,\mathbf{Q} - 1) \qquad (3.7.3)$$

Let **A** be a tensor of second order. The second order tensor $\mathbf{B} = \mathbf{Q}\,\mathbf{A}\,\mathbf{Q}^T$ is called the **Q**-*rotation* of the tensor **A**:

$$\mathbf{B} = \mathbf{Q}\,\mathbf{A}\,\mathbf{Q}^T \quad \text{the Q-rotation of the tensor A} \qquad (3.7.4)$$

We shall investigate the properties of the tensor **B** if the tensor **A** is either an orthogonal tensor or a symmetric tensor.

Theorem 3.1 *If* **A** *is an orthogonal tensor that represents a rotation* ϕ *with an axis of rotation parallel to a unit vector* **a**, *then the* **Q**-*rotation of* **A** *is a new orthogonal tensor* $\mathbf{B} = \mathbf{QAQ}^T$ *with the same rotation* ϕ *as* **A**, *and with an axis of rotation parallel to the unit vector* $\mathbf{b} = \mathbf{Q} \cdot \mathbf{a}$. *The proof of the theorem is given as Problem 3.5.*

Theorem 3.2 *If* **A** *is a symmetric tensor with principal values* α_k *and principal directions* $\mathbf{a}_k$. *Then the* **Q**-*rotation of* **A** *is a new symmetric tensor* $\mathbf{B} = \mathbf{QAQ}^T$ *with the same principal values* α_k *as* **A** *and with principal directions given by* $\mathbf{b}_k = \mathbf{Q} \cdot \mathbf{a}_k$. *The proof of the theorem is given as Problem 3.6.*

3.8   Polar Decomposition

A second order tensor **F** is called *non-singular* if the determinant of **F** is non-zero, i.e. det $\mathbf{F} \neq 0$.

The *polar decomposition theorem* states that:

A non-singular second order tensor **F** having a positive determinant, i.e. det $\mathbf{F} > 0$, can always be expressed as a composition of an orthogonal tensor **R** and a positive definite symmetric tensor **U**, or as a composition of a positive definite symmetric tensor **V** and the same orthogonal tensor **R**:

$$\mathbf{F} = \mathbf{RU} = \mathbf{VR} \qquad (3.8.1)$$

The tensors **R**, **U**, and **V** are uniquely determined by **F** through the relations:

$$\mathbf{U} = \sqrt{\mathbf{F}^T\mathbf{F}}, \quad \mathbf{V} = \sqrt{\mathbf{F}\mathbf{F}^T}, \quad \mathbf{R} = \mathbf{F}\mathbf{U}^{-1} \qquad (3.8.2)$$

An algebraic proof of the theorem will now be provided. In Sect. 4.5 on large deformations a geometrical proof of the polar decomposition theorem is presented.

The composition $\mathbf{F}^T\mathbf{F}$ is symmetric since: $(\mathbf{F}^T\mathbf{F})^T = \mathbf{F}^T\mathbf{F}$, and positive definite. The latter property is demonstrated as follows. For any vector $\mathbf{c}$:

$$\mathbf{c}\cdot(\mathbf{F}^T\mathbf{F})\cdot\mathbf{c} = c_i(F_{ki}F_{kj})c_j = (F_{ki}c_i)(F_{kj}c_j) = \text{sum of squares} > 0$$

Then according to the definition (3.3.33) $\mathbf{F}^T\mathbf{F}$ is a positive definite tensor. Now since $\mathbf{F}^T\mathbf{F}$ is a symmetric, positive definite tensor, we can through the relation (3.3.34) determine the symmetric and positive definite tensors:

$$\mathbf{U} = \sqrt{\mathbf{F}^T\mathbf{F}} \equiv (\mathbf{F}^T\mathbf{F})^{1/2}, \quad \mathbf{U}^{-1} = (\mathbf{F}^T\mathbf{F})^{-1/2} \qquad (3.8.3)$$

Next we will show that the composition $\mathbf{R} = \mathbf{F}\mathbf{U}^{-1}$ is an orthogonal tensor. First we find that, since $(\mathbf{U}^{-1})^T = \mathbf{U}^{-1}$:

$$\mathbf{R}\mathbf{R}^T = (\mathbf{F}\mathbf{U}^{-1})(\mathbf{U}^{-1}\mathbf{F}^T) = \mathbf{F}(\mathbf{U}^2)^{-1}\mathbf{F}^T = \mathbf{F}(\mathbf{F}^T\mathbf{F})^{-1}\mathbf{F}^T = (\mathbf{F}\mathbf{F}^{-1})(\mathbf{F}^{-T}\mathbf{F}^T)$$
$$= \mathbf{1}\mathbf{1} = \mathbf{1}$$

Hence:

$$\det(\mathbf{R}\mathbf{R}^T) = \det(\mathbf{R})\det(\mathbf{R}^T) = (\det\mathbf{R})^2 = 1 \Rightarrow \det\mathbf{R} = \pm 1$$

Now, since $\det\mathbf{F}$ by assumption is positive and the determinant of the positive definite tensor $\mathbf{U}^{-1}$ is positive, we get the result:

$$\det\mathbf{R} = (\det\mathbf{F})(\det\mathbf{U}^{-1}) > 0$$

Thus we have shown that $\det\mathbf{R} = +1$, and $\mathbf{R}$ is called a *proper orthogonal tensor*. If $\det\mathbf{F} < 0$, it follows that $\det\mathbf{R} = -1$, and $\mathbf{R}$ is called an *improper orthogonal tensor*. This completes the proof of the decomposition $\mathbf{F} = \mathbf{R}\mathbf{U}$.

We shall then prove that the decomposition $\mathbf{F} = \mathbf{R}\mathbf{U}$ is unique. Let us assume that two decompositions are possible:

$$\mathbf{F} = \mathbf{R}\mathbf{U} = \mathbf{R}_1\mathbf{U}_1$$

Then:

$$\mathbf{U}^2 = \mathbf{U}\mathbf{1}\mathbf{U} = \mathbf{U}(\mathbf{R}^T\mathbf{R})\mathbf{U} = (\mathbf{U}\mathbf{R}^T)(\mathbf{R}\mathbf{U}) = (\mathbf{R}\mathbf{U})^T(\mathbf{R}\mathbf{U}) = \mathbf{F}^T\mathbf{F}$$
$$= (\mathbf{R}_1\mathbf{U}_1)^T(\mathbf{R}_1\mathbf{U}_1) = \mathbf{U}_1^2$$
$$\Rightarrow \mathbf{U}_1 = \mathbf{U}$$

The implication follows from the fact that the square roots of the positive definite tensors $\mathbf{U}^2$ and $\mathbf{U}_1^2$ are unique tensors. Next we find that $\mathbf{R}_1 = \mathbf{F}\mathbf{U}_1^{-1} = \mathbf{F}\mathbf{U}^{-1} = \mathbf{R}$. Hence, the decomposition $\mathbf{F} = \mathbf{R}\mathbf{U}$ is unique.

The decomposition $\mathbf{F} = \mathbf{VR}$ and its uniqueness may be shown similarly. The relation between $\mathbf{U}$ and $\mathbf{V}$ is found as follows.

$$\mathbf{F}\mathbf{R}^T = (\mathbf{RU})\mathbf{R}^T = (\mathbf{VR})\mathbf{R}^T = \mathbf{V}(\mathbf{R}\mathbf{R}^T) = \mathbf{V} \Rightarrow$$
$$\mathbf{V} = \mathbf{RU}\mathbf{R}^T \Leftrightarrow \mathbf{U} = \mathbf{R}^T\mathbf{V}\mathbf{R}$$

(3.8.4)

We see that $\mathbf{V}$ is the $\mathbf{R}$-rotation of $\mathbf{U}$. The symmetric tensors $\mathbf{U}$ and $\mathbf{V}$ have the same principal values, and the principal directions of $\mathbf{V}$ are $\mathbf{R}$-rotations of the principal directions of $\mathbf{U}$. An application of the polar decomposition theorem and a geometrical interpretation of the properties of $\mathbf{R}$, $\mathbf{U}$, and $\mathbf{V}$ are presented in Sect. 4.5 on large deformations.

3.9   Isotropic Functions of Tensors

Let $\gamma[\mathbf{A}]$ be scalar-valued function of a second order tensor $\mathbf{A}$. The function $\gamma[\mathbf{A}]$ is called an *isotropic scalar-valued function* of $\mathbf{A}$ if:

$$\gamma[\mathbf{Q}\mathbf{A}\mathbf{Q}^T] = \gamma[\mathbf{A}] \quad \text{for all orthogonal tensors } \mathbf{Q}$$

(3.9.1)

If $\mathbf{S}$ is a second order symmetric tensor with principal values σ_k and principal invariants I, II, and III, the isotropic scalar-valued function $\gamma[\mathbf{S}]$ has the alternative representations:

$$\gamma[\mathbf{S}] = \gamma(\sigma_1, \sigma_2, \sigma_3), \quad \gamma[\mathbf{S}] = \gamma(I_1, I_2, I_3)$$

(3.9.2)

These results are found as follows. Let the orthogonal tensor $\mathbf{Q}$ be chosen such that the matrix of the tensor $\mathbf{Q}\mathbf{S}\mathbf{Q}^T$ is $(\sigma_i \delta_{ij})$ (no summation). Equation (3.9.1) then implies the representation $(3.9.2)_1$. The representation $(3.9.2)_2$ then follows from the fact that the principal values σ_k according to Eq. (3.3.25) are unique functions of the principal invariants I, II, III.

Let $\mathbf{B}[\mathbf{A}]$ be a second order tensor-valued function of a second order tensor $\mathbf{A}$. If:

$$\mathbf{B}[\mathbf{Q}\mathbf{A}\mathbf{Q}^T] = \mathbf{Q}\mathbf{B}[\mathbf{A}]\mathbf{Q}^T \quad \text{for all orthogonal tensors } \mathbf{Q}$$

(3.9.3)

then $\mathbf{B}[\mathbf{A}]$ is called an *isotropic second order tensor-valued function* of $\mathbf{A}$.

We now assume that both the argument tensor $\mathbf{A}$ and the tensor value $\mathbf{B}$ of the function $\mathbf{B}[\mathbf{A}]$ are symmetric tensors with principal values and principal direction given by α_k, $\mathbf{a}_k$ for $\mathbf{A}$ and β_k, $\mathbf{b}_k$ for $\mathbf{B}$. The property of isotropy (3.9.3) results in a special representation of the second tensor-valued function $\mathbf{B}[\mathbf{A}]$. First we shall prove the theorem.

Theorem 3.3 (a) *An isotropic, symmetric second tensor-valued function* $\mathbf{B}[\mathbf{A}]$ *of a symmetric second order tensor* $\mathbf{A}$ *is coaxial to the argument tensor* $\mathbf{A}$. (b) *If* $\mathbf{A}$ *is a* *symmetric tensor of second order, and* **B**[**A**] *is a symmetric second tensor-valued function of* **A***, then* **B**[**A**] *is an isotropic, symmetric second tensor-valued function of* **A***.*

Proof of Part (a) of Theorem 3.3 We need to show that any principal direction of **A**, here denoted $\mathbf{a}_3$, also is a principal direction of **B**. We choose a particular **Q**-rotation that has an axis of rotation parallel to $\mathbf{a}_3$ and the angle of rotation $\theta = 180°$. The **Q**-rotation of **A**, i.e. $\bar{\mathbf{A}} = \mathbf{Q}\,\mathbf{A}\,\mathbf{Q}^T$, has the same principal values as **A**, $\bar{\alpha}_k = \alpha_k$, and has the principal directions:

$$\bar{\mathbf{a}}_3 = \mathbf{Q} \cdot \mathbf{a}_3 = \mathbf{a}_3, \quad \bar{\mathbf{a}}_\alpha = \mathbf{Q} \cdot \mathbf{a}_\alpha = -\mathbf{a}_\alpha$$

The general representation (3.3.30) of a symmetric tensor of second order now shows that $\bar{\mathbf{A}} = \mathbf{A}$:

$$\bar{\mathbf{A}} = \sum_k \bar{\alpha}_k\, \bar{\mathbf{a}}_k \otimes \bar{\mathbf{a}}_k = \sum_k \alpha_k\, \mathbf{a}_k \otimes \mathbf{a}_k = \mathbf{A}$$

Then the property (3.9.3) implies that:

$$\mathbf{B} \equiv \mathbf{B}[\mathbf{A}] = \mathbf{B}\big[\bar{\mathbf{A}}\big] = \mathbf{Q}\,\mathbf{B}[\mathbf{A}]\mathbf{Q}^T \Rightarrow \mathbf{B}\mathbf{Q} = \mathbf{Q}\mathbf{B} \Rightarrow \mathbf{B}\mathbf{Q} \cdot \mathbf{a}_3 = \mathbf{Q}\mathbf{B} \cdot \mathbf{a}_3 \Rightarrow$$
$$\mathbf{B} \cdot \mathbf{a}_3 = \mathbf{Q}(\mathbf{B} \cdot \mathbf{a}_3)$$

This result shows that the vector $\mathbf{B} \cdot \mathbf{a}_3$ is not influenced by the special **Q**-rotation we have chosen. Thus the two vector $\mathbf{B} \cdot \mathbf{a}_3$ and $\mathbf{a}_3$ are parallel. Any principal direction of the argument tensor **A**, here represented by $\mathbf{a}_3$, is therefore also a principal direction of the tensor-valued function **B**[**A**]. This implies that **B** and **A** are coaxial tensors, and part (a) of Theorem 3.3 is thus proved.

Proof of Part (b) of Theorem 3.3 Since the two tensors **A** and **B**[**A**] are coaxial, they have the same principal directions $\mathbf{a}_k$, and the principal values of **B** must be scalar-valued functions of the principal values of **A**. Hence the tensor **B** may be expressed by:

$$\mathbf{B}[\mathbf{A}] = \sum_k \beta_k(\alpha)\, \mathbf{a}_k \otimes \mathbf{a}_k, \quad \beta_k(\alpha) \equiv \beta_k(\alpha_1, \alpha_2, \alpha_3)$$

Let **Q** be an orthogonal tensor. Then since the tensor $\mathbf{Q}\mathbf{A}\mathbf{Q}^T$ has the same principal values as the tensor **A** but principal directions that are **Q**-rotations of the principal directions of **A**, we obtain:

$$\mathbf{B}\big[\mathbf{Q}\,\mathbf{A}\,\mathbf{Q}^T\big] = \sum_k \beta_k(\alpha)(\mathbf{Q} \cdot \mathbf{a}_k) \otimes (\mathbf{Q} \cdot \mathbf{a}_k) = \mathbf{Q}\left(\sum_K \beta_k(\alpha)\mathbf{a}_k \otimes \mathbf{a}_k\right)\,\mathbf{Q}^T$$
$$= \mathbf{Q}\,\mathbf{B}[\mathbf{A}]\mathbf{Q}^T$$

The result proves that $\mathbf{B}[\mathbf{A}]$ is an isotropic, symmetric second order tensor-valued function of $\mathbf{A}$. This completes the proof of Theorem 3.3.

Next we shall prove an important theorem giving the most general representation of a symmetric tensor-valued function of a second order symmetric tensor.

Theorem 3.4 *Let $\mathbf{B}[\mathbf{A}]$ be a second order symmetric tensor-valued function of a second symmetric tensor $\mathbf{A}$. The function $\mathbf{B}[\mathbf{A}]$ is then isotropic if and only if the function has the representation*:

$$\mathbf{B}[\mathbf{A}] = \gamma_0\,\mathbf{1} + \gamma_1\,\mathbf{A} + \gamma_2\,\mathbf{A}^2, \tag{3.9.4}$$

where γ_0, γ_1, and γ_2 are isotropic scalar-valued function of $\mathbf{A}$.

Proof that: (3.9.4) $\Rightarrow$ (3.9.3). The functions γ_0, γ_1, and γ_2 are isotropic scalar-valued functions of $\mathbf{A}$. Thus according to the definition (3.9.3):

$$\gamma_i\left[\mathbf{QAQ}^T\right] = \gamma_i[\mathbf{A}] \quad \text{for } i = 0,\ 1,\ 2 \ \text{ and for all orthogonal tensors } \mathbf{Q}$$

Then from Eq. (3.9.4):

$$\mathbf{B}\left[\mathbf{Q\,A\,Q}^T\right] = \gamma_0[\mathbf{A}]\mathbf{1} + \gamma_1[\mathbf{A}]\mathbf{Q\,A\,Q}^T + \gamma_2[\mathbf{A}]\left(\mathbf{Q\,A\,Q}^T\right)\left(\mathbf{Q\,A\,Q}^T\right) \Rightarrow$$
$$= \mathbf{Q}\left(\gamma_0[\mathbf{A}]\mathbf{1} + \gamma_1[\mathbf{A}]\mathbf{A} + \gamma_2[\mathbf{A}]\mathbf{A}^2\right)\mathbf{Q}^T \Rightarrow \mathbf{B}\left[\mathbf{Q\,A\,Q}^T\right] = \mathbf{Q\,B}[\mathbf{A}]\mathbf{Q}^T \Rightarrow (3.9.3)$$

Proof that: (3.9.3) $\Rightarrow$ (3.9.4). Theorem 3.1 and Eq. (3.9.3) imply that the principal values β_k of $\mathbf{B}$ by the function $\mathbf{B}\ [\mathbf{A}]$ are functions of the principal values α_k of the argument tensor $\mathbf{A}$.

$$\beta_k = \beta_k(\alpha) \equiv \beta_k(\alpha_1, \alpha_2, \alpha_3) \tag{3.9.5}$$

Let us first assume that the principal values α_k are all unequal. The following three equations for the three unknown scalars γ_0, γ_1, and γ_2 will then have a unique solution.

$$\gamma_0 + \gamma_1\,\alpha_k + \gamma_2(\alpha_k)^2 = \beta_k \quad k = 1, 2, 3 \tag{3.9.6}$$

where β_k are given by Eq. (3.9.5). A unique solution of Eq. (3.9.6) is secured because the determinant of the coefficients is different from zero:

$$\det\begin{pmatrix} 1 & \alpha_1 & (\alpha_1)^2 \\ 1 & \alpha_2 & (\alpha_2)^2 \\ 1 & \alpha_3 & (\alpha_3)^2 \end{pmatrix} = (\alpha_1 - \alpha_2)(\alpha_2 - \alpha_3)(\alpha_3 - \alpha_1) \neq 0 \tag{3.9.7}$$

The scalars γ_i are according to Eq. (3.9.6) functions of the principal values α_k, alternatively of the principal invariants I, II, and III of $\mathbf{A}$. From the general representation of a tensor of second order (3.3.30) we now obtain:

$$\mathbf{B} = \sum_k \beta_k \mathbf{a}_k \otimes \mathbf{a}_k = \gamma_0 \sum_k \mathbf{a}_k \otimes \mathbf{a}_k + \gamma_1 \sum_k \alpha_k \mathbf{a}_k \otimes \mathbf{a}_k + \gamma_2 \sum_k (\alpha_k)^2 \mathbf{a}_k \otimes \mathbf{a}_k$$

which may be organized to:

$$\mathbf{B}[\mathbf{A}] = \gamma_0 \mathbf{1} + \gamma_1 \mathbf{A} + \gamma_2 \mathbf{A}^2 \qquad (3.9.8)$$

If any two principal values are equal, say $\alpha_2 = \alpha_3$, then any direction $\mathbf{a}$ normal to the principal direction $\mathbf{a}_1$ is a principal direction of $\mathbf{A}$, confer Sect. 2.3.1, and according to Theorem 3.1 also a principal direction of $\mathbf{B}$. This means that $\beta_2 = \beta_3$. The following two equations for the two unknown scalar-valued functions ϕ_0 and ϕ_1 of α_k have a unique solution:

$$\phi_0 + \phi_1 \alpha_\rho = \beta_\rho \quad \rho = 1, 2 \qquad (3.9.9)$$

The representation (3.9.5) now yields:

$$\mathbf{B}[\mathbf{A}] = \phi_0 \mathbf{1} + \phi_1 \mathbf{A} \qquad (3.9.10)$$

If all three principal values α_k are equal: $\alpha_1 = \alpha_2 = \alpha_3 = \alpha$, the argument tensor $\mathbf{A}$ and thus the function tensor $\mathbf{B}$ are both isotropic tensors, and we shall find:

$$\mathbf{B}[\mathbf{A}] = \psi_0 \mathbf{1} \quad \text{for } \mathbf{A} = \alpha \mathbf{1} \qquad (3.9.11)$$

where ψ_0 is a function of α. Equations (3.9.8, 3.9.10, 3.9.11) show that Eq. (3.9.3) implies Eq. (3.9.4). This completes the proof of Theorem 3.4.

It may be shown that:

$$\phi_0 = \gamma_0 - \alpha_1 \alpha_2 \gamma_2, \quad \phi_1 = \gamma_1 + (\alpha_1 + \alpha_2)\gamma_2 \quad \text{for } \alpha_2 = \alpha_3 \qquad (3.9.12)$$

$$\psi_0 = \gamma_0 + \gamma_1 \alpha + \gamma_2 \alpha^2 \quad \text{for } \alpha_1 = \alpha_2 = \alpha_3 = \alpha \qquad (3.9.13)$$

The proof of Theorem 3.4 is due to Serrin [3], who also shows that if the function $\mathbf{B}[\mathbf{A}]$ is three times differentiable with respect to $\mathbf{A}$, then the scalars γ_i are continuous functions of the principal invariants of $\mathbf{A}$.

An alternative form of the function (3.9.4) may be derived using the Cayley–Hamilton theorem (3.3.36), which gives:

$$\mathbf{A}^2 = I \, \mathbf{A} - II \, \mathbf{1} + III \, \mathbf{A}^{-1}$$

When this expression for $\mathbf{A}^2$ is substituted into the function (3.9.4), we obtain:

$$\mathbf{B}[\mathbf{A}] = \lambda_0 \mathbf{1} + \lambda_1 \mathbf{A} + \lambda_{-1} \mathbf{A}^{-1} \qquad (3.9.14)$$

where λ_i are isotropic scalar-valued functions of $\mathbf{A}$.

$$\lambda_0 = \gamma_0 - II\,\gamma_2, \quad \lambda_1 = \gamma_1 + I\,\gamma_2, \quad \lambda_{-1} = III\,\gamma_2 \qquad (3.9.15)$$

If $\mathbf{B}[\mathbf{A}]$ is a linear function of $\mathbf{A}$, it follows from the general expression (3.9.4) that the function takes the form:

$$\mathbf{B}[\mathbf{A}] = (\gamma + \lambda\,\mathrm{tr}\,\mathbf{A})\,\mathbf{1} + 2\mu\,\mathbf{A} \qquad (3.9.16)$$

where γ, λ, and μ are constant scalars. An alternative form of the expression (3.9.16) is:

$$\mathbf{B}[\mathbf{A}] = \gamma\,\mathbf{1} + \mathbf{I}_4^s : \mathbf{A} \qquad (3.9.17)$$

where $\mathbf{I}_4^s$ is the symmetric fourth order isotropic tensor presented in formula (3.2.45).

$$\mathbf{I}_4^s = 2\mu\,\mathbf{1}_4^s + \lambda\,\mathbf{1} \otimes \mathbf{1} \Leftrightarrow I_{4ijkl}^s = \mu\big(\delta_{ik}\delta_{jl} + \delta_{il}\delta_{jk}\big) + \lambda\,\delta_{ij}\delta_{kl} \qquad (3.9.18)$$

If each of the two symmetric tensors $\mathbf{A}$ and $\mathbf{B}$ are decomposed into trace–free deviators: $\mathbf{A}'$ and $\mathbf{B}'$, and isotrops: $\mathbf{A}^o$ and $\mathbf{B}^o$, as shown by Eqs. (3.3.39–3.3.41), the linear function (3.9.16) may be decomposed into:

$$\mathbf{B}' = 2\mu\,\mathbf{A}', \quad \mathbf{B}^o = 3\kappa\mathbf{A}^o + \gamma\mathbf{1} \qquad (3.9.19)$$

where μ, κ, and γ are scalars. The derivation of the result (3.9.19) is given as Problem 3.5. From the expressions (3.9.19) we obtain an alternative form of the function (3.9.16):

$$\mathbf{B}[\mathbf{A}] = 2\mu\,\mathbf{A} + \left(\kappa - \frac{2}{3}\mu\right)(\mathrm{tr}\,\mathbf{A})\mathbf{1} + \gamma\mathbf{1} \qquad (3.9.20)$$

The linear functions (3.9.16), (3.9.19), and (3.9.20) are very important in constitutive modelling of linear, isotropic materials. An example will be the generalized Hooke's law for isotropic linearly elastic materials, presented in Sect. 5.2.

Problems 3.1–3.5 with solutions see Appendix

References

1. Malvern LE (1969) Introduction to the mechanics of a continuous medium. Prentice-Hall Inc., Englewood Cliffs
2. Jaunzemis W (1967) Continuum mechanics. MacMillan, New York
3. Serrin J (1959) The derivation of the stress-deformation relations for a Stokesian fluid. J Math Mech 8:459–469

Chapter 4
Deformation Analysis

4.1 Strain Measures

The word strain is used about local deformation in a material, i.e. deformation in the neighbourhood of a particle. Strain represents changes of the lengths of material line elements, the angles between material line elements, and the volumes of material volume elements. Below we define three primary concepts of strain: *longitudinal strain* ε, *shear strain* γ, and *volumetric strain* ε_v. Strains are primarily due to mechanical stresses and temperature changes in the material. But strain may also have contributions from other effects. For instance, changes in the water content in wood and in some plastics lead to swelling or shrinking, which may introduce both strains and stresses in the material.

Figure 4.1 shows a body in the reference configuration K_0 at time t_0 and in the present configuration K at time t. The motion of the body is given by the place vector $\mathbf{r}(\mathbf{r}_0, t)$, where $\mathbf{r}_0$ (or X) represents an arbitrarily chosen particle. Alternatively modelling the motion may be given by the displacement vector $\mathbf{u}(\mathbf{r}_0, t)$:

$$\mathbf{u}(\mathbf{r}_0, t) = \mathbf{r}(\mathbf{r}_0, t) - \mathbf{r}_0 \tag{4.1.1}$$

At the particle $\mathbf{r}_0$ we select a material line element, which in K_0 is a straight line of length $s_0 = P_0 Q_0$, and with a direction given by the unit vector $\mathbf{e}$. In K the line element will in general have changed its length, which is denoted by $s = PQ$, and also have got a curved form. The *longitudinal strain ε in the direction $\mathbf{e}$ in a particle* $\mathbf{r}_0$ is defined by:

$$\varepsilon = \lim_{s_0 \to 0} \frac{s - s_0}{s_0} = \frac{ds - ds_0}{ds_0} = \frac{ds}{ds_0} - 1 \tag{4.1.2}$$

The strain ε represents change of length per unit length of undeformed line element in the direction of $\mathbf{e}$ in particle $\mathbf{r}_0$. The longitudinal strain is also called the *normal strain*.

© Springer Nature Switzerland AG 2019
F. Irgens, *Tensor Analysis*,
https://doi.org/10.1007/978-3-030-03412-2_4

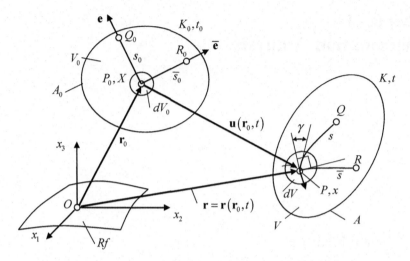

Fig. 4.1 General deformation of a body. Reference *Rf*. Reference configuration K_0, particle $\mathbf{r}_0$ (or X) at the position P_0, and underformed material lines P_0Q_0 and P_0R_0 at the time t_0. Volume V, volume element dV. Surface area A. Present configuration K and particle $\mathbf{r}_0$ (or X) at the position P (or place $\mathbf{r}$), and deformed material lines PQ and PR at the time t. Volume V_0, volume element dV_0. Surface area A_0

The *shear strain γ in a particle $\mathbf{r}_0$ with respect to two material elements*, which in K_0 are orthogonal, is defined as the change in the right angle between the line elements, and measured in radians. The shear strain is defined to be positive when the angle is reduced. Figure 4.1 illustrates the shear strain γ in the particle $\mathbf{r}_0$ between material line elements that in K_0 have the directions $\mathbf{e}$ and $\bar{\mathbf{e}}$.

A small body surrounding the particle $\mathbf{r}_0$ has the volume ΔV_0 and is deformed and obtains the volume ΔV in the configuration K. See Fig. 4.1. The *volumetric strain ε_v in a particle $\mathbf{r}_0$* is defined by:

$$\varepsilon_v = \lim_{\Delta V_o \to 0} \frac{\Delta V - \Delta V_0}{\Delta V_0} \qquad (4.1.3)$$

Volumetric strain ε_v represents change in volume per unit undeformed volume element surrounding the particle $\mathbf{r}_0$.

4.2   Green's Strain Tensor

In this section the primary measures of strain, introduced above, will be expressed in terms of the displacement vector $\mathbf{u}(\mathbf{r}_0, t)$. We look at the situation illustrated in Fig. 4.1 and again presented in Fig. 4.2. The length s_0 of the line element P_0Q_0 is now chosen to be a curve parameter for the straight material line P_0Q_0 and also for

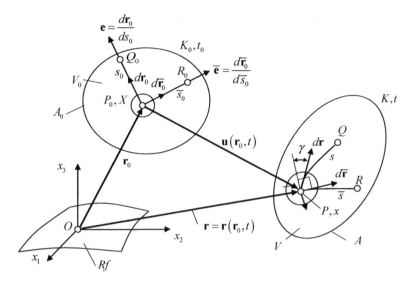

Fig. 4.2 General deformation of a body. Reference *Rf.* Orthogonal line elements $d\mathbf{r}_0$ and $d\bar{\mathbf{r}}_0$ from the particle $\mathbf{r}_0$ (or X) at the position P_0 in underformed material at the time t_0. Line elements $d\mathbf{r}$ and $d\bar{\mathbf{r}}$ from position P at the place $\mathbf{r}$ (or x) in the deformed material at the time t

the deformed material line PQ in K. The length s of the deformed line PQ then becomes a function of s_0. The unit vector $\mathbf{e}$ in the direction of P_0Q_0 has the components e_i in the Cartesian coordinate system Ox. The coordinates of the points Q_0 and Q are $X_i + s_0\,e_i$ and $x_i(X + s_0\,e, t)$ respectively. With s_0 as curve parameter, the arc length formula (1.4.5) gives for the length s of the line PQ:

$$s(\mathbf{r}_0, s_0) = \int_0^{s_0} \sqrt{\frac{\partial x_i}{\partial s_0}\frac{\partial x_i}{\partial s_0}}\,ds_0 \tag{4.2.1}$$

Let ds_0 be the length of the line element $d\mathbf{r}_0$ in P_0 and along P_0Q_0, and let dX_k be the components of $d\mathbf{r}_0$:

$$d\mathbf{r}_0 = \mathbf{e}\,ds_0 = dX_k\,\mathbf{e}_k \Rightarrow |d\mathbf{r}_0| = ds_0, \quad dX_k = e_k\,ds_0 \Leftrightarrow e_k = \frac{dX_k}{ds_0} \tag{4.2.2}$$

It follows that:

$$\frac{ds}{ds_0} = \sqrt{\frac{\partial x_i}{\partial s_0}\frac{\partial x_i}{\partial s_0}} \Rightarrow (ds)^2 = \left(\frac{\partial x_i}{\partial s_0}\frac{\partial x_i}{\partial s_0}\right)(ds_0)^2 \tag{4.2.3}$$

Note that e_k are components of the vector $\mathbf{e}$, while $\mathbf{e}_k$ are the base vector of the coordinate system, i.e. $\mathbf{e} = e_k\,\mathbf{e}_k$. The vector differential $d\mathbf{r}$, with components dx_i

represents a line element in P, which is a tangent vector to the curve PQ and defined by:

$$dr = \frac{\partial \mathbf{r}}{\partial s_0} ds_0 \Rightarrow dx_i = \frac{\partial x_i}{\partial s_0} ds_0 \qquad (4.2.4)$$

From Eqs. (4.2.2) and (4.2.4) it follows that ds is the length of the differential line element $d\mathbf{r}$:

$$|d\mathbf{r}| = ds \qquad (4.2.5)$$

The differential line elements $d\mathbf{r}_0$ and $d\mathbf{r}$ do not in general represent the same material line. That is only possible if the curve PQ is a straight line in the neighbourhood of point P.

The relation between the line elements $d\mathbf{r}_0$ in K_0 and dr in K is determined as follows.

$$d\mathbf{r} = \frac{\partial \mathbf{r}}{\partial \mathbf{r}_0} \cdot d\mathbf{r}_0 \Leftrightarrow dx_i = \frac{\partial x_i}{\partial X_k} dX_k \qquad (4.2.6)$$

The coordinate invariant form $\partial \mathbf{r}/\partial \mathbf{r}_0$ is a symbol for a tensor: Grad $\mathbf{r}$, with components $\partial x_i/\partial X_k$. This tensor is called the *deformation gradient tensor* and denoted by F.

$$\mathbf{F} = \text{Grad}\,\mathbf{r} = \frac{\partial \mathbf{r}}{\partial \mathbf{r}_o} \Leftrightarrow F_{ik} = \frac{\partial x_i}{\partial X_k} \qquad (4.2.7)$$

The relation (4.2.6) is now written as:

$$d\mathbf{r} = \mathbf{F} \cdot d\mathbf{r}_o \Leftrightarrow dx_i = F_{ik}\, dX_k = F_{ik}\, e_k\, ds_0 \qquad (4.2.8)$$

From the coordinate expressions $x_i(X + s_o e, t)$ and the definitions (4.2.7) we obtain:

$$\frac{\partial x_i(X+s_o e,t)}{\partial s_o}\bigg|_{s_o=0} = \frac{\partial x_i(X,t)}{\partial X_k}\frac{d(X_k+s_o e_k)}{ds_o}\bigg|_{s_o=0} \Rightarrow \frac{\partial x_i}{\partial s_o} = \frac{\partial x_i}{\partial X_k}\frac{dX_k}{ds_o} = F_{ik}\, e_k$$

Thus we have from Eq. (4.2.2) the result:

$$(ds)^2 = \left(\frac{\partial x_i}{\partial s_o}\frac{\partial x_i}{\partial s_o}\right)(ds_0)^2 = (F_{ik}\, e_k)(F_{il}\, e_l)(ds_0)^2 = (\mathbf{e}\cdot(\mathbf{F}^T\mathbf{F})\cdot\mathbf{e})(ds_0)^2 \qquad (4.2.9)$$
$$\Rightarrow ds = \sqrt{\mathbf{e}\cdot(\mathbf{F}^T\mathbf{F})\cdot\mathbf{e}}\, ds_o$$

We now define two symmetric tensors of second order: *Green's deformation tensor*
C and *Green's strain tensor* **E**, named after George Green [1793–1841]:

$$\mathbf{C} = \mathbf{F}^T\mathbf{F} \Leftrightarrow C_{kl} = F_{ik}F_{il} \qquad (4.2.10)$$

$$\mathbf{E} = \frac{1}{2}(\mathbf{C} - \mathbf{1}) \Leftrightarrow \mathbf{C} = \mathbf{1} + 2\mathbf{E} \qquad (4.2.11)$$

The result (4.2.9) may alternatively be expressed by:

$$ds = \sqrt{\mathbf{e} \cdot (\mathbf{F}^T\mathbf{F}) \cdot \mathbf{e}}\, ds_o = \sqrt{\mathbf{e} \cdot \mathbf{C} \cdot \mathbf{e}}\, ds_o = \sqrt{1 + 2\mathbf{e} \cdot \mathbf{E} \cdot \mathbf{e}}\, ds_o \qquad (4.2.12)$$

We are now ready to express the primary strain measures ε, γ, and ε_v, presented
in the previous section, by Green's strain tensor. According to Eqs. (4.1.2) and
(4.2.12) the longitudinal strain ε in direction **e** is given by:

$$\varepsilon = \frac{ds}{ds_o} - 1 = \sqrt{1 + 2\mathbf{e} \cdot \mathbf{E} \cdot \mathbf{e}} - 1 = \sqrt{1 + 2e_i E_{ik} e_k} - 1 \qquad (4.2.13)$$

In particular the longitudinal strain ε_{ii} (no summation) of a material line element
that in K_0 is parallel to the x_i − direction, is found from the formula (4.2.13) if **e** is
chosen to be the base vector $\mathbf{e}_i$.

$$\varepsilon_{ii} = \sqrt{1 + 2E_{ii}} - 1 \quad \text{(no summation w.r. to } i) \qquad (4.2.14)$$

In order to determine the shear strain γ with respect to two material line elements, which in K_0 are orthogonal and have directions $\bar{\mathbf{e}}$ and **e**, see Fig. 4.2, we
compute the scalar product of the tangent vectors $d\bar{\mathbf{r}}$ of *PR* and $d\mathbf{r}$ of *PQ*:

$$d\bar{\mathbf{r}} \cdot d\mathbf{r} = |d\bar{\mathbf{r}}||d\mathbf{r}| \sin\gamma$$
$$\Rightarrow \sin\gamma = \frac{d\bar{\mathbf{r}} \cdot d\mathbf{r}}{|d\bar{\mathbf{r}}||d\mathbf{r}|} \qquad (4.2.15)$$

Using Eqs. (4.2.9) and (4.2.12) we obtain:

$$|d\mathbf{r}| = ds = \sqrt{1 + 2\mathbf{e} \cdot \mathbf{E} \cdot \mathbf{e}}\, ds_o, \quad |d\bar{\mathbf{r}}| = d\bar{s} = \sqrt{1 + 2\bar{\mathbf{e}} \cdot \mathbf{E} \cdot \bar{\mathbf{e}}}\, d\bar{s}_o$$
$$d\bar{\mathbf{r}} \cdot d\mathbf{r} = (F_{ik}\bar{e}_k)(F_{il}e_l)d\bar{s}_0\, ds_o = \bar{\mathbf{e}} \cdot (\mathbf{F}^T\mathbf{F}) \cdot \mathbf{e}\, d\bar{s}_0\, ds_o = \bar{\mathbf{e}} \cdot (\mathbf{1} + 2\mathbf{E}) \cdot \mathbf{e}\, d\bar{s}_0\, ds_o$$
$$= 2\bar{\mathbf{e}} \cdot \mathbf{E} \cdot \mathbf{e}\, d\bar{s}_0\, ds_o$$

When these results are substituted into the formula (4.2.15), we obtain an expression for the shear strain γ:

$$\sin \gamma = \frac{2\bar{\mathbf{e}} \cdot \mathbf{E} \cdot \mathbf{e}}{\sqrt{(1+2\bar{\mathbf{e}} \cdot \mathbf{E} \cdot \bar{\mathbf{e}})(1+2\mathbf{e} \cdot \mathbf{E} \cdot \mathbf{e})}} = \frac{2\bar{\mathbf{e}} \cdot \mathbf{E} \cdot \mathbf{e}}{(1+\bar{\varepsilon})(1+\varepsilon)} \qquad (4.2.16)$$

$\bar{\varepsilon}$ is the longitudinal strain in the direction of $\bar{\mathbf{e}}$. The shear strain γ_{ij} for any two material line elements that in K_0 have directions $\bar{\mathbf{e}} = \mathbf{e}_i$ and $\mathbf{e} = \mathbf{e}_j$, is:

$$\sin \gamma_{ij} = \frac{2E_{ij}}{\sqrt{(1+2E_{ii})(1+2E_{jj})}} = \frac{2E_{ij}}{(1+\varepsilon_{ii})(1+\varepsilon_{jj})} \qquad i \neq j \qquad (4.2.17)$$

The longitudinal strains ε_{ii}, $i = 1, 2$, or 3, from Eq. (4.2.14) and the shear strains γ_{ij} from Eq. (4.2.17) will be called *coordinate strains*.

The volumetric strain ε_v may be expressed as follows. Figure 4.3 shows a material element surrounding a particle $\mathbf{r}_0$ in the position P_0 in the undeformed body and in position P in the deformed body. The volume dV_0 of the undeformed element is: $dV_0 = dX_1 \, dX_2 \, dX_3$. The volume dV of the deformed element is given by the box product of the three vectors $d\mathbf{r}_i$ in Fig. 4.3: $dV = [d\mathbf{r}_1 \, d\mathbf{r}_2 \, d\mathbf{r}_3]$. Using formula (4.2.4), we obtain:

$$d\mathbf{r}_1 = \frac{\partial \mathbf{r}}{\partial X_1} dX_1 = \frac{\partial x_i}{\partial X_1} \mathbf{e}_i \, dX_1, \quad d\mathbf{r}_2 = \frac{\partial \mathbf{r}}{\partial X_2} dX_2 = \frac{\partial x_j}{\partial X_2} \mathbf{e}_j \, dX_2,$$

$$d\mathbf{r}_3 = \frac{\partial \mathbf{r}}{\partial X_3} dX_3 = \frac{\partial x_k}{\partial X_3} \mathbf{e}_k \, dX_3$$

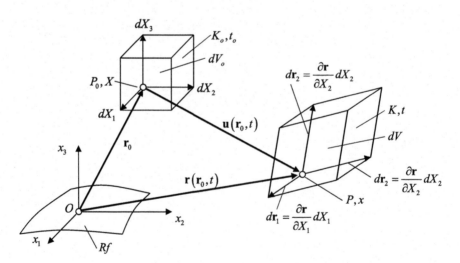

Fig. 4.3 Deformation of a material element of volume $dV_0 = dX_1 dX_2 dX_3$ surrounding a particle $\mathbf{r}_0$. The volume dV of deformed element is given by the box product of the line element $d\mathbf{r}_i = (\partial \mathbf{r}/\partial X_i)dX_i$

Therefore:

$$dV = [d\mathbf{r}_1\, d\mathbf{r}_2\, d\mathbf{r}_3] = e_{ijk}\left(\frac{\partial x_i}{\partial X_1}dX_1\right)\left(\frac{\partial x_j}{\partial X_2}dX_2\right)\left(\frac{\partial x_k}{\partial X_3}dX_3\right) = \det \mathbf{F}\, dV_0$$
$$= \sqrt{\det(\mathbf{1}+2\mathbf{E})}\, dV_o$$

Here we have used the formulas (4.2.10) and (4.2.11) for the development:

$$\det\left(\mathbf{F}\mathbf{F}^T\right) = \det \mathbf{F} \det \mathbf{F}^T = (\det \mathbf{F})^2 = \det \mathbf{C} = \det(\mathbf{1}+2\mathbf{E}) \quad \Rightarrow$$
$$\det \mathbf{F} = \sqrt{\det(\mathbf{1}+2\mathbf{E})}$$

The *volumetric strain* ε_v is then determined from the definition (4.1.3):

$$\varepsilon_v = \frac{dV - dV_o}{dV_o} = \det \mathbf{F} - 1 = \sqrt{\det(\mathbf{1}+2\mathbf{E})} - 1 \qquad (4.2.18)$$

It has now been demonstrated that the three primary strain measures ε, γ, and ε_v are all determined by the strain tensor $\mathbf{E}$, which again is determined by the deformation gradient $\mathbf{F}$. Most material models in Continuum Mechanics are defined by constitutive equations relating the stress tensor $\mathbf{T}$, through some functions or functionals, to the strain tensor $\mathbf{E}$ and other measures of deformation, which are derived from the deformation gradient $\mathbf{F}$.

It is sometimes convenient to express the strain measures in terms of the *displacement gradients* H_{ik} of the *displacement vector* $\mathbf{u}(\mathbf{r}_0, t)$ of the particle $\mathbf{r}_0$. From Fig. 4.2:

$$\mathbf{r} = \mathbf{r}(\mathbf{r}_0, t) = \mathbf{r}_0 + \mathbf{u}(\mathbf{r}_0, t) \Leftrightarrow x_i(X, t) = X_i + u_i(X, t) \qquad (4.2.19)$$

The displacement gradients H_{ik} represent a tensor $\mathbf{H}$ called the *displacement gradient tensor*. The tensor and its components are defined by:

$$\mathbf{H} = \frac{\partial \mathbf{u}}{\partial \mathbf{r}_o} \Leftrightarrow H_{ik} = \frac{\partial u_i}{\partial X_k} \qquad (4.2.20)$$

It follows from the formulas (4.2.11–4.2.12) that:

$$\frac{\partial x_i}{\partial X_k} = \frac{\partial X_i}{\partial X_k} + \frac{\partial u_i}{\partial X_k} \Leftrightarrow F_{ik} = \delta_{ik} + H_{ik} \Leftrightarrow \mathbf{F} = \mathbf{1} + \mathbf{H} \qquad (4.2.21)$$

From the formulas (4.2.10), (4.2.11), and (4.2.21) we obtain the results:

$$\mathbf{C} = \mathbf{F}^T\mathbf{F} = (\mathbf{1}+\mathbf{H}^T)(\mathbf{1}+\mathbf{H}) = \mathbf{1}+\mathbf{H}+\mathbf{H}^T+\mathbf{H}^T\mathbf{H} \qquad (4.2.22)$$

$$\mathbf{E} = \frac{1}{2}\left(\mathbf{H}+\mathbf{H}^T+\mathbf{H}^T\mathbf{H}\right) \Leftrightarrow E_{kl} = \frac{1}{2}\left(\frac{\partial u_k}{\partial X_l} + \frac{\partial u_l}{\partial X_k} + \frac{\partial u_i}{\partial X_k}\frac{\partial u_i}{\partial X_l}\right) \qquad (4.2.23)$$

A quantity related to the longitudinal strain ε is the *stretch* λ of a material line element and is defined as the ratio between the length ds of a deformed material line element to the length ds_0 of the undeformed element.

$$\lambda = \frac{ds}{ds_0} = 1 + \varepsilon \tag{4.2.24}$$

It follows by the definition that a stretch is always a positive quantity.

The stretch of a material line element that in K_0 is parallel to the x_i – direction will be called a *coordinate stretch* and is according to the definition (4.2.24) and Eq. (4.2.14):

$$\lambda_i = \sqrt{1 + 2E_{ii}} \tag{4.2.25}$$

All deformation effects and mechanical response related to the displacement gradient tensor $\mathbf{H}$, are most clearly exposed when we consider *homogeneous deformation*, which implies that the deformation gradient $\mathbf{F}$ and thus also the displacement gradient $\mathbf{H}$ are the same for all particles in the body, i.e. $\mathbf{F} = \mathbf{F}(t)$ and $\mathbf{H} = \mathbf{H}(t)$. Homogeneous deformations are given by the special motion: $\mathbf{r}(\mathbf{r}_0, t)$ or $\mathbf{u}(\mathbf{r}_0, t)$:

$$\mathbf{r}(\mathbf{r}_0, t) = \mathbf{u}_0(t) + \mathbf{F}(t) \cdot \mathbf{r}_0 \Leftrightarrow x_i(X, t) = u_{0i}(t) + F_{ik}(t) X_k \tag{4.2.26}$$

$$\mathbf{u}(\mathbf{r}_0, t) = \mathbf{r}(\mathbf{r}_0, t) - \mathbf{r}_0 = \mathbf{u}_0(t) + \mathbf{H}(t) \cdot \mathbf{r}_0$$
$$\Leftrightarrow u_i(X, t) = x_i(X, t) - X_i = u_{0i}(t) + H_{ik}(t) X_k \tag{4.2.27}$$

The displacement vector $\mathbf{u}_0(t)$ is the displacement of the particle $\mathbf{r}_0 = \mathbf{0}$, if that particle belongs to the body under consideration, but $\mathbf{u}_0(t)$ also represents a *translation* of the body. It may be shown, see Problem 4.1, that in a homogeneous deformation material planes and straight lines in K_0 deform into planes and straight lines in K.

4.3   Small Strains and Small Deformations

In most applications of structural materials like steel, aluminium, concrete, and wood the strains are small. For instance, elastic strains in mild steel under uniaxial stress are less than 0.001. We will characterize a deformation state in which the absolute values of all longitudinal strains are less than 0.01, as a state of *small strains*. According to Eq. (4.2.13) this implies that all components E_{ik} of the strain tensor $\mathbf{E}$ satisfy the inequality $E_{ik} \ll 1$. The formulas (4.2.16) and (4.2.18) imply that the absolute value of all shear strains is less than 0.02, and that the absolute value of the volumetric strain is less than 0.03.

Small strains do not imply that the displacements are small or that rotations of line element will be small. The following example will illustrate this.

Fig. 4.4 Ilustration of a case with small strains and large deformations

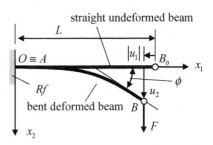

Figure 4.4 shows a thin horizontal steel bar *AB* fixed at the end *A* and loaded on the free *B* by a vertical force *F*. The bar is straight in the undeformed configuration AB_0. Due to the bending the bar experiences large elastic deformations and obtain the bent configuration *AB*.

However, as long as the bar remains elastic, and the strains in the bar do not exceed 0.001, the displacements $|u_1|$ and u_2 of the end *B* of the bar may be of the same magnitude as the length *L* of the bar. The angle of rotation ϕ of the tangent of the bar axis at *B* can be near 90°. Thus, this is a case of small strains and large deformations.

In ordinary engineering structures the displacements of particles and the rotations of line elements will be small quantities. Measured in radians we may assume the angle of rotation to be much less than 1. The displacement will be considerably less than a characteristic length of the structure, for instance the length *L* of the beam in Fig. 4.4. This means that the change of the geometry of the structure, due to the deformation, need not to be taken into account when the action of the forces is considered. If both the strains and rotations are small, we use the expression *state of small deformations*.

4.3.1 Small Strains

From formula (4.2.13) for the longitudinal strain ε in the direction **e** we obtain for small strains:

$$(\varepsilon + 1)^2 == \left(\sqrt{1 + 2\mathbf{e} \cdot \mathbf{E} \cdot \mathbf{e}}\right)^2 \Rightarrow \varepsilon^2 + 2\varepsilon + 1 = 1 + 2\mathbf{e} \cdot \mathbf{E} \cdot \mathbf{e}$$
$$\Rightarrow \varepsilon = \mathbf{e} \cdot \mathbf{E} \cdot \mathbf{e} = e_i E_{ij} e_j = e^T E e \tag{4.3.1}$$

The term ε^2 has been neglected when compared with the term ε. In formula (4.2.16) for the shear strain γ with respect to the two orthogonal directions $\bar{\mathbf{e}}$ and **e** we set $\sin \gamma \approx \gamma$ and replace the denominator by 1. Then we obtain for small strains:

$$\gamma = 2\bar{\mathbf{e}} \cdot \mathbf{E} \cdot \mathbf{e} = 2\bar{e}_i E_{ij} e_j = \bar{e}^T E e \tag{4.3.2}$$

For longitudinal strains and shear strains with respect to material line elements that in K_0 are parallel to the coordinate axes, we find:

$$\varepsilon_{ii} = E_{ii} \quad \text{(no summation)}, \quad \gamma_{ij} = 2E_{ij} \quad i \neq j \tag{4.3.3}$$

Because all the coordinate strains E_{ij} are small quantities, we get from Eq. (4.2.18):

$$(\varepsilon_v + 1)^2 = \det(\mathbf{1} + 2\mathbf{E}) \Rightarrow \varepsilon_v^2 + 2\varepsilon_v + 1 = 1 + 2(E_{11} + E_{22} + E_{33}) + \cdots$$
$$\Rightarrow \varepsilon_v = E_{11} + E_{22} + E_{33} = E_{ii} = \operatorname{tr} \mathbf{E} \tag{4.3.4}$$

It follows that the case of small strains is characterized by the inequalities: norm $\mathbf{E} \ll 1$.

4.3.2 Small Deformations

In order to make a basis for definition of what we shall mean by small deformations, we shall consider the deformation of the thin steel beam in Fig. 4.4. Let the reference coordinates of the end of the beam be $X = \{X_1 = L \ X_2 = 0 \ X_3 = 0\}$. For the displacement of the beam end we write: $u_1 = u_1(X_1, X_2)$ and $u_2 = u_2(X_1, X_2)$. The angle of rotation ϕ of the tangent of the bar axis at B is now expressed by: $\tan \phi = \partial u_2 / \partial X_1 = H_{21}$. For the deformation of the beam to be small it is natural to require the deformation gradient H_{21} to be small, i.e. $H_{21} \ll 1$. In a general case we define small deformations by the condition that all displacement gradients H_{ij} in absolute values must be small compared to 1:

$$\text{small deformations} \Leftrightarrow |H_{ij}| \equiv \left| \frac{\partial u_i}{\partial X_j} \right| \ll 1 \Leftrightarrow \text{norm } \mathbf{H} \ll 1 \tag{4.3.5}$$

Under the assumption of small deformations we get the following result for an arbitrary field $f(\mathbf{r}, t) = f(\mathbf{r}(\mathbf{r}_0, t), t)$, or $f(x, t) = f(x(X, t), t)$, using Eq. (4.2.21):

$$\frac{\partial f}{\partial X_i} = \frac{\partial f}{\partial x_k} \frac{\partial x_k}{\partial X_i} = \frac{\partial f}{\partial x_k} \left(\delta_{ki} + \frac{\partial u_k}{\partial X_i} \right) \approx \frac{\partial f}{\partial x_i} \equiv f_{,i} \tag{4.3.6}$$

Usually small deformations imply small displacements, and we may replace the particle reference $\mathbf{r}_0$ by the place vector $\mathbf{r}$ and the particle coordinates X_i by the place coordinates x_i as particle coordinates. For the same reason we will set:

$$\dot{f}(\mathbf{r}, t) = \dot{f}(x, t) = \partial_t f(\mathbf{r}, t) \equiv \frac{\partial f(\mathbf{r}, t)}{\partial t} \tag{4.3.7}$$

For small deformations the displacement gradients are denoted by:

$$H_{ij} = u_{i,j} \Leftrightarrow \mathbf{H} = \mathrm{grad}\,\mathbf{u} \qquad (4.3.8)$$

In the expressions for strains we take into account that the displacement gradients are small quantities. Products of H_{ik} – components are neglected compared to the components H_{ik}, and the components H_{ik} are neglected when compared with 1. The Green strain tensor (4.2.23) is thus reduced to:

$$\mathbf{E} = \frac{1}{2}\left(\mathbf{H} + \mathbf{H}^T\right) \Leftrightarrow E_{ij} = \frac{1}{2}\left(H_{ij} + H_{ji}\right) = \frac{1}{2}\left(u_{i,j} + u_{j,i}\right) \qquad (4.3.9)$$

This strain tensor is sometimes called the *small strain tensor*. We shall call the tensor $\mathbf{E}$ defined by the formulas (4.3.9) for the *strain tensor for small deformations*.

As we have seen small deformations imply small strains. In the literature the term *infinitesimal deformations* is sometime used about what is here called small deformations. As illustrated by the example in Fig. 4.4, small deformations also imply small strains.

The expression for the volumetric strain for small deformations is obtained from the formulas (4.3.4) and (4.3.9):

$$\varepsilon_v = \mathrm{tr}\,\mathbf{E} = E_{kk} = u_{k,k} = \mathrm{div}\,\mathbf{u} \qquad (4.3.10)$$

The divergence of the displacement field is thus an expression for the change in volume per unit volume when the material has been deformed from the reference configuration K_0 to the present configuration K. For the coordinate strains we get:

$$\varepsilon_{11} = E_{11} = u_{1,1}, \quad \gamma_{12} = 2E_{12} = u_{1,2} + u_{2,1} \quad \text{etc.} \qquad (4.3.11)$$

The results (4.3.11) may be illustrated directly as shown in Fig. 4.5, which for clarity illustrates only the two–dimensional case. Before deformation a particle is in the position P_0, and after deformation in the position P. Two orthogonal material line elements dx_1 and dx_2 are displaced and deformed to approximate new lengths $dx_1 + u_{1,1}\,dx_1$ and $dx_2 + u_{2,2}\,dx_2$. The longitudinal strains of the elements are then $\varepsilon_{11} = u_{1,1} = E_{11}$ and $\varepsilon_{22} = u_{2,2} = E_{22}$ respectively. The right angle between the material line elements dx_1 and dx_2 is diminished by the shear strain $\gamma_{12} = u_{1,2} + u_{2,1} = 2E_{12}$.

In the standard *xyz*–notation the formulas (4.3.11) are written as follows:

$$\varepsilon_x = \frac{\partial u_x}{\partial x}, \quad \gamma_{xy} = \frac{\partial u_x}{\partial y} + \frac{\partial u_y}{\partial x} \quad \text{etc.} \qquad (4.3.12)$$

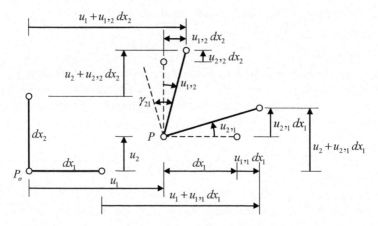

Fig. 4.5 Two—dimensional construction of the coordinate strains $\varepsilon_{11}, \varepsilon_{22}$, and γ_{12} in the case of small deformations

4.3.3 Principal Strains and Principal Directions for Small Deformations

The condition of small deformations implies that all components of the strain tensor for small deformations **E** are small in absolute values. The condition may be stated thus:

$$\text{Small deformations} \Leftrightarrow \left|E_{ij}\right| \ll 1 \Leftrightarrow \text{norm}\,\mathbf{E} \ll 1 \qquad (4.3.13)$$

Because the strain tensor **E** is symmetric, we may determine three orthogonal principal directions of strain and corresponding principal values. We shall interpret these quantities geometrically and follow the corresponding analysis for the stress tensor **T** in Sect. 2.3.1 and for a general symmetric tensor of second order in Sect. 3.3.1. For any direction in the reference configuration K_0 given by a unit vector **e**, we define the vector $\mathbf{E} \cdot \mathbf{e}$. The longitudinal strain ε in the direction **e** is the normal component of the strain tensor for the direction **e** and given by the formula (4.3.1). The shear strain γ with respect to two orthogonal directions **e** and $\bar{\mathbf{e}}$ is equal to twice the orthogonal shear component of the strain tensor for the directions **e** and $\bar{\mathbf{e}}$ and given by the formula (4.3.2). The *principal strains* ε_i and the *principal directions of strain* $\mathbf{a}_i$ are determined from the condition:

$$\mathbf{E} \cdot \mathbf{a} = \varepsilon \mathbf{a} \Leftrightarrow (\varepsilon \mathbf{1} - \mathbf{E}) \cdot \mathbf{a} = \mathbf{0} \Leftrightarrow \left(\varepsilon \delta_{ij} - E_{ij}\right) a_j = 0 \qquad (4.3.14)$$

The *characteristic equation* and the *principal invariants of the strain tensor* **E** are:

$$\varepsilon^3 - I\varepsilon^2 + II\varepsilon - III = 0$$

$$I = \mathrm{tr}\,\mathbf{E}, \quad II = \frac{1}{2}\left[(\mathrm{tr}\,\mathbf{E})^2 - (\mathrm{norm}\,\mathbf{E})^2\right], \quad III = \det\mathbf{E} \qquad (4.3.15)$$

The three principal strains ε_i are all real and the principal directions of strains are orthogonal. Through every material particle there exist three orthogonal material line elements before deformation that remain orthogonal after deformation. Material planes normal to the principal directions of strains are free of shear strains. The principal strains in a particle represent extreme values of longitudinal strain in the particle.

$$\varepsilon_3 \le \varepsilon_2 \le \varepsilon_1 \Rightarrow \varepsilon_{max} = \varepsilon_1, \quad \varepsilon_{min} = \varepsilon_3 \qquad (4.3.16)$$

The *maximum shear strain* in a particle is given by:

$$\gamma_{max} = \varepsilon_{max} - \varepsilon_{min}$$

$$\text{for the directions}: \mathbf{e} = \pm(\mathbf{a}_1 \pm \mathbf{a}_3)/\sqrt{2}, \quad \bar{\mathbf{e}} = \pm(\mathbf{a}_1 \mp \mathbf{a}_3)/\sqrt{2} \qquad (4.3.17)$$

4.3.4 Strain Deviator and Strain Isotrop for Small Deformations

In constitutive modelling of isotropic materials, it is convenient to decompose the state of small deformations into a *form invariant part* and a *volume invariant part*. This is done by decomposing the strain tensor for small deformations $\mathbf{E}$ into a *strain isotrop* $\mathbf{E}^o$, which is form invariant, and *strain deviator* $\mathbf{E}'$, which is volume invariant:

$$\mathbf{E} = \mathbf{E}^o + \mathbf{E}'$$

$$\mathbf{E}^o = \frac{1}{3}(\mathrm{tr}\,\mathbf{E})\mathbf{1} = \frac{1}{3}\varepsilon_v\,\mathbf{1} \Leftrightarrow \text{form invariant strain} \qquad (4.3.18)$$

$$\mathbf{E}' = \mathbf{E} - \mathbf{E}^o \Rightarrow \varepsilon_v = \mathrm{tr}\,\mathbf{E} = \mathrm{tr}\,\mathbf{E}' = 0 \Leftrightarrow \text{volume invariant strain}$$

Because the strains are assumed to be small, the strain tensors $\mathbf{E}^o$ and $\mathbf{E}'$ may be added commutatively.

The strain isotrop $\mathbf{E}^o$ represents a *state of form invariant strain* or an *isotropic state of strain*, because the angle between any two material lines does not change due to this deformation. All shear strains are zero.

The strain deviator $\mathbf{E}'$ is trace free, i.e. $\mathrm{tr}\,\mathbf{E}' = \mathrm{tr}\,\mathbf{E} - \mathrm{tr}\,\mathbf{E}^o = \mathrm{tr}\,\mathbf{E} - (1/3)(\mathrm{tr}\,\mathbf{E})(\mathrm{tr}\,\mathbf{1}) = 0$, and represents therefore a state of strain without change of volume. Therefore the strain deviator represents a *state of volume invariant strain* or an *isochoric state of strain*. The strain deviator $\mathbf{E}'$ and the strain tensor $\mathbf{E}$ are coaxial tensors. The principal strains of the two tensors are related through:

$$\varepsilon'_i = \varepsilon_i - \frac{1}{3}\varepsilon_v \qquad (4.3.19)$$

4.3.5   Rotation Tensor for Small Deformations

For small deformations it is convenient to decompose the displacement gradient tensor $\mathbf{H}$ into to distinct contributions: the strain tensor $\mathbf{E}$ defined by the formula (4.3.9) and the *rotation tensor for small deformations* $\widetilde{\mathbf{R}}$:

$$\mathbf{H} = \mathbf{E} + \widetilde{\mathbf{R}} \Leftrightarrow H_{ij} = u_{i,j} = E_{ij} + \widetilde{R}_{ij} \qquad (4.3.20)$$

It follows that $\widetilde{\mathbf{R}}$ is defined by:

$$\widetilde{\mathbf{R}} = \frac{1}{2}\left(\mathbf{H} - \mathbf{H}^T\right) \Leftrightarrow \widetilde{R}_{ij} = \frac{1}{2}\left(H_{ij} - H_{ji}\right) = \frac{1}{2}\left(u_{i,j} - u_{j,i}\right) \qquad (4.3.21)$$

The two tensors $\mathbf{E}$ and $\widetilde{\mathbf{R}}$ represent the symmetric part and the antisymmetric part of the displacement gradient tensor $\mathbf{H}$. The dual vector to the rotation tensor $\widetilde{\mathbf{R}}$ is the *rotation vector* $\mathbf{z}$ *for small deformations* and given by:

$$\mathbf{z} = -\frac{1}{2}\mathbf{P} : \widetilde{\mathbf{R}} = \frac{1}{2}\operatorname{rot}\mathbf{u} \Leftrightarrow z_i = -\frac{1}{2}e_{ijk}\widetilde{R}_{jk} = \frac{1}{2}e_{ijk}u_{k,j} \qquad (4.3.22)$$

Separately the strain tensor $\mathbf{E}$ represents pure strain in the following sense: Material line elements in the principal directions of the tensor $\mathbf{E}$ do not rotate. The rotation tensor $\widetilde{\mathbf{R}}$ and the rotation vector $\mathbf{z}$ represent the rotation of three orthogonal material line elements in the principal directions of the strain tensor $\mathbf{E}$.

The decomposition (4.3.20) will be illustrated as follows. We use the formulas (4.2.26), (4.2.27), (4.2.21), (4.3.20), and obtain:

$$\mathbf{r}(\mathbf{r}_0, t) = \mathbf{u}_0(t) + \mathbf{F}(t) \cdot \mathbf{r}_0 = \mathbf{u}_0(t) + (1 + \mathbf{H}(t)) \cdot \mathbf{r}_0 = \mathbf{u}_0(t) + \left(1 + \mathbf{E}(t) + \widetilde{\mathbf{R}}(t)\right) \cdot \mathbf{r}_0$$

$$\Rightarrow \mathbf{u} = \mathbf{r} - \mathbf{r}_0 = \mathbf{u}_0(t) + \mathbf{E} \cdot \mathbf{r}_0 + \widetilde{\mathbf{R}} \cdot \mathbf{r}_0 \Rightarrow d\mathbf{u} = d\mathbf{r} - d\mathbf{r}_0 = \mathbf{E} \cdot d\mathbf{r}_0 + \widetilde{\mathbf{R}} \cdot d\mathbf{r}_0$$

Let ε_i be the principal strains and $\mathbf{a}_i$ be the principal directions of the strain tensor $\mathbf{E}$, and let Ox be a Cartesian coordinate system with base vectors $\mathbf{e}_i \equiv \mathbf{a}_i$. We set: $d\mathbf{r} = dx_i\mathbf{e}_i$ and $d\mathbf{r}_0 = dX_i\mathbf{e}_i$. Figure 4.6a shows the displacement contribution from the term $\mathbf{E} \cdot d\mathbf{r}_0$. Figure 4.6b illustrates in two–dimensions the displacement contribution from the term $d\mathbf{r} \cong \widetilde{\mathbf{R}} \cdot d\mathbf{r}_0$ for two material line elements $d\mathbf{r}_{01} = \mathbf{e}_1 dX_1$ and $d\mathbf{r}_{02} = \mathbf{e}_2 dX_2$.

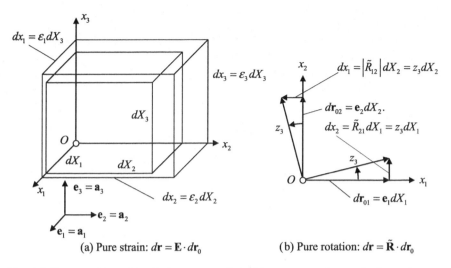

(a) Pure strain: $d\mathbf{r} = \mathbf{E} \cdot d\mathbf{r}_0$ (b) Pure rotation: $d\mathbf{r} = \tilde{\mathbf{R}} \cdot d\mathbf{r}_0$

Fig. 4.6 Decomposition of the displacement gradient $\mathbf{H}$ into a pure strain contribution and a pure rotation contribution. (a) Pure strain contribution: $d\mathbf{r} = \mathbf{E} \cdot d\mathbf{r}_0 \Rightarrow dx_i = E_{ij} dX_j = \varepsilon_i \delta_{ij} dX_j$. Material line elements in the principal directions $\mathbf{a}_i$ of the strain tensor $\mathbf{E}$ do not rotate. (b) Pure rotation contribution: $d\mathbf{r} = \tilde{\mathbf{R}} \cdot d\mathbf{r}_0 \Rightarrow dx_i = \tilde{R}_{ij} dX_j$. Material line elements in the principal directions $\mathbf{a}_i$ of the strain tensor $\mathbf{E}$ rotate according to the rotation vector $\mathbf{z} = -\mathbf{P} : \tilde{\mathbf{R}} = \mathrm{rot}\, \mathbf{u}/2$

A necessary and sufficient condition for a pure strain under small deformations, i.e. $\tilde{\mathbf{R}} = \mathbf{0}$, is that the displacement vector may be expressed by the gradient of a scalar field:

$$\mathbf{u} = \nabla \phi \Leftrightarrow \tilde{\mathbf{R}} = \mathbf{0} \Leftrightarrow \text{pure strain} \qquad (4.3.23)$$

The scalar field $\phi(\mathbf{r}, t)$ is called the *strain potential*. The proof of the bi–implication (4.3.23) is given by Theorem 9.13 in Chapter 9.

4.3.6 Small Deformations in a Material Surface

Indirect measurements of stresses in the surface of an elastic body are performed by measuring the longitudinal strains in the surface. The strain measurements may be done by using electrical strain gages or strain rosettes.

We shall now analyze the state of strain in a surface through a particle P and introduce a coordinate system Ox with the x_3-axis normal to the tangent plane of the surface. The other two axes are then in the tangent plane, Fig. 4.7. In Fig. 4.7 we have introduced two orthogonal unit directional vectors $\mathbf{e}$ and $\bar{\mathbf{e}}$ in the surface:

$$\mathbf{e} = [\cos \phi, \sin \phi, 0], \quad \bar{\mathbf{e}} = [\sin \phi, -\cos \phi, 0] \qquad (4.3.24)$$

Fig. 4.7 Orthogonal unit
vectors **e** and **ē** in a material
surface

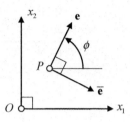

The longitudinal strain ε in the direction **e** in the material surface, and the shear strain γ with respect to the two orthogonal directions **e** and **ē** in the surface are given by:

$$\varepsilon = \varepsilon(\phi) = \mathbf{e} \cdot \mathbf{E} \cdot \mathbf{e} = e_\alpha E_{\alpha\beta}\, e_\beta = E_{11} \cos^2 \phi + E_{22} \sin^2 \phi + 2E_{12} \cos \phi \sin \phi$$
$$\gamma = \gamma(\phi) = 2\bar{\mathbf{e}} \cdot \mathbf{E} \cdot \mathbf{e} = 2\bar{e}_\alpha E_{\alpha\beta}\, e_\beta$$
$$= 2(E_{11} - E_{22}) \sin \phi \cos \phi - 2E_{12}\left(\cos^2 \phi - \sin^2 \phi\right)$$

$$(4.3.25)$$

Using the formulas (2.3.35) for $\sin 2\phi$ and $\cos 2\phi$, we may transform the formulas (4.3.25) to:

$$\varepsilon(\phi) = \frac{E_{11} + E_{22}}{2} + \frac{E_{11} - E_{22}}{2} \cos 2\phi + E_{12} \sin 2\phi$$
$$\gamma(\phi) = (E_{11} - E_{22}) \sin 2\phi - 2E_{12} \cos 2\phi$$

$$(4.3.26)$$

These formulas are analogous to those developed in Sect. 2.3.5 for the state of plane stress. The formulas for the principal strains ε_1 and ε_2 in the surface, and the angle ϕ_1 that the principal strain to ε_1 makes with the x_1 − axis, follow directly from the corresponding formulas for plane stress:

$$\begin{matrix} \varepsilon_1 \\ \varepsilon_2 \end{matrix} = \frac{E_{11} + E_{22}}{2} \pm \sqrt{\left(\frac{E_{11} - E_{22}}{2}\right)^2 + (E_{12})^2} \qquad (4.3.27)$$

$$\phi_1 = \arctan \frac{\varepsilon_1 - E_{11}}{E_{12}} \qquad (4.3.28)$$

Note that the formulas (4.3.29) do not necessarily represent the principal strains of the general state of strain. It is not assumed that the x_3 − direction, normal to the surface, is a principal direction of strain. However, very often this will be the case, especially when the surface is a free surface and with electrical strain gages for measuring strains attached to the surface.

From the analogous analysis of plane stress we may also conclude that the *maximum shear strain in the surface* is given by:

$$\gamma_{\max} = |\varepsilon_1 - \varepsilon_2| = 2\sqrt{\left(\frac{E_{11} - E_{22}}{2}\right)^2 + (E_{12})^2} \qquad (4.3.29)$$

4.3.7 Mohr–Diagram for Small Deformations in a Surface

We assume that the state of strain in a surface is given by the strain matrix:

$$E = \begin{pmatrix} E_{11} & E_{12} \\ E_{12} & E_{22} \end{pmatrix} \qquad (4.3.30)$$

The Mohr–diagram for strains in the surface is drawn in a Cartesian coordinate system with longitudinal strain ε and one half of shear strain γ as coordinates. The *Mohr strain circle* is constructed by the same principles as the *Mohr stress circle*. First the points $(E_{11}, -E_{12})$ and (E_{22}, E_{12}) are marked in the diagram. These two points are on a diameter of the circle and thus determine the circle. Figure 4.8 shows the complete Mohr–diagram.

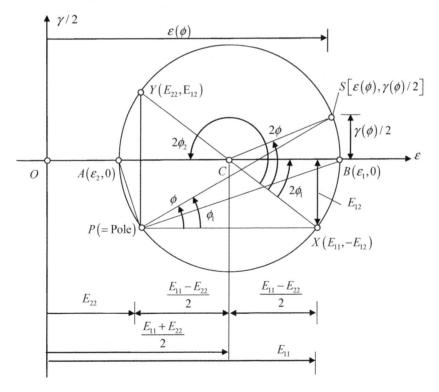

Fig. 4.8 Mohr—diagram for small deformations in a surface

4.4   Rates of Deformation, Strain, and Rotation

In order to describe the motion and deformation of fluids, and sometimes solids with fluid-like behaviour we choose to work with Eulerian coordinates and the present configuration K as reference configuration. The particles are now denoted by the place vector $\mathbf{r}$, or the place coordinates x. A particle P at the place $\mathbf{r}$ at the time t has the velocity $\mathbf{v}(\mathbf{r}, t)$ and is during a short time increment dt given the displacement $d\mathbf{u} = \mathbf{v}\, dt$. This displacement field leads to small deformations with a displacement gradient tensor $d\mathbf{H}$:

$$dH_{ik} = \frac{\partial v_i}{\partial x_k} dt \equiv v_{i,k}\, dt \tag{4.4.1}$$

and a strain tensor $d\mathbf{E}$ and a rotation tensor for small deformations $d\tilde{\mathbf{R}}$:

$$d\mathbf{E} = \frac{1}{2}\left(d\mathbf{H} + d\mathbf{H}^T\right), \quad d\tilde{\mathbf{R}} = \frac{1}{2}\left(d\mathbf{H} - d\mathbf{H}^T\right) \tag{4.4.2}$$

We now define three new tensors: the *velocity gradient tensor* $\mathbf{L}$ at the time t, the symmetric *rate of deformation tensor* $\mathbf{D}$ at the time t, and the antisymmetric *rate of rotation tensor* $\mathbf{W}$ at time t:

$$\mathbf{L} = \operatorname{grad} \mathbf{v} \equiv \frac{\partial \mathbf{v}}{\partial \mathbf{r}} \Leftrightarrow L_{ik} = v_{i,k} \tag{4.4.3}$$

$$\mathbf{D} = \frac{1}{2}\left(\mathbf{L} + \mathbf{L}^T\right) \Leftrightarrow D_{ik} = \frac{1}{2}\left(v_{i,k} + v_{k,i}\right) \tag{4.4.4}$$

$$\mathbf{W} = \frac{1}{2}\left(\mathbf{L} - \mathbf{L}^T\right) \Leftrightarrow W_{ik} = \frac{1}{2}\left(v_{i,k} - v_{k,i}\right) \tag{4.4.5}$$

$\mathbf{W}$ is also called the *spin tensor* and the *vorticity tensor*. It follows that:

$$d\mathbf{H} = \mathbf{L}\, dt, \quad d\mathbf{E} = \mathbf{D}\, dt, \quad d\tilde{\mathbf{R}} = \mathbf{W}\, dt, \quad \dot{\mathbf{E}} = \mathbf{D}, \quad \dot{\tilde{\mathbf{R}}} = \mathbf{W} \tag{4.4.6}$$

Note that $\mathbf{E}$ is here the strain tensor (4.3.9) for small deformations. As will be demonstrated in Sect. 4.5 the relation between the general Green's strain tensor $\mathbf{E}$ defined by formula (4.2.11) and the rate of deformation tensor $\mathbf{D}$ is more complex.

The change length of a material line element per unit length and per unit time is called the *rate of longitudinal strain* or for short the *rate of strain*. From the analysis of small deformations in Sect. 4.3 it follows the rate of strain in the direction $\mathbf{e}$ is, according to Eq. (4.3.1):

$$\dot{\varepsilon} = \mathbf{e} \cdot \mathbf{D} \cdot \mathbf{e} = e_i D_{ik} e_k = e^T De \tag{4.4.7}$$

In the coordinate direction $\mathbf{e}_i$ we get the *coordinate rates of longitudinal strain*:

$$\dot{\varepsilon}_{ii} = D_{ii} = v_{i,i} \quad \text{(no summation)} \tag{4.4.8}$$

The change per unit time of the angle between two material line elements that in K are orthogonal, is called the *rate of shear strain* or for short the *shear rate*. The rate of shear strain with respect to orthogonal line elements in the directions $\mathbf{e}$ and $\bar{\mathbf{e}}$ is, in analogy to Eq. (4.3.2):

$$\dot{\gamma} = 2\bar{\mathbf{e}} \cdot \mathbf{D} \cdot \mathbf{e} = \bar{e}_i D_{ik} e_k = e^T D e \tag{4.4.9}$$

For the coordinate directions $\mathbf{e}_1$ and $\mathbf{e}_2$ we get the *coordinate rate of shear strain*:

$$\dot{\gamma}_{21} = 2D_{21} = v_{2,1} + v_{2,1} \tag{4.4.10}$$

Figure 4.5 illustrates the coordinate rates of strain and shear rate when the displacement $\mathbf{u}$ is replaced by the increment in displacement $d\mathbf{u} = \mathbf{v}\,dt$.

The change in volume per unit of volume and per unit of time is called the *rate of volumetric strain*. By referring to the expression (4.3.4) for the volumetric strain, we may immediately write down the expression for the rate of volumetric strain.

$$\dot{\varepsilon}_v = D_{kk} = \text{tr}\,\mathbf{D} = \text{div}\,\mathbf{v} = v_{k,k} \tag{4.4.11}$$

The result (4.4.11) shows that the divergence of the velocity field $\mathbf{v}(\mathbf{r}, t)$ represents the rate of change of volume per unit volume about the particle under consideration.

The dual vector to the antisymmetric rate of rotation tensor $\mathbf{W}$ is the *angular velocity vector*:

$$\mathbf{w} = -\mathbf{P} : \mathbf{W} = \frac{1}{2}\text{rot}\,\mathbf{v} \Leftrightarrow w_i = \frac{1}{2}e_{ijk} W_{kj} = \frac{1}{2}e_{ijk} v_{k,j}$$
$$\Leftrightarrow \mathbf{W} = -\mathbf{P} \cdot \mathbf{w} \Leftrightarrow W_{ij} = -e_{ijk} w_k \tag{4.4.12}$$

The matrix of $\mathbf{W}$ may now be written as:

$$W = \begin{pmatrix} 0 & -w_3 & w_2 \\ w_3 & 0 & -w_1 \\ -w_2 & w_1 & 0 \end{pmatrix} \tag{4.4.13}$$

Note that angular velocity vector $\mathbf{w}$ is related to the rotation vector $\mathbf{z}$ for small deformation defined by Eq. (4.3.28):

$$\mathbf{w} = \dot{\mathbf{z}} \tag{4.4.14}$$

In Fluid Mechanics it is customary to introduce the concept of *vorticity* $\mathbf{c}$:

$$\mathbf{c} = 2\mathbf{w} = \mathrm{rot}\,\mathbf{v} \equiv \mathrm{curl}\,\mathbf{v} \equiv \nabla \times \mathbf{v} \qquad (4.4.15)$$

Due to the relationship between the rate of deformation tensor and the rate of strain tensor for small deformation, and the corresponding relationship between the angular velocity vector and the rotation vector for small deformation, we may conclude based on the discussion in Sect. 4.3.5 that material line elements parallel to the principal directions of the rate of deformation tensor **D** rotate instantaneously with the angular velocity **w**. In other words:

> The rate of rotation tensor W represents the instantaneous angular velocity of the three orthogonal material line elements that are oriented in the principal directions of the rate of deformation tensor **D**.

This fact is demonstrated in Fig. 4.9, in which the coordinate system is chosen to have base vectors $\mathbf{e}_i$ parallel to the principal directions $\mathbf{a}_i$ of **D**. The matrix of **D** in this coordinate system has the elements:

$$D_{ik} = \dot{\varepsilon}_i\,\delta_{ik} \qquad (4.4.16)$$

This implies that $D_{12} = v_{1,2} + v_{2,1} = 0$, and therefore $v_{1,2} = -v_{2,1}$. Thus: $w_3 = W_{21} = v_{2,1}$. In Fig. 4.9 we now see that $w_3\,dt$ represents, during the time interval dt, the angle of rotation about an axis parallel to the x_3-axis of material line elements in the principal directions $\mathbf{a}_i(= \mathbf{e}_i)$ of the rate of deformation tensor **D**. The quantity w_3 represents the angular velocity of those line elements about the axis of rotation.

To further illustrate the physical interpretation of the rate of rotation tensor, we consider the special motion of rigid—body rotation with angular velocity $\mathbf{w} = \mathbf{w}(t)$ about the x_3-axis through the origin O of the coordinate system Ox. The velocity field in this case is, see Eq. (3.5.26) and Fig. 4.10:

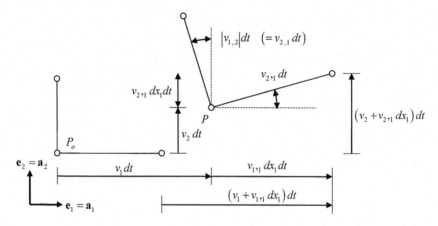

Fig. 4.9 Rotation about an axis parallel to the x_3-direction of material line elements in the principal directions $\mathbf{a}_i = \mathbf{e}_i$ of the rate of deformation tensor **D**

Fig. 4.10 Rigid—body rotation of a small material element about the x_3-axis through the origin O. Angular velocity w

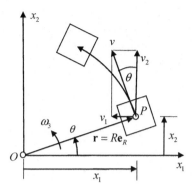

$$\mathbf{v} = \mathbf{W} \cdot \mathbf{r} = \mathbf{w} \times \mathbf{r}, \quad v = wR \Rightarrow v_1 = -v\sin\theta = -wx_2, \quad v_2 = v\cos\theta = wx_1, \quad v_3 = 0$$

In Fig. 4.10 a small material volume element is shown to rotate as a rigid body about the x_3-axis.

We easily find that the rate of deformation tensor $\mathbf{D} = \mathbf{0}$, and that matrix of the rate of rotation tensor is:

$$W = \begin{pmatrix} 0 & -w & 0 \\ w & 0 & 0 \\ 0 & 0 & 0 \end{pmatrix}$$

If the rate of rotation tensor is zero during the motion, i.e. $\mathbf{W} = \mathbf{0}$, the motion is called *irrotational motion* or *irrotational flow*. According to Theorem 9.13 the velocity field in the case of irrotational flow may be derived from a scalar field:

$$\mathbf{v} = \nabla\phi \tag{4.4.17}$$

The scalar field $\phi(\mathbf{r}, t)$ is called the *velocity potential*. Irrotational flow is for this reason also called *potential flow*. An important part of Fluid Mechanics is devoted to potential flow because we can in many practical flow cases show that the flow is in fact irrotational, or very close to be irrotational, and because potential flow introduces mathematical simplifications in Fluid Mechanics.

Example 4.1 Rectilinear Rotational Flow. Simple Shear Flow

A fluid flows between two parallel planes. One of the planes is at rest, while the other is moving with a constant velocity v parallel to the planes, as indicated in Fig. 4.11. We assume that the fluid particles move in straight parallel lines, and that the velocity field is:

$$v_1 = \frac{v}{h}x_2, \quad v_2 = v_3 = 0$$

This velocity field implies that the fluid sticks to the rigid boundary planes, a common assumption in Fluid Mechanics. The matrices of the velocity gradient tensor **L**, the rate of deformation tensor **D**, and the rate of rotation tensor **W** for this flow are:

$$L = \begin{pmatrix} 0 & 1 & 0 \\ 0 & 0 & 0 \\ 0 & 0 & 0 \end{pmatrix}\frac{v}{h}, \quad D = \begin{pmatrix} 0 & 1 & 0 \\ 1 & 0 & 0 \\ 0 & 0 & 0 \end{pmatrix}\frac{v}{2h}, \quad W = \begin{pmatrix} 0 & 1 & 0 \\ -1 & 0 & 0 \\ 0 & 0 & 0 \end{pmatrix}\frac{v}{2h}$$

Only one rate of deformation component is different from zero, which gives the shear rate $\dot{\gamma}_{12}$, and only rate of rotation component W_{12} is different from zero:

$$D_{12} = \frac{1}{2}\frac{v}{h} \Rightarrow \dot{\gamma}_{12} = 2D_{12} = \frac{v}{h}, \quad W_{12} = \frac{v}{2h}$$

The angular velocity vector **w** and the vorticity vector **c** are according to the formulas (4.4.12) and (4.4.15):

$$\mathbf{w} = w_3\,\mathbf{e}_3 \Rightarrow w_3 = -\frac{v}{2h}, \quad \mathbf{c} = 2\mathbf{w} = -\frac{v}{h}\mathbf{e}_3$$

Figure 4.11 illustrates the flow of two fluid elements during a short increment of time dt. Fluid element 1 has a shear rate $\dot{\gamma}_{12}$. The deformation of the fluid element 2 shows that the element sides remain orthogonal during the short time interval dt. The element shows no shear rate. This fact implies that the element is oriented according to the principal directions of the rate of deformation tensor **D**. The deformation also demonstrates that that the element 2 has an angular velocity component $w_3 = -v/2h$ about the x_3 – axis. Note that the *rate of volumetric strain*: $\dot{\varepsilon}_v = \text{div}\,\mathbf{v} = 0$ for this flow. We say that this is an *isochoric flow*, i.e. a volume preserving flow.

Because the fluid particles move in straight lines, the flow is called rectilinear, and because the flow give rise to rates of rotation, and thereby vorticity, it is called

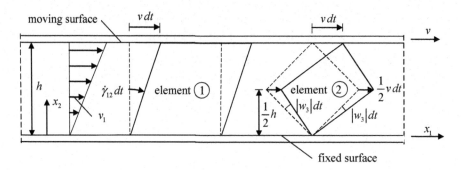

Fig. 4.11 Simple shear flow. Rectilinear flow with vorticity. Fluid element 1 has a shear rate γ_{12}. Fluid element 2 rotates with angular velocity w_3

Fig. 4.12 Circular
irrotational flow created by a
rotating cylinder of radius a in
a linear viscous fluid. Angular
velocity ω. Fluid particles
move in concentric circles

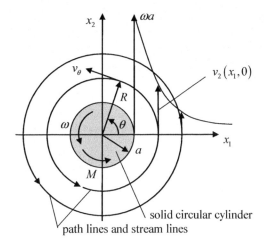

solid circular cylinder
path lines and stream lines

rotational flow. This particular kind of rectilinear, rotational flow is called *simple shear flow*.

Example 4.2 Circular Irrotational Flow. Potential vortex A solid circular cylinder with radius a that rotates about its axis with constant angular velocity ω in a linearly viscous fluid, creates an *irrotational vortex*, i.e. a vortex without vorticity, Fig. 4.12.

The fluid particles move in concentric circles and the velocity field is:

$$v_\theta = \frac{\alpha}{R}, \quad v_R = v_z = 0, \quad \alpha = \omega\, a^2$$

$$\Rightarrow v_1 = -v_\theta \sin\theta = -\frac{\alpha x_2}{x_1^2 + x_2^2}, \quad v_2 = v_\theta \cos\theta = \frac{\alpha x_1}{x_1^2 + x_2^2}, \quad v_3 = 0$$

For the non-zero velocity gradients we find:

$$v_{1,1} = \frac{2\alpha x_2 x_1}{\left(x_1^2 + x_2^2\right)^2}, \quad v_{2,2} = -\frac{2\alpha x_1 x_2}{\left(x_1^2 + x_2^2\right)^2}, \quad v_{1,2} = -\frac{\alpha}{x_1^2 + x_2^2} + \frac{2\alpha x_2^2}{\left(x_1^2 + x_2^2\right)^2} = \frac{\alpha\left(x_2^2 - x_1^2\right)}{\left(x_1^2 + x_2^2\right)^2}$$

$$v_{2,1} = \frac{\alpha}{x_1^2 + x_2^2} - \frac{2\alpha x_1^2}{\left(x_1^2 + x_2^2\right)^2} = \frac{\alpha\left(x_2^2 - x_1^2\right)}{\left(x_1^2 + x_2^2\right)^2}$$

The volumetric rate of strain rate $\dot{\varepsilon}_v = \operatorname{div} \mathbf{v} = v_{1,1} + v_{2,2} + v_{3,3} = 0$. Thus the flow is *isochoric*.

Since this is a planar flow only one non–trivial rate of rotation component has to be evaluated, i.e. W_{12}, for which we find:

$$W_{12} = -w_3 = -\frac{1}{2}c_3 = \frac{1}{2}(v_{1,2} - v_{2,1}) = 0$$

Hence we have the results:

$$\mathbf{W} = \mathbf{0}, \quad \mathbf{c} = 2\mathbf{w} = \operatorname{rot}\mathbf{v} = \nabla \times \mathbf{v} = \mathbf{0}$$

The velocity field may be derived from a velocity potential $\phi = \alpha\theta$ as follows:

$$\mathbf{v} = \nabla\phi \Rightarrow v_\theta = \frac{1}{R}\frac{\partial\phi}{\partial\theta} = \frac{\alpha}{R}, \quad v_R = \frac{\partial\phi}{\partial R} = 0, \quad v_z = \frac{\partial\phi}{\partial z} = 0$$

The flow has now three names: *circular irrotational flow*, a *potensial vortex flow*, and a *vorticity–free vortex*.

4.5 Large Deformations

In this section we shall analyze the deformation in the neighbourhood of a particle $\mathbf{r}_o$ that moves with a body from the reference configuration K_0 at time t_0 to the present configuration K at time t. The motion is given by, see Fig. 4.13:

$$\mathbf{r} = \mathbf{r}(\mathbf{r}_0, t) = \mathbf{r}_0 + \mathbf{u}(\mathbf{r}_0, t) \Leftrightarrow x_i = (X, t) = X_i + u_i(X, t) \tag{4.5.1}$$

$\mathbf{u}(\mathbf{r}_0, t)$ is the displacement vector. From the development in Sect. 4.2 we have seen that the three fundamental measures of strain ε, γ, and ε_v may be determined from the relation (4.2.8) between the material line element $d\mathbf{r}_0$ in the reference configuration K_0 and the corresponding material line element $d\mathbf{r}$ in the present configuration K:

$$d\mathbf{r} = \mathbf{F} \cdot d\mathbf{r}_0 \Leftrightarrow dx_i = F_{ik}\, dX_k \tag{4.5.2}$$

The tensor $\mathbf{F}$ is the *deformation gradient tensor* and defined by:

$$\mathbf{F} = \frac{\partial\mathbf{r}}{\partial\mathbf{r}_0} \Leftrightarrow F_{ik}(X, t) = \frac{\partial x_i(X, t)}{\partial X_k} \tag{4.5.3}$$

In order to simplify the following presentation we shall assume homogeneous deformation according to the formulas (4.2.26) and (4.2.27), in which the deformation is the same for all particles:

$$\mathbf{r} = \mathbf{u}_0 + \mathbf{F} \cdot \mathbf{r}_0, \quad \mathbf{u}_0 = \mathbf{u}_0(t), \quad \mathbf{F} = \mathbf{F}(t) = \frac{d\mathbf{r}}{d\mathbf{r}_0} \tag{4.5.4}$$

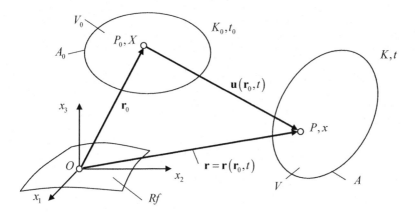

Fig. 4.13 General motion and deformation of a body. Reference *Rf*. Reference configuration K_0, particle $\mathbf{r}_0$ (or X) at the position P_0 at the time t_0. Present Configuration K and particle $\mathbf{r}_0$ (or X) at the position P (or place $\mathbf{r}$ (or x)) at the time t

The displacement vector $\mathbf{u}_0(t)$ represents a translation of the body. The deformation gradient $\mathbf{F}$ represents a linear transformation of a material line element $d\mathbf{r}_0$ in K_0 to the corresponding line element $d\mathbf{r}$ in K. The differentials $d\mathbf{r}_0$ and $d\mathbf{r}$ are the same material line element in K_0 and K respectively. Material planes and straight lines in K_0 are deformed into planes and straight lines in K, see Problem 4.1.

From Sect. 4.2 we import the general results expressing the longitudinal strain ε in the direction $\mathbf{e}$ in K_0, the shear strain γ with respect to two orthogonal directions $\bar{\mathbf{e}}$ and $\mathbf{e}$ in K_0, and the volumetric strain ε_v:

$$\varepsilon = \sqrt{1 + 2\mathbf{e} \cdot \mathbf{E} \cdot \mathbf{e}} - 1 = \sqrt{\mathbf{e} \cdot \mathbf{C} \cdot \mathbf{e}} - 1$$

$$\gamma = \arcsin \frac{2\bar{\mathbf{e}} \cdot \mathbf{E} \cdot \mathbf{e}}{\sqrt{(1 + 2\bar{\mathbf{e}} \cdot \mathbf{E} \cdot \bar{\mathbf{e}})(1 + 2\mathbf{e} \cdot \mathbf{E} \cdot \mathbf{e})}} = \arcsin \frac{\bar{\mathbf{e}} \cdot \mathbf{C} \cdot \mathbf{e}}{\sqrt{(\bar{\mathbf{e}} \cdot \mathbf{C} \cdot \bar{\mathbf{e}})(\mathbf{e} \cdot \mathbf{C} \cdot \mathbf{e})}}$$

$$\varepsilon_v = \det \mathbf{F} - 1 = \det(\mathbf{1} + \mathbf{H}) - 1 = \sqrt{\det(\mathbf{1} + 2\mathbf{E})} - 1 = \sqrt{\det \mathbf{C}} - 1$$

$$(4.5.5)$$

$\mathbf{H}$ is the *displacement gradient tensor*:

$$\mathbf{H} = \frac{\partial \mathbf{u}}{\partial \mathbf{r}_o} = \mathbf{F} - \mathbf{1} \Leftrightarrow H_{ik} = \frac{\partial u_i}{\partial X_k} = \frac{\partial x_i}{\partial X_k} - \delta_{ik} \qquad (4.5.6)$$

$\mathbf{E}$ is *Green's strain tensor*:

$$\mathbf{E} = \frac{1}{2}\left(\mathbf{H} + \mathbf{H}^T + \mathbf{H}^T\mathbf{H}\right) \Leftrightarrow E_{ik} = \frac{1}{2}\left(H_{ik} + H_{ki} + H_{ji}H_{jk}\right) \qquad (4.5.7)$$

$\mathbf{C}$ is the *Green's deformation tensor*:

$$\mathbf{C} = \mathbf{F}^T\mathbf{F} = \mathbf{1} + 2\mathbf{E} \tag{4.5.8}$$

When dealing with large deformations, it is more convenient to use the deformation gradient tensor $\mathbf{F}$ and the deformation tensor $\mathbf{C}$ rather than the displacement gradient tensor $\mathbf{H}$ and the strain tensor $\mathbf{E}$. For practical reasons we shall continue the discussion under the assumption of homogeneous deformations, as expressed by Eq. (4.5.4). Before we consider the general case, we will study two special cases, which the general case may be decomposed into. First we consider a *rigid–body motion*, as given by Eq. (3.5.33) and Fig. 3.4:

$$\mathbf{r} = \mathbf{u}_0 + \mathbf{R} \cdot \mathbf{r}_0, \quad \mathbf{u}_0 = \mathbf{u}_0(t), \quad \mathbf{R} = \mathbf{R}(t), \quad \mathbf{R}^T = \mathbf{R}^{-1}, \quad \mathbf{F} = \mathbf{R}, \quad \det \mathbf{F} = 1 \tag{4.5.9}$$

The displacement vector $\mathbf{u}_0(t)$ represents the displacement of a point $\overline{O}$ that moves with the body and a translation of the body, and the displacement gradient $\mathbf{F} = \mathbf{R}$ is an orthogonal tensor that represents a rigid–body rotation $\mathbf{R} \cdot \mathbf{r}_0$ about the point $\overline{O}$.

Next we consider a motion resulting in what is called *homogeneous pure strain*, in which the deformation gradient $\mathbf{F}$ is a *positive definite symmetric tensor* $\mathbf{U}$:

$$\mathbf{r} = \mathbf{u}_0 + \mathbf{F} \cdot \mathbf{r}_0, \quad \mathbf{F} = \mathbf{U}(t), \quad \mathbf{U}^T = \mathbf{U}, \quad \det \mathbf{U} > 0 \Rightarrow \mathbf{C} = \mathbf{F}^T\mathbf{F} = \mathbf{U}^2 \tag{4.5.10}$$

The symmetric tensors $\mathbf{C}$ and $\mathbf{U}$ are coaxial, i.e. both tensors have the same principal directions represented by the unit vectors $\mathbf{a}_i(t)$. If the principal values of the tensor $\mathbf{U}$ are $\lambda_i(t)$, the tensor $\mathbf{C}$ has the principal values $\lambda_i^2(t)$.

A material line element $d\mathbf{r}_0$ in K_0 parallel to one of the principal directions $\mathbf{a}_i$, such that $d\mathbf{r}_0 = ds_0\,\mathbf{a}_i$, will in configuration in K be represented by the line element $d\mathbf{r} = \mathbf{U} \cdot d\mathbf{r}_0$. By the property (3.3.22) for symmetric second order tensors, we obtain:

$$d\mathbf{r} = \mathbf{U} \cdot d\mathbf{r}_0 = \mathbf{U} \cdot (ds_0\,\mathbf{a}_i) = \lambda_i\,ds_0\,\mathbf{a}_i \Rightarrow ds = |d\mathbf{r}| = \lambda_i\,ds_0$$

The result shows that and the vectors $d\mathbf{r}$ and $d\mathbf{r}_0$ are parallel to the principal directions $\mathbf{a}_i$ of the tensor $\mathbf{U}$, and that λ_i are stretches, appropriately called the *principal stretches* of the deformation. By definition a stretch has to be positive. Therefore the tensor $\mathbf{U}$, which is now called a *stretch tensor*, has to be a *positive definite tensor*.

The property of the motion (4.5.10) that line elements parallel to the principal directions of the stretch tensor $\mathbf{U}$ do not rotate, is the reason for calling this a *motion of pure strain*. The property will now be further demonstrated in the following example illustrated in Fig. 4.14. For simplicity the translation $\mathbf{u}_0(t)$ is neglected in the example. We choose a coordinate system Ox with base vectors $\mathbf{e}_i$ that are parallel to the principal directions $\mathbf{a}_i$. The deformation gradient $\mathbf{F} = \mathbf{U}$, the deformation tensor $\mathbf{C}$, and the motion have now these representations:

$$F_{ik} = U_{ik} = \lambda_i \delta_{ik}, \quad C_{ik} = \lambda_i^2 \delta_{ik}, \quad \mathbf{r} = \mathbf{U} \cdot \mathbf{r}_o \Leftrightarrow x_i = \lambda_i X_i \tag{4.5.11}$$

In   Fig. 4.14   the   principal   stretches   have   been   chosen   to   be: $\lambda_1 = 3$, $\lambda_2 = 2$, and $\lambda_3 = 1$. The same material volume element is shown in the two configurations K_0 and K. In K_0 the element is chosen to have orthogonal edges parallel to the principal directions $\mathbf{a}_i$. The figure shows that the edges of the same material element in K are orthogonal, and have been stretched but not rotated. Note that the material element may have rotated on its motion from K_0 to K.

The general homogeneous deformation, represented by the motion (4.5.4), may be decomposed into a deformation of pure strain and a rigid–body motion. We are going to demonstrate the decomposition in the following way, using the two-dimensional case in Fig. 4.15 as illustration. For simplicity the translation $\mathbf{u}_0(t)$ is neglected in the demonstration.

The deformation from the reference configuration K_0 at the time t_0 to the present configuration K at the time t is assumed to be given by the deformation gradient $\mathbf{F}$. Then we can determine the deformation tensor $\mathbf{C} = \mathbf{F}^T\mathbf{F}$ and its principal directions $\mathbf{a}_i$. As indicated in Fig. 4.15 a material volume element with orthogonal edges parallel to $\mathbf{a}_i$ in K_0 will also in K have orthogonal edges. This result follows from the fact that according to the formula (4.5.5) the shear strain is zero for any two line elements parallel to the principal directions of the deformation tensor $\mathbf{C}$.

Without regard to the actual motion and deformation history of the material element between the configurations K_0 and K in Fig. 4.15, we may imagine two different alternatives to obtain the deformation represented by the deformation gradient $\mathbf{F}$, each alternative performed in two steps. The alternatives are shown in Fig. 4.16.

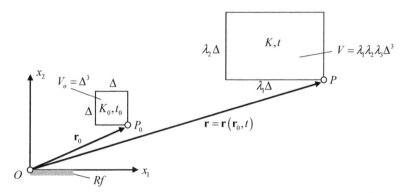

Fig. 4.14 Homogeneous pure strain, $\mathbf{F} = \mathbf{U}$. The cubic material element in the reference configuration K_0 are given principal stretches λ_i in the coordinate directions. The volume of the element is $V_o = \Delta^3$ in K_0 and $V = \lambda_1 \lambda_2 \lambda_3 \Delta^3$ in the present configuration K

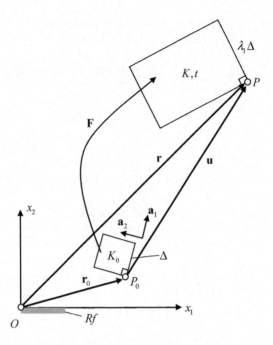

Fig. 4.15 General homogeneous deformation. Two-dimensional presentation of the result of the deformation tensor $\mathbf{C}$ from a reference configuration K_0 to the present configuration K. A cubic material element in K_0 has edges of length Δ that are oriented in the directions of the principal directions $\mathbf{a}_i$ of $\mathbf{C}$. The volume of the cubic element in K_0 is $V_0 = \Delta^3$. The particle $\mathbf{r}_0$, position P_0, moves to the place $\mathbf{r}$, in position P. The deformed material element has orthogonal edges with length $\lambda_i \Delta$, and the volume $V = \lambda_1 \lambda_2 \lambda_3 \Delta^3$. Figure 4.16 presents a polar decomposition of the deformation in two alternatives

Alternative I: The material is first deformed from K_0 to $\overline{K}$ by pure strain, $\mathbf{F} = \mathbf{U}$:

$$\bar{\mathbf{r}} = \mathbf{U} \cdot \mathbf{r}_o, \quad \mathbf{U}^T = \mathbf{U}$$

Then the material is given a rigid-body rotation from $\overline{K}$ to K:

$$\mathbf{r} = \mathbf{R} \cdot \bar{\mathbf{r}}, \quad \mathbf{R}^{-1} = \mathbf{R}^T, \quad \det \mathbf{R} = 1$$

The motion from the reference configuration K_0 to the present configuration K is then:

$$\mathbf{r} = \mathbf{F} \cdot \mathbf{r}_o \Rightarrow \mathbf{F} = \mathbf{RU}, \quad \mathbf{R}^{-1} = \mathbf{R}^T, \det \mathbf{R} = 1, \quad \mathbf{U}^T = \mathbf{U} \qquad (4.5.12)$$

$\mathbf{R}$ is called the *rotation tensor* and $\mathbf{U}$ is called the *right stretch tensor*.

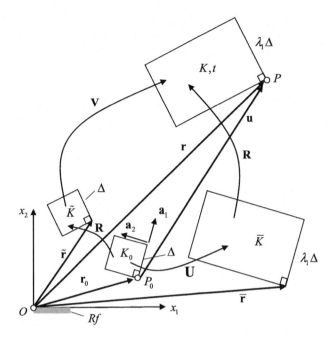

Fig. 4.16 Polar decomposition of a general homogeneous deformation in two alternatives. Alternative I: Deformation by pure strain, $\mathbf{F} = \mathbf{U}$, from K_0 to $\overline{K}$, followed by rigid—body motion, $\mathbf{F} = \mathbf{R}$, from $\overline{K}$ to K. Alternative II: Rigid—body motion, $\mathbf{F} = \mathbf{R}$, from K_0 to $\widetilde{K}$, followed by pure strain, $\mathbf{F} = \mathbf{V}$, from $\widetilde{K}$ to K

Alternative II: The material is first brought from K_0 to $\widetilde{K}$ by a rigid–body rotation:

$$\tilde{\mathbf{r}} = \mathbf{R} \cdot \mathbf{r}_0, \quad \mathbf{R}^{-1} = \mathbf{R}^T, \quad \det \mathbf{R} = 1$$

Next, the material is deformed from $\widetilde{K}$ to K by pure strain:

$$\mathbf{r} = \mathbf{V} \cdot \tilde{\mathbf{r}}, \quad \mathbf{V}^T = \mathbf{V}$$

The motion from the reference configuration K_0 to the present configuration K is then:

$$\mathbf{r} = \mathbf{F} \cdot \mathbf{r}_o \Rightarrow \mathbf{F} = \mathbf{V}\mathbf{R} \tag{4.5.13}$$

$\mathbf{R}$ is again the *rotation tensor* and $\mathbf{V}$ is called the *left stretch tensor*.

Note that the real motion and deformation between K_0 and K will in general be quite different from any of the two special motions presented above.

The important results of the demonstration related to Fig. 4.16 and the Eqs. (4.5.12) and (4.5.13) are that the general homogeneous deformation (4.5.4) may be considered to consist of a pure strain and a rigid–body motion, and furthermore, that the deformation gradient $\mathbf{F}$ may be decomposed into a composition of an orthogonal tensor **R**, which represents a rigid-body rotation, and positive definite symmetric tensors **U** or **V**, which represent pure strain. The decomposition of the deformation gradient **F** may be done in one of two alternatives ways:

$$\mathbf{F} = \mathbf{RU} = \mathbf{VR} \tag{4.5.14}$$

It also follows from how the decompositions (4.5.14) were developed above, that these decompositions are unique. The result (4.5.14) represents the *polar decomposition theorem* presented in Sect. 3.8, where an analytical proof is given.

A mathematical condition for the tensor **F** in the analytical proof is that **F** has to be a *non–singular tensor* with a positive determinant, $\det \mathbf{F} > O$. This condition also follows from the physical condition that the volume dV of a material element must be positive, which again according to the formula (4.2.18) leads to:

$$dV > 0 \Rightarrow \varepsilon_v = \frac{dV - dV_0}{dV_0} = \frac{dV}{dV_0} - 1 = \det \mathbf{F} - 1 > -1 \Rightarrow \det \mathbf{F} > 0 \tag{4.5.15}$$

The two tensors **U** and **V** represent the same state of strain. From Fig. 4.16 it is seen, and from Eq. (4.5.14) it follows, that the tensor **V** is the **R**–rotation of tensor **U**:

$$\mathbf{V} = \mathbf{RUR}^T \tag{4.5.16}$$

The principal directions of **V** are **R** – rotations of the principal directions $\mathbf{a}_i$ of **U**. The two tensors **V** and **U** have the same principal values $\lambda_i > 0$, i.e. the *principal stretches*. The principal stretch λ_i represents the ratio between the lengths in K and K_0 of a material line element that in K_0 is parallel to the principal direction $\mathbf{a}_i$ of **U**, see Fig. 4.15. In reference to their positions relative to the rotation tensor **R** in Eq. (4.5.14), **U** is the *right stretch tensor* and **V** is the *left stretch tensor*.

The right stretch tensor **U** is closely related to the Green deformation tensor **C**.

$$\mathbf{C} = \mathbf{F}^T\mathbf{F} = \mathbf{U}^2 \tag{4.5.17}$$

The two tensors **C** and **U** are coaxial. The principal values ζ_i of **C** and λ_i of **U** and **V** are related by:

$$\zeta_i = \lambda_i^2 \tag{4.5.18}$$

The principal strains are:

$$\varepsilon_i = \lambda_i - 1 \tag{4.5.19}$$

The principal strains ε_i are different from the principal values E_i of Green's strain tensor **E**. Formula (4.2.18) gives:

$$\varepsilon_i = \sqrt{1 + 2E_i} - 1 \qquad (4.5.20)$$

For the volumetric strain we get from Eq. (4.2.18):

$$\varepsilon_v = \det \mathbf{F} - 1 = \det \mathbf{U} - 1 = \det \mathbf{V} - 1 = \lambda_1 \lambda_2 \lambda_3 - 1 \qquad (4.5.21)$$

Green's deformation tensor $\mathbf{C}$ is also called the *right deformation tensor* due to its relation to the right stretch tensor $\mathbf{U}$. A *left deformation tensor* $\mathbf{B}$ related to the left stretch tensor $\mathbf{V}$ plays an important role in the modelling of isotropic materials, and is defined by:

$$\mathbf{B} = \mathbf{V}^2 = \mathbf{F}\mathbf{F}^T = \mathbf{R}\mathbf{C}\mathbf{R}^T \qquad (4.5.22)$$

We see that $\mathbf{B}$ is the $\mathbf{R}$–rotation of $\mathbf{C}$.

In a general, non–homogeneous deformation $\mathbf{r} = \mathbf{r}(\mathbf{r}_0, t)$ the deformation gradient $\mathbf{F} = \mathbf{F}(\mathbf{r}_0, t)$ is a tensor field defined by Eq. (4.2.7):

$$\mathbf{F} = \text{Grad}\,\mathbf{r} = \frac{\partial \mathbf{r}}{\partial \mathbf{r}_0} \mathbf{R}\mathbf{U} = \mathbf{V}\mathbf{R}$$
$$\mathbf{F} = \mathbf{F}(\mathbf{r}_0, t), \quad \mathbf{R} = \mathbf{R}(\mathbf{r}_0, t), \quad \mathbf{U} = \mathbf{U}(\mathbf{r}_0, t), \quad \mathbf{V} = \mathbf{V}(\mathbf{r}_0, t) \qquad (4.5.23)$$

The deformation of the material in the neighbourhood of the particle $\mathbf{r}_0$ is determined by a pure rotation $\mathbf{R}$ and a pure strain $\mathbf{U}$ or $\mathbf{V}$. These tensors fields may be determined as follows. First we compute the deformation gradient $\mathbf{F}$ from its definition in Eq. (4.2.7). Then the Green deformation tensor $\mathbf{C}$, the right stretch tensor $\mathbf{U}$, and the rotation tensor $\mathbf{R}$ may be determined from the formulas:

$$\mathbf{C} = \mathbf{F}^T\mathbf{F} = \mathbf{U}^2 \Rightarrow \mathbf{U} = \sqrt{\mathbf{C}}, \quad \mathbf{F} = \mathbf{R}\mathbf{U} \Rightarrow \mathbf{R} = \mathbf{F}\mathbf{U}^{-1} \qquad (4.5.24)$$

A deformation in a particle is called an *isotropic deformation*, also called a *form invariant deformation*, if any direction through the particle is a principal direction, and there is only one unique principal stretch λ.

$$\mathbf{U} = \mathbf{V} = \lambda \mathbf{1} \Leftrightarrow \mathbf{F} = \lambda \mathbf{R} \Leftrightarrow \mathbf{C} = \mathbf{B} = \mathbf{U} = \lambda^2 \mathbf{1} \qquad (4.5.25)$$

All the tensors $\mathbf{U}$, $\mathbf{V}$, $\mathbf{C}$, and $\mathbf{B}$ are now isotropic tensors. The volumetric strain is:

$$\varepsilon_v = \det \mathbf{F} - 1 = \lambda^3 - 1 \qquad (4.5.26)$$

If the deformation also is homogeneous the shape of any material volume element is unchanged by the deformation.

A deformation in which the volume is preserved is called an *isochoric deformation* or a *volume invariant deformation*. In this case:

$$\varepsilon_v = 0 \Leftrightarrow \det \mathbf{F} = \det \mathbf{U} = \det \mathbf{V} = 1 \qquad (4.5.27)$$

From the definitions of $\mathbf{F}$ in Eq. (4.2.7) and the velocity gradient tensor $\mathbf{L}$ in Eq. (4.4.3) we obtain the relationships:

$$\dot{\mathbf{F}} = \mathbf{L}\mathbf{F} \Leftrightarrow \mathbf{L} = \dot{\mathbf{F}}\mathbf{F}^{-1} \qquad (4.5.28)$$

The development of this result and the following relations (4.5.29–4.5.31) is given as Problem 4.2.

$$\mathbf{D} = \frac{1}{2}\mathbf{R}\big(\dot{\mathbf{U}}\mathbf{U}^{-T} + \mathbf{U}^{-1}\dot{\mathbf{U}}\big)\mathbf{R}^T \qquad (4.5.29)$$

$$\mathbf{W} = \dot{\mathbf{R}}\mathbf{R}^T + \frac{1}{2}\mathbf{R}\big(\dot{\mathbf{U}}\mathbf{U}^{-T} - \mathbf{U}^{-1}\dot{\mathbf{U}}\big)\mathbf{R}^T \qquad (4.5.30)$$

$$\dot{\mathbf{E}} = \frac{1}{2}\mathbf{F}^T\big(\mathbf{L} + \mathbf{L}^T\big)\mathbf{F} = \mathbf{F}^T\mathbf{D}\mathbf{F} \qquad (4.5.31)$$

Note that the motion corresponding to pure strain for large deformation, i.e. $\mathbf{F} = \mathbf{F}(\mathbf{r}_0, t) = \mathbf{U}(\mathbf{r}_0, t)$ and $\mathbf{R}(\mathbf{r}_0, t) = \mathbf{1}$, the motion is not irrotational in the sense that $\mathbf{W} = \mathbf{0}$. Furthermore, we see that "irrotational motion", in the sense that $\dot{\mathbf{R}} = \mathbf{0}$, does not imply that $\mathbf{W} = \mathbf{0}$. For large deformations $\dot{\mathbf{E}} \neq \mathbf{D}$, i.e. the rate of strain tensor is not equal to the rate of deformation tensor. For rigid–body motion $\dot{\mathbf{U}} = \mathbf{0}$, and formula (4.5.30) gives $\mathbf{W} = \dot{\mathbf{R}}\mathbf{R}^T$, which agrees with formula (3.5.22).

For later application we introduce the determinant J of the deformation gradient $\mathbf{F}$:

$$J = \det F = \det(F_{ij}) = \det\left(\frac{\partial x_i}{\partial X_j}\right) = \det \mathbf{F} \qquad (4.5.32)$$

J is called the *Jacobi determinant*, or for short the *Jacobian*, of the mapping from $\mathbf{r}_0$ to $\mathbf{r}(\mathbf{r}_0, t)$, or from X to $x(X, t)$. The name Jacobian is attributed to Carl Gustav Jacob Jacobi [1804–1851]. It may be shown that the material derivative of the Jacobian is given by, see Problem 4.3:

$$\dot{J} = J \operatorname{div} \mathbf{v} \qquad (4.5.33)$$

Problems 4.1–4.3 with solutions see Appendix.

Chapter 5
Constitutive Equations

5.1 Introduction

In this chapter two major theories in *continuum mechanics* are briefly presented. The purpose of presenting these theories in the present book is to show two important applications of Tensor Analysis. Firstly, the theory of linearly elastic materials shows how the stress tensor **T** and the strain tensor **E** for small deformations are connected through *constitutive equations*, to provide a complete theory of wide applications. Secondly, the theory of linearly viscous fluids illustrates how the stress tensor **T** and the velocity field through the rate of deformation tensor **D** are combined through constitutive equations, to provide some important solutions to fluid mechanics problems. The constitutive equations of both theories are tensor equations and will be presented as such, with the component forms in Cartesian coordinates.

In Chap. 7 the basic equations of continuum mechanics are presented in general curvilinear coordinates, and in particular cylindrical coordinates and spherical coordinates. The exposition includes equations of motion, deformation analysis, the constitutive equations for isotropic, linearly elastic materials and the constitutive equations for linearly viscous fluids.

The author's book Continuum Mechanics [1] provides a more extensive presentation of theories of elastic materials, visco-elastic material and advanced fluid models.

5.2 Linearly Elastic Materials

The classical theory of elasticity is primarily a theory of isotropic, linearly elastic materials subjected to small deformations. All governing equations of this theory are linear partial differential equations. This implies that the *principle of*

© Springer Nature Switzerland AG 2019
F. Irgens, *Tensor Analysis*,
https://doi.org/10.1007/978-3-030-03412-2_5

superposition may be applied: The sum of individual solutions to the set of equations also is a solution to the equations. The classical theory of elasticity has a *theorem of uniqueness of solution* and a *theorem of existence of solution*. The theorem of uniqueness insures that if a solution of the pertinent equations and the proper boundary conditions for a particular problem is found, then this solution is the only solution to the problem. The basic equations of the theory are presented. As an illustration of application of the theory some aspects of the propagation of stress waves in elastic materials are presented.

5.2.1   Generalized Hooke's Law

We assume small deformations and small displacements and replace the particle reference $\mathbf{r}_0$ by the place vector $\mathbf{r}$. In a Cartesian coordinate system Ox the particle coordinates X are replaced by the place coordinates x. The displacement vector field $\mathbf{u}(x, t)$ has the components $u_i(x, t)$ and the displacement gradient $H_{ik} = u_{i,k}$. The strains in the elastic material are given by the *strain tensor for small deformations* $\mathbf{E}$ with the components, i.e. coordinate strains:

$$E_{ik} = \frac{1}{2}(u_{i,k} + u_{k,i}) \qquad (5.2.1)$$

A material is called *elastic*, also called a *Cauchy-elastic material*, if the stresses in a particle $\mathbf{r} = x_i \, \mathbf{e}_i$ are functions only of the strains in the particle.

$$T_{ik} = T_{ik}(E, x) \Leftrightarrow \mathbf{T} = \mathbf{T}[\mathbf{E}, \mathbf{r}] \qquad (5.2.2)$$

Equation (5.2.2) are the basic *constitutive equations* for *Cauchy-elastic materials*.

If the elastic properties are the same in every particle in a material, the material is *elastically homogeneous*. If the elastic properties are the same in all directions through one and the same particle, the material is *elastically isotropic*. Metals, rocks, and concrete are in general considered to be both homogeneous and isotropic materials. The crystals in polycrystalline materials are assumed to be small and their orientations so random that the crystalline structure may be neglected when the material is considered to be a continuum. Each individual crystal is normally anisotropic. By milling or other forms of macro-mechanical forming of polycrystalline metals, an originally isotropic material may be anisotropic. Materials with fiber structure and well defined fiber directions have anisotropic elastic response. Wood and fiber reinforced plastic are typical examples. The elastic properties of wood are very different in the directions of the fibers and in the cross-fiber direction.

Isotropic elasticity implies that the principal directions of stress and strain coincide: *The stress tensor and the strain tensor are coaxial.*

Homogeneous elasticity implies that the stress tensor is independent of the particle coordinates. Therefore the constitutive equation of an elastic material should be of the form:

$$T_{ik} = T_{ik}(E) \Leftrightarrow \mathbf{T} = \mathbf{T}[\mathbf{E}] \qquad (5.2.3)$$

If the relation (5.2.2) is linear in **E**, the material is said to be *linearly elastic*. The six coordinate stresses T_{ij} with respect to a coordinate system Ox are now linear functions of the six coordinate strains E_{ij}. For a generally anisotropic material these linear relations contain $6 \times 6 = 36$ coefficients or material parameters, which are called *elasticities* or *stiffnesses*. For a homogeneously elastic material the stiffnesses are constant material parameters.

Figure 5.1 illustrates a bar of isotropic, linearly elastic material with cross-sectional area A and subjected to an axial force N. The state of stress in the bar will be uniaxial with the normal stress $\sigma = N/A$ in the direction of the axis of the bar on a cross-sectional. The bar experiences a small longitudinal strain ε in the direction of the stress and due to isotropy small negative longitudial strains ε_n in any direction normal to the stress. With respect to the Cartesian coordinate system Ox the stress matrix T and a strain matrix E are:

$$T = \begin{pmatrix} \sigma & 0 & 0 \\ 0 & 0 & 0 \\ 0 & 0 & 0 \end{pmatrix}, \quad E = \begin{pmatrix} \varepsilon & 0 & 0 \\ 0 & \varepsilon_n & 0 \\ 0 & 0 & \varepsilon_n \end{pmatrix}$$

For an isotropic, linearly elastic material the following constitutive relations may be stated:

$$\varepsilon = \frac{\sigma}{\eta}, \quad \varepsilon_n = -\nu\varepsilon = -\nu\frac{\sigma}{\eta} \qquad (5.2.4)$$

η is the *modulus of elasticity* and ν is the *Poisson's ratio*, after Siméon Denis Poisson [1781 − 1840].The symbol η for the modulus of elasticity is used rather then the more common symbol E to prevent confusion between the modulus of elasticity and the strain matrix E. The first of Eq. (5.2.4) is known as *Hooke's law*, named after Robert Hooke [1635–1703].

The constitutive relations (5.2.4) may be restated for a general situation of a state of uniaxial stress: $\sigma_1 \neq 0, \sigma_2 = \sigma_3 = 0$, in an isotropic, linearly elastic material:

Fig. 5.1 Isotropic, linearly elastic rod subjected to an axial force N. Normal stress σ on a cross-section

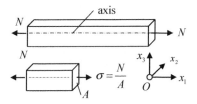

$$\varepsilon_1 = \frac{\sigma_1}{\eta} \quad , \quad \varepsilon_2 = \varepsilon_3 = -\nu\frac{\sigma_1}{\eta}$$

For a general state of triaxial stress with principal stresses σ_1, σ_2, and σ_3, the principal strains ε_i are given by:

$$\varepsilon_1 = \frac{\sigma_1}{\eta} - \frac{\nu}{\eta}(\sigma_2 + \sigma_3) = \frac{1+\nu}{\eta}\sigma_1 - \frac{\nu}{\eta}(\sigma_1 + \sigma_2 + \sigma_3) \text{ etc. for } \varepsilon_2 \text{ and } \varepsilon_3$$

The result follows from the fact that the relations between stresses and strains are linear such that the principal of superposition applies. The result may be rewritten to:

$$\varepsilon_i = \frac{1+\nu}{\eta}\sigma_i - \frac{\nu}{\eta}\text{tr}T, \text{tr}T = \sigma_1 + \sigma_2 + \sigma_3 = \text{tr}\mathbf{T}$$

The expression may also be presented in the matrix representation:

$$\varepsilon_i \delta_{ik} = \frac{1+\nu}{\eta}\sigma_i \delta_{ik} - \frac{\nu}{\eta}(\text{tr}T)\delta_{ik}$$

This is the matrix representation of the following tensor equation between the tensors $\mathbf{E}$ and $\mathbf{T}$ in a coordinate system with base vectors parallel to the principal directions of stress:

$$\mathbf{E} = \frac{1+\nu}{\eta}\mathbf{T} - \frac{\nu}{\eta}(\text{tr}\mathbf{T})\mathbf{1} \tag{5.2.5}$$

In any Cartesian coordinate system Ox Eq. (5.2.5) has the representation:

$$E_{ik} = \frac{1+\nu}{\eta}T_{ik} - \frac{\nu}{\eta}T_{jj}\delta_{ik} \tag{5.2.6}$$

From Eqs. (5.2.5) or equation (5.2.6) the inverse relations between the stress tensor and the strain tensor is obtained:

$$\mathbf{T} = \frac{\eta}{1+\nu}\left[\mathbf{E} + \frac{\nu}{1-2\nu}(\text{tr}\mathbf{E})\mathbf{1}\right] \Leftrightarrow T_{ik} = \frac{\eta}{1+\nu}\left[E_{ik} + \frac{\nu}{1-2\nu}E_{jj}\delta_{ik}\right] \tag{5.2.7}$$

The inversion procedure is given as Problem 5.1. The relations (5.2.5–5.2.7) represent the *generalized Hooke's law* and are the constitutive equations of the *Hookean material* or *Hookean solid*, which are names for an *isotropic, linearly elastic material*. Note that normal stresses T_{ii} only result in longitudinal strains E_{ii}, and visa versa, and that shear stresses T_{ij} only result in shear strains $\gamma_{ik} = 2E_{ik}$, and visa versa. This property is in general not the case for anisotropic materials.

For a state of pure shear stress, as in Example 2.2, the stress matrix T_{ik} related to the local coordinate system in Fig. 2.9 has only one non-zero element, i.e. the shear stress $T_{12} = \tau$. The generalized Hooke's Law (5.2.6) results in only one non-zero strain element: $E_{12} = (1+v)T_{12}/\eta$ for the strain matrix E_{ik}. According to formula (4.3.3) this element corresponds to a shear strain $\gamma = \gamma_{12} = 2E_{12}$. Hence we have found from Hooke's law (5.2.6):

$$\gamma = \frac{2(1+v)}{\eta} \tau \Leftrightarrow \tau = \mu\gamma \tag{5.2.8}$$

The material parameter μ is called the *shear modulus* and is given by:

$$\mu = \frac{\eta}{2(1+v)} \tag{5.2.9}$$

Table 5.1 presents the elasticities η, μ, and v for some characteristic materials. The values given in the table are at the temperature of 20 °C. The values vary some with the quality of the material and with temperature. We shall find that the relationship for the shear modulus in formula (5.2.9) is not exactly satisfied by the data given in Table 5.1. This is due to the fact that the values given in the table are standard values found from different sources.

The relationship between the elastic volumetric strain ε_v and the stresses is found by computing the trace of the strain matrix from Eq. (5.2.6).

$$\varepsilon_v = \mathrm{tr}\, E = E_{ii} = \frac{1+v}{\eta} T_{ii} - \frac{v}{\eta} T_{jj}\, \delta_{ii} = \frac{1-2v}{\eta} T_{ii} \tag{5.2.10}$$

The *mean normal stress* σ^o and the *bulk modulus of elasticity*, also called the *compression modulus of elasticity*, κ are introduced:

$$\sigma^o = \frac{1}{3} T_{ii} = \frac{1}{3} \mathrm{tr}\, \mathbf{T} \quad \text{the mean normal stress}$$

$$\kappa = \frac{\eta}{3(1-2v)} \quad \text{the bulk modulus} \tag{5.2.11}$$

Table 5.1 Properties of some elastic materials

Material	ρ [10^3 kg/m^3]	η [GPa]	μ [GPa]	v	α [10^{-6} °C^{-1}]
Steel	7.83	210	80	0.3	12
Aluminium	2.68	70	26	0.25	23
Concrete	2.35	20–40		0.15	10
Copper	8.86	118	41	0.33	17
Glass	2.5	80	24	0.23	3–9
Cast iron	7.75	103	41	0.25	11

Density ρ, modulus of elasticity η, shear modulus μ, Poisson's ratio v, and thermal expansion coefficient α

The result (5.2.10) may be presented as:

$$\varepsilon_v = \frac{1}{\kappa}\sigma^o \qquad (5.2.12)$$

For an isotropic state of stress $\mathbf{T} = \sigma\mathbf{1}$ the mean normal stress is equal to the normal stress: $\sigma^o = \sigma$.

Fluids are considered as linearly elastic materials when sound waves are analyzed. The only elasticity relevant for fluids is the bulk modulus κ. For water $\kappa = 2.1$ GPa, for mercury 27 GPa, and for alcohol 0.91 GPa.

It follows from Eq. (5.2.11) that a Poisson ratio $v > 0.5$ would have given $\kappa < 0$, which according to Eq. (5.2.12) would lead to the physically unacceptable result that the material increases its volume when subjected to isotropic pressure $p \Rightarrow \sigma^0 = -p$. Furthermore we may expect to find that $v \geq 0$ because a Poisson ratio $v < 0$ would give an expansion in the transverse direction when the material is subjected to uniaxial stress. Thus we may expect that:

$$0 \leq v \leq 0.5 \qquad (5.2.13)$$

The upper limit for the Poisson ratio, $v = 0.5$, which according to Eq. (5.2.11) gives $\kappa = \infty$, characterizes an *incompressible elastic material*. Among the real materials rubber, having $v = 0.49$, is considered to be (nearly) incompressible, while the other extreme, $v = 0$, is represented by cork, and is an advantageous property when corking bottles.

An interesting and sometimes convenient decomposition of Hooke's law (5.2.5-5.2.7) will now be demonstrated. First we decompose the stress tensor $\mathbf{T}$ and the strain tensor $\mathbf{E}$ into isotrops and deviators:

$$\mathbf{T} = \mathbf{T}^o + \mathbf{T}', \quad \mathbf{E} = \mathbf{E}^o + \mathbf{E}' \qquad (5.2.14)$$

$$\mathbf{T}^o = \frac{1}{3}(\mathrm{tr}\,\mathbf{T})\mathbf{1} = \sigma^o\mathbf{1}, \quad \mathbf{E}^o = \frac{1}{3}(\mathrm{tr}\,\mathbf{E})\mathbf{1} = \frac{1}{3}\varepsilon_v\mathbf{1} \qquad (5.2.15)$$

From Eq. (5.2.7) we find that:

$$\mathbf{T}^o = 3\kappa\mathbf{E}^o, \mathbf{T}' = 2\mu\mathbf{E}' \qquad (5.2.16)$$

The development of these results is given as Problem 5.2.

Alternative forms for Hooke's law, Eq. (5.2.8) and Eq. (5.2.6), are:

$$\mathbf{T} = 2\mu\left[\mathbf{E} - \frac{v}{1-2v}(\mathrm{tr}\,\mathbf{E})\mathbf{1}\right] = 2\mu\mathbf{E} + \left(\kappa - \frac{2}{3}\mu\right)(\mathrm{tr}\,\mathbf{E})\mathbf{1} \qquad (5.2.17)$$

$$\mathbf{E} = \frac{1}{2\mu}\left[\mathbf{T} - \frac{v}{1+v}(\mathrm{tr}\,\mathbf{T})\mathbf{1}\right] = \frac{1}{2\mu}\mathbf{T} - \frac{3\kappa - 2\mu}{18\mu\kappa}(\mathrm{tr}\,\mathbf{T})\mathbf{1} \qquad (5.2.18)$$

The parameters:

$$\mu \text{ and } \lambda \equiv \kappa - \frac{2}{3}\mu \tag{5.2.19}$$

are called the *Lamé constants* after Gabriel Lamé [1795–1870]. The parameter λ does not have any independent physical interpretation.

For an *incompressible elastic material*, i.e. $\varepsilon_v = \text{tr } \mathbf{E} = 0$, the mean normal stress σ^o cannot be determined from Hooke's law. For these materials it is customary to replace Eq. (5.2.17) by:

$$\mathbf{T} = -p\,\mathbf{1} + 2\mu\,\mathbf{E} \tag{5.2.20}$$

$p = p(\mathbf{r}, t)$ is an unknown pressure, which is an unknown tension if p is negative. The pressure p can only be determined from the equations of motion and the corresponding boundary conditions.

5.2.2   Some Basic Equations in Linear Elasticity. Navier's Equations

The primary objective of the theory of elasticity is to provide methods for calculating stresses, strains, and displacements in elastic bodies subjected to body forces and prescribed boundary conditions for contact forces and/or displacements on the surface of the bodies. The basic equations of the theory are:

The Cauchy equations of motion (2.2.37):

$$\text{div } \mathbf{T} + \rho\,\mathbf{b} = \rho\,\ddot{\mathbf{u}} \Leftrightarrow T_{ik,k} + \rho\,b_i = \rho\,\ddot{u}_i \tag{5.2.21}$$

Hooke's law for isotropic, linearly elastic materials, e.g. (5.2.17):

$$\mathbf{T} = 2\mu\left[\mathbf{E} + \frac{v}{1-2v}(\text{tr }\mathbf{E})\mathbf{1}\right] \Leftrightarrow T_{ik} = 2\mu\left[E_{ik} + \frac{v}{1-2v}E_{jj}\,\delta_{ik}\right] \tag{5.2.22}$$

The strain–displacement relations:

$$\mathbf{E} = \frac{1}{2}\left(\mathbf{H} + \mathbf{H}^T\right) \Leftrightarrow E_{ik} = \frac{1}{2}\left(u_{i,k} + u_{k,i}\right) \tag{5.2.23}$$

The component form of these equations applies only to Cartesian coordinate systems. Component versions in general coordinate systems of the basic equations are presented in Sect. 7.8.

Equations (5.2.21–5.2.23) represent 15 equations for the 15 unknown functions T_{ik}, E_{ik}, and u_i. The boundary conditions are expressed by the contact forces $\mathbf{t}$ and the displacements **u** on the surface A of the body. The part of the surface A on which **t** is prescribed is denoted A_σ. On the rest of the surface, denoted A_u, we assume that the displacement **u** is prescribed. For static problems ($\ddot{\mathbf{u}} = \mathbf{0}$) the boundary conditions are:

$$\mathbf{t} = \mathbf{T} \cdot \mathbf{n} = \mathbf{t}^* \text{on} A_\sigma, \quad \mathbf{u} = \mathbf{u}^* \text{on} A_u \qquad (5.2.24)$$

$\mathbf{t}^*$ and $\mathbf{u}^*$ are the prescribed functions, and **n** is the unit normal vector on A_σ. We may have cases where the boundary conditions on parts of A are given as combinations of prescribed components of the contact force **t** and displacement **u**. For dynamic problems conditions with respect to time must be added, for instance as the initial conditions on the displacement field $\mathbf{u}(\mathbf{r}, t)$:

$$\mathbf{u} = \mathbf{u}^{\#}(\mathbf{r}, 0) \quad \text{and} \, \dot{\mathbf{u}} = \dot{\mathbf{u}}^{\#}(\mathbf{r}, 0) \text{ in } V \qquad (5.2.25)$$

V Denotes the volume of the body. $\mathbf{u}^{\#}(\mathbf{r}, 0)$ and $\dot{\mathbf{u}}^{\#}(\mathbf{r}, 0)$ are prescribed functions

We will now let the displacement vector $\mathbf{u}(\mathbf{r}, t)$ be the primary unknown variable. The strain–displacement relation (5.2.23) is introduced in Hooke's law (5.2.22), and the result is:

$$T_{ik} = \mu \left[u_{i,k} + u_{k,i} + \frac{2v}{1 - 2v} u_{j,j} \, \delta_{ik} \right] \qquad (5.2.26)$$

When these expressions for the stresses are substituted into the Cauchy equations of motion (5.2.21), we obtain the equations of motion in terms of displacements.

$$\nabla^2 \mathbf{u} + \frac{1}{1 - 2v} \nabla(\nabla \cdot \mathbf{u}) + \frac{\rho}{\mu}(\mathbf{b} - \ddot{\mathbf{u}}) = \mathbf{0} \Leftrightarrow u_{i,kk} + \frac{1}{1 - 2v} u_{k,ki} + \frac{\rho}{\mu}(b_i - \ddot{u}_i) = 0$$

$$(5.2.27)$$

These three equations are *Navier's equations* named after Claude L. M. H. Navier [1785–1836] When the three displacement functions u_i – functions are found from Navier's equations, the strains and the stresses may be computed from Eqs. (5.2.23) and (5.2.26) respectively. The boundary conditions (5.2.24) and (5.2.25) will complete the solution of finding displacements, strains and stresses.

Navier's equations in cylindrical coordinates (R, θ, z) and in spherical coordianates (r, θ, ϕ) are presented in Sect. 7.8.

5.2.3 Stress Waves in Elastic Materials

In this section we discuss some relatively simple but fundamental aspects of propagation of *stress pulses* or *stress waves* in isotropic, linearly elastic materials. We may also consider these pulses or waves as displacement, deformation, or strain pulses or waves, according to which of these quantities we are interested in. In the general presentation we do not distinguish between pulses and waves, and since the materials are assumed to be elastic, we use the common name *elastic waves*.

In a large body of isotropic elastic material a relatively small region is subjected to a mechanical disturbance eminating from a point source. The disturbance may be considered to be a displacement field propagating into the undisturbed material as displacement waves. Sufficiently far away from the source the displacement propagates as approximately plane waves.

Figure 5.2 illustrates the situation. The point source is placed at the origin of a Cartesian coordinate system Ox. In the neighbourhood of the $x_3 -$ axis and at a long distance from the source we may consider the displacement propagates as approximately plane waves. assume the displacement field:

$$\mathbf{u} = \mathbf{u}(x_3, t) \Leftrightarrow u_i = u_i(x_3, t) \tag{5.2.28}$$

We shall see that the displacement field $\mathbf{u} = \mathbf{u}(x_3, t)$ really represents three different motions.

The motions are governed by the Navier Eq. (5.2.27). If we neglect the body forces, the equations are reduced to:

$$u_{i,kk} + \frac{1}{1 - 2v} u_{k,ki} = \frac{\rho}{\mu} \ddot{u}_i \tag{5.2.29}$$

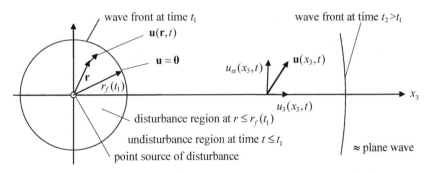

Fig. 5.2 Displacement disturbance $\mathbf{u}(r,t)$ from a point source at the origin O. Wave front $\mathbf{u} = \mathbf{0}$ on a sphere of radius $r = r_f$ and at time $t = t_1$. Approximately plane wave $\mathbf{u}(x_3, t) = u_\alpha(x_3, t)\mathbf{e}_\alpha + u_3(x_3, t)\mathbf{e}_3$ far away from the source at O. The displacement $u_3(x_3, t)$ represents a motion in the x_3-direction and is called a longitudinal wave propagating with the velocity c_l from formula (5.2.30). The displacements $u_3(x_3, t)$ represent motions in the x_α-directions and are called transverse waves propagating with the velocity c_t from formula (5.2.31)

The displacement field (5.2.28) is substituted into Eq. (5.2.29), and the result is three *one-dimensional wave equations*: one equation for the displacement $u_3(x_3, t)$ and two equations for the displacements $u_\alpha(x_3, t)$ for $\alpha = 1$ or 2.

$$c_l \frac{\partial^2 u_3}{\partial x_3^2} = \frac{\partial^2 u_3}{\partial t^2} \quad \text{where } c_l = \sqrt{\frac{2(1-v)}{1-2v}\frac{\mu}{\rho}} \equiv \sqrt{\frac{1-v}{(1+v)(1-2v)}\frac{\eta}{\rho}} \equiv \sqrt{\frac{\kappa + 4\mu/3}{\rho}}$$

(5.2.30)

$$c_t \frac{\partial^2 u_\alpha}{\partial x_3^2} = \frac{\partial^2 u_\alpha}{\partial t^2} \quad \text{where } \quad c_t = \sqrt{\frac{\mu}{\rho}} = \sqrt{\frac{1}{2(1+v)}\frac{\eta}{\rho}} \tag{5.2.31}$$

The displacement $u_3(x_3, t)$ represents a motion in the direction of the propagation, i.e. the x_3 − direction, and is called a *longitudinal wave*. The displacements $u_\alpha(x_3, t)$ represent motions in the directions normal to the direction of propagation, and are called *transversal waves*.

The general solutions of the wave Eqs. (5.2.30) and (5.2.31) are given as:

$$u_3(x_3, t) = f_3(c_l t + x_3) + g_3(c_l t - x_3) \tag{5.2.32}$$

$$u_\alpha(x_3, t) = f_\alpha(c_t t + x_3) + g_\alpha(c_t t - x_3) \tag{5.2.33}$$

The functions $f_i(k)$ and $g_i(k)$ are arbitrary functions of one variable k. Substitution of the displacements (5.2.32) and (5.2.33) into the wave Eqs. (5.2.30) and (5.2.31) will show that the equations are satisfied. It will be shown that the parameters c_l and c_t represent velocities of propagation of the displacement waves.

Figure 5.3 shall illustrate the propagation of the longitudinal wave $u_3(x_3, t) = g_3(c_l t - x_3)$. The graphs show the displacement $u_3(x_3, t) = g_3(c_l t - x_3)$ at two different times t and $t + \Delta t$. At the time t the displacement has reached the position $x_3 = x_{3f}$ which represents the *wave front*. At the later time $t + \Delta t$ the displacement has reached the position $x_3 = x_{3f} + \Delta x_3$ which represents the new position of the wave front. It follows that:

$u_3(x_{3f}, t) = g_3(c_l t - x_{3f}) = 0$
$u_3(x_{3f} + \Delta x_3, t + \Delta t) = g_3(c_l\{t + \Delta t\} - \{x_{3f} + \Delta x_3\}) = g_3(c_l t - x_{3f} + \{c_l \Delta t - \Delta x_3\}) = 0 \Rightarrow$
$c_l \Delta t - \Delta x_3 = 0 \Rightarrow \Delta x_3 = c_l \Delta t$

The result shows that the motion $u_3(x_3, t) = g_3(c_l t - x_3)$ can be characterized as a displacement wave that propagates in the positive x_3 − direction with the wave front velocity c_l. Using similar argument we may show that the motion $u_3(x_3, t) = f_3(c_l t + x_3)$ represents a displacement wave that propagates in the negative x_3 − direction with the *wave front velocity* c_l.

It also follows that the motions $u_\alpha(x_3, t) = f_\alpha(c_t t - x_3)$ and $u_\alpha + (x_3, t) = f_\alpha(c_t t + x_3)$ may represent motions in the x_α − directions propagating respectively in

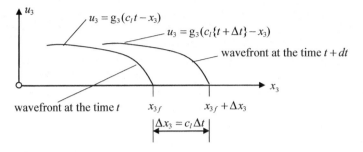

Fig. 5.3 Propagation of the longitudinal displacement wave $u_3(x_3, t) = g_3(c_l t - x_3)$. The positions of the wave are shown at the time t and the time $t + dt$. The wave fronts $x_3 = x_{3f}$ from $u_3(x_{3f}, t) = g_3(c_l t - x_{3f}) = 0$ and $x_3 = x_{3f} + \Delta x_3$ from $u_3(x_{3f} + \Delta x_3, t + \Delta t) = g_3(c_l t + \Delta t\} - x_{3f} + \Delta x_3) = 0$. c_l = velocity of propagation of longitudinal elastic waves

the positive x_3 – direction and in the neggative x_3 – direction with the wave front velocity c_t. These motions are therefore called transversal waves.

Because for known materials $0 \leq \nu \leq 0.5$ we will find that in general:

$$c_t < c_l \qquad (5.2.34)$$

For steel with $\nu = 0.3, \eta = 210\,\text{GPa}$, and $\rho = 7830\,\text{kg/m}^3$, we find $c_t = 3212$ m/s and $c_l = 6009$ m/s.

The longitudinal wave implies volume changes:$\varepsilon_v = u_{3,3} = E_{33}$, and the wave is therefore called a *dilatational wave* or *volumetric wave*. Because:

$$\text{rot}(u_3 \mathbf{e_3}) \equiv \text{curl}(u_3 \mathbf{e_3}) \equiv \nabla \times (u_3 \mathbf{e_3}) = \mathbf{0}$$

the displacement $u_3(x_3, t)$ is also called an *irrotational wave*. A physical consequence of an irrotational wave is that the principal directions of strain do not rotate. For the displacement field $u_3(x_3, t)$ the x_i – directions are the principal strain directions. According to Hooke's law (5.2.26) the stresses are determined by:

$$T_{33} = \frac{2(1 - \nu)\mu}{1 - 2\nu} u_{3,3}, \quad T_{11} = T_{22} = \frac{\nu}{1 - \nu} T_{33} \qquad (5.2.35)$$

The transverse waves $u_1(x_3, t)$ and $u_2(x_3, t)$ are isochoric, i.e. $\varepsilon_v = 0$, and are therefore also called *dilatation free waves* or *equivoluminal waves*. The stresses are according to Hooke's law (5.2.26):

$$T_{13} = \mu u_{1,3}, \quad T_{23} = \mu u_{2,3} \qquad (5.2.36)$$

Because these waves also represent shear stresses on planes normal to the direction of propagation, they are also called *shear waves*. Other names are *distortional waves* and *rotational waves*.

An earthquake initiates three elastic displacement waves. The fastest wave is a longitudinal wave called the *primary wave*, or the *P − wave*. The second fastest wave is a transverse wave, a shear wave, called the *secondary wave*, or the *S − wave*. Both waves propagate from the earth quake region in all directions and their intensities, or energy per unit area, decrease with the square of the distance from the earth quake. These two waves therefore are registered by relatively weak signal on a seismograph far away from the epicenter of the earth quake. The third, and generally the strongest wave propagates along the surface of the earth and is called a surface wave.

5.3   Linearly Viscous Fluids

5.3.1 Definition of Fluids

In continuum mechanics it is natural to define a fluid on the basis of what seems to be the most characteristic macromechanical aspects of liquids and gases as opposed to solid materials. Due to the fact that liquids and gases behave macroscopically similarly, the equations of motion have the same form and the most common constitutive models applied are in principle the same for liquids and gases. A fluid is thus a model for a liquid or a gas.

A common property of liquids and gases is that at rest they can only transmit pressure on solid boundaries or interfaces to other liquids. Figure 5.4a shows a small liquid volume element subjected to the pressure p. The liquid is in an isotropic state of stress. Shear stresses from the liquid on solid boundaries will only occur when there is relative motion between the boundary and the liquid. The capability to sustain shear stresses in a fluid is expressed by the viscosity of the fluid. In a fluid in motion shear stresses will always be present on material surfaces and the fluid is in an *anisotropic state of stress*, as illustrated in Fig. 5.4b. However, in some flow problems, especially with gasses, the shear stresses are very small and may be neglected in the part of the flow analysis. Based on these remarks we may choose that following general definition of fluids as a continuous material:

A fluid is a material that deforms continuously when subjected to anisotropic states of stress.

The constitutive equations of any fluid at rest relative to any reference must reduce to:

$$\mathbf{T} = -p\mathbf{1}, \quad p = p(\rho, \theta) \tag{5.3.1}$$

p is the *thermodynamic pressure*, which is a function of the *density* ρ and the *temperature* θ. The relationship for p in the formulas (5.3.1) is called an *equation of state*.

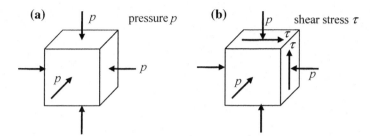

Fig. 5.4 a Stresses on a volume element in a fluid at rest. Isotrop state of stress **b** Stresses on a volume element in a fluid in motion. Anisotrop state of stress

An *ideal gas* is defined by the equation of state:

$$p = R\rho\theta \tag{5.3.2}$$

R is the *gas constant* for the gas, and θ is the absolute temperature, given in degrees Kelvin. The model ideal gas may be used with good results for many real gases, for example for air.

Due to the large displacements and the chaotic motions of the fluid particles it is in general impossible to follow the motion of the individual particles. Therefore the physical properties of the particles or quantities related to particles are observed or described at fixed positions in space. In other words we employ *spacial description* and *Eulerian coordinates*. The primary kinematic quantity in Fluid Mechanics is the velocity vector $\mathbf{v}(\mathbf{r},t)$.

An instrument that measures the viscosity in a liquid is called a *viscometer*. Figure 5.5 illustrates a *cylinder viscometer*. A test fluid is filled in the annular space between a cylindrical container and a cylinder. The container is at rest while the cylinder is subjected to a torque m_t and is made to rotate with a constant angular velocity omega. The rotation creates a flow in the annular space between the cylinder and the inner container wall. It is assumed that the distance h is very small as compared to the radius r and that the flow in the annular space may be considered as the flow between to parallel surfaces, represented by the moving wall of the rotating cylinder and the fixed inner wall of the container, as shown in the cut out in Fig. 5.5 It is also assumed that the fluid sticks to the solid wall. With respect to the local coordinate axes the velocity field in the annular space is assumed to be:

$$v_1(x_2) = \frac{v}{h}x_2, \quad v_2 = v_3 = 0, \quad v = \omega r \tag{5.3.3}$$

A small liquid element in the cut out in Fig. 5.5 is subjected to normal stresses and the shear stress τ. The rate of deformation in the flow is represented by the shear strain rate:

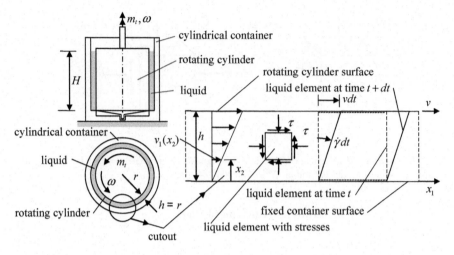

Fig. 5.5 Cylinder viscometer. Rotating cylinder of radius r in a cylindrical container. Test liquid in annular space of thickness $h \ll r$. The cylinder is subjected to a torque m_t and rotates with angular velocity ω

$$\dot{\gamma} = 2(D_{12}) = 2\left(\frac{1}{2}(v_{1,2} + v_{2,1})\right) = v_{1,2} = \frac{v}{h} \Rightarrow \dot{\gamma} = \frac{r}{h}\omega \qquad (5.3.4)$$

For the case of steady flow at constant angular velocity the torque m_t is balanced by the shear stress τ on the wall of the cylinder:

$$(\tau \cdot r) \cdot (2\pi r \cdot H) = M \quad \Rightarrow \quad \tau = \frac{m_t}{2\pi r^2 H} \qquad (5.3.5)$$

The viscometer records the relationship between the torque m_t and the angular velocity ω. Using the formulas (5.3.2) and (5.3.3) we obtain a relationship between the shear stress τ and the shear rate $\dot{\gamma}$.

A liquid is said to be purely viscous if the shear stress τ is a function of only $\dot{\gamma}$:

$$\tau = \tau(\dot{\gamma}) \Leftrightarrow \text{purely viscous fluid} \qquad (5.3.6)$$

A *Newtonian fluid* is a purely viscous fluid with a linear constitutive equation for the shear stress:

$$\tau = \mu\dot{\gamma} \Leftrightarrow \text{Newtonian fluid} \qquad (5.3.7)$$

The coefficient μ is called the *viscosity* of the fluid and has the unit Ns/m^2. The viscosity varies with temperature and to a certain extent with the pressure in the fluid. For water at $0\,^\circ C$ $\mu = 1.8 \times 10^{-3} Ns/m^2$ and at $20\,^\circ C$ $\mu = 1.0 \times 10^{-3} Ns/m^2$.

In some fluid flow problems the viscosity plays a minor role in describing the flow and may be neglected in the flow analysis. The fluid is then characterized as a *perfect fluid* or an *Eulerian fluid*.

5.3.2 The Continuity Equation

A small part of a fluid body of volume dV and density ρ has the mass $dm = \rho dV$. The principle of conservation of mass implies that the mass dm is constant, while both the density ρ and the volume dV may change with time. From Equation (4.4.11) we have found that the rate of change in volume per unit volume and per unit of time is given by div $\mathbf{v}$. Therefore we may write:

$$dm = \rho dV = \text{constant in time} \Rightarrow \frac{d(dm)}{dt} = 0 \Rightarrow$$

$$\dot{\rho}dV + \rho\frac{d(dV)}{dt} = 0 \Rightarrow \dot{\rho}dV + \rho\text{div }\mathbf{v}\,dV = 0 \Rightarrow$$

$$\dot{\rho} + \rho\,\text{div }\mathbf{v} = 0 \tag{5.3.8}$$

If we introduce the formula (2.1.17) for material derivative of an intensive quantity, Eq. (5.3.3) may be rewritten to:

$$\frac{\partial\rho}{\partial t} + \text{div}\,(\rho\mathbf{v}) = 0 \tag{5.3.9}$$

Equations (5.3.3) and (5.3.4) are both called the *continuity equation for a fluid particle*. If the fluid is incompressible the rate of volumetric strain is zero, and the continuity Eq. (5.3.3) is reduced to the *incompressibility condition*:

$$\text{div }\mathbf{v} = 0 \tag{5.3.10}$$

5.3.3 Constitutive Equations for Linearly Viscous Fluids

George Gabriel Stokes [1819–1903] presented the following four criteria for the relationships between the stresses and the velocity field in a viscous fluid:

1. The stress tensor $\mathbf{T}$ is a linear function of the rate of deformation tensor $\mathbf{D}$.
2. The stress tensor $\mathbf{T}$ is explicitly independent of the particle coordinates, which implies that the fluid is homogeneous.
3. When the fluid is not deforming, i.e. $\mathbf{D} = \mathbf{0}$, the stress tensor is: $\mathbf{T} = -p(\rho, \theta)\mathbf{1}$, where $p(\rho, \theta)$ is the *thermodynamic pressure*.
4. Viscosity is an isotropic property.

To fulfil these assumptions the constitutive equation must express the stress tensor **T** as a linear function of **D** and a function of the density ρ and the temperature θ. The following function satisfies the assumptions:

$$\mathbf{T} = -p(\rho, \theta)\mathbf{1} + 2\mu\mathbf{D} + \left(\kappa - \tfrac{2\mu}{3}\right)(\operatorname{tr}\mathbf{D})\mathbf{1} \Leftrightarrow$$
$$T_{ij} = -p(\rho, \theta)\delta_{ij} + 2\mu D_{ij} + \left(\kappa - \tfrac{2\mu}{3}\right)D_{kk}\delta_{ij} \tag{5.3.11}$$

The scalar parameters μ and κ will in general be functions of the density ρ and the temperature θ and are called the (shear) *viscosity* and the *bulk viscosity*. If we compare these equations with Eq. (5.2.17) for a Hookean solid, we understand that the isotropy assumption for the fluid is assured. Equation (5.3.11) are the *constitutive equations of linearly viscous fluids*, or what is called *Newtonian fluids*.

If the symmetric tensors **T** and **D** are decomposed into isotrops and deviators, the constitutive Eq. (5.3.7) take the alternative form:

$$\mathbf{T}^o = -p(\rho, \theta)\mathbf{1} + 2\mu\mathbf{D}^o \Leftrightarrow T_{ij}^o = -p(\rho, \theta)\delta_{ij} + 3\kappa D_{ij}^o \tag{5.3.12}$$

$$\mathbf{T}' = 2\mu\mathbf{D}' \Leftrightarrow T_{ij}' = 2\mu D_{ij}' \tag{5.3.13}$$

For isotropic states of stress, we may replace Eq. (5.3.7) by:

$$\mathbf{T} = -\tilde{p}\mathbf{1}, \quad \tilde{p} = p(\rho, \theta) - \kappa\dot{\varepsilon}_v, \quad \dot{\varepsilon}_v = \operatorname{tr}\mathbf{D} = \operatorname{div}\mathbf{v} = -\frac{\dot{\rho}}{\rho} \tag{5.3.14}$$

Note that the total pressure is not the same as the thermodynamic pressure $p(\rho, \theta)$.

The bulk viscosity κ expresses the resistance of the fluid toward rapid volume changes. Due to the fact that it is difficult to measure κ, values are hard to find in the literature. Kinetic theory of gasses shows that $\kappa = 0$ for monatomic gasses. But as shown by Truesdell in J. Rat. Mech. Anal. V.1 (1952) [2], this result is implied in the stress assumption that is the basis for the kinetic theory. Experiments show that for monatomic gasses it is reasonable to set $\kappa = 0$, while for other gasses and for all liquids the bulk viscosity κ, and values of λ, are larger than, and often much larger than μ. The assumption $\kappa = 0$, sometimes taken for granted in older literature on Fluid Mechanics, is called the *Stokes relation*, since it was introduced by him. However, Stokes did not really believe the relation to be relevant. Usually the deviator **D'** dominates over **D**^{o} such that the effects of the bulk viscosity are small. The bulk viscosity κ has dominating importance for the dissipation and absorption of sound energy.

In modern literature (5.3.6) is sometimes called the *Cauchy-Poisson law*. For an incompressible fluid, for which $\operatorname{tr}\mathbf{D} = 0$, Eq. (5.3.6) has to be replaced by:

$$\mathbf{T} = -p(\mathbf{r},t)\mathbf{1} + 2\mu\mathbf{D} \Leftrightarrow T_{ij} = -p(\mathbf{r},t)\delta_{ij} + 2\mu D_{ij} \qquad (5.3.15)$$

The pressure $p(\mathbf{r},t)$ is a function of position $\mathbf{r}$ and time t, and can only be determined from the equations of motion and the boundary conditions. An equation of state, $p = p(\rho,\theta)$, loses its meaning when incompressibility is assumed.

5.3.4   The Navier–Stokes Equations

The general equations of motion of a linearly viscous fluid are called the *Navier–Stokes equations*. These equations are obtained by the substitution of the constitutive Eq. (5.3.6) into the Cauchy equations of motion (2.2.37). If it is assumed that the viscosities μ and κ may be considered to be constant parameters, the resulting equations are:

$$\partial_t\mathbf{v} + (\mathbf{v}\cdot\nabla)\mathbf{v} = -\frac{1}{\rho}\nabla p + \frac{\mu}{\rho}\nabla^2\mathbf{v} + \frac{1}{\rho}\left(\kappa + \frac{\mu}{3}\right)\nabla(\nabla\cdot\mathbf{v}) + \mathbf{b} \qquad (5.3.16)$$

In a Cartesian coordinate system the Navier–Stokes equations are:

$$\partial_t v_i + v_k v_{i,k} = -\frac{1}{\rho}p_i + \frac{\mu}{\rho}v_{i,kk} + \frac{1}{\rho}\left(\kappa + \frac{\mu}{3}\right)v_{k,ki} + b_i \qquad (5.3.17)$$

For incompressible fluids $\nabla\cdot\mathbf{v} = 0$, and Eqs. (5.3.16) and (5.3.17) are reduced to:

$$\begin{aligned} \partial_t\mathbf{v} + (\mathbf{v}\cdot\nabla)\mathbf{v} &= -\tfrac{1}{\rho}\nabla p + \tfrac{\mu}{\rho}\nabla^2\mathbf{v} + \mathbf{b} \Leftrightarrow \\ \partial_t v_i + v_k v_{i,k} &= -\tfrac{1}{\rho}p_{,i} + \tfrac{\mu}{\rho}v_{i,kk} + b_i \end{aligned} \qquad (5.3.18)$$

The Navier–Stokes Eqs. (5.3.16–5.3.18) are the most important equations in the study of viscous fluids. The complexity of the equations indicates that analytical solutions in most cases require major simplifications and approximations. Modern computer codes make it possible to use the Navier–Stokes equations in numerical solutions of very complex fluid flow problems.

The Navier–Stokes equations in general coordinates and in particularly in cylindrical coordinates and spherical coordinates are presented in Sect. 7.9.

5.3.5   Film Flow

As an example of application of the basic equations of fluid mechanics, we shall determine the flow of a linearly viscous and incompressible fluid along an inclined conveyor belt presented in Fig. 5.6. The width of the belt is b and the angle of inclination is α. The conveyor belt moves with a constant velocity v_0. The fluid has the density ρ, the viscosity μ, and moves in a constant gravitational field given by the body force:

$$\mathbf{b} = -g \, \sin \alpha \, \mathbf{e}_1 - g \, \cos \alpha \, \mathbf{e}_2 \tag{5.3.19}$$

We assume stationary flow of the fluid in a film of constant height h and with the velocity field, as shown in Fig. 5.6:

$$v_1 = v_1(x_2), \quad v_2 = v_3 = 0 \tag{5.3.20}$$

The velocity field satisfies the incompressibility condition:

$$\operatorname{div} \mathbf{v} = 0 \Rightarrow v_{i,i} = \frac{\partial v_1}{\partial x_1} = 0 \tag{5.3.21}$$

With an *atmospheric pressure* of p_a at the free fluid surface at $x_2 = h$ the boundary conditions in the problem are:

$$
\begin{aligned}
v_1(0) &= v_0 \\
T_{22} &= -p_a, \quad T_{12} = 0 \quad \text{at} \, x_2 = h
\end{aligned}
\tag{5.3.22}
$$

The velocity field (5.3.20) results in the following non-zero rates of deformations and stresses in the fluid:

$$
\begin{aligned}
D_{ij} &= \tfrac{1}{2}\left(v_{i,j} + v_{j,i}\right), \quad T_{ij} = -p(\rho, \theta)\delta_{ij} + 2\mu D_{ij} + \left(\kappa - \tfrac{2\mu}{3}\right)D_{kk}\,\delta_{ij} \Rightarrow \\
D_{12} &= \tfrac{1}{2}\tfrac{dv_1}{dx_2}, \quad T_{11} = T_{22} = T_{33} = -p, \quad T_{12} = \mu \tfrac{dv_1}{dx_2}
\end{aligned}
\tag{5.3.23}
$$

The third of the boundary conditions (5.3.22) is now rewritten to:

$$T_{12} = 0 \, \text{at} \, x_2 = h \Rightarrow \frac{dv_1}{dx_2} = 0 \, \text{at} \, x_2 = h \tag{5.3.24}$$

Fig. 5.6 Flow of a viscous fluid film on a conveyor belt. The belt is moving with constant velocity v_0

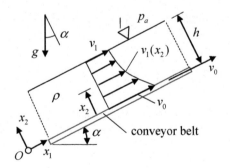

The velocity field (5.3.20) results in zero acceleration:

$$\partial_t v_1 + v_k v_{1,k} = 0, \quad \partial_t v_2 + v_k v_{2,k} = 0, \quad \partial_t v_3 + v_k v_{3,k} = 0 \tag{5.3.25}$$

The Navier–Stokes Eq. (5.3.18) for an incompressible fluid are reduced to:

$$0 = -\frac{1}{\rho}\frac{\partial p}{\partial x_1} + \frac{\mu}{\rho}\frac{d^2 v_1}{dx_2^2} - g\sin\alpha, \quad 0 = -\frac{1}{\rho}\frac{\partial p}{\partial x_2} - g\cos\alpha, \quad 0 = -\frac{1}{\rho}\frac{\partial p}{\partial x_3}$$
$$\tag{5.3.26}$$

First we conclude from these equations that the pressure can only be a function of x_1 and x_2, and furthermore we find:

$$\frac{\partial}{\partial x_1}\left(\frac{\partial p}{\partial x_1}\right) = 0, \quad \frac{\partial}{\partial x_1}\left(\frac{\partial p}{\partial x_2}\right) = \frac{\partial}{\partial x_2}\left(\frac{\partial p}{\partial x_1}\right) = 0 \Rightarrow \frac{\partial p(x_1, x_2)}{\partial x_1} = \text{constant} = c$$

Integrations of the last equation and the second of Eq. (5.3.26) yield:

$$p = p(x_1, x_2) = -(\rho g\cos\alpha)x_2 + cx_1 + C_1 \tag{5.3.27}$$

From the boundary conditions (5.3.22) and (5.3.24), and the result (5.3.27) we obtain:

$$-(\rho g\cos\alpha)h + cx_1 + C_1 = p_a \Rightarrow c = 0, \quad C_1 = p_a + (\rho g\cos\alpha)h \tag{5.3.28}$$

The general expression for the pressure is therefore:

$$p = p(x_2) = p_a + (h - x_2)\rho g\cos\alpha \tag{5.3.29}$$

The pressure is thus only a function of x_2.

With the expression for the pressure (5.3.29) Eq. (5.3.26)$_1$ gives:

$$\frac{d^2 v_1}{dx_2^2} = \frac{\rho g}{\mu}\sin\alpha \Rightarrow v_1(x_2) = \frac{\rho g}{\mu}\sin\alpha\frac{x_2^2}{2} + C_2 x_2 + C_3 \tag{5.3.30}$$

The boundary conditions (5.3.22) and (5.3.24) yield:

$$v_1(0) = v_0 = C_3$$
$$\frac{dv_1}{dx_2}\frac{dv_1}{dx_2}\Big|_{x_2=h} = 0 = \frac{\rho g}{\mu}\sin\alpha\frac{h^2}{2} + C_2 h \Rightarrow C_2 = -\frac{\rho g}{\mu}\sin\alpha\frac{h}{2}$$

Hence the velocity field for the film flow is given by:

$$v_1(x_2) = v_0 - \frac{\rho g h^2 \sin\alpha}{2\mu}\left[\frac{2x_2}{h} - \left(\frac{x_2}{h}\right)^2\right], \quad v_1(h) = v_0 - \frac{\rho g h^2 \sin\alpha}{2\mu}$$

The volumetric flow on the belt is:

$$Q = w \int_0^h v_z(y)\, dy = w \int_0^h \left\{ v_o - \frac{\rho g \sin \alpha\, h^2}{2\mu} \left[1 - \left(1 - \frac{y}{h}\right)^2 \right] \right\} dy \Rightarrow Q$$
$$= v_o wh - \frac{\rho g \sin \alpha\, w h^3}{3\mu}$$

Problems 5.1–5.2 with solutions see Appendix.

References

1. Irgens F (2008) Continuum Mechanics. Springer, Berlin
2. Truesdell C (1952) The mechanical foundations of elasticity and fluid dynamics. J Rational Mech Anal 1:125–300. Corrected reprint, Intl Sci Rev Ser. Gordon Breach 1965, New York

Chapter 6
General Coordinates in Euclidean Space E₃

6.1 Introduction

In the previous chapters we have almost exclusively used Cartesian coordinate systems in the three-dimensional physical space which we have called a three-dimensional Euclidean space E_3. Vectors and vector algebra were defined in Sect. 1.2, first geometrically, and then through their Cartesian components. Tensors were introduced via the Cauchy stress tensor in Sect. 2.2.4. A general definition of tensors in Euclidean space was presented in Chap. 3:

A tensor **A** of order n is a multilinear scalar-valued function of n argument vectors:

$$\mathbf{A}[\mathbf{a}, \mathbf{b}, \ldots] = \alpha.$$

Relative to a Cartesian coordinate system Ox with base vectors $\mathbf{e}_i$ the tensor **A** is represented by n components: $A_{ij\cdots k} = \mathbf{A}[\mathbf{e}_i, \mathbf{e}_j, \cdots, \mathbf{e}_k]$.

In the present chapter general coordinate systems will be introduced in the three-dimensional Euclidean space E_3. The geometrical definition of vectors leads to a vector being represented by two sets of components in each coordinate system. The definition of tensors as multilinear scalar-valued functions of vectors results to a tensor being represented by many related sets of components.

The tensor analysis in general coordinates will be applied to two special coordinate systems; the *cylindrical coordinate system* and the *spherical coordinate system*.

Equations of motion, analysis of stress, analysis of deformation, deformation kinematics, and basic equations for linear elasticity and for linearly viscous fluids in general coordinates will presented in Chap. 7.

© Springer Nature Switzerland AG 2019
F. Irgens, *Tensor Analysis*,
https://doi.org/10.1007/978-3-030-03412-2_6

6.2 General Coordinates. Base Vectors

In many applications of vector and tensor analysis it is convenient to describe the position of points in three-dimensional space using other parameters than the three coordinates x_i in a right-handed Cartesian Ox system. Three such parameters or *general coordinates* are necessary. These will be denoted:

$$y^1, y^2, y^3 \Leftrightarrow y^i \quad (i = 1, 2, 3) \tag{6.2.1}$$

Note that in Cartesian coordinate systems we have used sub indices for coordinates and components of vectors and tensors. In general coordinates it is, as we shall demonstrate convenient to use both sub indices and super indices.

A one-to-one correspondence between points in space and coordinate sets y^i for the points must exist, which implies that reversible dependencies must exist between the coordinates x_i and y^i:

$$x_i(y) \equiv x_i(y^1, y^2, y^3) \Leftrightarrow y^i(x) \equiv y^i \quad (x_1, x_2, x_3) \tag{6.2.2}$$

It may be shown, see Sokolnikoff [1], that sufficient conditions on the functions $y^i(x)$ for this to be the case are:

(a) The functions $y^i(x)$ are single-valued, continuous functions that have continuous partial derivatives of order one,
(b) The *Jacobian* to the mapping $y^i(x)$ must be non-zero:

$$(c) \qquad\qquad J_x^y \equiv \det\left(\frac{\partial y^i}{\partial x_j}\right) \neq 0 \tag{6.2.3}$$

We now assume that these conditions are satisfied in that part of the space E_3 that is of interest for the problem we are engaged in, except at isolated points, on isolated lines, or on isolated surfaces. Apart from these *singular regions*, the Jacobian J_x^y is different from zero. We shall see below that the *Jacobian either is positive everywhere or negative everywhere*.

According to the multiplication theorem for determinants, Eq. (1.1.23):

$$\det\left(\frac{\partial x_i}{\partial x_j}\right) = 1 \Rightarrow \det\left(\frac{\partial x_i}{\partial y^k}\frac{\partial y^k}{\partial x_j}\right) = \det\left(\frac{\partial x_i}{\partial y^k}\right)\det\left(\frac{\partial y^k}{\partial x_j}\right) = 1 \Rightarrow$$

$$J_y^x \equiv \det\left(\frac{\partial x_i}{\partial y^k}\right) = \left(\det\left(\frac{\partial y^k}{\partial x_j}\right)\right)^{-1} \equiv (J_x^y)^{-1} \tag{6.2.4}$$

When one of the y-coordinates is kept constant, the functions $x_i(y)$ describe a *coordinate surface*. For example, $y^1 = \text{constant} = \alpha_1$ represents a y^1-surface, Fig. 6.1:

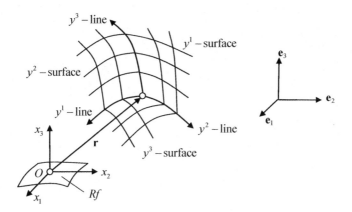

Fig. 6.1 Euclidean space with a reference Rf. Cartesian coordinate system Ox with origin O and base vectors $\mathbf{e}_i$. Coordinate surfaces and coordinate linesfor general coordinates y_i

$$y^1\text{-surface: } x_i = x_i(\alpha_1, y_2, y_3) \quad \text{is a oordinate surface for } y^1 = \alpha_1 \qquad (6.2.5)$$

The intersecting curve between two coordinate surfaces is called a *coordinate line*. The functions $x_i(y)$ describe a coordinate line when two of the y-coordinates are fixed. For example, with $y^1 = \alpha_1$ and $y^2 = \alpha_2$ the functions $x_i(y)$ represent a y^3-line, Fig. 6.1:

$$y^3\text{-line: } x_i = x_i(\alpha_1, \alpha_2, y_3) \quad \text{is a oordinate line for } y^1 = \alpha_1 \text{ and } y^2 = \alpha_2 \quad (6.2.6)$$

The coordinate lines are in general curved, and the coordinates y^i are therefore called *curvilinear coordinates*. We shall use the symbol y also to denote a *general coordinate system* or a *curvilinear coordinate system*. Likewise we shall, in this chapter, let the symbol x denote a right-handed Cartesian coordinate system Ox.

As will be demonstrated in the examples below, the dimensions of the y-coordinates may be different within one and the same coordinate system. If the coordinate difference between to arbitrary points on a coordinate line is equal to or proportional to the length of the line element between the two points, the coordinate is called a *metric coordinate*.

In *cylindrical coordinates* (R, θ, z), presented in Fig. 6.2, we set: $y^1 = R, y^2 = \theta$, $y^3 = z$. The mapping $x_i(y)$ is then:

$$x_1 = R\cos\theta, \quad x_2 = R\sin\theta, \quad x_3 = z \qquad (6.2.7)$$

The R-surfaces are cylindrical surfaces, and the θ-surfaces and the z-surfaces are planes. The θ-lines are circles, and the R-lines and the z-lines are straight lines. R and z are metric coordinates. The length of a line element between two points on a R-line, or a z-line is independent of the two other coordinates θ and z, for the R-line, or R and θ, for the z-line, and equal to the coordinate difference of the two points,

Fig. 6.2 Cylindrical coordinates: $y^1 = R$, $y^2 = \theta$, $y^3 = z$

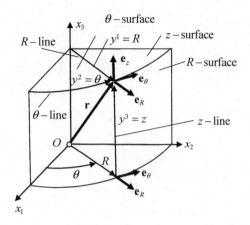

while the length of a line element between two points on a θ-line is dependent on the R-coordinate of the θ-line.

In *spherical coordinates* (r, θ, ϕ), presented in Fig. 6.3, we set: $y_1 = r, y_2 = \theta, y_3 = \phi$. The mapping $x_i(y)$ is then:

$$x_1 = r \sin \theta \cos \phi, \quad x_2 = r \sin \theta \sin \phi, \quad x_3 = r \cos \theta \qquad (6.2.8)$$

The r-surfaces are spherical surfaces, the θ-surfaces are cones, and the ϕ-surfaces are planes. The r-lines are straight lines, and the θ-lines and the ϕ-lines are circles. The coordinate r is a metric coordinate. The length of a line element between two points on a r-line is equal to the coordinate difference of the two points, while the length of a line element between two points on a θ-line is dependent on the coordinates r and ϕ of the θ-line. Likewise, the length of a line element between two points on a ϕ-line is dependent on the coordinates r and θ of the ϕ-line.

Any point P in space may be determined by a set of coordinates in any of the two coordinate systems x or y, or by the *place vector* **r** from a reference point O,

Fig. 6.3 Spherical coordinates: $y^1 = r$, $y^2 = \theta$, $y^3 = \phi$

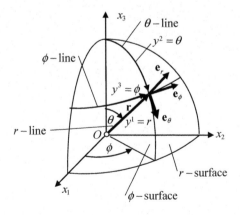

which in Fig. 6.4 is the origin of the Ox-system. The vector $\mathbf{r}$ may be consid-ered to be a function of either the x_i-coordinates or the y_i-coordinates. The *base vectors* in the Ox-system are the three unit vectors $\mathbf{e}_i$ in the positive directions of the coor-dinate axes:

$$\mathbf{r} = x_i\,\mathbf{e}_i \Rightarrow \frac{\partial \mathbf{r}}{\partial x_i} = \mathbf{e}_i \tag{6.2.9}$$

The *base vectors* in the y-system at the point P are defined as the tangent vectors to the coordinate lines y^i through P:

$$\mathbf{g}_i = \frac{\partial \mathbf{r}}{\partial y^i} \tag{6.2.10}$$

The relations between the base vectors $\mathbf{e}_i$ in the Ox-system and $\mathbf{g}_i$ in the y-system are given by:

$$\mathbf{g}_i = \frac{\partial \mathbf{r}}{\partial y^i} = \frac{\partial \mathbf{r}}{\partial x_k}\frac{\partial x_k}{\partial y^i} = \mathbf{e}_k \frac{\partial x_k}{\partial y^i}, \quad \mathbf{e}_i = \frac{\partial \mathbf{r}}{\partial x_i} = \frac{\partial \mathbf{r}}{\partial y^k}\frac{\partial y^k}{\partial x_i} = \mathbf{g}_k \frac{\partial y^k}{\partial x_i} \Rightarrow$$

$$\mathbf{g}_i = \frac{\partial \mathbf{r}}{\partial y^i} = \frac{\partial x_k}{\partial y^i}\mathbf{e}_k \Leftrightarrow \mathbf{e}_i = \frac{\partial \mathbf{r}}{\partial x_i} = \frac{\partial y^k}{\partial x_i}\mathbf{g}_k \tag{6.2.11}$$

The functions $\partial x_k(y)/\partial y^i$ represent the components in the x-system of the base vectors $\mathbf{g}_i$ for the y-system. These functions play the same role as the elements of the transformation matrix Q when transforming from the x-system to another Cartesian $\bar{x}$-system, confer the formulas (1.3.2).

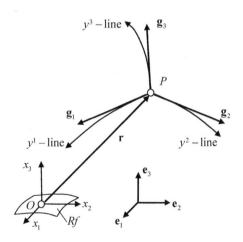

Fig. 6.4 Place vector $\mathbf{r}$. Base vectors $\mathbf{e}_i$ in the Cartesian coordinate sytem Ox. Base vectors $\mathbf{g}_i$ in the general coordinate system y

6.2.1 Covariant and Contravariant Transformations

Let the y-system be replaced by another general coordinate system $\bar{y}$ such that there is a one-to-one correspondence in the functional relationship between the sets of coordinates:

$$\bar{y}^i(y) \Leftrightarrow y^i(\bar{y}) \qquad (6.2.12)$$

The place vector $\mathbf{r}$ is now a function of either of the coordinate systems x, y or $\bar{y}$. We find the base vectors $\bar{\mathbf{g}}_i$ in the $\bar{y}$-system from the relations:

$$\bar{\mathbf{g}}_i = \frac{\partial \mathbf{r}}{\partial \bar{y}^i} = \frac{\partial \mathbf{r}}{\partial y^k}\frac{\partial y^k}{\partial \bar{y}^i} \Rightarrow$$
$$\bar{\mathbf{g}}_i = \frac{\partial y^k}{\partial \bar{y}^i}\mathbf{g}_k \Leftrightarrow \mathbf{g}_k = \frac{\partial \bar{y}^i}{\partial y^k}\bar{\mathbf{g}}_i \qquad (6.2.13)$$

The differential $d\mathbf{r}$ of the place vector $\mathbf{r}(y)$ or $\mathbf{r}(\bar{y})$ is expressed as follows:

$$d\mathbf{r} = \frac{\partial \mathbf{r}}{\partial y^i}dy^i = \mathbf{g}_i dy^i = \frac{\partial \mathbf{r}}{\partial \bar{y}^i}d\bar{y}^i = \bar{\mathbf{g}}_i d\bar{y}^i \qquad (6.2.14)$$

The relationships between the sets of differentials dy^i and $d\bar{y}^i$ are found as:

$$d\bar{y}^i = \frac{\partial \bar{y}_i}{\partial y_k}dy^k \Leftrightarrow dy^i = \frac{\partial y_i}{\partial \bar{y}_k}d\bar{y}^k \qquad (6.2.15)$$

The formulas (6.2.13) and (6.2.15) exemplify two kinds of linear transformations characteristic in general coordinate systems: *covariant transformation* in (6.2.13) and *contravariant transformation* in (6.2.15). When operating in general curvilinear coordinates, it is convenient to use indices in two positions. An index in the upper position, as for the differentials, is called a *super index*. When an index is in the lower position, as for the base vectors $\mathbf{g}_i$, it is called a *subindex*.

In coordinate transformations between two Cartesian systems x and $\bar{x}$:

$$\bar{x}_i = \bar{c}_i + Q_{ik}x_k, \quad x_k = -c_k + Q_{ik}\bar{x}_i \qquad (6.2.16)$$

where Q_{ik} are elements of the transformation matrix Q when transforming from the x-system to the $\bar{x}$-system, the two kinds of transformation coincide, as we can see from the relations:

$$\frac{\partial x_k}{\partial \bar{x}_i} = \frac{\partial \bar{x}_i}{\partial x_k} = Q_{ik} \qquad (6.2.17)$$

The base vector $\mathbf{e}_i$ in the x-system represents both a tangent vector to the coordinate line x_i and a normal vector to the coordinate surface $x_i = $ constant. In a general curvilinear coordinate system the tangent vectors and the normal vectors do not necessarily coincide. It becomes convenient to introduce two sets of base vectors: the tangent vectors $\mathbf{g}_i$ to the y^i-lines, and the normal vectors $\mathbf{g}^i$ to the coordinate surfaces $y^i = $ constant. The normal vectors $\mathbf{g}^i$ are defined by the relations:

$$\mathbf{g}^i \cdot \mathbf{g}_k = \delta^i_k \qquad (6.2.18)$$

where $\delta^i_k \equiv \delta_{ik}$ is a *Kronecker delta*. The normal vectors $\mathbf{g}^i$ are called the *reciprocal base vectors* or the *dual base vectors*.

Using Eq. (6.2.11)$_2$ and the formulas (6.2.18) we obtain:

$$\mathbf{g}^i \cdot \mathbf{e}_j = \mathbf{g}^i \cdot \left(\frac{\partial y^k}{\partial x_j} \mathbf{g}_k \right) = \frac{\partial y^k}{\partial x_j} \mathbf{g}^i \cdot \mathbf{g}_k = \frac{\partial y^k}{\partial x_j} \delta^i_k = \frac{\partial y^i}{\partial x_j}$$

The result shows that the symbols $\partial y^i / \partial x_j$ represent the components in the x-system of the reciprocal base vectors $\mathbf{g}^i$:

$$\mathbf{g}^i = \frac{\partial y^i}{\partial x_j} \mathbf{e}_j \Leftrightarrow \mathbf{e}_j = \frac{\partial x_j}{\partial y^i} \mathbf{g}^i \qquad (6.2.19)$$

The result on the right side of the bi-implication sign in Eq. (6.2.19) follows from the developments:

$$\mathbf{g}^i \frac{\partial x_j}{\partial y^i} = \left(\frac{\partial y^i}{\partial x_k} \mathbf{e}_k \right) \frac{\partial x_j}{\partial y^i} = \mathbf{e}_k \left(\frac{\partial x_j}{\partial y^i} \frac{\partial y^i}{\partial x_k} \right) \Rightarrow \mathbf{g}^i \frac{\partial x_j}{\partial y^i} = \mathbf{e}_k \delta^j_k = \mathbf{e}_j$$

Furthermore, with respect to a general $\bar{y}$-system:

$$\bar{\mathbf{g}}^i = \frac{\partial \bar{y}^i}{\partial x_j} \mathbf{e}_j = \frac{\partial \bar{y}^i}{\partial y^k} \frac{\partial y^k}{\partial x_j} \mathbf{e}_j$$

which with formula (6.2.19)$_1$ gives the results:

$$\bar{\mathbf{g}}^i = \frac{\partial \bar{y}^i}{\partial y^k} \mathbf{g}^k \Leftrightarrow \mathbf{g}^k = \frac{\partial y^k}{\partial \bar{y}^i} \bar{\mathbf{g}}^i \qquad (6.2.20)$$

The relationships (6.2.20) show that the reciprocal base vectors follow a contravariant transformation. For this reason it is also customary to call the two sets of base vectors $\mathbf{g}_i$ and $\mathbf{g}^i$ *covariant base vectors* and *contravariant base vectors*, respectively.

The vector product $\mathbf{g}_i \times \mathbf{g}_j$ of any two base vectors is a vector in the direction of the reciprocal base vector $\mathbf{g}^k$, where the index k is not equal to either i or j. Using the formulas (6.2.11)$_1$, (1.2.27), (6.2.19)$_2$ (1.1.21), and (6.2.4), we obtain:

$$\mathbf{g}_i \times \mathbf{g}_j = \frac{\partial x_r}{\partial y^i}\frac{\partial x_s}{\partial y^j}\,\mathbf{e}_r \times \mathbf{e}_s = \frac{\partial x_r}{\partial y^i}\frac{\partial x_s}{\partial y^j}\,e_{rst}\,\mathbf{e}_t = \frac{\partial x_r}{\partial y^i}\frac{\partial x_s}{\partial y^j}\frac{\partial x_t}{\partial y^k}\,e_{rst}\,\mathbf{g}^k = J_y^x\,e_{ijk}\,\mathbf{g}^k \Rightarrow$$

$$\mathbf{g}_i \times \mathbf{g}_j = \varepsilon_{ijk}\,\mathbf{g}^k$$

$$(6.2.21)$$

Here we have introduced the *permutation symbol*:

$$\varepsilon_{ijk} = J_y^x\,e_{ijk} \qquad (6.2.22)$$

Similarly we find:

$$\mathbf{g}^i \times \mathbf{g}^j = \varepsilon^{ijk}\,\mathbf{g}_k \qquad (6.2.23)$$

where we have introduced the *permutation symbol*:

$$\varepsilon^{ijk} = J_x^y\,e_{ijk} \qquad (6.2.24)$$

Using the formulas (6.2.21), (6.2.23), and (6.2.18), we obtain the box products:

$$\left[\mathbf{g}_i\mathbf{g}_j\mathbf{g}_k\right] = \varepsilon_{ijk}, \quad \left[\mathbf{g}^i\mathbf{g}^j\mathbf{g}^k\right] = \varepsilon^{ijk} \qquad (6.2.25)$$

Since it is assumed that the Jacobians J_y^x and J_x^y are different from zero, it follows that the base vectors $\mathbf{g}_i$ and $\mathbf{g}^i$ at a place cannot lie in one and the same plane. We say that the coordinate system y is a *right-handed/left-handed coordinate system* when J_x^y is positive/negative. Furthermore, we say that $\mathbf{g}_i$ ($i = 1, 2, 3$) and $\mathbf{g}^i$($i = 1, 2, 3$) represent *right-handed/left-handed systems of vectors* when J_x^y is positive/negative. Note that the results (6.2.21) and (6.2.23) are based on the assumption that the x-system is right-handed.

6.2.2 Fundamental Parameters of a General Coordinate System

The base vectors $\mathbf{g}_i$ may be decomposed along the $\mathbf{g}^j$-directions, and the base vectors $\mathbf{g}^i$ may be decomposed along the $\mathbf{g}_j$-directions:

$$\mathbf{g}_i = g_{ij}\,\mathbf{g}^j, \quad \mathbf{g}^i = g^{ij}\,\mathbf{g}_j \qquad (6.2.26)$$

Using the formulas (6.2.11)$_1$ and (6.2.19)$_1$ we obtain:

$$g_{ij} = \mathbf{g}_i \cdot \mathbf{g}_j = \frac{\partial x_k}{\partial y^i} \frac{\partial x_k}{\partial y^j}, \quad g^{ij} = \mathbf{g}^i \cdot \mathbf{g}^j = \frac{\partial y^i}{\partial x_k} \frac{\partial y^j}{\partial x_k} \tag{6.2.27}$$

The elements g_{ij} and g^{ij} are symmetric with respect to the indices and are respectively called the *fundamental parameters of the first order* and the *reciprocal fundamental parameters of first order* for the y-system. The elements g_{ii} and g^{ii}, $i = 1, 2$, or 3, represent the square of the magnitudes of the base vectors. The elements g_{ij} and $g^{ij}, i \neq j$, represent the angels between the base vectors.

$$|\mathbf{g}_i| = \sqrt{g_{ii}}, \quad |\mathbf{g}^i| = \sqrt{g^{ii}}$$

$$\cos(\mathbf{g}_i, \mathbf{g}_j) = \frac{g_{ij}}{\sqrt{g_{ii}\, g_{jj}}}, \quad \cos(\mathbf{g}^i, \mathbf{g}^j) = \frac{g^{ij}}{\sqrt{g^{ii}\, g^{jj}}}, \quad i \neq j \tag{6.2.28}$$

It will be demonstrated in Sect. 6.4.1 that the fundamental parameters are components of the *unit tensor of second order*, also called the *metric tensor*, in the y-system.

From the formulas (6.2.26) and (6.2.18) we obtain:

$$\mathbf{g}_i \cdot \mathbf{g}^j = g_{ik}\mathbf{g}^k \cdot \mathbf{g}^j = g_{ik}g^{kj} = \delta_i^j \Rightarrow$$
$$g_{ik}g^{kj} = \delta_i^j \tag{6.2.29}$$

The determinant of the matrix (g_{ij}) is denoted by the symbol g, and using the formulas (6.2.26), (1.1.23), (6.2.30), (6.2.28), (6.2.3) and (6.2.4), we obtain the results:

$$\det(g_{ik}g^{kj}) = \det(g_{ik})\det(g^{kj}) = \det(\delta_i^j) = 1 \Rightarrow g = \det(g_{ik}), \quad \det(g^{kj}) = \frac{1}{g}$$

$$g \equiv \det(g_{ij}) = \det\left(\frac{\partial x_k}{\partial y^i} \frac{\partial x_k}{\partial y^j}\right) = \left(\det\left(\frac{\partial x_k}{\partial y^i}\right)\right)^2 = (J_y^x)^2$$

$$\frac{1}{g} = \det(g^{ij}) = \det\left(\frac{\partial y^i}{\partial x_k} \frac{\partial y^j}{\partial x_k}\right) = \left(\det\left(\frac{\partial y^i}{\partial x_k}\right)\right)^2 = (J_x^y)^2 \Rightarrow$$

$$g \equiv \det(g_{ij}) = \left(\det\left(\frac{\partial x_k}{\partial y^i}\right)\right)^2 = (J_y^x)^2, \quad \frac{1}{g} = \det(g^{ij}) = \left(\det\left(\frac{\partial y^i}{\partial x_k}\right)\right)^2 = (J_x^y)^2 \tag{6.2.30}$$

With the cofactor, $\mathrm{Co}\, g_{ij}$, of the matrix element g_{ij} we use the formula (1.1.29) to write:

$$g^{ij} = \frac{\mathrm{Co}\, g_{ij}}{g} \tag{6.2.31}$$

6.2.3   Orthogonal Coordinates

In applications it is often convenient to use a coordinate system with orthogonal coordinate lines, which implies that:

$$\text{for } i \neq j: \quad \mathbf{g}_i \cdot \mathbf{g}_j = g_{ij} = 0 \quad \text{and} \quad \mathbf{g}^i \cdot \mathbf{g}^j = g^{ij} = 0 \tag{6.2.32}$$

Cylindrical coordinates (R, θ, z) in Fig. 6.2 and spherical coordinates (r, θ, ϕ) in Fig. 6.3 are the most common examples of orthogonal coordinates. The two sets of base vectors $\mathbf{g}_i$ and $\mathbf{g}^i$ in orthogonal coordinate systems are parallel sets of vectors, and (g_{ij}) and (g^{ij}) are diagonal matrices.

Let h_i denote the *magnitudes of the base vectors* $\mathbf{g}_i$ and let $\mathbf{e}_i^y$ be the unit tangent vectors to the coordinate lines. Then:

$$h_i = \sqrt{\mathbf{g}_i \cdot \mathbf{g}_i} = \sqrt{g_{ii}} \Rightarrow g = h_1 h_2 h_3, \quad \mathbf{e}_i^y = \frac{1}{h_i}\mathbf{g}_i \Leftrightarrow \mathbf{g}_i = h_i\,\mathbf{e}_i^y \tag{6.2.33}$$

From the definition (6.2.18) of the reciprocal base vectors and the formulas (6.2.33) it follows that:

$$\mathbf{g}^i = \frac{1}{h_i}\mathbf{e}_i^y = \frac{1}{h_i^2}\mathbf{g}_i, \quad g^{ii} = \frac{1}{g_{ii}} = \frac{1}{h_i^2} \tag{6.2.34}$$

In *cylindrical coordinates* (R, θ, z), Fig. 6.2, we introduce unit vectors $\mathbf{e}_R(\theta), \mathbf{e}_\theta(\theta)$, and $\mathbf{e}_z$ in the directions of the tangents to the coordinate lines, and write for the place vector:

$$\mathbf{r} = R\mathbf{e}_R(\theta) + z\mathbf{e}_z \tag{6.2.35}$$

From Fig. 6.2 we obtain:

$$\mathbf{e}_R(\theta) = \cos\theta\,\mathbf{e}_1 + \sin\theta\,\mathbf{e}_2, \quad \mathbf{e}_\theta(\theta) = -\sin\theta\,\mathbf{e}_1 + \cos\theta\,\mathbf{e}_2 \Rightarrow$$
$$\frac{d\mathbf{e}_R(\theta)}{d\theta} = \mathbf{e}_\theta(\theta), \quad \frac{d\mathbf{e}_\theta(\theta)}{d\theta} = -\mathbf{e}_R(\theta) \tag{6.2.36}$$

From the definitions of base vectors we now find:

$$\mathbf{g}_1 = \frac{\partial\mathbf{r}}{\partial R} = \mathbf{e}_R, \quad \mathbf{g}_2 = \frac{\partial\mathbf{r}}{\partial\theta} = R\frac{\partial\mathbf{e}_R}{\partial\theta} = R\mathbf{e}_\theta, \quad \mathbf{g}_3 = \frac{\partial\mathbf{r}}{\partial z} = \mathbf{e}_z$$
$$\mathbf{g}^1 = \mathbf{e}_R, \quad \mathbf{g}^2 = \frac{1}{R}\mathbf{e}_\theta, \quad \mathbf{g}^3 = \mathbf{e}_z, \quad h_1 = 1, \quad h_2 = R, \quad h_3 = 1, \quad g = R$$

$$\tag{6.2.37}$$

$$(g_{ij}) = (\mathbf{g}_i \cdot \mathbf{g}_j) = \begin{pmatrix} 1 & 0 & 0 \\ 0 & R^2 & 0 \\ 0 & 0 & 1 \end{pmatrix}, \quad g = \det(g_{ij}) = R^2 \tag{6.2.38}$$

$$(g^{ij}) = (\mathbf{g}^i \cdot \mathbf{g}^j) = \begin{pmatrix} 1 & 0 & 0 \\ 0 & 1/R^2 & 0 \\ 0 & 0 & 1 \end{pmatrix} \tag{6.2.39}$$

In *spherical coordinates* (r, θ, ϕ), Fig. 6.3, we introduce unit vectors $\mathbf{e}_r(\theta, \phi)$, $\mathbf{e}_\theta(\theta, \phi)$, and $\mathbf{e}_\phi(\phi)$ in the directions of the tangents to the coordinate lines, and write for the place vector:

$$\mathbf{r} = r\mathbf{e}_r(\theta, \phi) \tag{6.2.40}$$

From Fig. 6.3 we find:

$$\left. \begin{aligned} \mathbf{e}_r(\theta, \phi) &= \sin\theta\cos\phi\,\mathbf{e}_1 + \sin\theta\sin\phi\,\mathbf{e}_2 + \cos\theta\,\mathbf{e}_3 \\ \mathbf{e}_\theta(\theta, \phi) &= \cos\theta\cos\phi\,\mathbf{e}_1 + \cos\theta\sin\phi\,\mathbf{e}_2 - \sin\theta\,\mathbf{e}_3 \\ \mathbf{e}_\phi(\phi) &= -\sin\phi\,\mathbf{e}_1 + \cos\phi\,\mathbf{e}_2 \end{aligned} \right\} \Rightarrow$$

$$\left. \begin{aligned} \partial\mathbf{e}_r(\theta, \phi)/\partial\theta &= \cos\theta\cos\phi\,\mathbf{e}_1 + \cos\theta\sin\phi\,\mathbf{e}_2 - \sin\theta\,\mathbf{e}_3 \\ \partial\mathbf{e}_r(\theta, \phi)/\partial\phi &= -\sin\theta\sin\phi\,\mathbf{e}_1 + \sin\theta\cos\phi\,\mathbf{e}_2 \\ \partial\mathbf{e}_\theta(\theta, \phi)/\partial\theta &= -\sin\theta\cos\phi\,\mathbf{e}_1 - \sin\theta\sin\phi\,\mathbf{e}_2 - \cos\theta\,\mathbf{e}_3 \\ \partial\mathbf{e}_\theta(\theta, \phi)/\partial\phi &= -\cos\theta\sin\phi\,\mathbf{e}_1 + \cos\theta\cos\phi\,\mathbf{e}_2 \\ d\mathbf{e}_\phi(\phi)/d\phi &= -\cos\phi\,\mathbf{e}_1 - \sin\phi\,\mathbf{e}_2 \end{aligned} \right\} \Rightarrow \tag{6.2.41}$$

$$\frac{\partial\mathbf{e}_r(\theta,\phi)}{\partial\theta} = \mathbf{e}_\theta, \quad \frac{\partial\mathbf{e}_r(\theta,\phi)}{\partial\phi} = \sin\theta\,\mathbf{e}_\phi, \quad \frac{\partial\mathbf{e}_\theta(\theta,\phi)}{\partial\theta} = -\mathbf{e}_r, \quad \frac{\partial\mathbf{e}_\theta(\theta,\phi)}{\partial\phi} = \cos\theta\,\mathbf{e}_\phi$$

$$\frac{d\mathbf{e}_\phi(\phi)}{d\theta} = 0, \quad \frac{d\mathbf{e}_\phi(\phi)}{d\phi} = -\cos\phi\,\mathbf{e}_1 - \sin\phi\,\mathbf{e}_2 = -\cos\theta\,\mathbf{e}_\theta - \sin\theta\,\mathbf{e}_r$$

From the definitions of base vectors $\mathbf{g}_i$ and $\mathbf{g}^i$ we find:

$$\mathbf{g}_1 = \frac{\partial\mathbf{r}}{\partial r} = \mathbf{e}_r, \quad \mathbf{g}_2 = \frac{\partial\mathbf{r}}{\partial\theta} = r\frac{\partial\mathbf{e}_r}{\partial\theta} = r\,\mathbf{e}_\theta, \quad \mathbf{g}_3 = \frac{\partial\mathbf{r}}{\partial\phi} = r\frac{\partial\mathbf{e}_r}{\partial\phi} = r\sin\theta\,\mathbf{e}_\phi$$

$$\mathbf{g}^1 = \mathbf{e}_r, \quad \mathbf{g}^2 = \frac{1}{r}\mathbf{e}_\theta, \quad \mathbf{g}^3 = \frac{1}{r\sin\theta}\mathbf{e}_\phi$$

$$h_1 = 1, \quad h_2 = r, \quad h_3 = r\sin\theta, \quad h = r^2\sin\theta$$

$$\tag{6.2.42}$$

$$(g_{ij}) = (\mathbf{g}_i \cdot \mathbf{g}_j) = \begin{pmatrix} 1 & 0 & 0 \\ 0 & r^2 & 0 \\ 0 & 0 & r^2\sin^2\theta \end{pmatrix}, \quad g = \det(g_{ij}) = r^4\sin^2\theta$$

$$\tag{6.2.43}$$

$$(g^{ij}) = (\mathbf{g}^i \cdot \mathbf{g}^j) = \begin{pmatrix} 1 & 0 & 0 \\ 0 & 1/r^2 & 0 \\ 0 & 0 & 1/(r\sin\theta)^2 \end{pmatrix}$$

6.3   Vector Fields

Any vector field $\mathbf{a}(y)$ may be decomposed into components a^i along the directions of the base vectors $\mathbf{g}_i$ or into components a_i along the directions of the reciprocal base vectors $\mathbf{g}^i$, as illustrated in Fig. 6.5, which for simplicity does not show the third dimension. We write:

$$\mathbf{a} = a^i\, \mathbf{g}_i = a_i\, \mathbf{g}^i \qquad (6.3.1)$$

It follows that:

$$a^i = \mathbf{g}^i \cdot \mathbf{a} = \mathbf{g}^i \cdot \left(a_k\, \mathbf{g}^k\right) = g^{ik} a_k, \quad a_i = \mathbf{g}_i \cdot \mathbf{a} = \mathbf{g}_i \cdot \left(a^k\, \mathbf{g}_k\right) = g_{ik} a^k \qquad (6.3.2)$$

The parameters a^i are called the *contravariant components*, and the parameters a_i are called the *covariant components* of the vector field $\mathbf{a}$. From Eqs. (6.2.20) and (6.3.2) it follows that when changing from one general coordinate system y to another general coordinate system $\bar{y}$, the *contravariant components* and the *co-variant components* of the vector $\mathbf{a}$ transform according to the contravariant rule or covariant rule respectively:

$$\bar{a}^i = \frac{\partial \bar{y}^i}{\partial y^k} a^k, \quad \bar{a}_i = \frac{\partial y^k}{\partial \bar{y}^i} a_k \qquad (6.3.3)$$

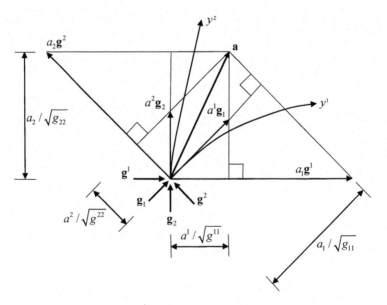

Fig. 6.5   Contravariant components a^i and covariant components a_i of a vector field $\mathbf{a}$. Coordinate lines y^1 and y^2 for general coordinates y. Base vectors $\mathbf{g}_i$ and reciprocal base vectors $\mathbf{g}_i$

The unit vectors in the directions of the base vectors $\mathbf{g}_i$ are $\mathbf{g}_i/\sqrt{g_{ii}}$ (no summation), and the magnitude of the vector components $a^i\mathbf{g}_i$ are $a^i\sqrt{g_{ii}}$ (no summation). The unit vectors in the directions of the reciprocal base vectors $\mathbf{g}^i$ are $\mathbf{g}^i/\sqrt{g^{ii}}$ (no summation), and the magnitude of the vector components $a_i\mathbf{g}^i$ are $a_i\sqrt{g^{ii}}$ (no summation).

The normal projections of the vector $\mathbf{a}$ onto the directions of the base vectors $\mathbf{g}_i$ and the reciprocal base vectors $\mathbf{g}^i$, respectively, become:

Normal projection of $\mathbf{a}$ onto the base vector $\mathbf{g}_i$:

$$\mathbf{a} \cdot \mathbf{g}_i/\sqrt{g_{ii}} \text{ (no summation)} = \sum_{k=1}^{k=3}\left(a_k\mathbf{g}^k \cdot \mathbf{g}_i/\sqrt{g_{ii}}\right) = a_i/\sqrt{g_{ii}} \text{ (no summation)}$$

(6.3.4)

Normal projection of $\mathbf{a}$ onto the reciprocal base vector $\mathbf{g}^i$:

$$\mathbf{a} \cdot \mathbf{g}^i/\sqrt{g^{ii}} \text{ (no summation)} = \sum_{k=1}^{k=3}\left(a^k\mathbf{g}_k \cdot \mathbf{g}^i/\sqrt{g^{ii}}\right) = a^i/\sqrt{g^{ii}} \text{ (no summation)}$$

(6.3.5)

These results are illustrated in Fig. 6.5. Note that the contravariant component a^i of the vector field $\mathbf{a}$ represents both the vector component $a^i\mathbf{g}_i$ and the projection of the vector field $\mathbf{a}$ onto the reciprocal base vector $\mathbf{g}^i$. Likewise, the covariant component a_i of the vector field $\mathbf{a}$ represents both the vector component $a_i\mathbf{g}^i$ and the projection of the vector field $\mathbf{a}$ onto the base vector $\mathbf{g}_i$.

The magnitudes $a^i\sqrt{g_{ii}}$ (no summation) of the vector components $a^i\mathbf{g}_i$ (no summation) of a vector field $\mathbf{a}$, are called the *physical components* of the vector field $\mathbf{a}$, and are denoted by $a(i)$:

$$a(i) \equiv a^i\sqrt{g_{ii}}, \quad \mathbf{a} = \sum_i a(i)\frac{\mathbf{g}_i}{\sqrt{g_{ii}}} = a(i)\mathbf{e}_i^y$$ (6.3.6)

In a Cartesian coordinate system we find that:

$$a(i) = a^i = a_i$$ (6.3.7)

In cylindrical and spherical coordinates the physical components of a vector $\mathbf{a}$ are related to the contravariant and covariant components of the vector through:

$$a_R = a^1 = a_1, \quad a_\theta = Ra^2 = \frac{1}{R}a_2, \quad a_z = a^3 = a_3$$

$$a_r = a^1 = a_1, \quad a_\theta = ra^2 = \frac{1}{r}a_2, \quad a_\phi = r\sin\theta\, a^3 = \frac{a_3}{r\sin\theta}$$

(6.3.8)

When adding or subtracting vectors, we find that:

$$\mathbf{a} + \mathbf{b} = \mathbf{c} \Leftrightarrow a^i + b^i = c^i \Leftrightarrow a_i + b_i = c_i \tag{6.3.9}$$

Adding/subtracting the covariant components of one vector to/from the contravariant components of another vector has no meaning.

The *scalar product* of two vectors $\mathbf{a}$ and $\mathbf{b}$ may be developed as follows.

$$\mathbf{a} \cdot \mathbf{b} = \left(a^i \, \mathbf{g}_i\right) \cdot \left(b_j \, \mathbf{g}^j\right) = a^i b_j \left(\mathbf{g}_i \cdot \mathbf{g}^j\right) = a^i b_j \, \delta_i^j = a^i b_i$$

Similarly we find: $\mathbf{a} \cdot \mathbf{b} = a_i \, b^i$. Thus:

$$\mathbf{a} \cdot \mathbf{b} = a^i \, b_i = a_i \, b^i \tag{6.3.10}$$

If we write:

$$a_i \, b^i = \sum_i \left[\frac{a_i}{\sqrt{g_{ii}}}\right] \left[b^i \sqrt{g_{ii}}\right] \tag{6.3.11}$$

we may interpret the scalar product $\mathbf{a} \cdot \mathbf{b}$ as the sum of the products of the normal projections of the vector $\mathbf{a}$ onto the base vectors $\mathbf{g}_i$, Formula (6.3.4), and the physical components of $\mathbf{b}$, formula (6.3.6).

The covariant and contravariant components of the *vector product* $\mathbf{b} \times \mathbf{c}$ of the vectors $\mathbf{b}$ and $\mathbf{c}$ may be found by using the formulas (6.2.22) and (6.2.24). The result is:

$$\mathbf{a} = \mathbf{b} \times \mathbf{c} \Leftrightarrow a_i = \varepsilon_{ijk} \, b^j \, c^k \Leftrightarrow a^i = \varepsilon^{ijk} b_j \, c_k \tag{6.3.12}$$

The *scalar triple product*, or *box product*, of three vectors $\mathbf{a}$, $\mathbf{b}$, and $\mathbf{c}$ may be computed from:

$$[\mathbf{abc}] \equiv \mathbf{a} \cdot (\mathbf{b} \times \mathbf{c}) = \varepsilon_{ijk} \, a^i \, b^j \, c^k = \varepsilon^{ijk} a_i \, b_j \, c_k \tag{6.3.13}$$

In some literature the definitions (6.2.23) and (6.2.25) are replaced by:

$$\varepsilon_{ijk} = \sqrt{g} \, e_{ijk}, \quad \varepsilon^{ijk} = \frac{1}{\sqrt{g}} e_{ijk} \tag{6.3.14}$$

When the expressions (6.3.14) are used for the symbols ε_{ijk} and ε^{ijk} in the formulas (6.3.12) for the components of the vector product, the vector product changes direction when transforming from a right-handed system to a left-handed system and the other way around. The vector product is now what is called an *axial vector*. We shall demonstrate this by transforming from a right-handed Cartesian system x to a *left-handed Cartesian system y*:

$$y^2 = -x_1, \quad y^2 = x_2, \quad y^3 = x_3 \quad \Rightarrow$$

$$\mathbf{g}_1 = \mathbf{g}^1 = -\mathbf{e}_1, \quad \mathbf{g}_2 = \mathbf{g}^2 = \mathbf{e}_2, \quad \mathbf{g}_3 = \mathbf{g}^3 = \mathbf{e}_3, \quad g_{ij} = g^{ij} = \delta_{ij}$$

Then:

$$\left(\frac{\partial x_i}{\partial y^j}\right) = \left(\frac{\partial y^i}{\partial x_j}\right) = \begin{pmatrix} -1 & 0 & 0 \\ 0 & 1 & 0 \\ 0 & 0 & 1 \end{pmatrix}, \quad J_y^x = \det\left(\frac{\partial x_i}{\partial y^j}\right) = -1,$$

$$g = \left(J_y^x\right)^2 = 1 \Rightarrow \sqrt{g} = 1$$

A vector $\mathbf{a}$ with components a_j^x in the x-system will have the following components in the y-system:

$$a^i = \frac{\partial y^i}{\partial x_j} a_j^x, \quad a_i = \frac{\partial x_j}{\partial y^i} a_j^x \Rightarrow a^1 = a_1 = -a_1^x, \quad a^2 = a_2 = a_2^x, \quad a^3 = a_3 = a_3^x$$

According to different presentations of vector algebra in the literature, the *vector product* $\mathbf{b} \times \mathbf{c}$ may be represented as either of the following two vectors. Using $\varepsilon_{ijk} = J_y^x e_{ijk}$ from the definition (6.2.23), we get the first alterative for the vector product $\mathbf{b} \times \mathbf{c}$:

$$\mathbf{a} = \mathbf{b} \times \mathbf{c} = \varepsilon_{ijk} b^j c^k \mathbf{g}^i = J_y^x e_{ijk} b^j c^k \mathbf{g}^i = -\det\begin{pmatrix} -\mathbf{e}_1 & \mathbf{e}_2 & \mathbf{e}_3 \\ -b_1^x & b_2^x & b_3^x \\ -c_1^x & c_2^x & c_3^x \end{pmatrix} = e_{ijk} b_j^x c_k^x \mathbf{e}_i$$

$$(6.3.15)$$

Using the definition (6.3.12)$_1$, we get the second alternative for the vector product $\mathbf{b} \times \mathbf{c}$:

$$\mathbf{a}_{\mathrm{alt}} = \mathbf{b} \times \mathbf{c} = \varepsilon_{ijk} b^j c^k \mathbf{g}^i = \sqrt{g} e_{ijk} b^j c^k \mathbf{g}^i = \det\begin{pmatrix} -\mathbf{e}_1 & \mathbf{e}_2 & \mathbf{e}_3 \\ -b_1^x & b_2^x & b_3^x \\ -c_1^x & c_2^x & c_3^x \end{pmatrix} = -e_{ijk} b_j^x c_k^x \mathbf{e}_i$$

$$(6.3.16)$$

The vector $\mathbf{a}_{\mathrm{alt}}$ has the opposite direction of the vector $\mathbf{a}$.

In the present presentation the vector product is defined geometrically in Sect. 1.2.2 by Fig. 1.9 and the formula (1.2.25) as a coordinate invariant quantity. This implies that if we do not intend to limit our choices of coordinate systems to only right-handed systems, or only to left-handed systems, we keep the formula (6.3.12) for the vector product and the definitions (6.2.23) and (6.2.25) for the symbols ε_{ijk} and ε^{ijk}. Then the vector $\mathbf{a}$ in Eq. (6.3.15) is the proper vector product of the vectors $\mathbf{b}$ and $\mathbf{c}$.

The scalar product and the vector product of the vectors **a** and **b** when the vectors are expressed in physical components, are:

$$\mathbf{a} \cdot \mathbf{b} = \sum_{i,k} \frac{a(i)\,b(k)\,g_{ik}}{\sqrt{g_{ii}\,g_{kk}}} \tag{6.3.17}$$

$$\mathbf{a} \times \mathbf{b} = \mathbf{c} \Leftrightarrow J_y^x \sum_{i,j,k} e_{ijk} a(i)\,b(\mathrm{j})\,g^{kl} \sqrt{\frac{g_{ll}}{g_{ii}\,g_{jj}}} = c(l) \tag{6.3.18}$$

The two formulas are derived as follows. For the scalar product **a** · **b** we write:

$$\mathbf{a} \cdot \mathbf{b} = \left(a^i \mathbf{g}_i\right) \cdot \left(b^k \mathbf{g}_k\right) = a^i b^k \mathbf{g}_i \cdot \mathbf{g}_k = \sum_{i,k} \frac{a(i)\,b(k)\,g_{ik}}{\sqrt{g_{ii}\,g_{kk}}} \Rightarrow 6.3.17$$

For the vector product **c** = **a** × **b** we write:

$$\mathbf{c} = c^l \mathbf{g}_l = \sum_l \frac{c(l)}{\sqrt{g_{ll}}} \mathbf{g}_l = \mathbf{a} \times \mathbf{b}$$

$$= \left(a^i \mathbf{g}_i\right) \times \left(b^j \mathbf{g}_j\right) = a^i b^j \mathbf{g}_i \times \mathbf{g}_j = a^i b^j \varepsilon_{ijk} \mathbf{g}^k = a^i b^j \varepsilon_{ijk} g^{kl} \mathbf{g}_l = J_y^x \sum_{i,j,k} e_{ijk} \frac{a(i)\,b(\mathrm{j})\,g^{kl}}{\sqrt{g_{ii}\,g_{jj}}} \mathbf{g}_l$$

$$\Rightarrow \quad \mathbf{a} \times \mathbf{b} = \mathbf{c} \Leftrightarrow J_y^x \sum_{i,j,k} e_{ijk} a(i)\,b(\mathrm{j})\,g^{kl} \sqrt{g_{ll}/\left(g_{ii}\,g_{jj}\right)} = c(l) \Rightarrow 6.3.18$$

In orthogonal coordinates the formulas (6.3.17) and (6.3.18) reduce to:

$$\mathbf{a} \cdot \mathbf{b} = a(i)\,b(i), \quad \mathbf{a} \times \mathbf{b} = \mathbf{c} \Leftrightarrow e_{ijk}\, a(j)\,b(k) = c(i) \tag{6.3.19}$$

6.4 Tensor Fields

6.4.1 Tensor Components. Tensor Algebra

In Sect. 3.1 a *tensor of order n* is defined as a *multilinear scalar-valued function of n argument vectors*. Let **C** be tensor of second order. Then the scalar value of **C** for the argument vectors **a** and **b** is:

$$\alpha = \mathbf{C}[\mathbf{a}, \mathbf{b}] = \mathbf{a} \cdot \mathbf{C} \cdot \mathbf{b} \tag{6.4.1}$$

In a Cartesian coordinate system x with base vectors $\mathbf{e}_i$ the tensor is represented by its components:

$$C_{ij} = \mathbf{C}[\mathbf{e}_i, \mathbf{e}_j] \quad \text{in a Cartesian coordinate system} \tag{6.4.2}$$

In a general coordinate system y with base vectors $\mathbf{g}_i$ and $\mathbf{g}^i$ the tensor $\mathbf{C}$ is represented by either of four associated sets of components:

$$
\begin{aligned}
C_{ij} &= \mathbf{C}[\mathbf{g}_i, \mathbf{g}_j] \quad \text{covariant components} \\
C^{ij} &= \mathbf{C}[\mathbf{g}^i, \mathbf{g}^j] \quad \text{contravariant components} \\
C^j_i &= \mathbf{C}[\mathbf{g}_i, \mathbf{g}^j], \quad C^i_j = \mathbf{C}[\mathbf{g}^i, \mathbf{g}_j] \quad \text{mixed components}
\end{aligned}
\tag{6.4.3}
$$

The formula (6.4.1) may now be presented as:

$$
\begin{aligned}
\alpha &= \mathbf{C}[\mathbf{a}, \mathbf{b}] = \mathbf{a} \cdot \mathbf{C} \cdot \mathbf{b} = a^i \mathbf{C}[\mathbf{g}_i, \mathbf{g}_j] b^j = a^i \mathbf{C}[\mathbf{g}_i, \mathbf{g}^j] b_j = a_i \mathbf{C}[\mathbf{g}^i, \mathbf{g}_j] b^j = a_i \mathbf{C}[\mathbf{g}^i, \mathbf{g}^j] b_j \Rightarrow \\
\alpha &= \mathbf{C}[\mathbf{a}, \mathbf{b}] = \mathbf{a} \cdot \mathbf{C} \cdot \mathbf{b} = a^i C_{ij} b^j = a^i C^j_i b_j = a_i C^i_j b^j = a_i C^{ij} b_j
\end{aligned}
$$
$$\tag{6.4.4}$$

Using the formulas (6.2.27) for the base vectors in the expressions (6.4.3), we find the relationships between the four sets of components of the tensor. For example:

$$
\left.
\begin{aligned}
C^i_j &= \mathbf{C}[\mathbf{g}^i, \mathbf{g}_j] = \mathbf{C}[g^{ik}\mathbf{g}_k, \mathbf{g}_j] = g^{ik}\mathbf{C}[\mathbf{g}_k, \mathbf{g}_j] = g^{ik}C_{kj} \\
C^{ij} &= \mathbf{C}[\mathbf{g}^i, \mathbf{g}^j] = \mathbf{C}[g^{ik}\mathbf{g}_k, g^{jl}\mathbf{g}_l] = g^{ik}g^{jl}\mathbf{C}[\mathbf{g}_k, \mathbf{g}_l] = g^{ik}g^{jl}C_{kl} \\
C^i_j &= g^{ik}C_{kj}, \quad C^{ij} = g^{ik}g^{jl}C_{kl}
\end{aligned}
\right\} \Rightarrow
\tag{6.4.5}
$$

Such operations are called "*raising and lowering of indices*".

A tensor $\mathbf{C}$ of second order is symmetric if $\mathbf{C}[\mathbf{a}, \mathbf{b}] = \mathbf{C}[\mathbf{b}, \mathbf{a}]$ and antisymmetric if $\mathbf{C}[\mathbf{a}, \mathbf{b}] = -\mathbf{C}[\mathbf{b}, \mathbf{a}]$. Symmetry implies that $C_{ij} = C_{ji}$, $C^{ij} = C^{ji}$, and $C^i_j = C^i_j$. For a symmetric tensor we introduce the notation:

$$C^i_j \equiv C^i_j \equiv C^i_j \tag{6.4.6}$$

The *unit tensor* $\mathbf{1}$ of second order is by formula (3.1.20) defined by the scalar product of the any to argument vectors $\mathbf{a}$ and $\mathbf{b}$:

$$\mathbf{1}[\mathbf{a}, \mathbf{b}] = \mathbf{a} \cdot \mathbf{b} \tag{6.4.7}$$

The components of the unit tensor $\mathbf{1}$ in a Cartesian coordinate system x with base vectors $\mathbf{e}_i$ are given by the Kronecker delta δ_{ij}:

$$\mathbf{1}[\mathbf{e}_i, \mathbf{e}_j] = \delta_{ij} \text{ in any Cartesian coordinate system} \tag{6.4.8}$$

From the definition (6.4.7) it follows that the components in a general coordinate system y of the second order unit tensor are equal to the fundamental parameters of first order in that coordinate system:

$$\mathbf{1}[\mathbf{g}_i, \mathbf{g}_j] = \mathbf{g}_i \cdot \mathbf{g}_j = g_{ij}, \quad \mathbf{1}[\mathbf{g}^i, \mathbf{g}^j] = \mathbf{g}^i \cdot \mathbf{g}^j = g^{ij}$$
$$\mathbf{1}[\mathbf{g}_i, \mathbf{g}^j] = \mathbf{g}_i \cdot \mathbf{g}^j = \mathbf{1}[\mathbf{g}^j, \mathbf{g}_i] = \mathbf{g}^j \cdot \mathbf{g}_i = \delta_i^j \tag{6.4.9}$$

Equation (6.4.5) may be considered as component expressions for the identities $\mathbf{C} = \mathbf{1C}$ and $\mathbf{C} = \mathbf{11C}$. The scalar product (6.4.7) may alternatively be expressed by:

$$\mathbf{a} \cdot \mathbf{b} = \mathbf{1}[\mathbf{a}, \mathbf{b}] = a^i b_j \mathbf{1}[\mathbf{g}_i, \mathbf{g}^j] = a^i b_j \delta_i^j = a^i b_i = a_i b^i \tag{6.4.10}$$

The results agree with the formulas (6.3.10).

When the components of a second order tensor $\mathbf{C}$ in one y-system are known, we may find the components of the tensor in another $\bar{y}$-system by the use of the formulas (6.2.13) and (6.2.21. For example:

$$\bar{C}_{ij} = \mathbf{C}[\bar{\mathbf{g}}_i, \bar{\mathbf{g}}_j] = \frac{\partial y^k}{\partial \bar{y}^i} \frac{\partial y^l}{\partial \bar{y}^j} \, \mathbf{C}[\mathbf{g}_k, \mathbf{g}_l] = \frac{\partial y^k}{\partial \bar{y}^i} \frac{\partial y^l}{\partial \bar{y}^j} \, C_{kl}$$

Using similar procedures, we obtain the results:

$$\bar{C}_{ij} = \frac{\partial y^k}{\partial \bar{y}^i} \frac{\partial y^l}{\partial \bar{y}^j} \, C_{kl} \quad \text{covariant transformation}$$

$$\bar{C}^{ij} = \frac{\partial \bar{y}^i}{\partial y^k} \frac{\partial \bar{y}^j}{\partial y^l} \, C^{kl} \quad \text{contravariant transformation} \tag{6.4.11}$$

$$\bar{C}_i^j = \frac{\partial y^k}{\partial \bar{y}^i} \frac{\partial \bar{y}^j}{\partial y^l} \, C_k^l, \quad \bar{C}_j^i = \frac{\partial \bar{y}^i}{\partial y^k} \frac{\partial y^l}{\partial \bar{y}^j} \, C_l^k \quad \text{mixed transformations}$$

A generalization of the results above to tensors of higher order is straight forward. A tensor of order n has in every y-system 2^n associated sets of components. The algebra of tensors using components in general coordinate systems is very similar to what applies in Cartesian coordinates. However, *addition* and *subtraction* of tensors of the same order must be performed with components of the same kind. For example, the sum of two tensors $\mathbf{A}$ and $\mathbf{B}$ of second order is a tensor $\mathbf{C}$ of second order, which may be represented by either of the following sets of components:

$$C_{ij} = A_{ij} + B_{ij}, \quad C^{ij} = A^{ij} + B^{ij}, \quad C_j^i = A_j^i + B_j^i, \quad C_i^j = A_i^j + B_i^j \tag{6.4.12}$$

The *tensor product* of a tensor $\mathbf{A}$ of order 2 and a tensor $\mathbf{B}$ of order 3 is a tensor $\mathbf{C}$ of order $2 + 3 = 5$:

$$\mathbf{A} \otimes \mathbf{B} = \mathbf{C} \Leftrightarrow A_{ij} B_{lp}^k = C_{ij\,lp}^k \quad \text{etc.} \tag{6.4.13}$$

Contraction of a tensor must be performed with respect to indices of opposite kind: one superindex and one subindex. For example, a vector **a** may be constructed from the tensor **C** of third order in the following manner:

$$a_i = C^k_{ik}, \quad a^i = C^{ik}_k \tag{6.4.14}$$

Scalar product, dot product, double dot product, linear mapping, and *composition* of tensors, defined in Cartesian coordinates by the formulas (3.2.12, 3.2.15, 3.2.16, 3.2.17, 3.2.20, 3.2.22, 3.2.23, 3.2.24), are illustrated by the following examples:

$$\alpha = \mathbf{A} : \mathbf{B} = A^{ij} B_{ij}, \quad \mathbf{a} = \mathbf{Ab} \equiv \mathbf{A} \cdot \mathbf{b} \Leftrightarrow a_i = A_{ik}b^k \tag{6.4.15}$$

$$\mathbf{d} = \mathbf{A} : \mathbf{C} \Leftrightarrow d_k = A_{ij} C^{ij}_k, \quad \mathbf{D} = \mathbf{AB} \Leftrightarrow D_{ij} = A_{ik}B^k_j \tag{6.4.16}$$

The component form of tensor equations in a Cartesian coordinate system x are readily translated into component equations in a general coordinate system y as shown by the following example. Let **a** and **b** be two vectors, **A** and **C** two tensors of second order, and **B** a tensor of third order with the following components in a Cartesian coordinate system x: a_i, b_j, A_{ij}, C_{jk}, and B_{ijk}. Suppose that these five tensor are related by the tensor equation:

$$\mathbf{a} = \mathbf{A} \cdot \mathbf{b} + \mathbf{B} : \mathbf{C} \Leftrightarrow a_i = A_{ij} b_j + B_{ijk} C_{jk} \quad (x\text{-system}) \tag{6.4.17}$$

Two possible representations of the tensor equation in a general coordinate system y are:

$$a_i = A_{ij} b^j + B_{ijk} C^{jk} \Leftrightarrow a^i = A^{ij} b_j + B^i_{jk} C^{jk}, \quad (y\text{-system}) \tag{6.4.18}$$

Note that all terms in a tensor equation must be tensors of the same order. In the tensor Eq. (6.4.14) all terms are first order tensors.

The *trace,* the *norm,* and the *determinant* of a second order tensor **B** are in a general coordinate system y given by:

$$\operatorname{tr} \mathbf{B} = B^i_i = B^i_i \tag{6.4.19}$$

$$\text{norm } \mathbf{B} \equiv \|\mathbf{B}\| = \sqrt{\mathbf{B}:\mathbf{B}} \equiv \sqrt{\operatorname{tr}\left(\mathbf{B}\mathbf{B}^T\right)} = \sqrt{\operatorname{tr}(BB^T)} = \sqrt{B_{ij}B_{ij}} \tag{6.4.20}$$

$$\det \mathbf{B} = \det\left(B_{ij}\right)/g = \det\left(B^j_i\right) = \det\left(B^i_j\right) = \det\left(B^{ij}\right)g \tag{6.4.21}$$

If $\det \mathbf{B} \neq 0$, the *inverse tensor* $\mathbf{B}^{-1}$ is found from:

$$\mathbf{B}^{-1}\,\mathbf{B} = \mathbf{1} \Leftrightarrow B_{ik}^{-1}\,B^{kj} = \delta_i^j \qquad (6.4.22)$$

The derivation of the formula (6.4.21) from the definition (3.3.10) of the determinant $\det \mathbf{B}$ of the tensor $\mathbf{B}$ is given as Problem 6.1.

It may be shown, Problem 6.2, that the symbols ε^{ijk} and ε_{ijk} are components of the *permutation tensor* $\mathbf{P}$ defined by formula (3.1.25), and that, Problem 6.3, the following identity holds:

$$\varepsilon^{ijk}\,\varepsilon_{rsk} = \delta_r^i\delta_s^j - \delta_s^i\delta_r^j \qquad (6.4.23)$$

6.4.2 Symmetric Tensors of Second Order

Let $\mathbf{S}$ be a symmetric second order tensor and $\mathbf{a}$ and $\mathbf{b}$ two orthogonal unit vectors. To find the principal values and the principal directions for the tensor we proceed as in Sect. 3.3.1. First we defined the *vector* $\mathbf{s}$ *of the tensor* $\mathbf{S}$ *for the direction* $\mathbf{a}$, *the normal component* σ *of the tensor* $\mathbf{S}$ *for the direction* $\mathbf{a}$, and *the orthogonal shear component* τ *of the tensor* $\mathbf{S}$ *for the orthogonal directions* $\mathbf{a}$ and $\mathbf{b}$:

$$\mathbf{s} = \mathbf{S} \cdot \mathbf{a} \Leftrightarrow s^i = S_k^i\,a^k$$
$$\sigma = \mathbf{a} \cdot \mathbf{s} = \mathbf{a} \cdot \mathbf{S} \cdot \mathbf{a} = a_i\,S_k^i\,a^k, \quad \tau = \mathbf{b} \cdot \mathbf{s} = \mathbf{b} \cdot \mathbf{S} \cdot \mathbf{a} = b_i\,S_k^i\,a^k \qquad (6.4.24)$$

The principal values $\sigma_i = \sigma$ and the principal directions $\mathbf{a}_i = \mathbf{a}$ for the tensor $\mathbf{S}$ are determined from the condition:

$$\mathbf{s} = \mathbf{S} \cdot \mathbf{a} = \sigma \mathbf{a} \Leftrightarrow \left(\sigma\,\delta_k^i - S_k^i\right)a_i = 0 \qquad (6.4.25)$$

The condition for a solution of these equations is that:

$$\det\left(\sigma\,\delta_k^i - S_k^i\right) = 0 \Leftrightarrow \sigma^3 - I\sigma^2 + II\sigma - III = 0 \qquad (6.4.26)$$

I, II, and *III* are the *principal invariants* of the tensor $\mathbf{S}$:

$$I = \operatorname{tr}\mathbf{S}, \quad III = \det\mathbf{S} = \det\left(S_i^i\right)$$
$$II = \frac{1}{2}\left[(\operatorname{tr}\mathbf{S})^2 - (\operatorname{norm}\mathbf{S})^2\right] = \frac{1}{2}\left[S_i^i\,S_k^k - S_k^i\,S_i^k\right] \qquad (6.4.27)$$

The three principal values σ_i are determined from Eq. (6.4.26), after which the principal directions $\mathbf{a}_i$ are computed from Eq. (6.4.25). As demonstrated in

Sect. 2.3.1 for the stress tensor $\mathbf{T}$, the principal directions are orthogonal. From the result expressed by the formula (3.3.26), we may write:

$$\left(\mathbf{S}[\mathbf{a}_i, \mathbf{a}_j]\right) = \begin{pmatrix} \sigma_1 & 0 & 0 \\ 0 & \sigma_2 & 0 \\ 0 & 0 & \sigma_3 \end{pmatrix} = (\sigma_i \delta_{ik}) \qquad (6.4.28)$$

6.4.3 Tensors as Polyadics

In Sect. 3.2.2 tensors were expressed as polyadics, i.e. linear combinations of polyads. For instance, a tensor $\mathbf{B}$ of second order with components B_{kl}^x in a Cartesian coordinate system x, may be expressed as:

$$\mathbf{B} = B_{kl}^x \, \mathbf{e}_k \otimes \mathbf{e}_l \qquad (6.4.29)$$

We use the formulas (6.4.11) to express the Cartesian components B_{kl}^x in terms of components in a general coordinate system y and the formulas (6.2.11) to express the base vectors $\mathbf{e}_i$ by the base vectors $\mathbf{g}_i$ and $\mathbf{g}^i$ in the y-system. We then obtain from the expression (6.4.29):

$$\mathbf{B} = B_{kl}^x \, \mathbf{e}_k \otimes \mathbf{e}_l = \left(\frac{\partial x_k}{\partial y^i}\frac{\partial x_l}{\partial y^j}B^{ij}\right)\left(\frac{\partial y^r}{\partial x_k}\mathbf{g}_r\right) \otimes \left(\frac{\partial y^s}{\partial x_l}\mathbf{g}_s\right) = \left(\frac{\partial y^r}{\partial x_k}\frac{\partial x_k}{\partial y^i}\right)\left(\frac{\partial y^s}{\partial x_l}\frac{\partial x_l}{\partial y^j}\right)B^{ij}\mathbf{g}_r \otimes \mathbf{g}_s \Rightarrow$$

$$\mathbf{B} = \delta_i^r \delta_j^s B^{ij}\mathbf{g}_r \otimes \mathbf{g}_s = B^{ij}\mathbf{g}_i \otimes \mathbf{g}_j$$

Similar expressions for the tensor $\mathbf{B}$ are obtained by raising and lowering of indices. Hence:

$$\mathbf{B} = B^{ij}\mathbf{g}_i \otimes \mathbf{g}_j = B_i^{\;j}\mathbf{g}^i \otimes \mathbf{g}_j = B_{\;j}^{i}\mathbf{g}_i \otimes \mathbf{g}^j = B_{ij}\mathbf{g}^i \otimes \mathbf{g}^j \qquad (6.4.30)$$

The results obtained for second order tensors are readily extended to higher order tensors. The operations discussed and presented in Sect. 3.2.2 may be generalized to apply in general coordinate systems. As an example the following operations are easily verified when $\mathbf{D}$ is a third order tensor, $\mathbf{B}$ is a second order tensor, and $\mathbf{a}$ is a vector, i.e. a first order tensor:

$$\mathbf{D} \cdot \mathbf{a} = \mathbf{B} \Leftrightarrow \left(D_{jk}^i \, \mathbf{g}_i \otimes \mathbf{g}^j \otimes \mathbf{g}^k\right) \cdot \left(a^l \mathbf{g}_l\right) = D_{jk}^i \, a^l \mathbf{g}_i \otimes \mathbf{g}^j (\mathbf{g}^k \cdot \mathbf{g}_l)$$

$$= D_{jk}^i \, a^l \mathbf{g}_i \otimes \mathbf{g}^j \delta_l^k = D_{jk}^i \, a^k \mathbf{g}_i \otimes \mathbf{g}^j = B_{\;j}^i \, \mathbf{g}_i \otimes \mathbf{g}^j \qquad (6.4.31)$$

6.5   Differentiation of Tensor Fields

6.5.1   Christoffel Symbols

The base vectors in a general curvilinear coordinate system are place functions: $\mathbf{g}_i(y)$ and $\mathbf{g}^i(y)$. We shall now find the changes of the base vectors along the coordinate lines. Using the comma notation when differentiating with respect to the coordinates y_i we write:

$$\frac{\partial \mathbf{g}_i}{\partial y^j} \equiv \mathbf{g}_{i,j} = \Gamma_{ijk}\mathbf{g}^k = \Gamma_{ij}^k \mathbf{g}_k \tag{6.5.1}$$

The components Γ_{ijk} and Γ_{ij}^k are called *Christoffel symbols of the first and the second kind* respectively. These symbols have their names after Elwin Bruno Christoffel [1829–1900]. Using the formulas (6.2.11) and (6.2.20) we obtain the results:

$$\mathbf{g}_{i,j} = \frac{\partial^2 x_r}{\partial y^j \partial y^i}\mathbf{e}_r = \frac{\partial^2 x_r}{\partial y^j \partial y^i}\frac{\partial x_r}{\partial y^k}\mathbf{g}^k \quad \text{or} \quad = \frac{\partial^2 x_r}{\partial y^j \partial y^i}\frac{\partial y^k}{\partial x_r}\mathbf{g}_k$$

By comparing the result with the definitions (6.5.1), we conclude that:

$$\Gamma_{ijk} = \frac{\partial^2 x_r}{\partial y^i \partial y^j}\frac{\partial x_r}{\partial y^k} = \Gamma_{jik}, \quad \Gamma_{ij}^k = \frac{\partial^2 x_r}{\partial y^i \partial y^j}\frac{\partial y^k}{\partial x_r} = \Gamma_{ji}^k \tag{6.5.2}$$

In Problem 6.4 the reader is asked to prove the following formulas:

$$\Gamma_{ij}^k = g^{kl}\Gamma_{ijl}, \quad \Gamma_{ijk} = g_{kl}\Gamma_{ij}^l \tag{6.5.3}$$

$$g_{ij,k} = \Gamma_{ikj} + \Gamma_{jki}, \quad \Gamma_{ijk} = \frac{1}{2}\left(g_{ik,j} + g_{jk,i} - g_{ij,k}\right) \tag{6.5.4}$$

$$\mathbf{g}_{,j}^k = -\Gamma_{ij}^k \mathbf{g}^i \tag{6.5.5}$$

In Problem 6.5 the reader is asked to use the formulas (6.5.1, 6.5.3, and 6.5.4) and (1.1.24) in order to show that:

$$\Gamma_{ik}^k = \frac{1}{2g}g_{,i} = \frac{1}{\sqrt{g}}(\sqrt{g})_{,i} \tag{6.5.6}$$

In orthogonal coordinate systems $g_{ij} = 0$ for $i \neq j$, and it follows from formulas (6.5.4)₂ and (6.5.3)₁ that:

$$\Gamma_{ijk} = \Gamma_{ij}^k = 0 \quad \text{when } i \neq j \neq k \neq i \qquad (6.5.7)$$

In *cylindrical coordinates* (R, θ, z) the non-zero Christoffel symbols are:

$$\Gamma_{122} = -\Gamma_{221} = R, \quad \Gamma_{12}^2 = \frac{1}{R}, \quad \Gamma_{22}^1 = -R \qquad (6.5.8)$$

In *spherical coordinates* (r, θ, ϕ) the non-zero Christoffel symbols are:

$$\Gamma_{122} = -\Gamma_{221} = -\Gamma_{22}^1 = r, \quad \Gamma_{133} = -\Gamma_{331} = -\Gamma_{33}^1 = r\sin^2\theta$$

$$\Gamma_{233} = -\Gamma_{322} = r^2\sin\theta\cos\theta, \quad \Gamma_{12}^2 = \Gamma_{13}^3 = \frac{1}{r}, \quad \Gamma_{23}^3 = \cot\theta \qquad (6.5.9)$$

$$\Gamma_{33}^2 = -\sin\theta\cos\theta$$

The Christoffel symbols Γ_{ijk} and $\Gamma_{ij}^{k\ k}$ do not represent a tensor. This fact follows from the following transformation formula relating $\overline{\Gamma}_{ij}^k$ and Γ_{ij}^k in two general coordinate systems $\bar{y}$ and y, see Problem 6.6:

$$\overline{\Gamma}_{ij}^k = \frac{\partial y^r}{\partial \bar{y}^i}\frac{\partial y^s}{\partial \bar{y}^j}\frac{\partial \bar{y}^k}{\partial y^t}\Gamma_{rs}^t + \frac{\partial^2 y^r}{\partial \bar{y}^i \partial \bar{y}^j}\frac{\partial \bar{y}^k}{\partial y^r} \qquad (6.5.10)$$

6.5.2 Absolute and Covariant Derivatives of Vector Components

Curves in three-dimensional space have been introduced in Sect. 1.4. A curve in space is described by the place vector $\mathbf{r} = \mathbf{r}(p)$ from the origin O in a Cartesian coordinate system Ox to a point on the curve identified by the curve parameter p. With respect to the coordinate system Ox with base vectors $\mathbf{e}_i$, the space vector is given as $\mathbf{r}(p) = x_i(p)\mathbf{e}_i$. The three coordinate functions $x_i(p)$ define the curve. In a general coordinate system the curve is defined by the coordinate functions $y_i(p)$. The length of the curve, from a point on the curve represented by the parameter value $p = p_0$ to the place $\mathbf{r}(p) = x_i(p)\,\mathbf{e}_i$, is given by the *arc length formula*, equation (1.4.5):

$$s(p) = \int_{p_o}^p \sqrt{\frac{d\mathbf{r}}{d\bar{p}} \cdot \frac{d\mathbf{r}}{d\bar{p}}}\, d\bar{p} = \int_{p_o}^p \sqrt{\frac{dx_i}{d\bar{p}}\frac{dx_i}{d\bar{p}}}\, d\bar{p}, \quad \text{arc length} \qquad (6.5.11)$$

To obtain the representation of the arc length formula in the y-system, we set:

$$\frac{d\mathbf{r}}{dp} = \frac{\partial \mathbf{r}}{\partial y^i}\frac{d(y^i)}{dp} = \mathbf{g}_i\frac{dy^i}{dp} \tag{6.5.12}$$

The differentials $dy^i \equiv d(y_i)$ are the contravariant components of the vector differential $d\mathbf{r}$:

$$d\mathbf{r} = \frac{\partial \mathbf{r}}{\partial y^i}dy^i = \mathbf{g}_i\,dy^i \tag{6.5.13}$$

The arc length formula (6.5.11) now takes the form:

$$s(p) = \int_{p_o}^{p}\sqrt{\frac{d\mathbf{r}}{d\bar{p}}\cdot\frac{d\mathbf{r}}{d\bar{p}}}\,d\bar{p} = \int_{p_o}^{p}\sqrt{g_{ij}\frac{dy^i}{d\bar{p}}\frac{dy^j}{d\bar{p}}}\,d\bar{p} \tag{6.5.14}$$

From this formula it follows that:

$$ds = |d\mathbf{r}| = \frac{ds(p)}{dp}dp = \sqrt{g_{ij}\frac{dy^i}{d\bar{p}}\frac{dy^j}{d\bar{p}}}\,dp = \sqrt{g_{ij}\,dy^i dy^j} \Rightarrow$$
$$(ds)^2 = g_{ij}\,dy^i\,dy^j \quad\text{and}\quad |d\mathbf{r}| = ds \tag{6.5.15}$$

The quadratic form $g_{ij}\,dy^i\,dy^j$, which is a scalar, is called the *metric of the space E_3* because the form relates the general coordinates y to length measurement. The result (6.5.15) may also be presented as:

$$(ds)^2 = d\mathbf{r}\cdot d\mathbf{r} = d\mathbf{r}\cdot\mathbf{1}\cdot d\mathbf{r} = dy^i g_{ij}\,dy^j \tag{6.5.16}$$

For this reason the unit tensor $\mathbf{1}$ is also called the *metric tensor* in the Euclidean space E_3.

With the arc length $s(p)$ as the curve parameter the curve is represented by the coordinate functions $y^i(s)$. The *tangent vector* $\mathbf{t}$ to the curve is defined as the unit vector:

$$\mathbf{t} = \frac{d\mathbf{r}}{ds} = \frac{\partial \mathbf{r}}{\partial y^i}\frac{dy^i}{ds} = \mathbf{g}_i t^i \Leftrightarrow t^i = \frac{dy^i}{ds} \tag{6.5.17}$$

Let $\alpha(\mathbf{r}, t)$ be a scalar field. The change of the field along the space curve $\mathbf{r}(y)$ where $y^i = y^i(s)$, may be expressed by the derivative:

$$\frac{\partial \alpha}{\partial s} = \frac{\partial \alpha}{\partial y_i}\frac{dy^i}{ds} \equiv \alpha_{,i}\frac{dy^i}{ds} = \alpha_{,i}\,t^i \tag{6.5.18}$$

Because $\partial\alpha/\partial s$ is a scalar the formula (6.5.18) shows that the functions $\alpha_{,i}$ represent the covariant components of a vector "grad α", the *gradient of the scalar field* $\alpha(y)$, defined in Cartesian coordinates by the formulas (1.6.6) and (1.6.9):

$$\text{grad } \alpha \equiv \frac{\partial\alpha}{\partial\mathbf{r}} \equiv \nabla\alpha = \mathbf{e}_i\frac{\partial\alpha}{\partial x_i} \equiv \mathbf{e}_i\,\alpha_{,i} \quad \text{in Cartesian Coordinate systems} \quad (6.5.19)$$

In a general coordinate system y the expression for the vector grad α is:

$$\text{grad } \alpha \equiv \frac{\partial\alpha}{\partial\mathbf{r}} \equiv \nabla\alpha = \mathbf{g}^i\frac{\partial\alpha}{\partial y^i} \equiv \mathbf{g}^i\,\alpha_{,i} \qquad (6.5.20)$$

The *del-operator* is a coordinate invariant scalar operator defined in a general coordinate system y by:

$$\nabla(\,) = \mathbf{g}^i\frac{\partial(\,)}{\partial y^i} \qquad (6.5.21)$$

The definition is in accordance with the del-operator (1.6.10) and with the left-operator (3.4.15) in Cartesian coordinates. To see that the operator is indeed coordinate invariant, we use the transformation $(6.2.21)_2$ for reciprocal base vectors and obtain:

$$\nabla(\,) = \mathbf{g}^i\frac{\partial(\,)}{\partial y^i} = \left[\frac{\partial y^i}{\partial \bar{y}^k}\,\bar{\mathbf{g}}^k\right]\frac{\partial(\,)}{\partial \bar{y}^j}\frac{\partial \bar{y}^j}{\partial y^i} = \bar{\mathbf{g}}^k\left[\frac{\partial \bar{y}^j}{\partial y^i}\frac{\partial y^i}{\partial \bar{y}^k}\right]\frac{\partial(\,)}{\partial \bar{y}^j} = \bar{\mathbf{g}}^k\delta^j_k\frac{\partial(\,)}{\partial \bar{y}^j} = \bar{\mathbf{g}}^k\frac{\partial(\,)}{\partial \bar{y}^k}$$

$$(6.5.22)$$

It is convenient to present the formula (6.5.20) as:

$$\text{grad } \alpha \equiv \frac{\partial\alpha}{\partial\mathbf{r}} \equiv \nabla\alpha = \alpha|_i\mathbf{g}^i = \alpha|^i\mathbf{g}_i \quad \text{where } \alpha|_i \equiv \alpha_{,i}\,, \quad \alpha|^i = g^{ik}\alpha_{,k} \qquad (6.5.23)$$

The functions $\alpha|^i$ are the contravariant components of the vector grad α. Equation (6.5.18) may now be presented as:

$$\frac{\partial\alpha}{\partial s} = (\text{grad } \alpha)\cdot\mathbf{t} \qquad (6.5.24)$$

This result is presented as formula (1.6.11) in Sect. 1.6, where it was shown that the vector grad al is a normal vector to the level surface $\alpha(\mathbf{r}, t) = \alpha_o(t)$.

For a space curve $s(y)$ along a y^i-coordinate line the formula (6.5.24) gives:

$$\frac{\partial \alpha}{\partial s} = \frac{\partial \alpha}{\partial y^i}\frac{dy^i}{ds} = (\text{grad } \alpha)\cdot \mathbf{t} = (\text{grad } \alpha)\cdot (\frac{dy^i}{ds}\mathbf{g}_i) \quad (\text{no summation w.r. to } i) \Rightarrow$$

$$\alpha_{,i} = \frac{\partial \alpha}{\partial y^i} = (\text{grad } \alpha)\cdot \mathbf{g}_i$$

$$(6.5.25)$$

This result is also obtained directly from the formula (6.5.20).

The derivative of vector field $\mathbf{a}(y,t)$ along the curve $y_i(s)$ is a new vector:

$$\frac{\partial \mathbf{a}}{\partial s} = \frac{\partial(a^i\mathbf{g}_i)}{\partial s} = \frac{\partial a^i}{\partial s}\mathbf{g}_i + a^i\mathbf{g}_{i,k}\frac{dy^k}{ds}, \quad \frac{\partial \mathbf{a}}{\partial s} = \frac{\partial(a_i\mathbf{g}^i)}{\partial s} = \frac{\partial a_i}{\partial s}\mathbf{g}^i + a_i\mathbf{g}^i_{,k}\frac{dy^k}{ds}$$

Using the formulas (6.5.1) and (6.5.5), we may present the vector $\partial\mathbf{a}/\partial s$ as:

$$\frac{\partial \mathbf{a}}{\partial s} = \left[\frac{\partial a^i}{\partial s} + a^j\Gamma^i_{jk}t^k\right]\mathbf{g}_i = \left[\frac{\partial a_i}{\partial s} - a_j\Gamma^j_{ik}t^k\right]\mathbf{g}^i \quad (6.5.26)$$

We now define *absolute derivatives of the vector components* a^i and a_i as the contravariant and covariant components of the vector $\partial\mathbf{a}/\partial s$:

$$\frac{\partial \mathbf{a}}{\partial s} = \frac{\delta a^i}{\delta s}\mathbf{g}_i = \frac{\delta a_i}{\delta s}\mathbf{g}^i \quad (6.5.27)$$

$$\frac{\delta a^i}{\delta s} = \frac{\partial \mathbf{a}}{\partial s}\cdot\mathbf{g}^i = \frac{\partial a^i}{\partial s} + a^j\Gamma^i_{jk}t^k, \quad \frac{\delta a_i}{\delta s} = \frac{\partial \mathbf{a}}{\partial s}\cdot\mathbf{g}_i = \frac{\partial a_i}{\partial s} - a_j\Gamma^j_{ik}t^k \quad (6.5.28)$$

We write:

$$\frac{\partial a^i}{\partial s} = a^i_{,k}\frac{dy^k}{ds} = a^i_{,k}t^k, \quad \frac{\partial a_i}{\partial s} = a_{i,k}\frac{dy^k}{ds} = a_{i,k}t^k \quad (6.5.29)$$

Then the expressions for the absolute derivatives of the vector components in Eq. (6.5.28) take the forms:

$$\frac{\delta a^i}{\delta s} = a^i|_k t^k, \quad \frac{\delta a_i}{\delta s} = a_i|_k t^k \quad (6.5.30)$$

where

$$a^i|_k = a^i_{,k} + a^j\Gamma^i_{jk}, \quad a_i|_k = a_{i,k} - a_j\Gamma^j_{ik} \quad (6.5.31)$$

The expressions $a^i|_k$ and $a_i|_k$ are called *covariant derivatives* of the *vector components* a^i and a_i respectively.

In Cartesian coordinate systems the Christoffel symbols are all zero and the absolute derivatives of vector components reduce to partial derivatives with respect to the curve parameter s and the covariant derivatives reduce to partial derivatives with respect to the general coordinates y:

$$\frac{\delta a^i}{\delta s} = \frac{\delta a_i}{\delta s} = \frac{\partial a_i}{\partial s} \text{ and } a^i|_k = a^i{}_{,k} = a_i|_k = a_{i,k} \text{ in Cartesian coordinate systems}$$

(6.5.32)

The covariant derivatives of vector components are components of the tensor we in Sect. 3.4.1 have called the *gradient of the vector field* $\mathbf{a}(y, t)$. Equations (6.5.30) is now seen to express the coordinate invariant relation between the vector $\partial \mathbf{a}/\partial s$, the second order tensor grad $\mathbf{a}$, and the tangent vector $\mathbf{t}$:

$$\frac{\partial \mathbf{a}}{\partial s} = (\text{grad } \mathbf{a}) \cdot \mathbf{t}$$

(6.5.33)

We may express grad $\mathbf{a}$ as the polyadic, confer the formulas (3.4.6):

$$\text{grad } \mathbf{a} = a^i|_k \mathbf{g}_i \otimes \mathbf{g}^k = a_i|_k \mathbf{g}^i \otimes \mathbf{g}^k$$

(6.5.34)

For a space curve $s(y_i)$ along a y_i-coordinate line the formula (6.5.33) gives:

$$\frac{\partial \mathbf{a}}{\partial s} = \frac{\partial \mathbf{a}}{\partial y_i} \frac{dy^i}{ds} = (\text{grad } \mathbf{a}) \cdot \mathbf{t} = (\text{grad } \mathbf{a}) \cdot (\frac{dy^i}{ds} \mathbf{g}_i) \quad (\text{no summation w.r. to } i) \Rightarrow$$

$$\mathbf{a}_{,i} \equiv \frac{\partial \mathbf{a}}{\partial y_i} = (\text{grad } \mathbf{a}) \cdot \mathbf{g}_i = a^k|_i \mathbf{g}_k = a_k|_i \mathbf{g}^k$$

(6.5.35)

This result is also obtained by differentiating the formulas: $\mathbf{a} = a^k \mathbf{g}_k = a_k \mathbf{g}^k$ with respect to y_i.

A vector field $\mathbf{a}(y, t)$ is said to be *uniform* at the time t along a curve $y_i(s)$ if the vector $\mathbf{a}(y, t)$ has the same direction and magnitude at all points on the curve at the time t. This implies that:

$$\frac{\partial \mathbf{a}}{\partial s} = \mathbf{0} \Leftrightarrow \frac{\delta a^i}{\delta s} = 0 \text{ along the space curve } y_i(s)$$

(6.5.36)

If the vector field $\mathbf{a}(y, t)$ is uniform everywhere in a volume V in space, then:

$$\text{grad } \mathbf{a} = \mathbf{0} \Leftrightarrow a^i|_k = 0 \text{ everywhere in the volume } V$$

(6.5.37)

6.5.3   The Frenet-Serret Formulas for Space Curves

Curves in space were introduced in Sect. 1.4. We shall present the components of the properties of a space curve given by the coordinate functions $y_i(s)$, where s represents the arc length from some reference point on the curve to the curve point given by the coordinates $y_i(s)$, or by a place vector $\mathbf{r}(s)$ from some fixed point O in space to the curve point $y_i(s)$. The *tangent vector* $\mathbf{t}$ to the curve is given by formula (6.5.17). The *principal normal* to the curve is defined by the unit vector $\mathbf{n}$ in formula (1.4.10), and now presented as:

$$\mathbf{n} = \frac{1}{\kappa}\frac{d\mathbf{t}}{ds} = \frac{1}{\kappa}\frac{d^2\mathbf{r}}{ds^2} \Leftrightarrow n^i = \frac{1}{\kappa}\frac{\delta t^i}{\delta s} \tag{6.5.38}$$

κ is the *curvature* to the space curve and defined by:

$$\kappa = \left|\frac{d\mathbf{t}}{ds}\right| = \sqrt{g_{ij}\frac{\delta t^i}{\delta s}\frac{\delta t^j}{\delta s}} \tag{6.5.39}$$

The *binormal* $\mathbf{b}$ to the curve is defined by the unit vector $\mathbf{b}$ in formula (1.4.11), and now presented as:

$$\mathbf{b} = \mathbf{t} \times \mathbf{n} \Leftrightarrow b_i = \varepsilon_{ijk}t^j n^k \tag{6.5.40}$$

The *torsion* τ of the curve is defined by the formula (1.4.12):

$$\frac{d\mathbf{b}}{ds} = -\tau\mathbf{n} \Rightarrow \tau = -\mathbf{n}\cdot\frac{d\mathbf{b}}{ds} \tag{6.5.41}$$

From the formulas (1.4.14) we obtain the formula for the torsion of the space curve in general curvilinear coordinates y:

$$\tau = \mathbf{b}\cdot\frac{d\mathbf{n}}{ds} = (\mathbf{t}\times\mathbf{n})\cdot\frac{d\mathbf{n}}{ds} = \left[\mathbf{t}\mathbf{n}\frac{d\mathbf{n}}{ds}\right] = \varepsilon_{ijk}t^i n^j \frac{\delta n^k}{\delta s} \tag{6.5.42}$$

The three *Frenet-Serret formulas* (1.4.13) for a space curve are in general coordinates:

$$\begin{aligned}
\frac{d\mathbf{t}}{ds} &= \kappa\mathbf{n} \Leftrightarrow \frac{\delta t^i}{\delta s} = \kappa n^i \\
\frac{d\mathbf{n}}{ds} &= \tau\mathbf{b} - \kappa\mathbf{t} \Leftrightarrow \frac{\delta n^i}{\delta s} = \tau b^i - \kappa t^i \\
\frac{d\mathbf{b}}{ds} &= -\tau\mathbf{n} \Leftrightarrow \frac{\delta b^i}{\delta s} = -\tau n^i
\end{aligned} \tag{6.5.43}$$

Example 6.1. Kinematics of a Particle Moving on a Space Curve
A mass particle is at the time t at the place $\mathbf{r}(t)$. The path of the particle is given by the space curve $y_i\,(t)$. Let $s(t)$ be the path length along the curve, measured from some reference point on the path to the place $\mathbf{r}(t)$. The velocity of the particle is given by:

$$\mathbf{v} = \frac{d\mathbf{r}}{dt} = \frac{d\mathbf{r}}{ds}\frac{ds}{dt} = \dot{s}\,\mathbf{t} \Leftrightarrow v^i = \frac{dy^i}{dt} = \frac{dy^i}{ds}\frac{ds}{dt} = \dot{s}\,t^i \qquad (6.5.44)$$

The acceleration of the particle is:

$$\mathbf{a} = \frac{d\mathbf{v}}{dt} = \ddot{s}\,\mathbf{t} + \dot{s}\frac{d\mathbf{t}}{ds}\frac{ds}{dt} \Rightarrow \mathbf{a} = \frac{d\mathbf{v}}{dt} = \ddot{s}\,\mathbf{t} + \kappa\dot{s}^2\mathbf{n} \qquad (6.5.45)$$

The first term in the expression for the acceleration $\mathbf{a}$, $\ddot{s}\,\mathbf{t}$, is the *tangential acceleration* and the second term, $\kappa\dot{s}^2\mathbf{n}$, is the *normal acceleration*. The contravariant components of the acceleration vector $\mathbf{a} = d\mathbf{v}/dt$ are, according to the formulas (6.5.45), (6.5.29), and (6.5.30):

$$a^i = \frac{\delta v^i}{\delta t} = \frac{\delta v^i}{\delta s}\frac{ds}{dt} = v^i\big|_k t^k \dot{s} = v^i\big|_k v^k = \left[v^i_{,k} + v^j\Gamma^i_{jk}\right]v^k = \frac{dv^i}{dt} + v^j\Gamma^i_{jk}v^k \quad (6.5.46)$$

The contravariant components and the physical components of the velocity vector $\mathbf{v}$ in cylindrical coordinates $(y_1, y_2, y_3) \equiv (R, \theta, z)$ are obtained from the formulas (6.5.44), (6.3.6), (6.5.46), and (6.5.8):

$$\mathbf{v} = \frac{d\mathbf{r}}{dt} = \frac{\partial\mathbf{r}}{\partial y_i}\frac{dy^i}{dt} = \mathbf{g}_1\frac{dR}{dt} + \mathbf{g}_2\frac{d\theta}{dt} + \mathbf{g}_3\frac{dz}{dt} = \mathbf{e}_R\dot{R} + R\mathbf{e}_\theta\dot{\theta} + \mathbf{e}_z\dot{z} \Rightarrow$$

$$\mathbf{v} = v^i\mathbf{g}_i \Rightarrow (v^i) = \left\{\dot{R}\ \dot{\theta}\ \dot{z}\right\}\left(\dot{R}, \dot{\theta}, \dot{z}\right), \quad \mathbf{v} = v(i)\mathbf{e}_i^y \Rightarrow (v(i)) = \left(v^i\sqrt{g_{ii}}\right) = \left\{\dot{R}\ R\dot{\theta}\ \dot{z}\right\}$$

$$(6.5.47)$$

The contravariant components and the physical components of the acceleration vector $\mathbf{a}$ in cylindrical coordinates $(y_1, y_2, y_3) \equiv (R, \theta, z)$ are obtained from the formulas (6.5.46) and (6.5.8):

$$a^1 = \ddot{R} + \dot{\theta}(-R)\dot{\theta}, \quad a^2 = \ddot{\theta} + \dot{R}\frac{1}{R}\dot{\theta} + \dot{\theta}\frac{1}{R}\dot{R} = \ddot{\theta} + 2\frac{\dot{R}\dot{\theta}}{R}, \quad a^3 = \ddot{z} \Rightarrow$$

$$\mathbf{a} = a^i\mathbf{g}_i \Rightarrow (a^i) = \left\{\ddot{R} - R\dot{\theta}^2\ \ddot{\theta} + 2\frac{\dot{R}\dot{\theta}}{R}\ \ddot{z}\right\}\left(\ddot{R} - R\dot{\theta}^2\ \ddot{\theta} + 2\frac{\dot{R}\dot{\theta}}{R}\ \ddot{z}\right)$$

$$\mathbf{a} = a(i)\mathbf{e}_i^y \Rightarrow (a(i)) = \left(a^i\sqrt{g_{ii}}\right) = \left\{\ddot{R} - R\dot{\theta}^2\ R\ddot{\theta} + 2\dot{R}\dot{\theta}\ \ddot{z}\right\}$$

$$(6.5.48)$$

6.5.4 Divergence and Rotation of Vector Fields

The divergence and the rotation of a vector field $\mathbf{a}(y, t)$ are defined by the formulas
(1.6.12) and (1.6.15) respectively. The expression for the *divergence of a vector
field* $\mathbf{a}(y, t)$ in general coordinates y is obtained by application of the formula
(6.5.21) for the del-operator ∇ and the formulas (6.5.35):

$$\operatorname{div} \mathbf{a} = \nabla \cdot \mathbf{a} = \mathbf{g}^i \cdot \frac{\partial \mathbf{a}}{\partial y^i} = \mathbf{g}^i \cdot (a^k|_i \mathbf{g}_k) = \mathbf{g}^i \cdot (a_k|_i \mathbf{g}^k) = a^i|_i = a_k|_i g^{ik} = a_k|^k \Rightarrow$$

$$\operatorname{div} \mathbf{a} = \nabla \cdot \mathbf{a} = a^i|_i = a_k|^k$$

$$(6.5.49)$$

The *rotation of a vector field* $\mathbf{a}(y, t)$ in general coordinates y is also obtained by
application of the formula (6.5.20) for the del-operator ∇ and the formulas (6.5.34):

$$\operatorname{rot} \mathbf{a} \equiv \operatorname{curl} \mathbf{a} \equiv \nabla \times \mathbf{a} = \mathbf{g}^j \times \frac{\partial \mathbf{a}}{\partial y_j} = \mathbf{g}^j \times \left(a_k|_j \mathbf{g}^k \right) = \varepsilon^{ijk} a_k|_j \mathbf{g}_i \Rightarrow$$

$$\operatorname{rot} \mathbf{a} \equiv \operatorname{curl} \mathbf{a} \equiv \nabla \times \mathbf{a} = \varepsilon^{ijk} a_k|_j \mathbf{g}_i = \varepsilon^{ijk} a_{k,j}\, \mathbf{g}_i = \frac{1}{\sqrt{g}} \det \begin{pmatrix} \mathbf{g}_1 & \mathbf{g}_2 & \mathbf{g}_3 \\ \frac{\partial}{\partial y^1} & \frac{\partial}{\partial y^2} & \frac{\partial}{\partial y^3} \\ a_1 & a_2 & a_3 \end{pmatrix}$$

$$(6.5.50)$$

In Problem 6.7 the reader is asked to prove the last two equalities in Eq. (6.5.50). In
Problem 6.8 the reader is asked to use formulas (6.5.6) and $(6.5.31)_1$ to derive the
formula:

$$\operatorname{div} \mathbf{a} \equiv \nabla \cdot \mathbf{a} = \frac{1}{\sqrt{g}} \left(\sqrt{g}\, a^i \right)_{,i} \tag{6.5.51}$$

This formula may be utilized to develop expression for $\nabla \cdot \mathbf{a}$ in a particular coordinate system.

6.5.5 Orthogonal Coordinates

In an orthogonal coordinate system we introduce the unit tangent vectors $\mathbf{e}_i^y$ and the
symbols h_i defined in Eq. (6.2.34). We also apply the symbol:

$$h = \sqrt{g} = h_1 h_2 h_3 \tag{6.5.52}$$

A vector **a** may be represented by contravariant components a^i, covariant components a_i, or by physical components $a(i)$. From the formulas (6.3.4) and (6.2.34) we obtain:

$$a(i) = a^i h_i = a_i / h_i \tag{6.5.53}$$

To express the del-operator (6.5.21) we use the formula $(6.2.35)_1$:

$$\nabla(\ ) = \mathbf{g}^i \frac{\partial(\ )}{\partial y^i} = \sum_i \frac{1}{h_i} \mathbf{e}_i^y \frac{\partial(\ )}{\partial y^i} \tag{6.5.54}$$

From the general formulas (6.5.54), (6.5.51), and (6.5.50) we obtain for *orthogonal coordinates*:

$$\text{grad } \alpha \equiv \nabla\alpha = \sum_i \frac{1}{h_i} \alpha_{,i} \, \mathbf{e}_i^y \tag{6.5.55}$$

$$\text{div } \mathbf{a} \equiv \nabla \cdot \mathbf{a} = \frac{1}{h} \sum_i \frac{\partial}{\partial y^i} \left(h \frac{a(i)}{h_i} \right) \tag{6.5.56}$$

$$\text{rot } \mathbf{a} \equiv \text{curl } \mathbf{a} \equiv \nabla \times \mathbf{a} = \frac{1}{h} \sum_{i,j,k} e_{ijk} \frac{\partial}{\partial y^j} (h_k \, a(k)) h_i \, \mathbf{e}_i^y$$

$$= \frac{1}{h} \det \begin{pmatrix} h_1 \, \mathbf{e}_1^y & h_2 \, \mathbf{e}_2^y & h_3 \, \mathbf{e}_3^y \\ \frac{\partial}{\partial y^1} & \frac{\partial}{\partial y^2} & \frac{\partial}{\partial y^3} \\ h_1 \, a(1) & h_2 \, a(2) & h_3 \, a(3) \end{pmatrix} \tag{6.5.57}$$

$$\text{div grad } \alpha \equiv \nabla \cdot \nabla\alpha \equiv \nabla^2\alpha = \frac{1}{h} \sum_i \frac{\partial}{\partial y^i} \left(\frac{h}{h_i^2} \frac{\partial\alpha}{\partial y^i} \right) \tag{6.5.58}$$

In *cylindrical coordinates* (R, θ, z) we have from the formulas (6.2.38) and (6.5.52): $h_1 = h_3 = 1$, $h_2 = R$, $h = R$. The formulas (6.5.55)–(6.5.58) provide the results:

$$\text{grad } \alpha \equiv \nabla\alpha = \frac{\partial\alpha}{\partial R} \mathbf{e}_R + \frac{1}{R} \frac{\partial\alpha}{\partial\theta} \mathbf{e}_\theta + \frac{\partial\alpha}{\partial z} \mathbf{e}_z \tag{6.5.59}$$

$$\text{div } \mathbf{a} = \frac{1}{R} \frac{\partial}{\partial R} (R a_R) + \frac{1}{R} \frac{\partial a_\theta}{\partial\theta} + \frac{\partial\alpha}{\partial z} \tag{6.5.60}$$

$$\text{rot } \mathbf{a} = \mathbf{e}_R \left[\frac{1}{R} \frac{\partial a_z}{\partial\theta} - \frac{\partial a_\theta}{\partial z} \right] + \mathbf{e}_\theta \left[\frac{\partial a_R}{\partial z} - \frac{\partial a_z}{\partial R} \right] + \mathbf{e}_z \left[\frac{1}{R} \frac{\partial}{\partial R} (R a_\theta) - \frac{1}{R} \frac{\partial a_R}{\partial\theta} \right] \tag{6.5.61}$$

$$\nabla^2 \alpha = \frac{1}{R}\frac{\partial}{\partial R}\left(R\frac{\partial \alpha}{\partial R}\right) + \frac{1}{R^2}\frac{\partial^2 \alpha}{\partial \theta^2} + \frac{\partial^2 \alpha}{\partial z^2} = \frac{\partial^2 \alpha}{\partial R^2} + \frac{1}{R}\frac{\partial \alpha}{\partial R} + \frac{1}{R^2}\frac{\partial^2 \alpha}{\partial \theta^2} + \frac{\partial^2 \alpha}{\partial z^2} \quad (6.5.62)$$

In *spherical coordinates* (r, θ, ϕ) we have from the formulas (6.2.42) and (6.5.52): $h_1 = 1$, $h_2 = r$, $h_3 = r \sin \theta$, $h = r^2 \sin \theta$. The formulas (6.5.55)–(6.5.58) provide the results:

$$\mathrm{grad}\,\alpha \equiv \nabla \alpha = \frac{\partial \alpha}{\partial r}\,\mathbf{e}_r + \frac{1}{r}\frac{\partial \alpha}{\partial \theta}\,\mathbf{e}_\theta + \frac{1}{r \sin \theta}\frac{\partial \alpha}{\partial \phi}\,\mathbf{e}_\phi \quad (6.5.63)$$

$$\mathrm{div}\,\mathbf{a} = \frac{1}{r^2}\frac{\partial}{\partial r}\left(r^2 a_r\right) + \frac{1}{r \sin \theta}\frac{\partial}{\partial \theta}\left(\sin \theta a_\theta\right) + \frac{1}{r \sin \theta}\frac{\partial a_\phi}{\partial \phi} \quad (6.5.64)$$

$$\mathrm{rot}\,\mathbf{a} = \mathbf{e}_r\left[\frac{1}{r \sin \theta}\frac{\partial}{\partial \theta}\left(\sin \theta a_\phi\right) - \frac{1}{r \sin \theta}\frac{\partial a_\theta}{\partial \phi}\right]$$
$$+ \mathbf{e}_\theta\left[\frac{1}{r \sin \theta}\frac{\partial a_r}{\partial \phi} - \frac{\partial}{\partial r}\left(r a_\phi\right)\right] + \mathbf{e}_\phi\left[\frac{1}{r}\frac{\partial}{\partial r}\left(r a_\theta\right) - \frac{1}{r}\frac{\partial a_r}{\partial \theta}\right] \quad (6.5.65)$$

$$\mathrm{div\,grad}\,\alpha = \nabla \cdot \nabla \alpha = \nabla^2 \alpha$$
$$= \frac{1}{r^2}\frac{\partial}{\partial r}\left(r^2 \frac{\partial \alpha}{\partial r}\right) + \frac{1}{r^2 \sin \theta}\frac{\partial}{\partial \theta}\left(\sin \theta \frac{\partial \alpha}{\partial \theta}\right) + \frac{1}{r^2 \sin^2 \theta}\frac{\partial^2 \alpha}{\partial \phi^2} \quad (6.5.66)$$

In order to find expressions for the Christoffel symbols of the first kind we first develop from the formulas (6.5.4) some special results:

$$\Gamma_{ijk} = \Gamma_{jik} = \frac{1}{2}\left(g_{ik,j} + g_{jk,i} - g_{ij,k}\right), \quad g_{ij} = h_i^2 \delta_j^i \Rightarrow$$
$$\Gamma_{111} = \frac{1}{2}g_{11,1} = \frac{1}{2}\left(h_1^2\right)_{,1} = h_1 h_{1,1}, \quad \Gamma_{123} = 0, \quad \Gamma_{112} = -\frac{1}{2}g_{11,2} = -\frac{1}{2}\left(h_1^2\right)_{,2} = -h_1 h_{1,2}$$
$$\Gamma_{122} = \frac{1}{2}g_{22,1} = \frac{1}{2}\left(h_2^2\right)_{,1} = h_2 h_{2,1}, \quad \Gamma_{121} = \frac{1}{2}g_{11,2} = \frac{1}{2}\left(h_1^2\right)_{,2} = h_1 h_{1,2}$$

Keeping in mind the symmetry $\Gamma_{ijk} = \Gamma_{jik}$, the above results are generalized to the formulas:

$$\begin{aligned}
\Gamma_{ijk} &= 0 \quad \text{for } i \neq j \neq k \neq i \\
\Gamma_{iik} &= -h_i h_{i,k} \quad \text{no summation w.r. to } i \text{ and } k \neq i) \\
\Gamma_{ikk} &= \Gamma_{kik} = h_k h_{k,i} \quad \text{no summation w.r. to } k \\
\Gamma_{iki} &= \Gamma_{kii} = h_i h_{i,k} \quad \text{no summation w.r. to } i
\end{aligned} \quad (6.5.67)$$

For the Christoffel symbols of the second kind we obtain from Eqs. $(6.5.3)_1$ and (6.5.67) first some special results:

$$\Gamma_{ij}^k = \Gamma_{ji}^k = g^{kl}\Gamma_{ijl}, \quad g^{kl} = \frac{1}{h_i^2}\delta_j^i \Rightarrow$$

$$\Gamma_{23}^1 = g^{1l}\Gamma_{23l} = g^{11}\Gamma_{231} = 0,$$

$$\Gamma_{22}^1 = g^{1l}\Gamma_{22l} = g^{11}\Gamma_{221} = \frac{1}{h_1^2}(-h_2 h_{2,1}) = -\frac{h_2}{h_1^2}h_{2,1}$$

$$\Gamma_{12}^1 = \Gamma_{21}^1 = g^{1l}\Gamma_{12l} = g^{11}\Gamma_{121} = \frac{1}{h_1^2}h_1 h_{1,2} = \frac{1}{h_1}h_{1,2},$$

$$\Gamma_{11}^1 = g^{1l}\Gamma_{11l} = g^{11}\Gamma_{111} = \frac{1}{h_1^2}h_1 h_{1,1} = \frac{1}{h_1}h_{1,1}$$

Keeping in mind the symmetry $\Gamma_{ij}^k = \Gamma_{ji}^k$, we may generalize the above results to the formulas:

$$\Gamma_{jk}^i = 0 \text{ for } i \neq j \neq k \neq i, \quad \Gamma_{ii}^k = -\frac{h_i}{h_k^2}\frac{\partial h_i}{\partial y^k} \quad \text{no summation w.r. to } i, \text{ and } k \neq i$$

$$\Gamma_{ik}^i = \Gamma_{ki}^i = \frac{1}{h_i}\frac{\partial h_i}{\partial y^k} \quad \text{no summation w.r. to } i \tag{6.5.68}$$

The covariant derivatives of the vector components a^i and a_i are developed from Eq. (6.5.31). First we develop some special results for the components $a^i|_k$:

$$a^i|_k = a^i{}_{,k} + a^j\Gamma_{jk}^i, \quad a^i = \frac{a(i)}{\sqrt{g_{ii}}} = \frac{a(i)}{h_i} \Rightarrow$$

$$a^1|_1 = a^1{}_{,1} + a^j\Gamma_{j1}^1 = a^1{}_{,1} + a^1\Gamma_{11}^1 + a^2\Gamma_{21}^1 + a^3\Gamma_{31}^1$$

$$= a^1{}_{,1} + a^1\frac{1}{h_1}h_{1,1} + a^2\frac{1}{h_1}h_{1,2} + a^3\frac{1}{h_1}h_{1,3}$$

$$a^1|_1 = \frac{\partial}{\partial y^1}\left(\frac{a(1)}{h_1}\right) + \sum_k \frac{a(k)}{h_k}\frac{1}{h_1}\frac{\partial h_1}{\partial y^k}$$

$$a^1|_2 = a^1{}_{,2} + a^j\Gamma_{j2}^1 = a^1{}_{,2} + a^1\Gamma_{12}^1 + a^2\Gamma_{22}^1 + a^3\Gamma_{32}^1 = a^1{}_{,2} + a^1\frac{1}{h_1}h_{1,2} - a^2\frac{h_2}{h_1^2}h_{2,1}$$

$$= \frac{1}{h_1}\frac{\partial a(1)}{\partial y^2} - a(1)\frac{1}{h_1^2}h_{1,2} + \frac{a(1)}{h_1}\frac{1}{h_1}h_{1,2} - \frac{a(2)}{h_2}\frac{h_2}{h_1^2}h_{2,1} = a^1|_2 = \frac{1}{h_1}\frac{\partial a(1)}{\partial y^2} - \frac{a(2)}{h_1^2}\frac{\partial h_2}{\partial y^1}$$

The above results are generalized to the formulas:

$$a^i|_i = \frac{\partial}{\partial y^i}\left(\frac{a(i)}{h_i}\right) + \sum_k \frac{a(k)}{h_k}\frac{1}{h_i}\frac{\partial h_i}{\partial y^k} \quad \text{no summation w.r. to } i$$

$$\tag{6.5.69}$$

$$a^i|_k = \frac{1}{h_i}\frac{\partial a(i)}{\partial y^k} - \frac{a(k)}{h_i^2}\frac{\partial h_k}{\partial y^i} \quad i \neq k$$

Next we shall derive the formulas for the components $a_i|_k = g_{il}a^l|_k$:

$$a_i|_k = h_i^2 a^i|_k = h_i \frac{\partial a(i)}{\partial y^k} - a(k)\frac{\partial h_k}{\partial y^i} \quad i \neq k$$

$$a_i|_i = h_i^2 a^i|_i = h_i^2 \frac{\partial}{\partial y^i}\left(\frac{a(i)}{h_i}\right) + \sum_k \frac{a(k)}{h_k} h_i \frac{\partial h_i}{\partial y^k} \quad \text{no summation w.r. to } i$$

$$(6.5.70)$$

Example 6.2. Covariant Derivatives of Vector Components in Cylindrical Coordinates

A vector $\mathbf{a}$ is expressed in physical, contravariant, and covariant components in cylindrical coordinates (R, θ, z) in the formulas (6.3.8) and we write:

$$\mathbf{a} = a_R \mathbf{e}_R + a_\theta \mathbf{e}_\theta + a_z \mathbf{e}_z = a^i \mathbf{g}_i = a_i \mathbf{g}^i \Leftrightarrow [a(1), a(2), a(3)] \equiv [a_R, a_\theta, a_z]$$

$$[a^1, a^2, a^3] \equiv \left[a_R, \frac{1}{R}a_\theta, a_z\right], \quad [a_1, a_2, a_3] \equiv [a_R, Ra_\theta, a_z]$$

$$(6.5.71)$$

We shall derive expressions for the covariant derivatives of the contravariant vector components a^k and the covariant vector components a_k in cylindrical coordinates (R, θ, z). From the formulas (6.2.37), (6.5.69), and (6.5.70) we obtain the following matrices:

$$(a^i|_k) = \begin{pmatrix} \frac{\partial a_R}{\partial R} & \frac{\partial a_R}{\partial \theta} - a_\theta & \frac{\partial a_R}{\partial z} \\ \frac{1}{R}\frac{\partial a_\theta}{\partial R} & \frac{1}{R}\frac{\partial a_\theta}{\partial \theta} + \frac{a_R}{R} & \frac{1}{R}\frac{\partial a_\theta}{\partial z} \\ \frac{\partial a_z}{\partial R} & \frac{\partial a_z}{\partial \theta} & \frac{\partial a_z}{\partial z} \end{pmatrix}, \quad (a_i|_k) = \begin{pmatrix} \frac{\partial a_R}{\partial R} & \frac{\partial a_R}{\partial \theta} - a_\theta & \frac{\partial a_R}{\partial z} \\ R\frac{\partial a_\theta}{\partial R} & R\frac{\partial a_\theta}{\partial \theta} + Ra_R & R\frac{\partial a_\theta}{\partial z} \\ \frac{\partial a_z}{\partial R} & \frac{\partial a_z}{\partial \theta} & \frac{\partial a_z}{\partial z} \end{pmatrix}$$

$$(6.5.72)$$

6.5.6　Absolute and Covariant Derivatives of Tensor Components

The results derived in Sect. 6.5.2 for scalar and vector fields, i.e. for tensors of order 0 and 1, will now be generalized to tensor fields of any order.

Let $\mathbf{A}(y, t)$ be a tensor field of order n. The components of the two new tensors: $\partial \mathbf{A}/ds$ of order n and grad $\mathbf{A}$ of order $(n + 1)$, are well-defined in any Cartesian coordinate system Ox, as presented in Sect. 3.4.1:

$$\frac{\partial \mathbf{A}}{\partial s} \text{ is defined through the components: } \frac{\partial A_{i \cdots j}}{\partial s} \text{ in Cartesian coordinates} \quad (6.5.73)$$

grad **A** is defined through the Cartesian components: $\dfrac{\partial A_{i\cdot\cdot j}}{\partial x_k} \equiv A_{i\cdot\cdot j,k}$ (6.5.74)

The following tensor relation is valid in a Cartesian coordinate system, confer the formulas (3.3.8):

$$\frac{\partial \mathbf{A}}{\partial s} = (\text{grad } \mathbf{A}) \cdot \mathbf{t} \Leftrightarrow \frac{\partial A_{i\cdot\cdot j}}{\partial s} = A_{i\cdot\cdot j,k}\, t_k \quad \text{in Cartesian coordinates} \qquad (6.5.75)$$

In a general coordinate system y we write:

$$\frac{\partial \mathbf{A}}{\partial s} = (\text{grad } \mathbf{A}) \cdot \mathbf{t} \Leftrightarrow \frac{\delta A_{i\cdot\cdot j}}{\delta s} = A_{i\cdot\cdot j}|_k t^k \qquad (6.5.76)$$

Confer the formulas (6.5.33) and (6.5.30) for tensors of first order. The tensor $\partial \mathbf{A}/\partial s$ is now represented by the *absolute derivatives* $\delta A_{i\cdot\cdot j}/\delta s$ of the tensor components $A_{i\cdot\cdot j}$, and the tensor grad **A** is represented by the *covariant derivatives* $A_{i\cdot\cdot j}|_k$ of the tensor components $A_{i\cdot\cdot j}$.

It will be demonstrated below how we may obtain the expressions for the covariant derivatives of the tensor components. Analogous to the development of the formulas (6.5.24) and (6.5.35) for scalar fields and vector fields, we obtain in general for tensor fields of any order:

$$\mathbf{A}_{,i} = (\text{grad } \mathbf{A}) \cdot \mathbf{g}_i \qquad (6.5.77)$$

From the definitions above it also follows that the rules of ordinary and partial differentiation also apply to absolute and covariant differentiation. For example, from the tensor equation:

$$\mathbf{A} = \mathbf{b} \otimes \mathbf{c} \Leftrightarrow A_j^i = b^i c_j \qquad (6.5.78)$$

we may compute:

$$\frac{\partial \mathbf{A}}{\partial s} = \frac{\partial \mathbf{b}}{\partial s} \otimes \mathbf{c} + \mathbf{b} \otimes \frac{\partial \mathbf{c}}{\partial s} \Leftrightarrow \frac{\delta A_j^i}{\delta s} = \frac{\delta b^i}{\delta s} c_j + b^i \frac{\delta c_j}{\delta s} \qquad (6.5.79)$$

$$\text{grad } \mathbf{A} = (\text{grad } \mathbf{b}) \otimes \mathbf{c} + \mathbf{b} \otimes (\text{grad } \mathbf{c}) \Leftrightarrow A_j^i|_k = b^i|_k c_j + b^i c_j|_k \qquad (6.5.80)$$

Equations (6.5.79) and (6.5.80) are true in Cartesian coordinate systems, and as tensor equations are thus true in general coordinates.

We now seek the expressions for covariant derivatives of the component sets of a tensor field, which by definition represent the component sets of the gradient of the tensor field. In order to simplify the presentation, we take as an example a tensor field **C** of second order, and we shall find the expressions:

$$C^i_j\big|_k = C^i_{j,k} + C^l_j\,\Gamma^i_{lk} - C^i_l\,\Gamma^l_{jk} \tag{6.5.81}$$

$$C_{ij}\big|_k = C_{ij,k} - C_{lj}\,\Gamma^l_{ik} - C_{il}\,\Gamma^l_{jk} \tag{6.5.82}$$

$$C^{ij}\big|_k = C^{ij}{}_{,k} + C^{lj}\,\Gamma^i_{lk} + C^{il}\,\Gamma^j_{lk} \tag{6.5.83}$$

$$C^j_i\big|_k = C^j_{i,k} - C^j_l\,\Gamma^l_{ik} + C^l_i\,\Gamma^j_{lk} \tag{6.5.84}$$

With these expressions for the covariant derivatives of tensor components, the absolute derivatives of the tensor components may be found from the formulas (6.5.76).

To derive the formulas (6.5.81)–(6.5.84) we may use either one of three methods:

1. by application of the transformation rules for tensor components from the partial derivatives in a Cartesian x-system to the covariant derivatives in a general y-system,
2. by partial and covariant differentiations of the component form of the scalar-valued function that the tensor represents,
3. by partial differentiation of the tensor presented as a polyadic.

The first method is straightforward but leads to some lengthy manipulations, and for that reason the method will not be demonstrated here. The second method shall now be applied to find the covariant derivative (6.5.81) as follows. Let **a** and **b** be two vector fields, such that according to the formulas (6.5.31):

$$a_i\big|_k = a_{i,k} - a_l\,\Gamma^l_{ik}, \quad b^j\big|_k = b^j{}_{,k} + b^l\,\Gamma^j_{lk} \tag{6.5.85}$$

The scalar value α of the second order tensor **C** for the argument vectors **a** and **b** is:

$$\alpha = \mathbf{C}[\mathbf{a}, \mathbf{b}] = C^i_j\,a_i\,b^j$$

Partial differentiation of this equation yields:

$$\alpha_{,k} = C^i_{j,k}\,a_i\,b^j + C^i_j\,a_{i,k}\,b^j + C^i_j\,a_i\,b^j{}_{,k} \tag{6.5.86}$$

Since partial differentiation coincides with covariant differentiation in a Cartesian coordinate system, Eq. (6.5.86) may also be considered to be the Cartesian form of the more general tensor equation:

$$\alpha\big|_k = C^i_j\big|_k a_i\,b^j + C^i_j\,a_i\big|_k b^j + C^i_j\,a_i\,b^j\big|_k \tag{6.5.87}$$

Now since $\alpha\big|_k \equiv \alpha_{,k}$, we obtain from Eqs. (6.5.87) and (6.5.86) that:

$$C^i_j|_k a_i\, b^j + \left[C^i_j\, a_i|_k b^j + C^i_j\, a_i\, b^j|_k\right] - \alpha|_k = 0 \quad \Rightarrow$$

$$C^i_j|_k a_i\, b^j + \left[C^i_j\left(a_{i,k} - a_l\, \Gamma^l_{ik}\right)b^j + C^i_j\, a_i\left(b^j{}_{,k} + b^l\, \Gamma^j_{lk}\right)\right]$$

$$- \left[C^i_{j,k}\, a_i\, b^j + C^i_j\, a_{i,k}\, b^j + C^i_j\, a_i\, b^j{}_{,k}\right] = 0 \Rightarrow$$

$$C^i_j|_k a_i\, b^j - \left[C^i_{j,k}\, a_i\, b^j + C^i_j\, a_l\, \Gamma^l_{ik}\, b^j - C^i_j\, a_i\, b^l\, \Gamma^j_{lk}\right] = 0$$

From the last result we obtain by renaming indices appropriately:

$$\left\{C^i_j|_k - \left[C^i_{j,k} + C^l_j\, \Gamma^i_{lk} - C^i_l\, \Gamma^l_{jk}\right]\right\}a_i\, b^j = 0$$

Because the vector fields **a** and **b** may be chosen arbitrarily, it follows that the coefficients in the brackets {} must be zero. Hence:

$$C^i_j|_k = C^i_{j,k} + C^l_j\, \Gamma^i_{lk} - C^i_l\, \Gamma^l_{jk} \Rightarrow \quad (6.5.81)$$

The result indicates a general recipe on how to obtain covariant derivatives of tensor components as a sum of the partial derivative of the component and contracted products of the components and the Christoffel symbols, one product for each component index. The products are added with a positive/negative sign according to whether the contracted tensor index is a superscript or a subscript. This rule applies to tensors of any order.

The third method for finding the covariant derivatives of tensor components will now be illustrated. The tensor **C** and grad **C** are presented as:

$$\mathbf{C} = C^i_j\, \mathbf{g}_i \otimes \mathbf{g}^j, \quad \text{grad } \mathbf{C} = C^i_j|_l \mathbf{g}_i \otimes \mathbf{g}^j \otimes \mathbf{g}^l \qquad (6.5.88)$$

Substituting the expression for grad **C** into formula (6.5.77), we get:

$$\mathbf{C}_{,k} = (\text{grad } \mathbf{C}) \cdot \mathbf{g}_k = \left(C^i_j|_l \mathbf{g}_i \otimes \mathbf{g}^j \otimes \mathbf{g}^l\right) \cdot \mathbf{g}_k = \left(C^i_j|_l \mathbf{g}_i \otimes \mathbf{g}^j\right)\delta^l_k = C^i_j|_k \mathbf{g}_i \otimes \mathbf{g}^j \Rightarrow$$

$$\mathbf{C}_{,k} = (\text{grad } \mathbf{C}) \cdot \mathbf{g}_k = C^i_j|_k \mathbf{g}_i \otimes \mathbf{g}^j$$

$$(6.5.89)$$

The expression for $\mathbf{C}_{,k}$ is also obtained by differentiation of equation in $(6.5.88)_1$:

$$\mathbf{C}_{,k} = C^i_{j,k}\, \mathbf{g}_i \otimes \mathbf{g}^j + C^i_j\, \mathbf{g}_{i,k} \otimes \mathbf{g}^j + C^i_j\, \mathbf{g}_i \otimes \mathbf{g}^j{}_{,k}$$

By use of the formulas (6.5.1) and (6.5.5), the above equation is rewritten to:

$$\mathbf{C}_{,k} = \left[C^i_{j,k} + C^l_j \, \Gamma^i_{lk} - C^i_l \, \Gamma^l_{jk} \right] \mathbf{g}_i \otimes \mathbf{g}^j \tag{6.5.90}$$

By comparing the two expressions for $\mathbf{C}_{,k}$ in Eqs. (6.5.89) and (6.5.90), we get the expression for the covariant derivatives presented in Eq. (6.5.81).

A tensor field $\mathbf{A}(y,t)$ is said to be *uniform* at the time t along a space curve $y_i(s)$ if its components in a Cartesian coordinate system are constants at all points on the curve at the time t. This implies that:

$$\text{along the curve } y_i(s) : \frac{\partial \mathbf{A}}{\partial s} = \mathbf{0} \text{ and grad } \mathbf{A} = \mathbf{0} \Leftrightarrow \frac{\delta A^i_{..j}}{\delta s} = 0 \text{ and } A^i_{..j}|_k = 0 \tag{6.5.91}$$

If the tensor field $\mathbf{A}(y,t)$ is uniform everywhere in a volume V in space, then:

$$\text{grad } \mathbf{A} = \mathbf{0} \Leftrightarrow A^i_{..j}|_k = 0 \text{ everywhere in the volume } V \tag{6.5.92}$$

The unit tensor $\mathbf{1}$ and the permutation tensor $\mathbf{P}$ are uniform tensor fields everywhere in E_3. Therefore:

$$\frac{\delta g_{ij}}{\delta s} = \frac{\delta g^{ij}}{\delta s} = 0, \quad g_{ij}|_k = g^{ij}|_k = 0 \tag{6.5.93}$$

$$\frac{\delta \varepsilon_{ijk}}{\delta s} = \frac{\delta \varepsilon^{ijk}}{\delta s} = 0, \quad \varepsilon_{ijk}|_l = \varepsilon^{ijk}|_l = 0 \tag{6.5.94}$$

Second covariant derivatives of the tensor components C^i_j are presented by:

$$C^i_j|_k|_l \equiv C^i_j|_{kl} = C^i_j|_{lk} \tag{6.5.95}$$

The symmetry with respect to the last two indices follows from the fact that this symmetry is true in Cartesian coordinate systems. This type of symmetry, which in fact shows that the order of differentiation is immaterial, is an inherent property of Euclidean spaces. As will be demonstrated in Sect. 8.5, this property does not in general hold in the two-dimensional Riemannian space R_2.

The second covariant derivatives of a scalar field $\alpha(y,t)$ are components of a tensor of second order appropriately called the *gradient of the gradient of the scalar field* $\alpha(y,t)$:

$$\text{grad grad } \alpha = \alpha|_{ij} \mathbf{g}^i \otimes \mathbf{g}^j \tag{6.5.96}$$

The symmetry implies that we may write:

$$\alpha|_j^i = \alpha|_j^i = \alpha|_j^i \qquad (6.5.97)$$

For div grad $\alpha = \nabla^2 \alpha$, we obtain, using the formulas (6.5.22) and (6.5.55):

$$\text{div grad } \alpha = \nabla^2 \alpha \equiv \nabla \cdot \nabla \alpha = \left(\alpha|^i\right)|_i = \alpha|^i|_i = \alpha|_i^i \qquad (6.5.98)$$

In Problem 6.9 the reader is asked to use formulas (6.5.6) and (6.5.51) to derive the formula:

$$\nabla^2 \alpha = \frac{1}{\sqrt{g}} \left(\sqrt{g}\, g^{ij} \alpha_{,j}\right)_{,i} \qquad (6.5.99)$$

This formula may be utilized to develop the expression for $\nabla^2 \alpha$ in a particular coordinate system.

The *divergence* of a tensor field $\mathbf{A}(y, t)$ of order n is a new tensor field div $\mathbf{A}$ of order $(n-1)$ and defined by the components $A_{i..}{}^k|_k$, confer Eq. (3.4.9):

$$\text{div } \mathbf{A} = A_{i..}{}^k|_k \mathbf{g}^i \otimes .. \qquad (6.5.100)$$

The *rotation* of a tensor field $\mathbf{A}(y, t)$ of order n is a new tensor rot $\mathbf{A}$ of order n and defined by the components $\varepsilon^{ijk} A_{r..k}|_j$, confer Eq. (3.4.13):

$$\text{rot } \mathbf{A} \equiv \text{curl } \mathbf{A} = \varepsilon^{ijk} A_{r..k}|_j \mathbf{g}^r \otimes .. \otimes \mathbf{g}_i \qquad (6.5.101)$$

The rotation of a vector field $\mathbf{a}(y, t)$ is the following vector rot $\mathbf{a}$, confer the formulas (1.6.15) and (3.4.14):

$$\text{rot } \mathbf{a} \equiv \text{curl } \mathbf{a} = \varepsilon^{ijk} a_k|_j \mathbf{g}_i \qquad (6.5.102)$$

The *Laplace-operator* ∇^2 is used to express the divergence of the gradient of a tensor $\mathbf{A}$ of order n, confer formulas (3.4.11):

$$\nabla^2 \mathbf{A} \equiv \text{div grad } \mathbf{A} = A^i{}_{..j}|_k^k \mathbf{g}_i \otimes .. \otimes \mathbf{g}^j \qquad (6.5.103)$$

In the Navier equations (5.2.27) in linear theory of elasticity and in the Navier-Stokes equations (5.3.16) for linearly viscous fluids the divergence of the gradient of a vector field and the gradient of the divergence of a vector field appear. For a vector field $\mathbf{a}(y, t)$:

$$\text{div grad } \mathbf{a} \equiv \nabla^2 \mathbf{a} = a^i|_k^k \mathbf{g}_i, \quad \text{grad div } \mathbf{a} = a^k|_k^i \mathbf{g}_i \qquad (6.5.104)$$

We shall present formulas for these two vector fields. With the help of the formulas (6.5.20) and (6.5.51) we obtain the result:

$$\text{grad div } \mathbf{a} = \nabla(\nabla \cdot \mathbf{a}) = a^k|^i_k \mathbf{g}_i = \frac{\partial}{\partial y^i}\left[\frac{1}{\sqrt{g}}\frac{\partial}{\partial y^k}\left(\sqrt{g}\, a^k\right)\right]\mathbf{g}^i \qquad (6.5.105)$$

The following expression for div grad $\mathbf{a} = \nabla^2\mathbf{a}$ shall be derived:

$$\nabla^2\mathbf{a} = \left\{\frac{\partial}{\partial y^j}\left[\frac{1}{\sqrt{g}}\frac{\partial}{\partial y^k}\left(\sqrt{g}\, a^k\right)\right]g^{ij} - \frac{1}{\sqrt{g}}\frac{\partial}{\partial y^k}\left[\sqrt{g}g^{kr}g^{is}\left(\frac{\partial a_r}{\partial y^s} - \frac{\partial a_s}{\partial y^r}\right)\right]\right\}\mathbf{g}_i$$

$$(6.5.106)$$

We need the result, see Problem 6.10:

$$\nabla \times (\nabla \times \mathbf{a}) = \frac{1}{\sqrt{g}}\frac{\partial}{\partial y^k}\left[\sqrt{g}g^{kr}g^{is}\left(\frac{\partial a_r}{\partial y^s} - \frac{\partial a_s}{\partial y^r}\right)\right]\mathbf{g}_i \qquad (6.5.107)$$

The formula $(1.6.17)_2$ is rewritten to:

$$\text{div grad } \mathbf{a} = \text{grad div } \mathbf{a} - \text{rot rot } \mathbf{a} \Leftrightarrow \nabla^2\mathbf{a} = \nabla(\nabla \cdot \mathbf{a}) - \nabla \times (\nabla \times \mathbf{a})$$

$$(6.5.108)$$

The expression (6.5.106) is now obtained by substitutions of the result (6.5.105) for $\nabla(\nabla \cdot \mathbf{a})$ and the result (6.5.107) for $\nabla \times (\nabla \times \mathbf{a})$ into the formula (6.5.108).

In *orthogonal coordinates*:

$$\sqrt{g} = h, \quad \mathbf{g}_i = h_i\mathbf{e}^y_i, \quad \mathbf{g}^i = \frac{\mathbf{e}^y_i}{h_i}, \quad \mathbf{g}_i = h_i\mathbf{e}^y_i, \quad g^{ij} = \frac{1}{h^2_i}\delta^{ij}, \quad a^k = \frac{a(k)}{h_i},$$

$$a_k = h_k a(k)$$

From the formulas (6.5.99), (6.5.105), (6.5.107), and (6.5.108) we obtain:

$$\nabla^2 = \sum_i\left[\frac{1}{h}\frac{\partial}{\partial y^i}\left(\frac{h}{h^2_i}\frac{\partial}{\partial y^i}\right)\right] \qquad (6.5.109)$$

$$\text{grad div } \mathbf{a} = \nabla(\nabla \cdot \mathbf{a}) = \sum_{i,k}\frac{1}{h_i}\frac{\partial}{\partial y^i}\left[\frac{1}{h}\frac{\partial}{\partial y^k}\left(h\frac{a(k)}{h_k}\right)\right]\mathbf{e}^y_i \qquad (6.5.110)$$

$$\nabla \times (\nabla \times \mathbf{a}) = \sum_{i,k}\frac{h_i}{h}\frac{\partial}{\partial y^k}\left[\frac{h}{(h_k h_i)^2}\left(\frac{\partial}{\partial y^i}(h_k a(k)) - \frac{\partial}{\partial y^k}(h_i a(i))\right)\right]\mathbf{e}^y_i \qquad (6.5.111)$$

$$\nabla^2 \mathbf{a} = \nabla(\nabla \cdot \mathbf{a}) - \nabla \times (\nabla \times \mathbf{a}) = \sum_{i,k} \left\{ \frac{1}{h_i} \frac{\partial}{\partial y^i} \left[\frac{1}{h} \frac{\partial}{\partial y^k} \left(h \frac{a(k)}{h_k} \right) \right] \right\} \mathbf{e}_i^y$$
$$- \sum_{i,k} \left\{ \frac{h_i}{h} \frac{\partial}{\partial y^k} \left[\frac{h}{(h_k h_i)^2} \left(\frac{\partial}{\partial y^i} (h_k a(k)) - \frac{\partial}{\partial y^k} (h_i a(i)) \right) \right] \right\} \mathbf{e}_i^y \tag{6.5.112}$$

In the following two examples we record the formulas for the operators presented in the formulas (6.5.109)–(6.5.112) in cylindrical coordinates (R, θ, z) and spherical coordinates (r, θ, ϕ). These formulas will be applied in the Navier equations for linearly elastic materials in Sect. 7.8 and in the Navier–Stokes equations for linearly viscous fluids in Sect. 7.9.

Example 6.3. Vector Operators in Cylindrical Coordinates (R, θ, z).
In cylindrical coordinates: $h = R$, $h_1 = h_3 = 1$, $h_2 = R$. Formula (6.5.109) yields:

$$\nabla^2 = \frac{\partial^2}{\partial R^2} + \frac{1}{R} \frac{\partial}{\partial R} + \frac{1}{R} \frac{\partial^2}{\partial \theta^2} + \frac{\partial^2}{\partial z^2} \tag{6.5.113}$$

Formula (6.5.110) gives:

$$\text{grad div } \mathbf{a} \equiv \nabla(\nabla \cdot \mathbf{a}) = \mathbf{b} = b_R \mathbf{e}_R + b_\theta \mathbf{e}_\theta + b_z \mathbf{e}_z \quad \Rightarrow$$
$$b_R = \frac{\partial^2 a_R}{\partial R^2} + \frac{1}{R} \frac{\partial a_R}{\partial R} - \frac{a_R}{R^2} + \frac{1}{R} \frac{\partial^2 a_\theta}{\partial R \partial \theta} - \frac{1}{R^2} \frac{\partial a_\theta}{\partial \theta} + \frac{\partial^2 a_z}{\partial R \partial z}$$
$$b_\theta = \frac{1}{R} \frac{\partial^2 a_R}{\partial \theta \partial R} + \frac{1}{R^2} \frac{\partial a_R}{\partial \theta} + \frac{1}{R^2} \frac{\partial^2 a_\theta}{\partial \theta^2} + \frac{1}{R} \frac{\partial^2 a_z}{\partial \theta \partial z} \tag{6.5.114}$$
$$b_z = \frac{\partial^2 a_R}{\partial z \partial R} + \frac{1}{R} \frac{\partial a_R}{\partial z} + \frac{1}{R} \frac{\partial^2 a_\theta}{\partial z \partial \theta} + \frac{\partial^2 a_z}{\partial z^2}$$

Formula (6.5.111) gives:

$$\nabla \times (\nabla \times \mathbf{a}) = \mathbf{c} = c_R \mathbf{e}_R + c_\theta \mathbf{e}_\theta + c_z \mathbf{e}_z \Rightarrow$$
$$c_R = \frac{1}{R} \frac{\partial^2 a_\theta}{\partial \theta \partial R} + \frac{1}{R^2} \frac{\partial a_\theta}{\partial \theta} - \frac{1}{R^2} \frac{\partial^2 a_R}{\partial \theta^2} + \frac{\partial^2 a_z}{\partial z \partial R} - \frac{\partial^2 a_R}{\partial z^2}$$
$$c_\theta = \frac{1}{R} \frac{\partial^2 a_R}{\partial R \partial \theta} - \frac{1}{R^2} \frac{\partial a_R}{\partial \theta} - \frac{\partial^2 a_\theta}{\partial R^2} - \frac{1}{R} \frac{\partial a_\theta}{\partial R} + \frac{1}{R^2} a_\theta + \frac{1}{R} \frac{\partial^2 a_z}{\partial z \partial \theta} - \frac{\partial^2 a_\theta}{\partial z^2} \tag{6.5.115}$$
$$c_z = \frac{\partial^2 a_R}{\partial R \partial z} + \frac{1}{R} \frac{\partial a_R}{\partial z} - \frac{1}{R} \frac{\partial a_z}{\partial R} - \frac{\partial^2 a_z}{\partial R^2} + \frac{1}{R} \frac{\partial^2 a_\theta}{\partial \theta \partial z} - \frac{1}{R^2} \frac{\partial^2 a_z}{\partial \theta^2}$$

Formulas (6.5.112) gives:

$$\nabla^2 \mathbf{a} = \nabla(\nabla \cdot \mathbf{a}) - \nabla \times (\nabla \times \mathbf{a}) = \mathbf{d} = d_R \mathbf{e}_R + d_\theta \mathbf{e}_\theta + d_z \mathbf{e}_z = \mathbf{b} - \mathbf{c}$$

$$= \left[\nabla^2 a_R - \frac{a_R}{R^2} - \frac{2}{R^2} \frac{\partial a_\theta}{\partial \theta} \right] \mathbf{e}_r = \left[\nabla^2 a_\theta - \frac{1}{R^2} a_\theta + \frac{2}{R^2} \frac{\partial a_R}{\partial \theta} \right] \mathbf{e}_\theta + \nabla^2 a_z \mathbf{e}_\phi$$

$$(6.5.116)$$

Example 6.4. Vector Operators in Spherical Coordinates (r, θ, ϕ).

In spherical coordinates: $h = r^2 \sin \theta$, $h_1 = 1, h_2 = r$, $h_3 = r \sin \theta$.   Formula (6.5.109) yields:

$$\nabla^2 = \frac{\partial^2}{\partial r^2} + \frac{2}{r} \frac{\partial}{\partial r} + \frac{1}{r^2} \frac{\partial^2}{\partial \theta^2} + \frac{1}{r^2} \cot \theta \frac{\partial}{\partial \theta} + \frac{1}{r^2 \sin^2 \theta} \frac{\partial^2}{\partial \phi^2} \qquad (6.5.117)$$

Formula (6.5.110) gives:

$$\text{grad div } \mathbf{a} \equiv \nabla(\nabla \cdot \mathbf{a}) = \mathbf{b} = b_r \mathbf{e}_r + b_\theta \mathbf{e}_\theta + b_\phi \mathbf{e}_\phi \quad \Rightarrow$$

$$b_r = \frac{\partial^2 a_r}{\partial r^2} + \frac{2}{r} \frac{\partial a_r}{\partial r} - \frac{2}{r^2} a_r + \frac{1}{r} \frac{\partial^2 a_\theta}{\partial r \partial \theta} - \frac{1}{r^2} \frac{\partial a_\theta}{\partial \theta} + \frac{1}{r \sin \theta} \frac{\partial^2 a_\phi}{\partial r \partial \phi} - \frac{1}{r^2 \sin \theta} \frac{\partial a_\phi}{\partial \phi}$$

$$b_\theta = \frac{1}{r^2} \frac{\partial^2 a_\theta}{\partial \theta^2} + \frac{\cot \theta}{r^2} \frac{\partial a_\theta}{\partial \theta} - \frac{1}{r^2 \sin^2 \theta} a_\theta + \frac{1}{r} \frac{\partial^2 a_r}{\partial \theta \partial r} + \frac{2}{r^2} \frac{\partial a_r}{\partial \theta} - \frac{\cot \theta}{r^2 \sin \theta} \frac{\partial a_\phi}{\partial \phi} + \frac{1}{r^2 \sin \theta} \frac{\partial^2 a_\phi}{\partial \theta \partial \phi}$$

$$b_\phi = \frac{1}{r^2 \sin^2 \theta} \frac{\partial^2 a_\phi}{\partial \phi^2} + \frac{2}{r^2 \sin \theta} \frac{\partial a_r}{\partial \phi} + \frac{1}{r^2 \sin \theta} \frac{\partial^2 a_\theta}{\partial \phi \partial \theta} + \frac{1}{r \sin \theta} \frac{\partial^2 a_r}{\partial \phi \partial r} + \frac{\cot \theta}{r^2 \sin \theta} \frac{\partial a_\theta}{\partial \phi}$$

$$(6.5.118)$$

Formula (6.5.111) gives:

$$\nabla \times (\nabla \times \mathbf{a}) = \mathbf{c} = c_r \mathbf{e}_r + c_\theta \mathbf{e}_\theta + c_\phi \mathbf{e}_\phi \Rightarrow$$

$$c_r = \frac{1}{r^2} \cot \theta a_\theta + \frac{1}{r} \cot \theta \frac{\partial a_\theta}{\partial r} - \frac{1}{r^2} \cot \theta \frac{\partial a_r}{\partial \theta} + \frac{1}{r^2} \frac{\partial a_\theta}{\partial \theta}$$

$$+ \frac{1}{r} \frac{\partial^2 a_\theta}{\partial \theta \partial r} - \frac{1}{r^2} \frac{\partial^2 a_r}{\partial \theta^2} + \frac{1}{r^2 \sin \theta} \frac{\partial a_\phi}{\partial \phi} + \frac{1}{r \sin \theta} \frac{\partial^2 a_\phi}{\partial \phi \partial r} - \frac{1}{r^2 \sin^2 \theta} \frac{\partial^2 a_r}{\partial \phi^2}$$

$$c_\theta = \frac{1}{r \sin \theta} \frac{\partial^2 a_r}{\partial r \partial \theta} - \frac{2}{r} \frac{\partial a_\theta}{\partial r} - \frac{\partial^2 a_\theta}{\partial r^2} + \frac{\cot \theta}{r^2 \sin \theta} \frac{\partial a_\phi}{\partial \phi} + \frac{1}{r^2 \sin \theta} \frac{\partial^2 a_\phi}{\partial \phi \partial \theta} \qquad (6.5.119)$$

$$- \frac{1}{r^2 \sin^2 \theta} \frac{\partial^2 a_\theta}{\partial \phi^2}$$

$$c_\phi = \frac{1}{r \sin \theta} \frac{\partial^2 a_r}{\partial r \partial \phi} - \frac{2}{r} \frac{\partial a_\phi}{\partial r} - \frac{\partial^2 a_\phi}{\partial r^2} - \frac{\cos \theta}{r^2 \sin^2 \theta} \frac{\partial a_\theta}{\partial \phi} + \frac{1}{r^2 \sin \theta} \frac{\partial^2 a_\theta}{\partial \theta \partial \phi}$$

$$+ \frac{1}{r^2 \sin^2 \theta} a_\phi - \frac{\cot \theta}{r^2} \frac{\partial a_\phi}{\partial \theta} - \frac{1}{r^2} \frac{\partial^2 a_\phi}{\partial \theta^2}$$

Formula (6.5.112) gives:

$$\nabla^2 \mathbf{a} = \nabla(\nabla \cdot \mathbf{a}) - \nabla \times (\nabla \times \mathbf{a}) = \mathbf{b} - \mathbf{c}$$
$$= \left[\nabla^2 a_r - \frac{2}{r^2} a_r - \frac{2}{r^2} \frac{\partial a_\theta}{\partial \theta} - \frac{2 \cot \theta}{r^2} a_\theta - \frac{2}{r^2 \sin \theta} \frac{\partial a_\phi}{\partial \phi} \right] \mathbf{e}_r$$
$$+ \left[\nabla^2 a_\theta - \frac{1}{r^2 \sin^2 \theta} a_\theta + \frac{2}{r^2} \frac{\partial a_r}{\partial \theta} - \frac{\cos \theta}{r^2 \sin^2 \theta} \frac{\partial a_\phi}{\partial \phi} \right] \mathbf{e}_\theta$$
$$+ \left[\nabla^2 a_\phi - \frac{1}{r^2 \sin^2 \theta} a_\phi + \frac{2}{r^2 \sin^2 \theta} \frac{\partial a_r}{\partial \phi} + \frac{2}{r^2 \sin^2 \theta} \cos \theta \frac{\partial a_\theta}{\partial \phi} \right] \mathbf{e}_\phi$$

$$(6.5.120)$$

6.6   Two-Point Tensor Components

The content in this section will be applied in the deformation analysis in Sect. 7.3. With reference to Fig. 6.6 we introduce three coordinate systems: y and Y are general curvilinear coordinate systems, while x is a Cartesian system. The coordinate sets y and Y may represent two different places in E_3. The place y has the position vector $\mathbf{r}$ and Cartesian coordinates x_i, while the place Y has the position vector $\mathbf{r}_0$ and Cartesian coordinates X_i. The base vectors in the Y-system are $\mathbf{g}_K$ and $\mathbf{g}^K$, and in the y-system $\mathbf{g}_i$ and $\mathbf{g}^i$. Upper case and lower case letter indices shall refer to the Y-system and the y-system respectively.

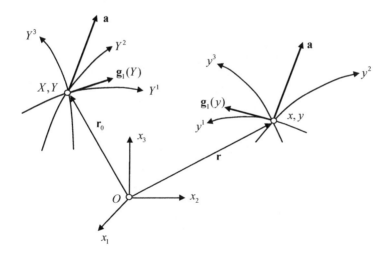

Fig. 6.6 Threes coordinate systems:general coordinates y and Y, and Cartesian coordinates

The base vectors $\mathbf{g}_K$ and $\mathbf{g}^K$ at the place Y are moved to the place y and decomposed there. We write:

$$\mathbf{g}_K = g_K^i\,\mathbf{g}_i = g_{Ki}\,\mathbf{g}^i, \quad \mathbf{g}^K = g^{Ki}\,\mathbf{g}_i = g_i^K\,\mathbf{g}^i \tag{6.6.1}$$

The components g_K^i, g_{Ki}, g^{Ki}, and g_i^K are called *Euclidean shifters* and represent *two-point components of the unit tensor* $\mathbf{1}$ of second order:

$$g_K^i = \mathbf{g}_K \cdot \mathbf{g}^i = \mathbf{1}\left[\mathbf{g}_K,\ \mathbf{g}^i\right] = \frac{\partial X_j}{\partial Y^K}\frac{\partial y^i}{\partial x_j} \tag{6.6.1}$$

$$g_{Ki} = g_{iK} = \mathbf{g}_K \cdot \mathbf{g}_i = \mathbf{1}[\mathbf{g}_K,\ \mathbf{g}_i] = \frac{\partial X_j}{\partial Y^K}\frac{\partial x_j}{\partial y^i} \tag{6.6.2}$$

$$g_i^K = \mathbf{g}^K \cdot \mathbf{g}_i = \mathbf{1}\left[\mathbf{g}^K,\ \mathbf{g}_i\right] = \frac{\partial Y^K}{\partial X_j}\frac{\partial x_j}{\partial y^i} \tag{6.6.3}$$

$$g^{Ki} = g^{iK} = \mathbf{g}^K \cdot \mathbf{g}^i = \mathbf{1}\left[\mathbf{g}^K,\ \mathbf{g}^i\right] = \frac{\partial Y^K}{\partial X_j}\frac{\partial y^i}{\partial x_j} \tag{6.6.4}$$

When the base vectors $\mathbf{g}_i$ and $\mathbf{g}^i$ at the place y are shifted (moved) to the place Y, we get:

$$\mathbf{g}_i = g_i^K\,\mathbf{g}_K = g_{iK}\,\mathbf{g}^K, \quad \mathbf{g}^i = g^{iK}\,\mathbf{g}_K = g_K^i\,\mathbf{g}^K \tag{6.6.5}$$

A vector $\mathbf{a}$ may be decomposed at the places Y or y with respect to either set of base vectors:

$$\mathbf{a} = a^K\,\mathbf{g}_K = a_K\,\mathbf{g}^K = a^i\,\mathbf{g}_i = a_i\,\mathbf{g}^i \tag{6.6.6}$$

From these expressions we get the following relations between the component sets:

$$\begin{aligned} a^K &= g_i^K\,a^i = g^{Ki}a_i, \quad a_K = g_K^i\,a_i = g_{Ki}\,a^i \\ a^i &= g_K^i\,a^K = g^{iK}a_K, \quad a_i = g_i^K\,a_K = g_{iK}\,a^K \end{aligned} \tag{6.6.7}$$

We conclude that the Euclidean shifters have the property of moving a vector from one place in space to another place in space. The Euclidean shifters are functions of both y-and Y-coordinates.

Let $\mathbf{C}(y,t)$ be a tensor field of second order and $\mathbf{a}$ and $\mathbf{b}$ two argument vectors fields. The scalar field $\mathbf{C}[\mathbf{a}, \mathbf{b}]$ may now be calculated from alternative formulas, as for instance:

$$\alpha = \mathbf{C}[\mathbf{a}, \mathbf{b}] = C_{Ki}\,a^K\,b^i = C_i^K\,a_K\,b^i = C_{iK}\,a^i\,b^K \text{ etc.} \tag{6.6.8}$$

The components:

$$C_{Ki} = \mathbf{C}[\mathbf{g}_K, \mathbf{g}_i], \quad C_i^K = \mathbf{C}[\mathbf{g}^K, \mathbf{g}_i] \text{ etc.} \tag{6.6.9}$$

are called *two-point components* of the tensor $\mathbf{C}$. The component sets are related through formulas of the type:

$$C_{Ki} = g_K^j\, C_{ji}, \quad C_i^K = g_i^L\, C_L^K \tag{6.6.10}$$

Two-point components of a second order tensor $\mathbf{C}$ may be used to transform a vector $\mathbf{a}$ at one place to a new vector $\mathbf{b}$ at another place. For example, let $\mathbf{a}$ be a vector at the place $\mathbf{r}_0$. Then:

$$b^i = C_K^i a^K \tag{6.6.11}$$

are the components of a vector $\mathbf{b}$ at the place $\mathbf{r}$. For each distinct value of the index K the components C_K^i and C_{iK} represent a vector at the place $\mathbf{r}$. Likewise, for each distinct value of the index i the components C_K^i and C_{iK} represent a vector at the place $\mathbf{r}_0$. In the literature, the two-point tensor components are therefore often presented as the components of a *two-point tensor* or a *double vector*. In the present exposition, we shall not distinguish between tensors with argument vectors at one and the same place or at two different places.

A vector field $\mathbf{a} = \mathbf{a}(y, Y)$ is called a *two-point vector field*, or a *two-point tensor field of order 1*. The components of the vector in both the y-system and the Y-system are in general functions of the six variables y^i and Y^K, with special cases when $\mathbf{a} = \mathbf{a}(y)$ or $\mathbf{a} = \mathbf{a}(Y)$.

$$
\begin{aligned}
\mathbf{a}(y, Y) &\Leftrightarrow a_i(y, Y) \text{ and } a_K(y, Y) \\
\mathbf{a}(y) &\Leftrightarrow a_i(y) \text{ and } a_K(y, Y) \\
\mathbf{a}(Y) &\Leftrightarrow a_i(y, Y) \text{ and } a_K(Y)
\end{aligned}
\tag{6.6.12}
$$

Two− point tensor fields of higher order are defined similarly.

Let $\mathbf{C}(y, Y)$ be a two-point tensor field of second order. Since the components C_K^i behave as vector components at the places y and Y, we may calculate covariant derivatives at these places from Eq. (6.5.31):

$$C_K^i|_j = C_{K,j}^i + C_K^k\, \Gamma_{kj}^i, \quad C_K^i|_L = C_{K,L}^i - C_N^i\, \Gamma_{KL}^N \tag{6.6.13}$$

The Christoffel symbols $\Gamma_{kj}^{\;i}$ and Γ_{KL}^N refer to the systems y and Y respectively. The expressions (6.6.14) are called *partial-covariant derivatives* of the two-point components of $\mathbf{C}$. If the coordinate systems y and Y are chosen to be identical to the Cartesian system Ox, the partial-covariant derivatives reduce to:

$$C_K^i|_j = \frac{\partial C_K^i}{\partial x_j}, \quad C_K^i|_L = \frac{\partial C_K^i}{\partial X_L} \tag{6.6.14}$$

If a one-to-one mapping between the places y and Y is given:

$$y = y(Y) \Leftrightarrow Y = Y(y) \tag{6.6.15}$$

the *total-covariant derivatives* of tensor components may be defined:

$$A_K^i||_j = A_K^i|_j + A_K^i|_L \frac{\partial Y^L}{\partial y^j}, \quad A_K^i||_L = A_K^i|_L + A_K^i|_j \frac{\partial y^j}{\partial Y^L} \tag{6.6.16}$$

6.7   Relative Tensors

The tensor fields presented so far in this book will in the present section be called *absolute tensors fields*, *absolute vectors fields*, or *absolute scalar fields*. In the extended definition to *relative tensor fields* the tensors lose some of their characteristic coordinate invariance. Since relative tensors are used in the literature to some extend, they will be given a brief introduction here.

An *NP-scalar field* is defined as a quantity that in every coordinate systems y is represented by a magnitude α, such that:

$$\alpha = \left| J_y^x \right|^N \left(\text{sign } J_y^x \right)^P \alpha_0 \tag{6.7.1}$$

N is an integer, $P = 0$ or $= 1$, and α_0 is a scalar field. If y is a right-handed Cartesian system x, it follows that $\alpha = \alpha_0$. In the case $P = 0$, the quantity is called a *relative scalar field of weight N*. A relative scalar field of weight $N = 1$ is called a *scalar density*. From Eq. (6.2.31) it follows that the quantity $g = \det(g_{ij})$ is a relative scalar of weight 2, and that $\sqrt{g}$ is a scalar density. For $N = 0$ and $P = 1$ the quantity α in Eq. (6.7.1) is called an *axial scalar*. Note that the magnitude α of an axial scalar changes sign by a transformation from a right-handed/left-handed coordinate system and to a left-handed/right-handed coordinate system. For $N = P = 0$ the *NP*-scalar is an *absolute scalar*. From Eq. (6.7.1) it finally follows that the magnitudes α and $\bar{\alpha}$ in two coordinate systems y and $\bar{y}$ respectively, are related through the formula:

$$\bar{\alpha} = \left| J_{\bar{y}}^y \right|^N \left(\text{sign } J_{\bar{y}}^y \right)^P \alpha \tag{6.7.2}$$

An *NP-vector field* **b** is defined as a linear *NP*-scalar-valued function of a vector **a**:

$$\alpha = \mathbf{b}[\mathbf{a}] \tag{6.7.3}$$

In the coordinate system y the NP-vector field $\mathbf{b}$ is represented by the component sets:

$$b_i = \mathbf{b}[\mathbf{g}_i], \quad b^i = \mathbf{b}[\mathbf{g}^i] \tag{6.7.4}$$

It now follows from Eqs. (6.7.2)–(6.7.4) that the components sets of $\mathbf{b}$ in two coordinate system y and $\bar{y}$ are related through the formulas:

$$\bar{b}_i = \left|J_{\bar{y}}^y\right|^N \left(\mathrm{sign}\, J_{\bar{y}}^y\right)^P \frac{\partial y^k}{\partial \bar{y}^i}\, b_k, \quad \bar{b}^i = \left|J_{\bar{y}}^y\right|^N \left(\mathrm{sign}\, J_{\bar{y}}^y\right)^P \frac{\partial \bar{y}^i}{\partial y^k}\, b^k \tag{6.7.5}$$

For $P = 0$ the NP-vector is called a *relative vector of weight N*, and for $N = 0$ and $P = 1$ the NP-vector is called an *axial vector*. For $N = P = 0$ the NP-vector is an *absolute vector*. If the vector product of two absolute vectors $\mathbf{a}$ and $\mathbf{b}$ are defined by the components $\sqrt{g}\, e_{ijk}\, a^j b^k$ rather then by the components $\varepsilon_{ijk}\, a^j b^k$ as in Sect. 6.3, the vector product becomes an axial vector.

An *NP-tensor of order n* is defined as a multilinear NP-scalarvalued function of n absolute vectors. In the coordinate system y the NP-tensor is represented by component sets defined similarly to the components of absolute tensors. For $P = 0$ the NP-tensor is called a *relative tensor of weight N*, and for $N = 0$ and $P = 1$ the NP-tensor is called an *axial tensor*. For $N = P = 0$ the NP-tensor is an *absolute tensor*. The components $\sqrt{g}\, e_{ijk}$ define an axial tensor of third order. Confer the discussion in Sect. 6.3.

The algebra of NP-tensors follows the rules applying for absolute tensors. Addition has only meaning for tensors of equal weight N and of the same value of P. By tensor multiplication the weights of the tensors are added as are the P-values.

Problems 6.1–6.10 with solutions see Appendix

Reference

1. Sokolnikoff IS (1939) Advanced calculus. McGraw-Hill, New York

Chapter 7
Elements of Continuum Mechanics in General Coordinates

7.1 Introduction

In this chapter the basic equations of continuum mechanics are presented in general curvilinear coordinates. Section 7.2 presents the kinematics and the material derivative of intensive quantities. In Sect. 7.3 the deformation analysis presented in Chap. 4 is extended to applications in curvilinear coordinates. The analysis of large deformations introduced in Sect. 4.5, is generalized in Sect. 7.4. The concept of convected coordinates is presented in Sect. 7.5. Section 7.6 introduces the concept of convected derivatives. The components of the Cauchy stress tensor **T** and the Cauchy equations of motion, introduced in Chap. 2, are presented in curvilinear coordinates in Sect. 7.7. Some basic equations of linear elasticity and linear viscous fluids, from Chap. 5, are the subject matter of Sects. 7.8 and 7.9.

7.2 Kinematics

Figure 7.1 illustrates a reference Rf to which motion and deformation of a material body will be referred, a *reference configuration* K_0 representing the body at a *reference time* t_0 and the *present configuration* K representing the same body at the time t, which we call the *present time*. The reference configuration will usually be chosen to be real configuration of the body, such that the configurations K and K_0 coincide at $t = t_0$.

Figure 7.1 also introduces three coordinate systems fixed in the reference Rf: (1) an orthogonal Cartesian coordinate system Ox with base vectors $\mathbf{e}_i$, (2) a general coordinate system y with base vectors $\mathbf{g}_i$, and (3) a general coordinate system Y with base vectors $\mathbf{g}_K$. In Sect. 7.5 the coordinate system Y will be used as a *material coordinate system* imbedded in the continuum.

© Springer Nature Switzerland AG 2019
F. Irgens, *Tensor Analysis*,
https://doi.org/10.1007/978-3-030-03412-2_7

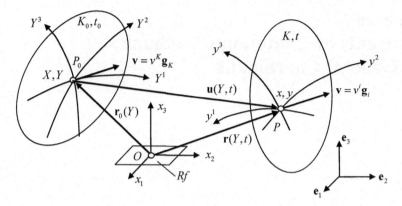

Fig. 7.1 Body of a continuum in a reference configuration K_0 at the time t_0 and in a present configuration K at the time t. Three coordinate systems fixed relative to the reference Rf: Cartesian coordinate system Ox with base vectors e_i. General coordinate system y with base vectors g_i. General coordinate system Y with base vectors g_K. Particle vector r_0 (Y). Place vector $r(Y, t)$. Displacement vector $u(Y, t)$. Velocity vector $v = v^i g_i = v^k g_K$

A *particle* in the body is a material point which at the reference time t_0 is denoted by P_0 and localized by the Cartesian coordinates X_i, the general coordinates Y^K, or the place vector $r_0(Y)$, which we call the *particle vector*. At the present time t the particle is denoted by P and localized by the Cartesian coordinates x_i, the general coordinates y_i, or by the *place vector* $r(Y, t)$. The coordinates X_i and Y_K are called *particle coordinates*, while the coordinates x_i and y_i are called *place coordinates*. We use the alternative notations for a particle and for the place of the particle at the present time t, confer the notations (2.1.1) and (2.1.2):

$$\text{particle: } r_0, \ [X_1, X_2, X_3], \ X_i, \ X \equiv \begin{pmatrix} X_1 \\ X_2 \\ X_3 \end{pmatrix}, \ Y^K, \ Y \equiv \begin{pmatrix} Y^1 \\ Y^2 \\ Y^3 \end{pmatrix}$$

$$\text{place: } r, \ [x_1, x_2, x_3], \ x_i, \ x \equiv \begin{pmatrix} x_1 \\ x_2 \\ x_3 \end{pmatrix}, \ y^i, \ y \equiv \begin{pmatrix} y^1 \\ y^2 \\ y^3 \end{pmatrix}$$

$$(7.2.1)$$

Note that lower case Latin letters are used as indices in the x-system and the y-system, while upper case Latin letters are used as indices in the Y-system.

The motion of the body at the present time t may be given by the functional relationship between the place vector r and the place vector r_0, or the functional relationships between the place coordinates y_i and the particle coordinates Y^K, confer the expressions (2.1.3):

$$\mathbf{r} = \mathbf{r}(\mathbf{r}_0, t) \quad \text{or} \quad \mathbf{r} = \mathbf{r}(Y, t), \quad y^i = y^i(Y^1, Y^2, Y^3, t) = y^i(Y, t) \qquad (7.2.2)$$

The motion of the body may also be represented by the *displacement vector* $\mathbf{u}(\mathbf{r}_0, t)$, or $\mathbf{u}(Y, t)$:

$$\mathbf{u} = \mathbf{u}(\mathbf{r}_0, t) = \mathbf{r}(\mathbf{r}_0, t) - \mathbf{r}_0 \quad \text{or} \quad \mathbf{u} = \mathbf{u}(Y, t) = \mathbf{r}(Y, t) - \mathbf{r}_0(Y) \qquad (7.2.3)$$

Intensive physical quantities are expressed either by *particle functions* or *place functions*. Particle functions are field functions of the particle vector $\mathbf{r}_0$, or particle coordinates Y, at the reference time t_0, and the present time t. Place functions are field functions of the place vector $\mathbf{r}$, or the place coordinates y, and the present time t.

$$
\begin{aligned}
&\text{particle function: } f(\mathbf{r}_0, t) = f(Y, t) \equiv f(Y^1, Y^2, Y^3, t) \\
&\text{place function: } f(\mathbf{r}, t) = f(y, t) \equiv f(y^1, y^2, y^3, t)
\end{aligned}
\qquad (7.2.4)
$$

In accordance with the presentation in Sect. 2.1.1 the coordinates $(Y, t) \equiv (Y^1, Y^2, Y^3, t)$ are called *Lagrangian coordinates*, *material coordinates*, or *reference coordinates*. The application of these coordinates is called *Lagrangian description* or *reference description*. The coordinates $(y, t) \equiv (y^1, y^2, y^3, t)$ are called *Eulerian coordinates* or *space coordinates*, and their application is *Eulerian description* or *spacial description*.

7.2.1 Material Derivative of Intensive Quantities

Let the particle function $f(\mathbf{r}_0, t) = f(Y, t)$ represent an arbitrary intensive physical quantity. For a particular choice of the particle $\mathbf{r}_0$ or the particle coordinates Y the function $f(\mathbf{r}_0, t) = f(Y, t)$ is connected to that particle at all times. The *material derivative* of the particle function f is defined as the time rate of change of the function when it is attached to the particle:

$$
\begin{aligned}
\dot{f} = \dot{f}(\mathbf{r}_0, t) &\equiv \left.\frac{df}{dt}\right|_{\mathbf{r}_0 = \text{constant}} = \frac{\partial f(\mathbf{r}_0, t)}{\partial t} \equiv \partial_t f(\mathbf{r}_0, t) \Leftrightarrow \\
\dot{f} = \dot{f}(Y, t) &\equiv \left.\frac{df}{dt}\right|_{Y = \text{constant}} = \frac{\partial f(Y, t)}{\partial t} \equiv \partial_t f(Y, t)
\end{aligned}
\qquad (7.2.5)
$$

This definition is in accordance with the definition (2.1.8) in Sect. 2.1.1.

The *velocity* $\mathbf{v}$ of the particle $\mathbf{r}_0$ is given by:

$$
\begin{aligned}
\mathbf{v} = \mathbf{v}(\mathbf{r}_0, t) = \dot{\mathbf{r}} = \partial_t \mathbf{r}(\mathbf{r}_0, t) = \partial_t \mathbf{u}(\mathbf{r}_0, t) &\Leftrightarrow \mathbf{v} = \mathbf{v}(Y, t) = \dot{\mathbf{r}} = \partial_t \mathbf{r}(Y, t) \\
= \partial_t \mathbf{u}(Y, t)&
\end{aligned}
\qquad (7.2.6)
$$

Another expression for the particle velocity $\mathbf{v}$ is found as follows.

$$\mathbf{v} = \mathbf{v}(Y,t) = \dot{\mathbf{r}} = \frac{\partial \mathbf{r}}{\partial y^i} \frac{\partial y^i(Y,t)}{\partial t} = \mathbf{g}_i v^i = v^K \mathbf{g}_K, \ v^i = \frac{\partial y^i(Y,t)}{\partial t}, \ v^K = v^i g_i^K \quad (7.2.7)$$

The symbols $g_i^K = \mathbf{g}^K \cdot \mathbf{g}_i$ are *Euclidian shifters* as defined by the formulas (6.6.3), and v^K are components of the velocity vector $\mathbf{v}$ with respect to the Y-system at the place $\mathbf{r}_0$.

The material derivative of a tensor field $\mathbf{A}(\mathbf{r}_0,t) = \mathbf{A}(Y,t) = A_{K..}{}^L \mathbf{g}_K \otimes .. \otimes \mathbf{g}^L$ of order n is a new tensor field of the same order and defined by, confer with the formulas (3.4.27):

$$\dot{\mathbf{A}} = \frac{\partial \mathbf{A}(Y,t)}{\partial t} = \dot{A}_{K..}{}^L \mathbf{g}_K \otimes .. \otimes \mathbf{g}^L \Rightarrow \dot{A}_{K..}{}^L = \frac{\partial}{\partial t}\left(A_{K..}{}^L\right) = \dot{\mathbf{A}}\left[\mathbf{g}_K, .., \mathbf{g}^L\right] \quad (7.28)$$

Let the place function $f(\mathbf{r},t) = f(y,t)$ represent an arbitrary intensive physical quantity. The local change of the function f per unit time is:

$$\frac{\partial f(\mathbf{r},t)}{\partial t} \equiv \partial_t f(\mathbf{r},t) \Leftrightarrow \frac{\partial f(y,t)}{\partial t} \equiv \partial_t f(y,t) \quad (7.2.9)$$

In order to find the material derivative of the place function $f(\mathbf{r},t) = f(y,t)$, we attach the function to the particle $\mathbf{r}_0$, or Y, that takes the place $\mathbf{r}$, or y, at the time t. Thus we write:

$$f(\mathbf{r},t) = f(\mathbf{r}(\mathbf{r}_0,t),t) \Leftrightarrow f(y,t) = f(y(Y,t),t) \quad (7.2.10)$$

The definition (7.2.5) of the material derivative of an intensive quantity leads us to the result:

$$\dot{f} \equiv \left.\frac{df}{dt}\right|_{\mathbf{r}_0 = \text{constant}} = \frac{\partial f(y,t)}{\partial t} + \frac{\partial f(y,t)}{\partial y^i}\frac{\partial y^i(Y,t)}{\partial t} = \partial_t f(y,t) + f(y,t)_{,i} v^i \Rightarrow$$
$$\dot{f} = \partial_t f(y,t) + f(y,t)_{,i} v^i = \partial_t f(y,t) + (\text{grad} f) \cdot \mathbf{v}$$
$$(7.2.11)$$

The material derivative of a tensor field $\mathbf{A}(y,t)$ of order n is a new tensor field of the same order and defined by the following formulas, confer with the formulas (3.4.28):

$$\dot{\mathbf{A}} = \frac{\partial}{\partial t}\mathbf{A}(y(Y,t),t) = \partial_t\mathbf{A}(y,t) + \text{grad}\,\mathbf{A}(y,t) \cdot \mathbf{v}(y,t) \Leftrightarrow$$
$$\dot{A}_{i..}{}^j \equiv \dot{\mathbf{A}}\left[\mathbf{g}_i, .., \mathbf{g}^j\right] = \partial_t A_{i..}{}^j + A_{i..}{}^j|_k v^k$$
$$(7.2.12)$$

Note that $\dot{A}_{i..}{}^{j}$ are components of the tensor $\dot{\mathbf{A}}$ and not the material derivatives of the tensor components $A_{i..}{}^{j}$. The latter are given by:

$$\frac{\partial}{\partial t}\left[A_{i..}{}^{j}(y(Y,t),t)\right] = \frac{\partial}{\partial t}\left[A_{i..}{}^{j}(y,t)\right] + \frac{\partial}{\partial y^{k}}\left[A_{i..}{}^{j}(y,t)\right]v^{k} \qquad (7.2.13)$$

and do not represent a tensor in general coordinates y, only when the coordinate system y is Cartesian.

Material differentiation follows the standard differentiation rules. For example, the material derivative of the vector $\mathbf{a} = \mathbf{B}\mathbf{c}$, where $\mathbf{B}$ is a second order tensor and $\mathbf{c}$ is a vector, has the components:

$$\dot{a}_i = \dot{B}_i^k\, c_k + B_i^k\, \dot{c}_k \qquad (7.2.14)$$

This result is easy to prove by use of Eq. (7.2.12) and the differentiation rules for covariant derivatives.

Because the *unit tensor* $\mathbf{1}$ and the *permutation tensor* $\mathbf{P}$ are time independent and uniform tensors it follows that:

$$\dot{g}_{ij} = \dot{g}^{ij} = 0, \quad \dot{\varepsilon}_{ijk} = \dot{\varepsilon}^{ijk} = 0 \qquad (7.2.15)$$

The definition (7.2.11) may be used to find the following expression for the *particle acceleration* $\mathbf{a}(y, t)$:

$$\mathbf{a} = \dot{\mathbf{v}} = \partial_t\mathbf{v} + (\text{grad } \mathbf{v})\cdot\mathbf{v} = \partial_t\mathbf{v} + (\mathbf{v}\cdot\nabla)\mathbf{v} \Leftrightarrow a^i = \partial_t v^i + v^i|_k v^k \qquad (7.2.16)$$

If $\mathbf{A}$ is a steady tensor field: $\mathbf{A} = \mathbf{A}(y)$, the material derivative of $\mathbf{A}$ may be expressed by the absolute derivative of tensor components with respect to time. From the formulas (7.2.12) and (6.5.76) we obtain:

$$\dot{A}_{i..}{}^{j} = A_{i..}{}^{j}|_k v^k = A_{i..}{}^{j}|_k\, t^k \frac{ds}{dt} = \frac{\delta A_{i..}{}^{j}}{\delta t} \qquad (7.2.17)$$

$$\dot{\mathbf{A}} = \frac{d\mathbf{A}}{dt} \Leftrightarrow \dot{A}_{i..}{}^{j} = \frac{\delta A_{i..}{}^{j}}{\delta t} \qquad (7.2.18)$$

7.3   Deformation Analysis

It has been shown in Sect. 4.2 that the deformation in the neighborhood of a particle Y, or X, is determined by the *deformation gradient tensor* $\mathbf{F}$ defined by the formulas (4.2.7) and (4.2.8):

$$d\mathbf{r} = \mathbf{F} \cdot d\mathbf{r}_0, \quad \mathbf{F} = \text{Grad } \mathbf{r} = \frac{\partial \mathbf{r}}{\partial \mathbf{r}_0} \tag{7.3.1}$$

$d\mathbf{r}_0$ and $d\mathbf{r}$ represent a material line element respectively in the reference configuration K_0 and in the present configuration K. In the Cartesian coordinate system x the relations (7.3.1) are represented by:

$$dx_i = F_{ik} dX_k, \quad F_{ik} = \frac{\partial x_i}{\partial X_k} \tag{7.3.2}$$

In a general curvilinear coordinate system y the motion of the particles in body of continuous material is given by $y\,(Y, t)$ and the relation (7.3.1) has the component representation:

$$dy^i = F^i_K dY^K, \quad F^i_K = \frac{\partial y^i}{\partial Y^K} \tag{7.3.3}$$

F^i_K are *two-point components of the deformation gradient tensor* **F**.

The lengths ds_0 and ds of the material line element in K_0 and K are found from the formulas:

$$(ds_0)^2 = d\mathbf{r}_0 \cdot d\mathbf{r}_0 = g_{KL} dY^K dY^L, \quad (ds)^2 = d\mathbf{r} \cdot d\mathbf{r} = g_{ij} dy^i dy^j \tag{7.3.4}$$

We introduce the unit vector **e** in the direction of the material line element $d\mathbf{r}_0$, see Fig. 4.2:

$$\mathbf{e} = \frac{d\mathbf{r}_0}{ds_0} \tag{7.3.5}$$

Using Eqs. (7.3.1) and (7.3.5) we may write:

$$(ds)^2 = d\mathbf{r} \cdot d\mathbf{r} = (\mathbf{F} \cdot d\mathbf{r}_0) \cdot (\mathbf{F} \cdot d\mathbf{r}_0) = \mathbf{e} \cdot (\mathbf{F}^T \mathbf{F}) \cdot \mathbf{e}(ds_0)^2 \quad \Rightarrow$$
$$(ds)^2 = \mathbf{e} \cdot (\mathbf{F}^T \mathbf{F}) \cdot \mathbf{e}(ds_0)^2 = \mathbf{e} \cdot \mathbf{C} \cdot \mathbf{e}(ds_0)^2 \tag{7.3.6}$$

C is *Green's deformation tensor* and is defined by:

$$\mathbf{C} = \mathbf{F}^T \mathbf{F} \Leftrightarrow C_{KL} = F^i_K F_{iL} \tag{7.3.7}$$

As an alternative to the result (7.3.6) we write:

$$(ds)^2 = d\mathbf{r} \cdot d\mathbf{r} = (\mathbf{F} \cdot d\mathbf{r}_0) \cdot (\mathbf{F} \cdot d\mathbf{r}_0) = d\mathbf{r}_0 \cdot (\mathbf{F}^T \mathbf{F}) \cdot d\mathbf{r}_0 = d\mathbf{r}_0 \cdot \mathbf{C} \cdot d\mathbf{r}_0 \quad \Rightarrow$$
$$(ds)^2 = d\mathbf{r}_0 \cdot \mathbf{C} \cdot d\mathbf{r}_0 = C_{KL} dY^K dY^L$$

$$\tag{7.3.8}$$

It follows that we can write:

$$(ds_0)^2 = d\mathbf{r}_0 \cdot \mathbf{1} \cdot d\mathbf{r}_0 = g_{KL}dY^K dY^L \tag{7.3.9}$$

$$(ds)^2 - (ds_0)^2 = d\mathbf{r}_0 \cdot (\mathbf{C} - \mathbf{1}) \cdot d\mathbf{r}_0 = 2d\mathbf{r}_0 \cdot \mathbf{E} \cdot d\mathbf{r}_0 \tag{7.3.10}$$

$\mathbf{E}$ is *Green's strain tensor* and defined by:

$$\mathbf{E} = \frac{1}{2}(\mathbf{C} - \mathbf{1}) \Leftrightarrow E_{KL} = \frac{1}{2}(C_{KL} - g_{KL}) \tag{7.3.11}$$

If we imagine that the Y-system is imbedded in the continuum and moves and deforms with the body as a *convected coordinate system*, the result (7.3.8) shows that the components C_{KL} of the Green deformation tensor also represent the fundamental parameters of first order for the convected Y-system. This interpretation of C_{KL} will be utilized below in Sect. 7.5.

7.3.1   Strain Measures

The following strain measures are defined in Sect. 4.1: the *longitudinal strain* ε in the direction $\mathbf{e}$, formula (4.2.13), the *volumetric strain* ε_v, formula (4.2.18), and the *shear strain* γ with respect to two material line elements, which in the reference configuration are orthogonal and have the directions $\bar{\mathbf{e}}$ and $\mathbf{e}$, formula (4.2.16). The expressions for these strain measures in general coordinates are:

$$\varepsilon = \frac{ds - ds_o}{ds_o} = \sqrt{\mathbf{e} \cdot \mathbf{C} \cdot \mathbf{e}} - 1 = \sqrt{e^K C_{KL} e^L} - 1$$
$$\varepsilon = \sqrt{1 + 2\mathbf{e} \cdot \mathbf{E} \cdot \mathbf{e}} - 1 = \sqrt{1 + 2e^K E_{KL} e^L} - 1 \tag{7.3.12}$$

$$\varepsilon_v = \frac{dV - dV_o}{dV_o} = \det \mathbf{F} - 1 = \sqrt{\det \mathbf{C}} - 1 = \sqrt{\det(\mathbf{1} + 2\mathbf{E})} - 1$$
$$= \sqrt{\det(C_K^L)} - 1 = \sqrt{\det(\delta_K^L + 2E_K^L)} - 1 \tag{7.3.13}$$

$$\sin \gamma = \frac{\bar{\mathbf{e}} \cdot \mathbf{C} \cdot \mathbf{e}}{\sqrt{(\bar{\mathbf{e}} \cdot \mathbf{C} \cdot \bar{\mathbf{e}})(\mathbf{e} \cdot \mathbf{C} \cdot \mathbf{e})}} = \frac{\bar{e}^K C_{KL} e^L}{\sqrt{(\bar{e}^M C_{MN} \bar{e}^N)(e^P C_{PQ} e^Q)}}$$
$$\sin \gamma = \frac{2\bar{\mathbf{e}} \cdot \mathbf{E} \cdot \mathbf{e}}{\sqrt{(1 + 2\bar{\mathbf{e}} \cdot \mathbf{E} \cdot \bar{\mathbf{e}})(1 + 2\mathbf{e} \cdot \mathbf{E} \cdot \mathbf{e})}} = \frac{2\,\bar{e}^K E_{KL} e^L}{\sqrt{(1 + \bar{\varepsilon})(1 + \varepsilon)}} \tag{7.3.14}$$

$\bar{\varepsilon}$ is the longitudinal strain in the direction $\bar{\mathbf{e}}$.

When the motion is expressed by the displacement vector: $\mathbf{u}(\mathbf{r}_0, t) = \mathbf{r}(\mathbf{r}_0, t) - \mathbf{r}_0$, the deformation will be defined in terms of the *displacement gradient tensor* $\mathbf{H}$:

$$\mathbf{H} = \frac{\partial \mathbf{u}}{\partial \mathbf{r}_0} = \mathbf{F} - \mathbf{1}, \quad H_{KL} \equiv u_K|_L = F_{KL} - g_{KL} \tag{7.3.15}$$

It now follows that:

$$\mathbf{C} = \mathbf{1} + \mathbf{H} + \mathbf{H}^T + \mathbf{H}^T \mathbf{H} \Leftrightarrow C_{KL} = g_{KL} + u_K|_L + u_L|_K + u^N|_K u_N|_L \tag{7.3.16}$$

$$\mathbf{E} = \frac{1}{2}\left(\mathbf{H} + \mathbf{H}^T + \mathbf{H}^T \mathbf{H}\right) \Leftrightarrow E_{KL} = \frac{1}{2}\left(u_K|_L + u_L|_K + u^N|_K u_N|_L\right) \tag{7.3.17}$$

7.3.2   Small Strains and Small Deformations

The special but very important case of small strains and small deformations has been discussed in Sect. 4.3. Small strains may be characterized by the inequality: norm $\mathbf{E} \ll 1$. The expressions (7.3.12–7.3.14) for the primary strain measures are reduced to:

$$\varepsilon = \mathbf{e} \cdot \mathbf{E} \cdot \mathbf{e} = e^K E_{KL} e^L, \gamma = 2\bar{\mathbf{e}} \cdot \mathbf{C} \cdot \mathbf{e} = 2\bar{e}^K E_{KL} e^L, \varepsilon_v = \operatorname{tr} \mathbf{E} = E_K^K \tag{7.3.18}$$

Small deformations imply small strains and small rotations. The condition of small deformations is defined by the formulas (4.3.5) and (7.3.15):

$$\operatorname{norm} \mathbf{H} = \sqrt{\operatorname{tr}\left(\mathbf{H}\mathbf{H}^T\right)} = \sqrt{u^K|_L u^L|_K} \ll 1 \tag{7.3.19}$$

Note that the components $u^K|_L$ are dimensionless quantities. The expressions (7.3.16–7.3.17) for the Green deformation tensor $\mathbf{C}$ and the Green strain tensor $\mathbf{E}$ may be approximated by:

$$\mathbf{C} = \mathbf{1} + \mathbf{H} + \mathbf{H}^T \Leftrightarrow C_{KL} = g_{KL} + u_K|_L + u_K|_L \tag{7.3.20}$$

$$\mathbf{E} = \frac{1}{2}\left(\mathbf{H} + \mathbf{H}^T\right) \Leftrightarrow E_{KL} = \frac{1}{2}\left(u_K|_L + u_K|_L\right) \tag{7.3.21}$$

If we further can assume that the displacements are small, such that the K_0 and K are configurations close to one another, the place coordinates y_i may be used as particle coordinates. The expressions (7.3.21) are now written as:

$$\mathbf{E} = \frac{1}{2}\left(\mathbf{H} + \mathbf{H}^T\right) \Leftrightarrow E_{ij} = \frac{1}{2}\left(u_i|_j + u_j|_i\right) \tag{7.3.22}$$

The tensor $\mathbf{E}$ defined by the formulas (7.3.21) is the *strain tensor for small deformation*.

The three characteristic strain measures at a particle Y: the *longitudinal strain* ε in the direction $\mathbf{e}$, the *shear strain* γ with respect to two orthogonal directions $\mathbf{e}$ and $\bar{\mathbf{e}}$, and the *volumetric strain* ε_v are now expressed by:

$$\varepsilon = \mathbf{e} \cdot \mathbf{E} \cdot \mathbf{e} = e^i u_i|_j e^j, \ \gamma = 2\bar{\mathbf{e}} \cdot \mathbf{E} \cdot \mathbf{e} = \bar{e}^i\left(u_i|_j + u_j|_i\right)e^j, \ \varepsilon_v = \mathrm{tr}\,\mathbf{E} = u^i|_i \tag{7.3.23}$$

The *rotation tensor for small deformations* $\tilde{\mathbf{R}}$ is defined by formula (4.3.20) and now by:

$$\tilde{\mathbf{R}} = \frac{1}{2}\left(\mathbf{H} - \mathbf{H}^T\right) \Leftrightarrow \tilde{R}_{ij} = \frac{1}{2}\left(u_i|_j - u_j|_i\right) \tag{7.3.24}$$

The principal directions of the strain tensor $\mathbf{E}$ rotates a small angle determined by the rotation vector in the formula (4.3.21):

$$\mathbf{z} = \frac{1}{2}\nabla \times \mathbf{u} \Leftrightarrow z^i = \frac{1}{2}\varepsilon^{ijk}u_k|_j = \frac{1}{2}\varepsilon^{ijk}u_{k,j} = \frac{1}{2}\varepsilon^{ijk}\tilde{R}_{kj} \tag{7.3.25}$$

We shall now present coordinate strains for small deformations in *orthogonal coordinates*. First we present the *physical components* $u(i)$ of the displacement vector $\mathbf{u}$, according to the formulas (6.5.53):

$$\mathbf{u} = u(i)\mathbf{e}_i^y \Leftrightarrow u(i) = u^i h_i = \frac{u_i}{h_i}, \mathbf{e}_i^y = \frac{1}{h_i}\mathbf{g}_i = h_i\mathbf{g}^i \tag{7.3.26}$$

$\mathbf{e}_i^y$ are the unit tangent vectors to coordinate lines in the orthogonal system y. The *physical components* of the strain tensor for small deformations $\mathbf{E}$ are defined by:

$$\varepsilon_{ii} = E(ii) = \mathbf{e}_i^y \cdot \mathbf{E} \cdot \mathbf{e}_i^y = \frac{1}{h_i^2}E_{ii} = \frac{1}{h_i^2}u_i|_i \quad \text{no summation w.r.t } i$$
$$\gamma_{ij} = 2E(ij) = 2\mathbf{e}_i^y \cdot \mathbf{E} \cdot \mathbf{e}_j^y = \frac{2}{h_i h_j}E_{ij} = \frac{1}{h_i h_j}\left(u_i|_j + u_j|_i\right) \tag{7.3.27}$$

Using the formulas (6.5.70) we obtain from the formulas (7.3.27):

$$\varepsilon_{ii} = E(ii) = \frac{1}{h_i^2} E_{ii} = \left[\frac{\partial}{\partial y^i} \left(\frac{u(i)}{h_i} \right) + \sum_j \frac{\partial h_i}{\partial y^j} \frac{u(j)}{h_i h_j} \right] \quad \text{no summation w. r. to } i$$

$$\gamma_{ij} = 2E(ij) = \frac{2}{h_i h_j} E_{ij} = \frac{h_i}{h_j} \frac{\partial}{\partial y^j} \left(\frac{u(i)}{h_i} \right) + \frac{h_j}{h_i} \frac{\partial}{\partial y^i} \left(\frac{u(j)}{h_j} \right) \quad i \neq j$$

$$\tag{7.3.28}$$

$$\varepsilon_v = \text{tr} \, \mathbf{E} = E(ii) = \frac{1}{h} \sum_i \frac{\partial}{\partial y^i} \left(\frac{h u(i)}{h_i} \right) \tag{7.3.29}$$

The physical components of the angle of rotation $\mathbf{z}$ in formula (7.3.25) becomes according to the formula (6.5.57):

$$z(i) = \frac{h_i}{2h} \sum_{j,k} e_{ijk} \frac{\partial}{\partial y^j} [h_k u(k)] \tag{7.3.30}$$

The results of applying the formulas (7.3.28–7.3.29) in cylindrical coordinates and spherical coordinates are presented in the examples below.

Example 7.1 Physical Components of E in Cylindrical Coordinates.
(R, θ, z): $\mathbf{u} = [u_R, u_\theta, u_z]$, $h_1 = 1$, $h_2 = R$, $h_3 = 1$, $h = R$. The formulas (7.3.28–7.3.29) yield:

$$(E(ij)) = \begin{pmatrix} \varepsilon_R & \gamma_{R\theta}/2 & \gamma_{Rz}/2 \\ \gamma_{\theta R}/2 & \varepsilon_\theta & \gamma_{\theta z}/2 \\ \gamma_{zR}/2 & \gamma_{z\theta}/2 & \varepsilon_z \end{pmatrix}$$

$$= \begin{pmatrix} \frac{\partial u_R}{\partial R} & \frac{1}{2R}\frac{\partial u_R}{\partial \theta} + \frac{R}{2}\frac{\partial}{\partial R}\left(\frac{u_\theta}{R}\right) & \frac{1}{2}\frac{\partial u_z}{\partial R} + \frac{1}{2}\frac{\partial u_R}{\partial z} \\ \frac{1}{2R}\frac{\partial u_R}{\partial \theta} + \frac{R}{2}\frac{\partial}{\partial R}\left(\frac{u_\theta}{R}\right) & \frac{1}{R}\frac{\partial u_\theta}{\partial \theta} + \frac{u_R}{R} & \frac{1}{2}\frac{\partial u_\theta}{\partial z} + \frac{1}{2R}\frac{\partial u_z}{\partial \theta} \\ \frac{1}{2}\frac{\partial u_z}{\partial R} + \frac{1}{2}\frac{\partial u_R}{\partial z} & \frac{1}{2}\frac{\partial u_\theta}{\partial z} + \frac{1}{2R}\frac{\partial u_z}{\partial \theta} & \frac{\partial u_z}{\partial z} \end{pmatrix}$$

$$\varepsilon_v = \text{tr} \, \mathbf{E} = E(ii) = \varepsilon_R + \varepsilon_\theta + \varepsilon_z = \frac{\partial u_R}{\partial R} + \frac{u_R}{R} + \frac{1}{R}\frac{\partial u_\theta}{\partial \theta} + \frac{\partial u_z}{\partial z}$$

Example 7.2 Physical Components of E in Spherical Coordinates.
(r, θ, ϕ) : $\mathbf{u} = \left[u_r, u_\theta, u_\phi\right]$, $h_1 = 1, h_2 = r, h_3 = r \sin\theta, h = r^2 \sin\theta$. The formulas (7.3.38–7.3.43) yield:

$$(E(ij)) = \begin{pmatrix} \varepsilon_r & \gamma_{r\theta}/2 & \gamma_{r\phi}/2 \\ \gamma_{\theta r}/2 & \varepsilon_\theta & \gamma_{\theta\phi}/2 \\ \gamma_{\phi r}/2 & \gamma_{\phi\theta}/2 & \varepsilon_\phi \end{pmatrix}$$

$$\varepsilon_r = \frac{\partial u_r}{\partial r}, \varepsilon_\theta = \frac{1}{r}\frac{\partial u_\theta}{\partial \theta} + \frac{u_r}{r}, \varepsilon_\phi = \frac{1}{r \sin \theta}\frac{\partial u_\phi}{\partial \phi} + \frac{u_r}{r} + \frac{\cot \theta}{r}u_\theta$$

$$\gamma_{r\theta} = \frac{1}{r}\frac{\partial u_r}{\partial \theta} + r\frac{\partial}{\partial r}\left(\frac{u_\theta}{r}\right), \gamma_{\theta\phi} = \frac{1}{r \sin \theta}\frac{\partial u_\theta}{\partial \phi} + \frac{\sin \theta}{r}\frac{\partial}{\partial \theta}\left(\frac{u_\phi}{\sin \theta}\right),$$

$$\gamma_{\phi r} = r\frac{\partial}{\partial r}\left(\frac{u_\phi}{r}\right) + \frac{1}{r \sin \theta}\frac{\partial u_r}{\partial \phi}$$

$$\varepsilon_v = \operatorname{tr}\mathbf{E} = E(ii) = \varepsilon_r + \varepsilon_\theta + \varepsilon_\phi = \frac{\partial u_R}{\partial R} + \frac{u_R}{R} + \frac{1}{R}\frac{\partial u_\theta}{\partial \theta} + \frac{\partial u_z}{\partial z}$$

7.3.3   Rates of Deformation, Strain, and Rotation

The *velocity gradient tensor* $\mathbf{L}$, the *rate of deformation tensor* $\mathbf{D}$, and the *rate of rotation tensor* $\mathbf{W}$ at the time t are introduced and defined in Sect. 4.4. The components of these tensors in a general curvilinear coordinate system y are directly transformed from the corresponding expressions in a Cartesian coordinate system Ox:

$$\mathbf{L} = \operatorname{grad}\mathbf{v} \equiv \frac{\partial \mathbf{v}}{\partial \mathbf{r}} \Leftrightarrow L_{ij} = v_i|_j \tag{7.3.31}$$

$$\mathbf{D} = \frac{1}{2}\left(\mathbf{L} + \mathbf{L}^T\right) \Leftrightarrow D_{ij} = \frac{1}{2}\left(v_i|_j + v_j|_i\right) \tag{7.3.32}$$

$$\mathbf{W} = \frac{1}{2}\left(\mathbf{L} - \mathbf{L}^T\right) \Leftrightarrow W_{ik} = \frac{1}{2}\left(v_i|_j - v_j|_i\right) = \frac{1}{2}\left(v_{i,j} - v_{j,i}\right) \tag{7.3.33}$$

The last equality follows from the fact that the *rate of rotation tensor* is antisymmetric.

The *longitudinal strain rate* $\dot{\varepsilon}$ in the direction $\mathbf{e}$, the *shear strain rate* $\dot{\gamma}$ with respect to two orthogonal directions $\mathbf{e}$ and $\bar{\mathbf{e}}$, and the *volumetric strain rate* $\dot{\varepsilon}_v$ have been defined in Sect. 4.4. The expressions in general coordinates are:

$$\dot{\varepsilon} = \mathbf{e}\cdot\mathbf{D}\cdot\mathbf{e} = e^i D_{ij} e^j = e^i v_i|_j e^j \tag{7.3.34}$$

$$\dot{\gamma} = 2\bar{\mathbf{e}}\cdot\mathbf{D}\cdot\mathbf{e} = 2\bar{e}^i D_{ij} e^j = \bar{e}^i\left(v_i|_j + v_j|_i\right) e^j \tag{7.3.35}$$

$$\dot{\varepsilon}_v = \operatorname{tr}\mathbf{D} = \operatorname{div}\mathbf{v} = D_i^i = v^i|_i \tag{7.3.36}$$

The *principal directions* of the tensor $\mathbf{D}$ rotate with the angular velocity $\mathbf{w}$:

$$\mathbf{w} = \frac{1}{2}\nabla \times \mathbf{v} \Leftrightarrow w^i = \frac{1}{2}\varepsilon^{ijk}v_{k,j} = \frac{1}{2}\varepsilon^{ijk}W_{kj} \tag{7.3.37}$$

To find the physical components of the rate tensors in *orthogonal coordinates* we first present the physical components of the velocity vector **v**:

$$\mathbf{v} = v(i)\mathbf{e}_i^y \Leftrightarrow v(i) = v^i h_i = \frac{v_i}{h_i}, \mathbf{e}_i^y = \frac{1}{h_i}\mathbf{g}_i = h_i\mathbf{g}^i \tag{7.3.38}$$

The *physical components* of the rate of deformation tensor **D** are defined by:

$$\dot{\varepsilon}_{ii} = D(ii) = \mathbf{e}_i^y \cdot \mathbf{D} \cdot \mathbf{e}_i^y = \frac{1}{h_i^2}D_{ii} = \frac{1}{h_i^2}v_i|_i \quad \text{no summation w.r. to } i$$

$$\dot{\gamma}_{ij} = 2D(ij) = 2\mathbf{e}_i^y \cdot \mathbf{D} \cdot \mathbf{e}_j^y = \frac{2}{h_i h_j}D_{ij} = \frac{1}{h_i h_j}\left(v_i|_j + v_j|_i\right) \quad i \neq j \tag{7.3.39}$$

Using the formulas (6.5.70) we obtain from the formulas (7.3.39):

$$\dot{\varepsilon}_{ii} = D(ii) = \frac{1}{h_i^2}D_{ii} = \left[\frac{\partial}{\partial y^i}\left(\frac{v(i)}{h_i}\right) + \sum_j \frac{v(j)}{h_i h_j}\frac{\partial h_i}{\partial y^j}\right] \quad \text{no summation w.r. to } i$$

$$\dot{\gamma}_{ij} = 2D(ij) = \frac{2}{h_i h_j}D_{ij} = \frac{h_i}{h_j}\frac{\partial}{\partial y^j}\left(\frac{v(i)}{h_i}\right) + \frac{h_j}{h_i}\frac{\partial}{\partial y^i}\left(\frac{v(j)}{h_j}\right) \quad i \neq j$$

$$\tag{7.3.40}$$

$$\dot{\varepsilon}_v = \operatorname{tr}\mathbf{D} = D(ii) = \frac{1}{h}\sum_i \frac{\partial}{\partial y^i}\left(\frac{hv(i)}{h_i}\right) \tag{7.3.41}$$

The physical components of the *vorticity vector* $\mathbf{c} = \operatorname{rot}\mathbf{v}$ are according to the formulas (6.5.57):

$$c(i) = \frac{h_i}{h}\sum_{j,k}e_{ijk}\frac{\partial}{\partial y^j}[h_k v(k)] \Rightarrow c(1) = \frac{1}{h_2 h_3}\left(\frac{\partial}{\partial y^2}[h_3 v(3)] - \frac{\partial}{\partial y^3}[h_2 v(2)]\right) \text{ etc.}$$

$$\tag{7.3.42}$$

In Cartesian coordinate systems the vorticity vector **c** and the rate of rotation tensor **W** are according to Eqs. (4.4.12) and (4.4.15) related through:

$$c_i = e_{ijk}W_{kj} \Leftrightarrow W_{ij} = -\frac{1}{2}e_{ijk}c_k \text{ in Cartesian coordinate systems} \tag{7.3.43}$$

In orthogonal coordinate systems we define the *physical components* of **c** and **W** through relations similar to the relations (7.3.43):

$$c(i) = e_{ijk} W(kj) \Leftrightarrow W(ij) = -\frac{1}{2} e_{ijk} c(k) \tag{7.3.44}$$

From Eqs. (7.3.44) and (7.3.42) we obtain:

$$W(23) = -\frac{1}{2} e_{231} c(1) = \frac{1}{2} \frac{1}{h_2 h_3} \left([h_2 v(2)]_{,3} - [h_3 v(3)]_{,3}\right) \text{ etc. } \Rightarrow$$

$$W(ij) = \frac{1}{2} \frac{1}{h_i h_j} \left([h_i v(i)]_{,j} - [h_j v(j)]_{,i}\right) \tag{7.3.45}$$

The results of applying the formulas (7.3.40–7.3.45) in cylindrical coordinates and spherical coordinates are presented in the examples below.

Example 7.3 Physical Components of D, W and c in Cylindrical Coordinates.
$(R, \theta, z) : \mathbf{v} = [v_R, v_\theta, v_z], h_1 = 1, h_2 = R, h_3 = 1, h = R$. The formulas (7.3.40–7.3.45) yield:

$$(D(ij)) = \begin{pmatrix} \dot{\varepsilon}_R & \dot{\gamma}_{R\theta}/2 & \dot{\gamma}_{Rz}/2 \\ \dot{\gamma}_{\theta R}/2 & \dot{\varepsilon}_\theta & \dot{\gamma}_{\theta z}/2 \\ \dot{\gamma}_{zR}/2 & \dot{\gamma}_{z\theta}/2 & \dot{\varepsilon}_z \end{pmatrix}$$

$$= \begin{pmatrix} \frac{\partial v_R}{\partial R} & \frac{1}{2R}\frac{\partial v_R}{\partial \theta} + \frac{R}{2}\frac{\partial}{\partial R}\left(\frac{v_\theta}{R}\right) & \frac{1}{2}\frac{\partial v_z}{\partial R} + \frac{1}{2}\frac{\partial v_R}{\partial z} \\ \frac{1}{2R}\frac{\partial v_R}{\partial \theta} + \frac{R}{2}\frac{\partial}{\partial R}\left(\frac{v_\theta}{R}\right) & \frac{1}{R}\frac{\partial v_\theta}{\partial \theta} + \frac{v_R}{R} & \frac{1}{2}\frac{\partial v_\theta}{\partial z} + \frac{1}{2R}\frac{\partial v_z}{\partial \theta} \\ \frac{1}{2}\frac{\partial v_z}{\partial R} + \frac{1}{2}\frac{\partial v_R}{\partial z} & \frac{1}{2}\frac{\partial v_\theta}{\partial z} + \frac{1}{2R}\frac{\partial v_z}{\partial \theta} & \frac{\partial v_z}{\partial z} \end{pmatrix}$$

$$\dot{\varepsilon}_v = \text{tr}\,\mathbf{D} = D(ii) = \dot{\varepsilon}_R + \dot{\varepsilon}_\theta + \dot{\varepsilon}_z = \frac{\partial v_R}{\partial R} + \frac{v_R}{R} + \frac{1}{R}\frac{\partial v_\theta}{\partial \theta} + \frac{\partial v_z}{\partial z}$$

$$(W(ij)) = \begin{pmatrix} 0 & W_{R\theta} & W_{Rz} \\ W_{\theta R} & 0 & W_{\theta z} \\ W_{zR} & W_{z\theta} & 0 \end{pmatrix}$$

$$= \begin{pmatrix} 0 & \frac{1}{2R}\left[\frac{\partial v_R}{\partial \theta} - \frac{\partial (Rv_\theta)}{\partial R}\right] & -\frac{1}{2}\left[\frac{\partial v_z}{\partial R} - \frac{\partial v_R}{\partial z}\right] \\ -\frac{1}{2R}\left[\frac{\partial v_R}{\partial \theta} - \frac{\partial (Rv_\theta)}{\partial R}\right] & 0 & \frac{1}{2}\left[\frac{\partial v_\theta}{\partial z} - \frac{1}{R}\frac{\partial v_z}{\partial \theta}\right] \\ \frac{1}{2}\left[\frac{\partial v_z}{\partial R} - \frac{\partial v_R}{\partial z}\right] & -\frac{1}{2}\left[\frac{\partial v_\theta}{\partial z} - \frac{1}{R}\frac{\partial v_z}{\partial \theta}\right] & 0 \end{pmatrix}$$

$$c(1) \equiv c_R = 2W_{z\theta} = \frac{1}{R}\frac{\partial v_z}{\partial \theta} - \frac{\partial v_\theta}{\partial z}, c(2) \equiv c_\theta = 2W_{Rz} = \frac{\partial v_R}{\partial z} - \frac{\partial v_z}{\partial R}$$

$$c(3) \equiv c_z = 2W_{\theta R} = \frac{1}{R}\left[\frac{\partial v_R}{\partial \theta} - \frac{\partial (Rv_\theta)}{\partial R}\right]$$

Example 7.4 Physical Components of D, W and c in Spherical Coordinates.
$(r, \theta, \phi) : \mathbf{v} = [v_r, v_\theta, v_\phi], h_1 = 1, h_2 = r, h_3 = r \sin \theta, h = r^2 \sin \theta.$ The formulas
(7.3.38–7.3.43) yield:

$$(D(ij)) = \begin{pmatrix} \dot{\varepsilon}_r & \dot{\gamma}_{r\theta}/2 & \dot{\gamma}_{r\phi}/2 \\ \dot{\gamma}_{\theta r}/2 & \dot{\varepsilon}_\theta & \dot{\gamma}_{\theta\phi}/2 \\ \dot{\gamma}_{\phi r}/2 & \dot{\gamma}_{\phi\theta}/2 & \dot{\varepsilon}_\phi \end{pmatrix}$$

$$\dot{\varepsilon}_r = \frac{\partial v_r}{\partial r}, \dot{\varepsilon}_\theta = \frac{1}{r}\frac{\partial v_\theta}{\partial \theta} + \frac{v_r}{r}, \dot{\varepsilon}_\phi = \frac{1}{r \sin \theta}\frac{\partial v_\phi}{\partial \phi} + \frac{v_r}{r} + \frac{\cot \theta}{r} v_\theta$$

$$\dot{\gamma}_{r\theta} = \frac{1}{r}\frac{\partial v_r}{\partial \theta} + r\frac{\partial}{\partial r}\left(\frac{v_\theta}{r}\right), \dot{\gamma}_{\theta\phi} = \frac{1}{r \sin \theta}\frac{\partial v_\theta}{\partial \phi} + \frac{\sin \theta}{r}\frac{\partial}{\partial \theta}\left(\frac{v_\phi}{\sin \theta}\right),$$

$$\dot{\gamma}_{\phi r} = r\frac{\partial}{\partial r}\left(\frac{v_\phi}{r}\right) + \frac{1}{r \sin \theta}\frac{\partial v_r}{\partial \phi}$$

$$(W(ij)) = \begin{pmatrix} 0 & W_{r\theta} & W_{r\phi} \\ W_{\theta r} & 0 & W_{\theta\phi} \\ W_{\phi r} & W_{\phi\theta} & 0 \end{pmatrix}, W_{r\theta} = -W_{\theta r} = \frac{1}{2r}\left[\frac{\partial v_r}{\partial \theta} - \frac{\partial}{\partial r}(r v_\theta)\right]$$

$$W_{\theta\phi} = -W_{\phi\theta} = \frac{1}{2r \sin \theta}\left[\frac{\partial v_\theta}{\partial \phi} - \frac{\partial}{\partial \theta}\left(\sin \theta\, v_\phi\right)\right],$$

$$W_{\phi r} = -W_{r\phi} = \frac{1}{2r}\left[\frac{\partial}{\partial r}(r v_\phi) - \frac{1}{\sin \theta}\frac{\partial v_r}{\partial \phi}\right]$$

$$c(1) \equiv c_r = 2W_{\phi\theta} = \frac{1}{r \sin \theta}\left[\frac{\partial}{\partial \theta}\left(\sin \theta\, v_\phi\right) - \frac{\partial v_\theta}{\partial \phi}\right],$$

$$c(2) \equiv c_\theta = 2W_{r\phi} = \frac{1}{r}\left[\frac{1}{\sin \theta}\frac{\partial v_r}{\partial \phi} - \frac{\partial}{\partial r}(r v_\phi)\right]$$

$$c(3) \equiv c_\phi = 2W_{\theta r} = \frac{1}{r}\left[\frac{\partial}{\partial r}(r v_\theta) - \frac{\partial v_r}{\partial \theta}\right]$$

7.4   General Analysis of Large Deformations

The deformation of a differential material line element $d\mathbf{r}_0$ from the reference
configuration K_0 to the line element $d\mathbf{r}$ in the present configuration K may be
decomposed into a deformation of pure strain and a rigid-body motion as discussed
in Sect. 4.5 and illustrated in the Figs. 4.15 and 4.16. In addition to the deformation
the line element may experience a displacement $\mathbf{u}$. The decomposition of the
deformation, which does not necessarily represent the actual deformation of the
material, may be considered in two alternative ways. We applied the coordinate
systems presented in Fig. 7.1.

In the first alternative let the material be subjected to pure strain through the *right stretch tensor* **U**, transforming the line element $d\mathbf{r}_0 = dY^K \mathbf{g}_K$ emanating from the particle Y at the place $\mathbf{r}_0$ to the line element $d\bar{\mathbf{r}}$:

$$d\bar{\mathbf{r}} = \mathbf{U} \cdot d\mathbf{r}_0 \Leftrightarrow d\bar{Y}^L = U_K^L dY^K \tag{7.4.1}$$

Then the line element $d\bar{\mathbf{r}}$ is rotated to give the element $d\tilde{\mathbf{r}}$:

$$d\tilde{\mathbf{r}} = \mathbf{R} \cdot d\bar{\mathbf{r}} \Leftrightarrow d\tilde{Y}^N = R_L^N d\bar{Y}^L \tag{7.4.2}$$

Finally, the line element $d\tilde{\mathbf{r}}$ is given the displacement **u** from the place $\mathbf{r}_0$ to the place $\mathbf{r}$ with coordinates y, and becomes the element $d\mathbf{r} = dy^i \mathbf{g}_i$:

$$
\begin{aligned}
d\mathbf{r} &= d\tilde{\mathbf{r}} \Leftrightarrow dy^i = g_N^i d\tilde{Y}^N \Rightarrow \\
d\mathbf{r} &= d\tilde{\mathbf{r}} = \mathbf{R}\mathbf{U} \cdot d\mathbf{r}_0 \Leftrightarrow dy^i = g_N^i R_L^N U_K^L dY^K
\end{aligned}
\tag{7.4.3}
$$

The symbols are *Euclidean shifters* as defined by the formulas (6.6.1).

In the second alternative of deformation decomposition the material line element $d\mathbf{r}_0$ is given a displacement **u** and moved from the reference place $\mathbf{r}_0$ to the final place $\mathbf{r}$. The element $d\mathbf{r}_0$ is then represented by the element $d\tilde{\mathbf{r}}_0$ emanating from the place $\mathbf{r}$:

$$d\tilde{\mathbf{r}}_0 = d\mathbf{r}_0 \Leftrightarrow dy_0^k = g_N^k \, dY^N \tag{7.4.4}$$

Then the line element $d\tilde{\mathbf{r}}_0$ is rotated according to the rotation tensor **R** to give the element $d\tilde{\mathbf{r}}$:

$$d\tilde{\mathbf{r}} = \mathbf{R} \cdot d\tilde{\mathbf{r}}_0 \Leftrightarrow d\tilde{y}^j = R_k^j dy_0^k \tag{7.4.5}$$

The element is finally subjected to pure strain through the left stretch tensor **V**, transforming the element from $d\tilde{\mathbf{r}}$ to $d\mathbf{r}$:

$$
\begin{aligned}
d\mathbf{r} &= \mathbf{V} \cdot d\tilde{\mathbf{r}} \Leftrightarrow dy^i = V_j^i d\tilde{y}^j \Rightarrow \\
d\mathbf{r} &= \mathbf{V}\mathbf{R} \cdot d\tilde{\mathbf{r}}_0 = \mathbf{V}\mathbf{R} \cdot d\mathbf{r}_0 \Rightarrow dy^i = V_j^i R_k^j g_N^k dY^N
\end{aligned}
\tag{7.4.6}
$$

The *displacement gradient tensor* **F** is defined by the relation (4.2.7) and expresses the relation between the line elements $d\mathbf{r}_0$ and $d\mathbf{r}$:

$$\mathbf{F} = \operatorname{Grad} \mathbf{r} = \frac{\partial \mathbf{r}}{\partial \mathbf{r}_0}, \, d\mathbf{r} = \mathbf{F} \cdot d\mathbf{r}_0 \tag{7.4.7}$$

The results (7.4.3) and (7.4.6) show that the displacement gradient tensor **F** may be decomposed in either of two ways:

$$\mathbf{F} = \mathbf{RU} = \mathbf{VR} \Leftrightarrow F_N^i = g_N^i R_L^N U_K^L = V_j^i R_k^j g_N^k \tag{7.4.8}$$

The decomposition (7.4.8) represents the *polar decomposition theorem* presented in Sect. 3.8 with an analytical proof, and in Sect. 4.5 with a proof equivalent to the proof presented above.

When the present configuration K is used as reference configuration it is convenient to introduce the *inverse deformation gradient tensor* $\mathbf{F}^{-1}$:

$$\mathbf{F}^{-1} \equiv \frac{\partial \mathbf{r}_0}{\partial \mathbf{r}} \Leftrightarrow F_i^{-1K} = \frac{\partial Y^K}{\partial y^i} \Rightarrow d\mathbf{r}_0 = \mathbf{F}^{-1} \cdot d\mathbf{r} \tag{7.4.9}$$

The *inverse deformation tensor*, also called *Cauchy's deformation tensor*, is defined by:

$$\mathbf{B}^{-1} = \mathbf{F}^{-T}\mathbf{F}^{-1} \Leftrightarrow B_{ij}^{-1} = F_i^{-1K} F_{Kj}^{-1} \tag{7.4.10}$$

It follows that:

$$(ds_0)^2 = d\mathbf{r}_0 \cdot d\mathbf{r}_0 = d\mathbf{r} \cdot \mathbf{B}^{-1} \cdot d\mathbf{r} = dy^i B_{ij}^{-1} dy^j \tag{7.4.11}$$

Euler's strain tensor $\tilde{\mathbf{E}}$, also called *Almansi's strain tensor* after Emilio Almansi [1869–1948], is a symmetric tensor defined by the expression:

$$(ds)^2 - (ds_0)^2 = 2 d\mathbf{r} \cdot \tilde{\mathbf{E}} \cdot d\mathbf{r} \tag{7.4.12}$$

From the relation $(ds)^2 = d\mathbf{r} \cdot d\mathbf{r} = d\mathbf{r} \cdot \mathbf{1} \cdot d\mathbf{r}$ and the formulas (7.4.11–7.4.12) we obtain:

$$\tilde{\mathbf{E}} = \frac{1}{2}(\mathbf{1} - \mathbf{B}^{-1}), \tilde{E}_{ij} = \frac{1}{2}\left(\delta_{ij} - B_{ij}^{-1}\right) \tag{7.4.13}$$

The following formulas are to be derived in Problem 7.1:

$$\tilde{E}_{ij} = \frac{1}{2}\left(\mathbf{g}_i \cdot \mathbf{u}_{,j} + \mathbf{g}_j \cdot \mathbf{u}_{,i} - \mathbf{u}_{,i} \cdot \mathbf{u}_{,j}\right) = \frac{1}{2}\left(u_i|_j + u_j|_i - u^k|_i u_k|_j\right) \tag{7.4.14}$$

$$\mathbf{E} = \mathbf{F}^T \tilde{\mathbf{E}} \mathbf{F} \Leftrightarrow E_{KL} = \tilde{E}_{ij} F_K^i F_L^j \tag{7.4.15}$$

7.5   Convected Coordinates

The coordinate system Y presented in Fig. 7.1 to represent the reference configuration K_0 of a body of continuous material is now assumed to be imbedded in the continuum, which implies that the coordinate system moves and deforms with the material. The system is called a *convected coordinate system*. Convected coordinate systems are used in *Rheology* to describe constitutive models for non-Newtonian fluids, see the author's book Rheology and Non-Newtonian Fluids [1].

The base vectors and the fundamental parameters for the Y-system are place and time functions. The base vectors of the convected Y coordinate system are denoted by $\mathbf{c}_K$ and defined by:

$$\mathbf{c}_K(Y,t) = \frac{\partial \mathbf{r}(Y,t)}{\partial Y^K} \Rightarrow \mathbf{c}_K(Y,t_o) = \mathbf{g}_K(Y) \tag{7.5.1}$$

The reciprocal base vectors $\mathbf{c}^K(Y,t)$ are defined by:

$$\mathbf{c}^K \cdot \mathbf{c}_L = \delta_L^K \tag{7.5.2}$$

The fundamental parameters for the Y-system are:

$$C_{KL} = C_{KL}(Y,t) = \mathbf{c}_K \cdot \mathbf{c}_L, \ C^{KL} = C^{KL}(Y,t) = \mathbf{c}^K \cdot \mathbf{c}^L \tag{7.5.3}$$

The length ds_0 and ds of a material line element in the two configurations K_0 and K are respectively given by:

$$(ds_0)^2 = g_{KL}\, dY^K\, dY^L, (ds)^2 = C_{KL}\, dY^K\, dY^L \tag{7.5.4b}$$

The components $C_{KL}(Y,t)$ now represent two tensors in the Y-system: In the formulas (7.3.7) and (7.3.8) $C_{KL}(Y,t)$ are components in K_0 of the deformation tensor **C**. In Eq. (7.5.4)$_2$ the components $C_{KL}(Y,t)$ are fundamental parameters in the convected Y-system and components of the unit tensor **1** in K. In particular:

$$C_{KL}(Y,t_0) = g_{KL}(Y) \tag{7.5.5}$$

The components $C^{KL}(Y,t)$ also represent two tensors in the Y-system: contravariant components in K_0 of the deformation tensor **C** and fundamental parameters in the convected Y-system in K, and components of the unit tensor **1** in the Y-system in K. In particular:

$$C^{KL}(Y,t_0) = g^{KL}(Y) \tag{7.5.6}$$

It follows from the expressions (7.3.11) that in convected coordinates the components E_{KL} of the strain tensor represent half of the change in the fundamental parameters from g_{KL} in K_0 to C_{KL} in K.

If the y-system in K is chosen such that the coordinate system coincides with the convective Y-system at time t then:

$$F^i_K = \frac{\partial y^i}{\partial Y^K} = \delta^i_K \qquad (7.5.7)$$

The Eqs. (7.5.7) and (7.4.15) show that the matrix of *Green's strain tensor* and the matrix of *Euler's strain tensor* become identical for this choice of the y-system.

If the motion from K_0 to K is given by the displacement vector $\mathbf{u}(Y, t)$ in Eq. (7.2.3), then:

$$\mathbf{c}_K = \frac{\partial \mathbf{r}}{\partial Y^K} = \frac{\partial \mathbf{r}_0}{\partial Y^K} + \frac{\partial \mathbf{u}}{\partial Y^K} = \mathbf{g}_K + \frac{\partial \mathbf{u}}{\partial Y^K} \qquad (7.5.8)$$

$$C_{KL} = \mathbf{c}_K \cdot \mathbf{c}_L = \mathbf{g}_K \cdot \mathbf{g}_L + \mathbf{g}_K \cdot \frac{\partial \mathbf{u}}{\partial Y^L} + \frac{\partial \mathbf{u}}{\partial Y^K} \cdot \mathbf{g}_L + \frac{\partial \mathbf{u}}{\partial Y^K} \cdot \frac{\partial \mathbf{u}}{\partial Y^L} \Rightarrow$$
$$C_{KL} = g_{KL} + u_K|_L + u_L|_K + u^N|_K u_N|_L \qquad (7.5.9)$$

Note that the displacement components u_K and u^K are related to the Y-system in K_0:

$$u_K = \mathbf{u} \cdot \mathbf{g}_K, \quad u^K = \mathbf{u} \cdot \mathbf{g}^K \qquad (7.5.10)$$

and that covariant differentiation is to be performed in K_o with Christoffel symbols based on the fundamental parameters g_{KL}.

Alternatively we may use the displacement components u_K and u^K related to the Y-system in K:

$$u_K = \mathbf{u} \cdot \mathbf{c}_K, \quad u^K = \mathbf{u} \cdot \mathbf{c}^K \qquad (7.5.11)$$

For simplicity we use the same symbols for the displacement components here. Using the result (7.5.8), we obtain:

$$g_{KL} = \mathbf{g}_K \cdot \mathbf{g}_L = (\mathbf{c}_K - \mathbf{u}_{,K}) \cdot (\mathbf{c}_L - \mathbf{u}_{,L})$$
$$= \mathbf{c}_K \cdot \mathbf{c}_L - \mathbf{c}_K \cdot \mathbf{u}_{,L} - \mathbf{u}_{,K} \cdot \mathbf{c}_L + \mathbf{u}_{,K} \cdot \mathbf{u}_{,L} \Rightarrow \qquad (7.5.12)$$
$$C_{KL} = g_{KL} + u_K\|_L + u_L\|_K - u^N\|_K u_N\|_L$$

Covariant differentiation is now marked by a double vertical line to indicate that it is to be performed in K and with Christoffel symbols computed from the fundamental parameters C_{KL}. The strain components E_{KL} in the formulas (7.3.17) will have two different forms:

$$E_{KL} = \frac{1}{2} \left(u_K|_L + u_L|_K + u^N|_K u_N|_L \right) \qquad (7.5.13)$$

$$E_{KL} = \frac{1}{2} \left(u_K\|_L + u_L\|_K - u^N\|_K u_N\|_L \right) \qquad (7.5.14)$$

Note the two sets of the strain components E_{KL} in the formulas (7.5.13) and (7.5.14) are identical, while the displacement components u_K and the covariant differentiations are not the same. Comparing the formulas (7.5.14) for the components of Green's strain tensor with formulas (7.4.14) for the components of Euler's strain tensor, we see that the two components set are identical if we choose a y-system that coincides with the Y-system at the present time t.

Example 7.5 Simple Shear

Figure 7.2 illustrates the deformation of a material block from the reference configuration K_0 at the time t_0 to the present configuration K at the present time t. Ox is a Cartesian coordinate system. The coordinate system Y is attached to the body and moves and deforms with the body as a convected coordinate system. The deformation is called *simple shear* and is described by:

$$x_1(Y^1, Y^2, t) = Y^1 + \beta(t)\, Y_2, \quad x_2(Y^2) = Y_2, \quad x_3(Y^3) = Y^3 \tag{7.5.15}$$

$\beta(t)$ is a scalar function of time and $\beta(t_0) = 0$. Compare the deformation simple shear with simple shear flow in Example 4.1.

The inverse mapping of the mapping (7.5.15) is:

$$Y^1(x_1, x_1, t) = x_1 - \beta(t)x_2, \quad Y^2(x_2) = x_2, \quad Y^3(x_3) = x_3 \tag{7.5.16}$$

The matrices for Green's deformation tensor $\mathbf{C}$ and Green's strain tensor $\mathbf{E}$ will be computed. First we compute:

$$\left(\frac{\partial x_i}{\partial Y^K}\right)_{t_0} = \begin{pmatrix} 1 & 0 & 0 \\ 0 & 1 & 0 \\ 0 & 0 & 1 \end{pmatrix}, \quad \left(\frac{\partial x_i}{\partial Y^K}\right)_t = \begin{pmatrix} 1 & \beta & 0 \\ 0 & 1 & 0 \\ 0 & 0 & 1 \end{pmatrix}$$

The base vectors $\mathbf{c}_K$ and the fundamental parameters C_{KL} for the Y-system are found from the formulas (6.2.11) and (6.2.27):

Fig. 7.2 Deformation of a material block from configuration K_0 at the time t_0 to configuration K at the time t. Cartesian coordinate system Ox. Convected coordinate system Y

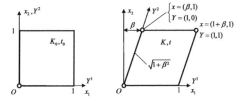

$$\mathbf{c}_K(Y,t) = \frac{\partial \mathbf{r}}{\partial Y^K} = \frac{\partial \mathbf{r}}{\partial x_i}\frac{\partial x_i}{\partial Y^K} = \frac{\partial x_i}{\partial Y^K}\mathbf{e}_i, \quad \mathbf{c}_K(Y,t_0) = \mathbf{e}_K$$

$$C_{KL}(Y,t) = \mathbf{c}_K \cdot \mathbf{c}_L \Leftrightarrow (C_{KL}) = \begin{pmatrix} 1 & \beta & 0 \\ \beta & 1+\beta^2 & 0 \\ 0 & 0 & 1 \end{pmatrix},$$

$$C_{KL}(Y,t_0) = g_{KL}(Y) = \delta_{KL}$$

The components of Green's strain tensor become are obtained from the formula (7.3.11):

$$(E_{KL}) = \left(\frac{1}{2}(C_{KL} - g_{KL})\right) = \frac{1}{2}\begin{pmatrix} 0 & \beta & 0 \\ \beta & \beta^2 & 0 \\ 0 & 0 & 0 \end{pmatrix}$$

Because the Y-system in K_0 is Cartesian the non-zero coordinate strains ε_{22} and γ_{12} are obtained from the Eqs. (4.2.13) and (4.2.16):

$$\varepsilon_{22} = \sqrt{1 + 2\mathbf{e}_2 \cdot \mathbf{E} \cdot \mathbf{e}_2} - 1 = \sqrt{1 + 2E_{22}} - 1 = \sqrt{1 + \beta^2} - 1$$

$$\sin\gamma_{12} = \frac{2\mathbf{e}_1 \cdot \mathbf{E} \cdot \mathbf{e}_2}{\sqrt{1 + 2\mathbf{e}_1 \cdot \mathbf{E} \cdot \mathbf{e}_1}\sqrt{1 + 2\mathbf{e}_2 \cdot \mathbf{E} \cdot \mathbf{e}_2}}$$

$$= \frac{2E_{12}}{\sqrt{1 + 2E_{11}}\sqrt{1 + 2E_{22}}} = \frac{\beta}{\sqrt{1 + \beta^2}}$$

These results are easily obtained directly from Fig. 7.2. For small deformations: $\beta \ll 1 \Rightarrow \beta^2 \ll \beta$, the strain matrix and the non-zero coordinate strains are:

$$(E_{KL}) = \frac{1}{2}\begin{pmatrix} 0 & \beta & 0 \\ \beta & 0 & 0 \\ 0 & 0 & 0 \end{pmatrix}, \gamma_{12} = \beta$$

7.6   Convected Derivatives of Tensors

A tensor quantity that is independent of the choice of reference, is called a *reference invariant tensor* or *objective tensor*. Tensor quantities dependent of the choice of reference, are called *reference related tensors*.

In this section we apply convected coordinates Y with base vectors $\mathbf{c}_K(Y,t)$. The material derivatives of these base vectors are:

$$\dot{\mathbf{c}}_K = \frac{\partial}{\partial t}\frac{\partial \mathbf{r}}{\partial Y^K} = \frac{\partial}{\partial Y^K}\frac{\partial \mathbf{r}}{\partial t} = \frac{\partial \mathbf{v}}{\partial Y^K} \equiv \mathbf{v}_{,K} \tag{7.6.1}$$

$\mathbf{v}$ is the particle velocity vector. The components of the material derivatives of the base vectors are covariant derivatives of the velocity components:

$$\dot{\mathbf{c}}_K = v^L||_K \, \mathbf{c}_L \tag{7.6.2}$$

It is straight forward to show that, see Problem 7.2:

$$\dot{\mathbf{c}}^K = -v^K||_L \, \mathbf{c}^L \tag{7.6.3}$$

Let $\mathbf{a}(Y,\, t)$ be a convected vector field, i.e. a vector field associated with the particles in the material we are considering:

$$\mathbf{a} = a^K \, \mathbf{c}_K = a_K \, \mathbf{c}^K \tag{7.6.4}$$

The material derivative of $\mathbf{a}$ is:

$$\dot{\mathbf{a}} = \dot{a}^K \, \mathbf{c}_K = \left[\frac{\partial a^K}{\partial t} + a^L v^K||_L\right]\mathbf{c}_K, \quad \dot{\mathbf{a}} = \dot{a}_K \, \mathbf{c}^K = \left[\frac{\partial a_K}{\partial t} - a_L v^L||_K\right]\mathbf{c}^K \tag{7.6.5}$$

$\dot{\mathbf{a}}$ is a reference related vector field, while the two vector fields defined by the components:

$$\partial_c a^K \equiv \frac{\partial a^K}{\partial t}, \; \partial_c a_K \equiv \frac{\partial a_K}{\partial t} \tag{7.6.6}$$

are objective. The expressions (7.6.6) are called *convected differentiated vector components*. It follows from the expressions (7.6.5) that the vectors defined by the components (7.6.6) are two different vector fields. In order to determine the components of these vectors in a reference fixed coordinate system y, we rearrange Eq. (7.6.5) to:

$$\partial_c a^K = \dot{a}^K - a^L v^K||_L, \quad \partial_c a_K = \dot{a}_K + a_L v^L||_K \tag{7.6.7}$$

Because these equations are tensor equations, they may directly be transformed to the fixed y-system:

$$\partial_c a^i = \dot{a}^i - a^k v^i|_k = \frac{\partial a^i}{\partial t} + a^i|_k v^k - a^k v^i|_k \tag{7.6.8}$$

$$\partial_c a_i = \dot{a}_i + a_k v^k|_i = \frac{\partial a_i}{\partial t} + a_i|_k v^k + a_k v^k|_i \tag{7.6.9}$$

Note that the convective derivatives of the contravariant and the covariant vector components do not result in one and the same vector. The vector field defined by Eq. (7.6.8) is called the *upper-convected derivative* of the vector field **a**, and the vector field defined by Eq. (7.6.9) is called the *lower-convected derivative* of the vector field **a**. The two derivatives are also presented as:

$$\mathbf{a}^\nabla = \dot{\mathbf{a}} - \mathbf{L}\mathbf{a} \Leftrightarrow a^{\nabla i} = \dot{a}^i - v^i|_k a^k \text{ upper-convected derivate of } \mathbf{a} \tag{7.6.10}$$

$$\mathbf{a}^\Delta = \dot{\mathbf{a}} + \mathbf{L}^T\mathbf{a} \Leftrightarrow a^\Delta_i = \dot{a}_i + v^k|_i a_k \text{ lower-convected derivate of } \mathbf{a} \tag{7.6.11}$$

Convected differentiated tensor components are defined by their representation in a convected Y-system in which they are given directly by the material derivatives of the tensor components. As an example we consider a second order tensor **B**. The convected derivatives of the components B^K_L in the Y-system are defined by:

$$\partial_c B^K_L \equiv \frac{\partial}{\partial t} B^K_L \tag{7.6.12}$$

If **B** is an objective tensor then so is the tensor defined by the components (7.6.12). One set of components of this tensor in the y-system is denoted $\partial_c B^i_j$ and is found as follows. Let **a** and **b** be two objective vectors. The scalar $\alpha = \mathbf{B}[\mathbf{a}, \mathbf{b}]$ may alternatively be computed from:

$$\alpha = B^K_L a_K b^L, \ \alpha = B^i_j a_i b^j \tag{7.6.13}$$

Then we may write:

$$\dot{\alpha} = \partial_c B^K_L a_K b^L + B^K_L \partial_c a_K b^L + B^K_L a_K \partial_c b^L \tag{7.6.14}$$

$$\dot{\alpha} = \dot{B}^i_j a_i b^j + B^i_j \dot{a}_i b^j + B^i_j a_i \dot{b}^j \tag{7.6.15}$$

Equation (7.6.14) is transformed to the y-system:

$$\dot{\alpha} = \partial_c B^i_j a_i b^j + B^i_j \partial_c a_i b^j + B^i_j a_i \partial_c b^j \tag{7.6.16}$$

The formulas (7.6.8–7.6.9) are used for $\partial_c a_i$ and $\partial_c b^j$ in Eq. (7.6.16), and then Eq. (7.6.15) is subtracted from Eq. (7.6.16). The result is:

$$\left[\partial_c B^i_j - \dot{B}^i_j + B^k_j v^i|_k - B^i_k v^k|_j\right] a_i b^j = 0$$

Because the vectors **a** and **b** may be chosen arbitrarily, the expression in the brackets must be zero. Thus we have the result:

$$\partial_c B^i_j = \dot{B}^i_j - B^k_j v^i|_k + B^i_k v^k|_j \tag{7.6.17}$$

When the material derivative and the covariant derivatives in Eq. (7.6.17) are written out in detail, we shall see that the terms with Christoffel symbols are eliminated, and that the result is, see Problem 7.3:

$$\partial_c B^i_j = \frac{\partial}{\partial t} B^i_j + B^i_{j,k} v^k - B^k_j v^i_{,k} + B^i_k v^k_{,j} \tag{7.6.18}$$

Oldroyd [2] derived the formula (7.6.18) by evaluating the material derivative $\partial_c B^K_L$ from the transformation equation:

$$B^K_L = \frac{\partial Y^K}{\partial y^i} \frac{\partial y^j}{\partial Y^L} B^i_j \tag{7.6.19}$$

The result is then substituted into the transformation equation, see Problem 7.4:

$$\partial_c B^i_j = \frac{\partial y^i}{\partial Y^K} \frac{\partial Y^L}{\partial y^j} \partial_c B^K_L \tag{7.6.20}$$

Now we define two tensors by their components in the fixed y-system and the corresponding components in the convected Y-system:

$$\partial_c B_{ij} \text{ in the } y-\text{system} \Leftrightarrow \frac{\partial}{\partial t} B_{KL} \text{ in the } Y-\text{system} \tag{7.6.21}$$

$$\partial_c B^{ij} \text{ in the } y-\text{system} \Leftrightarrow \frac{\partial}{\partial t} B^{KL} \text{ in the } Y-\text{system} \tag{7.6.22}$$

Starting with the representations $\alpha = B_{ij} a^i b^j = B^{ij} a_i b_j$ for the scalar $\alpha = \mathbf{B}[\mathbf{a}, \mathbf{b}]$ and following the procedure that gave the result (7.6.17) we can derive the formulas, see Problem 7.5:

$$\partial_c B_{ij} = \dot{B}_{ij} + B_{kj} v^k|_i + B_{ik} v^k|_j, \quad \partial_c B^{ij} = \dot{B}^{ij} - B^{kj} v^i|_k - B^{ik} v^j|_k \tag{7.6.23}$$

The results (7.6.8), (7.6.9), (7.6.17), and (7.6.23) show a pattern for constructing the convected derivatives of objective tensors of any order. Note that the convective derivatives of the tensor components of different types do not result in one and the same tensor. For a second order tensor $\mathbf{B}$ it is customary to let the components $(7.6.23)_1$ define the *lower-convected derivative* of the tensor $\mathbf{B}$, while the components $(7.6.23)_2$ define the *upper-convected derivative* of the tensor $\mathbf{B}$. Special symbols are introduced for these two tensors:

$$\mathbf{B}^{\Delta} = \dot{\mathbf{B}} + \mathbf{L}^T\mathbf{B} + \mathbf{B}\mathbf{L} \Leftrightarrow$$
$$B^{\Delta}_{ij} = \dot{B}_{ij} + v^k\big|_i B_{kj} + B_{ik}v^k\big|_j \qquad \text{lower-convected derivative of } \mathbf{B} \qquad (7.6.24)$$

$$\mathbf{B}^{\nabla} = \dot{\mathbf{B}} - \mathbf{L}\mathbf{B} - \mathbf{B}\mathbf{L}^T \Leftrightarrow$$
$$B^{\nabla ij} = \dot{B}^{ij} - v^i\big|_k B^{kj} - B^{ik}v^j\big|_k \qquad \text{upper-convected derivative of } \mathbf{B} \qquad (7.6.25)$$

These tensors play an important role in constitutive modeling of non-Newtonian fluids, see the author's book Rheology and Non-Newtonian Fluids [4].

7.7 Cauchy's Stress Tensor. Equations of Motion

The stress vector $\mathbf{t}$ on a material surface through a particle P and with unit normal $\mathbf{n}$ is determined by the Cauchy stress theorem (2.2.27) and is in a general coordinate system y represented by the three sets of components $T^{ik}, T_{ik},$ and T^i_k:

$$\mathbf{t} = \mathbf{T} \cdot \mathbf{n} \Leftrightarrow t^i = T^{ik}n_k = T^i_k n^k, \; t_i = T^k_i n_k = T_{ik} n^k \qquad (7.7.1)$$

$\mathbf{T}$ is the Cauchy stress tensor. The normal stress σ and the shear stress τ on the surface are given by (Fig. 7.3):

$$\sigma = \mathbf{n} \cdot \mathbf{t} = \mathbf{n} \cdot \mathbf{T} \cdot \mathbf{n} = n_i T^i_k n^k$$
$$\tau = \sqrt{\mathbf{t} \cdot \mathbf{t} - \sigma^2} = \sqrt{t^k t_k - \sigma^2} \qquad (7.7.2)$$

We shall develop expressions for the stress vector $\mathbf{t}_i$, the normal stress σ_i, and the shear stress τ_i on a coordinate surface $y^i = $ constant. The unit normal vector to the surface is:

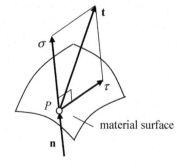

Fig. 7.3 Stress vector $\mathbf{t}$ from a particle P on a material surface with normal $\mathbf{n}$. Normal stress σ and shear stress τ

$$\mathbf{n}^i = \frac{\mathbf{g}^i}{\sqrt{g^{ii}}} \tag{7.7.3}$$

In order to simplify the development we start by showing the stresses on the material coordinate surface $y^3 = $ constant in Fig. 7.4. From the formulas (7.7.1–7.7.3) we obtain:

$$\mathbf{t}_3 = \mathbf{T} \cdot \mathbf{n}^3 \Rightarrow t_3^k = T^{kj} n_j^3 = \frac{T^{k3}}{\sqrt{g^{33}}} \tag{7.7.4}$$

$$\sigma_3 = \mathbf{n}^3 \cdot \mathbf{T} \cdot \mathbf{n}^3 = \frac{1}{\sqrt{g^{33}}} T^{33} \frac{1}{\sqrt{g^{33}}} = \frac{T^{33}}{g^{33}}, \ \tau_3 = \sqrt{t_3^k t_{3k} - (\sigma_3)^2}$$

$$= \sqrt{\frac{T^{k3} T_k^3}{g^{33}} - (\sigma_3)^2} \tag{7.7.5}$$

On a coordinate surface $y^i = $ constant, we write:

$$\mathbf{t}_i = \mathbf{T} \cdot \mathbf{n}^i \Rightarrow t_i^k = \frac{T^{ki}}{\sqrt{g^{ii}}}, \sigma_i = \frac{T^{ii}}{g^{ii}}, \ \tau_i = \sum_k \sqrt{\frac{T^{ki} T_k^i}{g^{ii}} - (\sigma_i)^2} \tag{7.7.6}$$

The shear stress τ_i has two components on the material coordinate surface $y^i = $ constant:

$$(\tau_i)_k = \frac{\mathbf{g}_k}{\sqrt{g_{kk}}} \cdot \mathbf{t}_i = \frac{\mathbf{g}_k}{\sqrt{g_{kk}}} \cdot \mathbf{T} \cdot \frac{\mathbf{g}^i}{\sqrt{g^{ii}}} = \frac{T_k^i}{\sqrt{g_{kk} g^{ii}}} \tag{7.7.7}$$

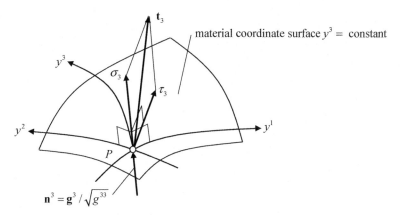

material coordinate surface $y^3 = $ constant

Fig. 7.4 Stress vector $\mathbf{t}_3$ from a particle P on a material coordinate surface $y^3 = $ constant. Unit normal vector $\mathbf{n}^3 = \mathbf{g}^3 / \sqrt{g^{33}}$. Normal stress σ_3 and shear stress τ_3. General coordinate system y. Coordinate lines $y^1, y^2,$ and y^3

7.7.1 Physical Stress Components

The physical components of the stress vector $\mathbf{t}_i$ are denoted by τ_{ki}, and are by Green and Zerna [3] called the *physical stress components*. Figure 7.5 shows the physical components τ_{13}, τ_{23}, and τ_{33} of the stress vector $\mathbf{t}_3$. Note that while τ_{13} and τ_{23} are shear stresses, the component τ_{33} represents in general both a shear stress and a normal stress.

According to the general definition (6.3.4) of physical components of a vector $\mathbf{a}$: $a(k) = a^k \sqrt{g_{kk}}$, we get:

$$\tau_{ki} = T^{ki} \sqrt{\frac{g_{kk}}{g^{ii}}} = t_i^k \sqrt{g_{kk}} = \frac{T^{ki}}{\sqrt{g^{ii}}} \sqrt{g_{kk}} \Rightarrow$$

$$\tau_{ki} = T^{ki} \sqrt{\frac{g_{kk}}{g^{ii}}} \quad \text{physical stress components according to Green and Zerna [3]}$$

(7.7.8)

Truesdell [4] defined *physical stress components* differently from Green and Zerna [3]. First the stress vector $\mathbf{t}$ and the unit normal surface vector $\mathbf{n}$ are expressed in physical components:

$$t(k) = t^k \sqrt{g_{kk}}, \quad n(i) = n^i \sqrt{g_{ii}} \tag{7.7.9}$$

From the Cauchy stress theorem (7.7.1) we now obtain:

$$t^k = T_i^k n^i \Rightarrow t(k) \frac{1}{\sqrt{g_{kk}}} = T_i^k \frac{n(i)}{\sqrt{g_{ii}}} \Rightarrow t(k) = \left(T_i^k \sqrt{\frac{g_{kk}}{g_{ii}}} \right) n(i) \quad \Rightarrow$$

$$t(k) = T(ki)n(i) \quad \text{Cauchy' stress theorem in terms of physical components}$$

(7.7.10)

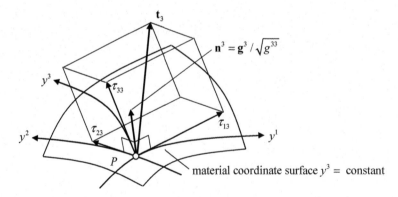

Fig. 7.5 Physical stress components τ_{13}, τ_{23}, and τ_{33} of a stress vector $\mathbf{t}_3$ from a particle P on a material coordinate surface $y_3 = $ constant. Unit normal vector $\mathbf{n}^3 = \mathbf{g}^3 / \sqrt{g^{33}}$ General coordinate system y. Coordinate lines y^1, y^2, and y^3

$$T(ki) = T_i^k \sqrt{\frac{g_{kk}}{g_{ii}}} \quad \text{Truesdell's physical stress components} \qquad (7.7.11)$$

Since the physical components $n(i)$ are dimensionless and the physical components $t(k)$ have dimension force per unit area, the dimension of the components $T(ki)$ is also force per unit area. $T(ki)$ are called *physical stress components* because they are the proper coefficients in the linear relations (7.7.10) between the physical components of the unit normal vector $\mathbf{n}$ and the physical components of the stress vector $\mathbf{t}$. Note that the components $T(ki)$ are not in general symmetric. It may be shown, see Problem 7.6, that the two sets of physical components defined by Eqs. (7.7.8) and (7.7.11) are related through:

$$\tau_{ki} = \sum_j T(kj) \, g^{ij} \sqrt{\frac{g_{jj}}{g^{ii}}} \qquad (7.7.12)$$

In orthogonal coordinate systems:

$$g_{ij} = g^{ij} = 0 \quad \text{for } i \neq j, \quad g_{ii} = \frac{1}{g^{ii}} = h_i^2$$

Now formulas (7.7.11) and (7.7.10) yield:

$$\left.\begin{aligned}
(7.7.11) &\Rightarrow T(ki) = T_i^k \sqrt{\frac{g_{kk}}{g_{ii}}} = \left(h_i^2 T^{ki}\right)\sqrt{\frac{h_k^2}{h_i^2}} = h_k h_i T^{ki} \\
(7.7.12) &\Rightarrow \tau_{ki} = \sum_j T(kj)\, g^{ij}\sqrt{\frac{g_{jj}}{g^{ii}}} = T(ki)\frac{1}{h_i^2}\sqrt{\frac{h_i^2}{1/h_i^2}} = T(ki) \\
&\quad \tau_{ki} = T(ki) = \frac{h_k}{h_i} T_i^k = h_k h_i T^{ki}
\end{aligned}\right\} \Rightarrow \quad (7.7.13)$$

$T(ki)$ for $k = i$ is the normal stress σ_i on the coordinate surface $y^i = $ constant, while $T(ki)$ for $k \neq i$ are orthogonal components of the shear stress τ_i on the same surface. In the following the physical stress components are only used in orthogonal coordinates, and for practical reasons we shall use the notation $T(ki)$. Note that, with the unit tangent vectors to the coordinate lines $\mathbf{e}_i^y$:

$$T(ki) = \mathbf{e}_k^y \cdot \mathbf{T} \cdot \mathbf{e}_i^y \qquad (7.7.14)$$

Example 7.6 Physical Stress Components in Cylindrical Coordinates, Fig. 7.6
The physical components $T(ki)$ of the stress tensor $\mathbf{T}$ in cylindrical coordinates are expressed in Fig. 7.6 and with alternative symbols in the formulas (7.7.15). The

formulas also present the mixed tensor components and the contravariant tensor components.

$$(T(ij)) \equiv \begin{pmatrix} \sigma_R & \tau_{R\theta} & \tau_{Rz} \\ \tau_{\theta R} & \sigma_\theta & \tau_{\theta z} \\ \tau_{zR} & \tau_{z\theta} & \sigma_z \end{pmatrix} \equiv \begin{pmatrix} T_{RR} & T_{R\theta} & T_{Rz} \\ T_{\theta R} & T_{\theta\theta} & T_{\theta z} \\ T_{zR} & T_{z\theta} & T_{zz} \end{pmatrix} \quad \text{(Fig. 7.6)}$$

$$= \begin{pmatrix} T_1^1 & T_2^1/R & T_3^1 \\ RT_1^2 & T_2^2 & RT_3^2 \\ T_1^3 & T_2^3/R & T_3^3 \end{pmatrix} = \begin{pmatrix} T^{11} & RT^{12} & T^{13} \\ RT^{21} & R^2T^{22} & RT^{23} \\ T^{31} & RT^{31} & T^{33} \end{pmatrix} \qquad (7.7.15)$$

7.7.2 Cauchy's Equations of Motion

The Cauchy equations of motion (2.2.35) have the following representation in a general curvilinear coordinates:

$$T^{ik}\big|_k + \rho b^i = \rho a^i \Leftrightarrow T^{ik}{}_{,k} + T^{lk}\Gamma^i_{lk} + T^{il}\Gamma^k_{lk} + \rho b^i = \rho a^i \qquad (7.7.16)$$

By using the result (6.5.6), we obtain the alternative form:

$$\frac{1}{\sqrt{g}} \frac{\partial}{\partial y^k}\left(\sqrt{g}T^{ik}\right) + T^{lk}\Gamma^i_{lk} + \rho b^i = \rho a^i \qquad (7.7.17)$$

In *orthogonal coordinates* we use the formulas (6.5.51), (7.7.12), and (6.5.51) and can rewrite Eq. (7.7.17) to:

$$\sum_k \left[\frac{1}{h}\frac{\partial}{\partial y^k}\left(\frac{h}{h_k}T(ik)\right) + \frac{1}{h_i h_k}\frac{\partial h_i}{\partial y^k}T(ik) - \frac{1}{h_i h_k}\frac{\partial h_k}{\partial y^i}T(kk) \right] + \rho b(i) = \rho a(i)$$

$$(7.7.18)$$

Fig. 7.6 Physical stress components in cylindrical coordinates

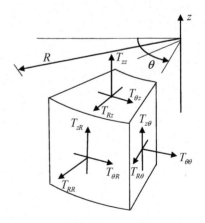

Example 7.7 Cauchy's Equations in Cylindrical Coordinates.
See Problem 7.7

$$\frac{\partial \sigma_R}{\partial R} + \frac{\sigma_R - \sigma_\theta}{R} + \frac{1}{R}\frac{\partial \tau_{R\theta}}{\partial \theta} + \frac{\partial \tau_{Rz}}{\partial z} + \rho b_R = \rho a_R$$

$$\frac{1}{R^2}\frac{\partial}{\partial R}\left(R^2 \tau_{\theta R}\right) + \frac{1}{R}\frac{\partial \sigma_\theta}{\partial \theta} + \frac{\partial \tau_{\theta z}}{\partial z} + \rho b_\theta = \rho a_\theta \qquad (7.7.19)$$

$$\frac{1}{R}\frac{\partial}{\partial R}\left(R\tau_{zR}\right) + \frac{1}{R}\frac{\partial \tau_{\theta z}}{\partial \theta} + \frac{\partial \sigma_z}{\partial z} + \rho b_z = \rho a_z$$

Polar coordinates and rotational symmetry:

$$\frac{d\sigma_R}{dR} + \frac{\sigma_R - \sigma_\theta}{R} + \rho b_R = \rho a_R \Leftrightarrow \frac{d}{dR}(R\sigma_R) - \sigma_\theta + \rho R b_R = \rho R a_R \qquad (7.7.20)$$

Example 7.8 Cauchy's Equations in Spherical Coordinates. See Problem 7.8

$$\frac{\partial \sigma_r}{\partial r} + \frac{2\sigma_r - \sigma_\theta - \sigma_\phi}{r} + \frac{1}{r\sin\theta}\left[\frac{\partial}{\partial \theta}(\sin\theta\,\tau_{r\theta}) + \frac{\partial \tau_{r\phi}}{\partial \phi}\right] + \rho b_r = \rho a_r$$

$$\frac{1}{r^3}\frac{\partial}{\partial r}\left(r^3 \tau_{\theta r}\right) + \frac{1}{r\sin\theta}\left[\frac{\partial}{\partial \theta}(\sin\theta\,\sigma_\theta) + \frac{\partial \tau_{\theta\phi}}{\partial \phi}\right] - \frac{\cot\theta}{r}\sigma_\phi + \rho b_\theta = \rho a_\theta \quad (7.7.21)$$

$$\frac{1}{r^3}\frac{\partial}{\partial r}\left(r^3 \tau_{\phi r}\right) + \frac{1}{r\sin^2\theta}\frac{\partial}{\partial \theta}(\sin^2\theta\,\tau_{\phi\theta}) + \frac{1}{r\sin\theta}\frac{\partial \sigma_\phi}{\partial \phi} + \rho b_\phi = \rho a_\phi$$

Point symmetry:

$$\sigma_\theta = \sigma_\phi, \tau_{r\theta} = \tau_{\theta\phi} = \tau_{\phi r} = 0 \Rightarrow$$

$$\frac{d\sigma_r}{dr} + \frac{2(\sigma_r - \sigma_\phi)}{r} + \rho b_r = \rho a_r \Leftrightarrow \frac{1}{r^2}\frac{d}{dr}(r^2\sigma_r) - \frac{2\sigma_\phi}{r} + \rho b_r = \rho a_r$$

$$(7.7.22)$$

7.8 Basic Equations in Linear Elasticity

The classical theory of elasticity has been presented in Sect. 7.5.2. The most important equations of the theory will now be transformed to general coordinates.

In the Cauchy equations of motion represented above by Eqs. (7.7.16–7.7.18) the particle acceleration will be expressed through the displacement vector $\mathbf{u}(y, t)$:

$$\mathbf{a} = \ddot{\mathbf{u}} \Leftrightarrow a^i = \ddot{u}^i \qquad (7.8.1)$$

The constitutive equations for isotropic, linearly elastic material are represented by the *generalized Hooke's law*, which in tensorial form, i.e. invariant form, and in the Cartesian form, is represented by Eq. (5.2.7), and repeated here:

$$\mathbf{T} = \frac{\eta}{1+v}\left[\mathbf{E} + \frac{v}{1-2v}(\operatorname{tr}\mathbf{E})\mathbf{1}\right] \Leftrightarrow T_{ik} = \frac{\eta}{1+v}\left[E_{ik} + \frac{v}{1-2v}E_{jj}\delta_{ik}\right] \qquad (7.8.2)$$

In general coordinates the generalized Hooke's law is expressed by the formula:

$$T_k^i = \frac{\eta}{1+v}\left[E_k^i + \frac{v}{1-2v}E_j^j\delta_k^i\right] \qquad (7.8.3)$$

The element E_k^i of the strain tensor $\mathbf{E}$ for small deformation may be expressed by the displacement gradients:

$$E_k^i = \frac{1}{2}\left(u^i|_k + u_k|^i\right) \qquad (7.8.4)$$

Navier's equations, i.e. the equations of motion expressed through the displacement $\mathbf{u}(y,t)$, are in invariant form and Cartesian form given by Eq. (5.2.27). The invariant form and the form in general coordinates are:

$$\nabla^2\mathbf{u} + \frac{1}{1-2v}\nabla(\nabla\cdot\mathbf{u}) + \frac{\rho}{\mu}(\mathbf{b}-\ddot{\mathbf{u}}) = 0 \Leftrightarrow u^i|_k^k + \frac{1}{1-2v}u^k|_k^i + \frac{\rho}{\mu}(b^i-\ddot{u}^i) = 0$$
$$(7.8.5)$$

The any orthogonal coordinate system Navier's equations may be expanded in details by application of formula (6.5.104) for $\nabla^2\mathbf{u}$ and formula (6.5.102) for $\nabla(\nabla\cdot\mathbf{u})$. The result is:

$$\sum_k\left\{\frac{1}{h_i}\frac{\partial}{\partial y^i}\left[\frac{1}{h}\frac{\partial}{\partial y^k}\left(h\frac{u(k)}{h_k}\right)\right] - \frac{h_i}{h}\frac{\partial}{\partial y^k}\left[\frac{h}{(h_kh_i)^2}\left(\frac{\partial}{\partial y^i}(h_ku(k)) - \frac{\partial}{\partial y^k}(h_iu(i))\right)\right]\right\}$$
$$+ \sum_k\left\{\frac{1}{1-2v}\frac{1}{h_i}\frac{\partial}{\partial y^i}\left[\frac{1}{h}\frac{\partial}{\partial y^k}\left(h\frac{u(k)}{h_k}\right)\right]\right\} + \frac{\rho}{\mu}(b(i)-\ddot{u}(i)) = 0 \qquad (7.8.6)$$

Example 7.9 Navier's Equations in Cylindrical Coordinates
Applying the formula (6.5.114) for $\nabla^2\mathbf{u}$ and formula (6.5.112) for $\nabla(\nabla\cdot\mathbf{u})$ we obtain:

$$\nabla^2 u_R - \frac{u_R}{R^2} - \frac{2}{R^2}\frac{\partial u_\theta}{\partial\theta} + \frac{1}{1-2v}\frac{\partial}{\partial R}(\nabla\cdot\mathbf{u}) + \frac{\rho}{\mu}(b_R-\ddot{u}_R) = 0$$
$$\nabla^2 u_\theta + \frac{2}{R^2}\frac{\partial u_R}{\partial\theta} - \frac{u_\theta}{R^2} + \frac{1}{1-2v}\frac{1}{R}\frac{\partial}{\partial\theta}(\nabla\cdot\mathbf{u}) + \frac{\rho}{\mu}(b_\theta-\ddot{u}_\theta) = 0 \qquad (7.8.7)$$
$$\nabla^2 u_z + \frac{1}{1-2v}\frac{\partial}{\partial z}(\nabla\cdot\mathbf{u}) + \frac{\rho}{\mu}(b_z-\ddot{u}_z) = 0$$

where:

$$\nabla^2 = \frac{1}{R}\frac{\partial}{\partial R}\left(R\frac{\partial}{\partial R}\right) + \frac{1}{R^2}\frac{\partial^2}{\partial \theta^2} + \frac{\partial^2}{\partial z^2}, \nabla \cdot \mathbf{u} = \frac{1}{R}\frac{\partial}{\partial R}(Ru_R) + \frac{1}{R}\frac{\partial u_\theta}{\partial \theta} + \frac{\partial u_z}{\partial z}$$

$$(7.8.8)$$

Example 7.10 Navier's Equations in Spherical Coordinates

Applying the formula (6.5.117) for $\nabla^2\mathbf{u}$ and formula (6.5.115) for $\nabla(\nabla \cdot \mathbf{u})$ we obtain:

$$\nabla^2 u_r - \frac{2}{r^2}u_r - \frac{2}{r^2}\frac{\partial u_\theta}{\partial \theta} - \frac{2\cot\theta}{r^2}u_\theta - \frac{2}{r^2\sin\theta}\frac{\partial u_\phi}{\partial \phi} + \frac{1}{1-2\nu}\frac{\partial}{\partial r}(\nabla \cdot \mathbf{u})$$

$$+ \frac{\rho}{\mu}(b_r - \ddot{u}_r) = 0$$

$$(7.8.9)$$

$$\nabla^2 u_\theta - \frac{1}{r^2\sin^2\theta}u_\theta + \frac{2}{r^2}\frac{\partial u_r}{\partial \theta} - \frac{\cos\theta}{r^2\sin^2\theta}\frac{\partial u_\phi}{\partial \phi} + \frac{1}{1-2\nu}\frac{1}{r}\frac{\partial}{\partial \theta}(\nabla \cdot \mathbf{u})$$

$$+ \frac{\rho}{\mu}(b_\theta - \ddot{u}_\theta) = 0$$

$$\nabla^2 u_\phi - \frac{1}{r^2\sin^2\theta}u_\phi + \frac{2}{r^2\sin^2\theta}\frac{\partial u_r}{\partial \phi} + \frac{2}{r^2\sin^2\theta}\cos\theta\frac{\partial u_\theta}{\partial \phi}$$

$$+ \frac{1}{1-2\nu}\frac{1}{r\sin\theta}\frac{\partial}{\partial \phi}(\nabla \cdot \mathbf{u}) + \frac{\rho}{\mu}(b_\phi - \ddot{u}_\phi) = 0$$

$$(7.8.10)$$

where:

$$\nabla^2 = \frac{\partial^2}{\partial r^2} + \frac{2}{r}\frac{\partial}{\partial r} + \frac{1}{r^2}\frac{\partial^2}{\partial \theta^2} + \frac{1}{r^2}\cot\theta\frac{\partial}{\partial \theta} + \frac{1}{r^2\sin^2\theta}\frac{\partial^2}{\partial \phi^2}$$

$$\nabla \cdot \mathbf{u} = \frac{1}{r^2}\frac{\partial}{\partial r}(r^2 u_r) + \frac{1}{r\sin\theta}\frac{\partial}{\partial \theta}(\sin\theta u_\theta) + \frac{1}{r\sin\theta}\frac{\partial u_\phi}{\partial \phi}$$

$$(7.8.11)$$

7.9   Basic Equations for Linearly Viscous Fluids

This section will only present the most important equations pertaining to the mechanics of linearly viscous fluids. The equations have been presented in invariant forms and in terms of Cartesian components in Sect. 5.3 and will now be given the general versions in general curvilinear coordinates.

The equation of continuity introduced in Eq. (5.3.9) is represented by:

$$\frac{\partial \rho}{\partial t} + \text{div}(\rho\mathbf{v}) = 0 \Leftrightarrow \frac{\partial \rho}{\partial t} + (\rho v^i)|_i = 0 \tag{7.9.1}$$

By application of formula (6.5.51) we may rewrite the component form to:

$$\frac{\partial \rho}{\partial t} + \frac{1}{\sqrt{g}} \left(\sqrt{g} \, \rho \, v^i \right)_{,i} = 0 \tag{7.9.2}$$

The *linearly viscous fluid* also called the *Newtonian fluid* is defined by the constitutive equation (5.3.11), which are now generalized to:

$$\mathbf{T} = -p(\rho, \theta)\, \mathbf{1} + 2\mu \mathbf{D} + \left(\kappa - \frac{2\mu}{3} \right)(\mathrm{tr}\mathbf{D})\mathbf{1} \Leftrightarrow$$

$$T^i_j = -p(\rho, \theta)\, \delta^i_j + 2\mu D^i_j + \left(\kappa - \frac{2\mu}{3} \right) D^k_k \, \delta^i_j \Leftrightarrow \tag{7.9.3}$$

$$T^i_j = -p(\rho, \theta)\, \delta^i_j + \mu \left(v^i|_j + v_j|^i \right) + \left(\kappa - \frac{2\mu}{3} \right) v^k|_k \, \delta^i_j$$

When these constitutive equations are substituted into the Cauchy equation of motion (2.2.37) we obtain the *Navier-Stokes' equations*, which are presented in invariant form and Cartesian form by Eqs. (5.3.16) and (5.3.17). These equations are repeated here for reference:

$$\frac{\partial \mathbf{v}}{\partial t} + (\mathbf{v} \cdot \nabla)\mathbf{v} = -\frac{1}{\rho}\nabla p + \frac{\mu}{\rho}\nabla^2 \mathbf{v} + \frac{1}{\rho}\left(\frac{\mu}{3} + \kappa \right)\nabla(\nabla \cdot \mathbf{v}) + \mathbf{b} \Leftrightarrow$$

$$\frac{\partial v_i}{\partial t} + v_k v_{i,k} = -\frac{1}{\rho}p_{,i} + \frac{\mu}{\rho}v_{i,kk} + \frac{1}{\rho}\left(\frac{\mu}{3} + \kappa \right)v_{k,ki} + b_i \text{ in Cartesian coordinates}$$

$$\tag{7.9.4}$$

In general curvilinear coordinates y we obtain the contravarient components of $(\mathbf{v} \cdot \nabla)\mathbf{v}$ using formula $(6.5.31)_1$, of ∇p using formula (6.5.23), of $\nabla^2 \mathbf{v}$ using formula (6.5.106), and of $\nabla(\nabla \cdot \mathbf{v})$ using formula (6.5.105). The *Navier-Stokes equations* in general coordinates y now become:

$$\frac{\partial v^i}{\partial t} + v^k v^i|_k = -\frac{1}{\rho}p|^i + \frac{\mu}{\rho}v^i|^k_k + \frac{1}{\rho}\left(\frac{\mu}{3} + \kappa \right)v^k|^i_k \Leftrightarrow$$

$$\frac{\partial v^i}{\partial t} + v^k \left(v^i{}_{,k} + v^l \Gamma^i_{lk} \right) = -\frac{1}{\rho}g^{ik}p_{,k}$$

$$+ \frac{\mu}{\rho}\left\{ g^{ij}\frac{\partial}{\partial y^j}\left[\frac{1}{\sqrt{g}}\frac{\partial}{\partial y^k}\left(\sqrt{g}v^k \right) \right] - \frac{1}{\sqrt{g}}\frac{\partial}{\partial y^k}\left[\sqrt{g}g^{kr}g^{is}\left(\frac{\partial v_r}{\partial y^s} - \frac{\partial v_s}{\partial y^r} \right) \right] \right\} \tag{7.9.5}$$

$$+ \frac{1}{\rho}\left(\frac{\mu}{3} + \kappa \right)g^{ij}\frac{\partial}{\partial y^j}\left[\frac{1}{\sqrt{g}}\left(\sqrt{g}v^k \right)_{,k} \right] + b^i$$

7.9.1 Basic Equations in Orthogonal Coordinates

The velocity vector $\mathbf{v}$ will now be represented by the physical components $v(i) = v^i h_i$.

The equation of continuity expressed in orthogonal coordinates is obtained from Eq. (7.9.2) and with $\sqrt{g} = h$:

$$\frac{\partial \rho}{\partial t} + \frac{1}{h} \sum_i \frac{\partial}{\partial y^i} \left(\frac{h}{h_i} \rho \, v(i) \right) = 0 \qquad (7.9.6)$$

We may use the expressions (6.5.69–6.5.70) for the covariant derivatives of vector components in orthogonal coordinates to obtain the formulas for the physical components of the particle acceleration in the Navier-Stokes' equations, see Problem 7.9:

$$a(i) = \frac{Dv(i)}{Dt} + \sum_k \frac{v(k)}{h_i h_k} \left[\frac{\partial h_i}{\partial y^k} v(i) - \frac{\partial h_k}{\partial y^i} v(k) \right] \qquad (7.9.7)$$

where we have introduced the operator:

$$\frac{D}{Dt} = \frac{\partial}{\partial t} + \sum_k \frac{v(k)}{h_k} \frac{\partial}{\partial y^k} \qquad (7.9.8)$$

In cylindrical coordinates the physical acceleration components become:

$$
\begin{aligned}
a_R &= \frac{\partial v_R}{\partial t} + v_R \frac{\partial v_R}{\partial R} + \frac{v_\theta}{R} \frac{\partial v_R}{\partial \theta} + v_z \frac{\partial v_R}{\partial z} - \frac{v_\theta^2}{R} \\
a_\theta &= \frac{\partial v_\theta}{\partial t} + v_R \frac{\partial v_\theta}{\partial R} + \frac{v_\theta}{R} \frac{\partial v_\theta}{\partial \theta} + v_z \frac{\partial v_\theta}{\partial z} + \frac{v_R v_\theta}{R} \\
a_z &= \frac{\partial v_z}{\partial t} + v_R \frac{\partial v_z}{\partial R} + \frac{v_\theta}{R} \frac{\partial v_z}{\partial \theta} + v_z \frac{\partial v_z}{\partial z}
\end{aligned}
\qquad (7.9.9)
$$

In spherical coordinates the physical acceleration components become:

$$
\begin{aligned}
a_r &= \frac{\partial v_r}{\partial t} + v_r \frac{\partial v_r}{\partial r} + \frac{v_\theta}{r} \frac{\partial v_r}{\partial \theta} + \frac{v_\phi}{r \sin\theta} \frac{\partial v_r}{\partial \phi} - \frac{v_\theta^2 + v_\phi^2}{r} \\
a_\theta &= \frac{\partial v_\theta}{\partial t} + v_r \frac{\partial v_\theta}{\partial r} + \frac{v_\theta}{r} \frac{\partial v_\theta}{\partial \theta} + \frac{v_\phi}{r \sin\theta} \frac{\partial v_\theta}{\partial \phi} + \frac{v_r v_\theta}{r} - \frac{\cot\theta}{r} v_\phi^2 \\
a_\phi &= \frac{\partial v_\phi}{\partial t} + v_r \frac{\partial v_\phi}{\partial r} + \frac{v_\theta}{r} \frac{\partial v_\phi}{\partial \theta} + \frac{v_\phi}{r \sin\theta} \frac{\partial v_\phi}{\partial \phi} + \frac{v_r v_\phi}{r} + \frac{v_\theta v_\phi}{r} \cot\theta
\end{aligned}
\qquad (7.9.10)
$$

The constitutive equations of the Newton fluid in terms of physical components are:

$$T(ij) = -p(\rho, \theta)\, \delta_{ij} + 2\mu D(ij) + \left(\kappa - \frac{2\mu}{3}\right) D(kk)\delta_{ij} \tag{7.9.11}$$

The physical components $T(ij)$ and $D(ij)$ are given by the formulas (7.7.11) and (7.3.38) respectively.

In order to obtain the physical components of the Navier-Stokes equations we need the expressions for $\nabla(\nabla \cdot \mathbf{v})$ and $\nabla^2\mathbf{v}$ presented respectively by the formulas (6.5.109) and (6.5.111) in the general case and the formulas (6.5.112) for $\nabla(\nabla \cdot \mathbf{v})$ and the formulas (6.5.114) for $\nabla^2\mathbf{v}$ in cylindrical coordinates and the formulas (6.5.115) for $\nabla(\nabla \cdot \mathbf{v})$ and the formulas (6.5.117) for $\nabla^2\mathbf{v}$ in spherical coordinates. The results are:

General orthogonal coordinates:

$$\text{grad div }\mathbf{v} = \nabla(\nabla \cdot \mathbf{v}) = \sum_{i,k} \frac{1}{h_i}\frac{\partial}{\partial y_i}\left[\frac{1}{h}\frac{\partial}{\partial y^k}\left(h\frac{v(k)}{h_k}\right)\right]\mathbf{e}_i^y \tag{7.9.12}$$

$$\nabla^2\mathbf{v} = \nabla(\nabla \cdot \mathbf{v}) - \nabla \times (\nabla \times \mathbf{v}) = \sum_{i,k}\left\{\frac{1}{h_i}\frac{\partial}{\partial y^i}\left[\frac{1}{h}\frac{\partial}{\partial y^k}\left(h\frac{v(k)}{h_k}\right)\right]\right\}\mathbf{e}_i^y$$

$$- \sum_{i,k}\left\{\frac{h_i}{h}\frac{\partial}{\partial y^k}\left[\frac{h}{(h_k h_i)^2}\left(\frac{\partial}{\partial y^i}(h_k v(k)) - \frac{\partial}{\partial y^k}(h_i v(i))\right)\right]\right\}\mathbf{e}_i^y \tag{7.9.13}$$

Example 7.11 Expressions for $\nabla(\nabla \cdot \mathbf{v})$ and $\nabla^2\mathbf{v}$ in Cylindrical Coordinates (R, θ, z)

$$\mathbf{b} = \text{grad div }\mathbf{v} = \nabla(\nabla \cdot \mathbf{v}) = b_R\mathbf{e}_R + b_\theta\mathbf{e}_\theta + b_z\mathbf{e}_z \quad \Rightarrow$$

$$b_R = \frac{\partial^2 v_R}{\partial R^2} + \frac{1}{R}\frac{\partial v_R}{\partial R} - \frac{v_R}{R^2} + \frac{1}{R}\frac{\partial^2 v_\theta}{\partial R\partial\theta} - \frac{1}{R^2}\frac{\partial v_\theta}{\partial\theta} + \frac{\partial^2 v_z}{\partial R\partial z}$$

$$b_\theta = \frac{1}{R}\frac{\partial^2 v_R}{\partial\theta\partial R} + \frac{1}{R^2}\frac{\partial v_R}{\partial\theta} + \frac{1}{R^2}\frac{\partial^2 v_\theta}{\partial\theta^2} + \frac{1}{R}\frac{\partial^2 v_z}{\partial\theta\partial z} \tag{7.9.14}$$

$$b_z = \frac{\partial^2 v_R}{\partial z\partial R} + \frac{1}{R}\frac{\partial v_R}{\partial z} + \frac{1}{R}\frac{\partial^2 v_\theta}{\partial z\partial\theta} + \frac{\partial^2 v_z}{\partial z^2}$$

$$\nabla^2 \mathbf{v} = \nabla(\nabla \cdot \mathbf{v}) - \nabla \times (\nabla \times \mathbf{v})$$

$$= \left[\nabla^2 v_R - \frac{v_R}{R^2} - \frac{2}{R^2}\frac{\partial v_\theta}{\partial \theta}\right]\mathbf{e}_r + \left[\nabla^2 v_\theta - \frac{1}{R^2}v_\theta + \frac{2}{R^2}\frac{\partial v_R}{\partial \theta}\right]\mathbf{e}_\theta + \nabla^2 v_z \mathbf{e}_\phi \quad (7.9.15)$$

$$\nabla^2 = \frac{\partial^2}{\partial r^2} + \frac{2}{r}\frac{\partial}{\partial r} + \frac{1}{r^2}\frac{\partial^2}{\partial \theta^2} + \frac{1}{r^2}\cot\theta\frac{\partial}{\partial \theta} + \frac{1}{r^2\sin^2\theta}\frac{\partial^2}{\partial \phi^2}$$

Example 7.12 Expressions for $\nabla(\nabla \cdot \mathbf{v})$ and $\nabla^2\mathbf{v}$ in Spherical Coordinates (r, θ, ϕ)

$$\mathbf{b} = \mathrm{grad}\,\mathrm{div}\,\mathbf{v} = \nabla(\nabla \cdot \mathbf{v}) = b_r\mathbf{e}_r + b_\theta\mathbf{e}_\theta + b_\phi\mathbf{e}_\phi \Rightarrow \quad (7.9.16)$$

$$b_r = \frac{\partial^2 v_r}{\partial r^2} + \frac{2}{r}\frac{\partial v_r}{\partial r} - \frac{2}{r^2}v_r + \frac{1}{r}\frac{\partial^2 v_\theta}{\partial r\partial\theta} + \frac{\cot\theta}{r}\frac{\partial v_\theta}{\partial r} - \frac{1}{r^2}\frac{\partial v_\theta}{\partial \theta} - \frac{\cot\theta}{r^2}v_\theta$$

$$+ \frac{1}{r\sin\theta}\frac{\partial^2 v_\phi}{\partial r\partial\phi} - \frac{1}{r^2\sin\theta}\frac{\partial v_\phi}{\partial \phi}$$

$$b_\theta = \frac{1}{r^2}\frac{\partial^2 v_\theta}{\partial \theta^2} + \frac{\cot\theta}{r^2}\frac{\partial v_\theta}{\partial \theta} - \frac{1}{r^2\sin^2\theta}v_\theta + \frac{1}{r}\frac{\partial^2 v_r}{\partial\theta\partial r} + \frac{2}{r^2}\frac{\partial v_r}{\partial \theta} - \frac{\cos\theta}{r^2\sin^2\theta}\frac{\partial v_\phi}{\partial \phi}$$

$$+ \frac{1}{r^2\sin\theta}\frac{\partial^2 v_\phi}{\partial\theta\partial\phi}$$

$$b_\phi = \frac{1}{r^2\sin^2\theta}\frac{\partial^2 v_\phi}{\partial \phi^2} + \frac{2}{r^2\sin^2\theta}\frac{\partial v_r}{\partial \phi} + \frac{1}{r^2\sin\theta}\frac{\partial^2 v_r}{\partial\theta\partial\phi} + \frac{1}{r^2\sin^2\theta}\frac{\partial^2 v_r}{\partial\phi\partial r}$$

$$+ \frac{1}{r^2\sin^2\theta}\cos\theta\frac{\partial v_\theta}{\partial \phi}$$

$$\nabla^2\mathbf{v} = \nabla(\nabla \cdot \mathbf{v}) - \nabla \times (\nabla \times \mathbf{v})$$

$$= \left[\nabla^2 v_r - \frac{2}{r^2}v_r - \frac{2}{r^2}\frac{\partial v_\theta}{\partial \theta} - \frac{2\cot\theta}{r^2}v_\theta - \frac{2}{r^2\sin\theta}\frac{\partial v_\phi}{\partial \phi}\right]\mathbf{e}_r$$

$$+ \left[\nabla^2 v_\theta - \frac{1}{r^2\sin^2\theta}v_\theta + \frac{2}{r^2}\frac{\partial v_r}{\partial \theta} - \frac{\cos\theta}{r^2\sin^2\theta}\frac{\partial v_\phi}{\partial \phi}\right]\mathbf{e}_\theta \quad (7.9.17)$$

$$+ \left[\nabla^2 v_\phi - \frac{1}{r^2\sin^2\theta}v_\phi + \frac{2}{r^2\sin^2\theta}\frac{\partial v_r}{\partial \phi} + \frac{2}{r^2\sin^2\theta}\cos\theta\frac{\partial v_\theta}{\partial \phi}\right]\mathbf{e}_\phi$$

$$\nabla^2 = \frac{\partial^2}{\partial r^2} + \frac{2}{r}\frac{\partial}{\partial r} + \frac{1}{r^2}\frac{\partial^2}{\partial \theta^2} + \frac{1}{r^2}\cot\theta\frac{\partial}{\partial \theta} + \frac{1}{r^2\sin^2\theta}\frac{\partial^2}{\partial \phi^2} \quad (7.9.18)$$

The Navier-Stokes equations for orthogonal coordinates are obtained by applying the formulas (7.9.7), (7.9.12), and (7.9.13) in the general expression (7.9.4)$_1$ The result is:

$$\frac{\partial \mathbf{v}}{\partial t} + (\mathbf{v} \cdot \nabla)\mathbf{v} = -\frac{1}{\rho}\nabla p + \frac{\mu}{\rho}\nabla^2 \mathbf{v} + \frac{1}{\rho}\left(\frac{\mu}{3} + \kappa\right)\nabla(\nabla \cdot \mathbf{v}) + \mathbf{b} \quad \Leftrightarrow$$

$$\frac{Dv(i)}{Dt} + \sum_k \frac{v(k)}{h_i h_k}\left[\frac{\partial h_i}{\partial y^k}v(i) - \frac{\partial h_k}{\partial y^i}v(k)\right] = -\frac{1}{\rho h_i}\frac{\partial p}{\partial y^i} + b(i)$$

$$+ \frac{\mu}{\rho}\sum_k\left\{\frac{1}{h_i}\frac{\partial}{\partial y^i}\left[\frac{1}{h}\frac{\partial}{\partial y^k}\left(h\frac{v(k)}{h_k}\right)\right]\right\}$$ (7.9.19)

$$-\frac{\mu}{\rho}\sum_k\left\{\frac{h_i}{h}\frac{\partial}{\partial y^k}\left[\frac{h}{(h_k h_i)^2}\left(\frac{\partial}{\partial y^i}(h_k v(k)) - \frac{\partial}{\partial y^k}(h_i v(i))\right)\right]\right\}$$

$$+ \frac{1}{\rho}\left(\frac{\mu}{3} + \kappa\right)\sum_k\frac{1}{h_i}\frac{\partial}{\partial y^i}\left[\frac{1}{h}\frac{\partial}{\partial y^k}\left(h\frac{v(k)}{h_k}\right)\right]$$

Example 7.13 Navier-Stokes Equations in Cylindrical Coordinates

In cylindrical coordinates: $h_1 = 1$, $h_2 = R$, $h_3 = 1$, and $h = R$. Applying the formulas (7.9.9), (7.9.14), and (7.9.15), we obtain:

$$\frac{\partial \mathbf{v}}{\partial t} + (\mathbf{v} \cdot \nabla)\mathbf{v} = -\frac{1}{\rho}\nabla p + \frac{\mu}{\rho}\nabla^2 \mathbf{v} + \frac{1}{\rho}\left(\frac{\mu}{3} + \kappa\right)\nabla(\nabla \cdot \mathbf{v}) + \mathbf{b} \quad \Leftrightarrow$$

$$\frac{\partial v_R}{\partial t} + v_R\frac{\partial v_R}{\partial R} + \frac{v_\theta}{R}\frac{\partial v_R}{\partial \theta} + v_z\frac{\partial v_R}{\partial z} - \frac{v_\theta^2}{R} = -\frac{1}{\rho}\frac{\partial p}{\partial R} + \frac{\mu}{\rho}\left(\nabla^2 a_R - \frac{a_R}{R^2} - \frac{2}{R^2}\frac{\partial a_\theta}{\partial \theta}\right)$$

$$\frac{1}{\rho}\left(\frac{\mu}{3} + \kappa\right)\left(\frac{\partial^2 v_R}{\partial R^2} + \frac{1}{R}\frac{\partial v_R}{\partial R} - \frac{a_v}{R^2} + \frac{1}{R}\frac{\partial^2 v_\theta}{\partial R\partial \theta} - \frac{1}{R^2}\frac{\partial v_\theta}{\partial \theta} + \frac{\partial^2 v_z}{\partial R\partial z}\right) + b_R$$

$$\frac{\partial v_\theta}{\partial t} + v_R\frac{\partial v_\theta}{\partial R} + \frac{v_\theta}{R}\frac{\partial v_\theta}{\partial \theta} + v_z\frac{\partial v_\theta}{\partial z} + \frac{v_R v_\theta}{R} = -\frac{1}{\rho R}\frac{\partial p}{\partial \theta} + \frac{\mu}{\rho}\left(\nabla^2 v_\theta - \frac{1}{R^2}v_\theta + \frac{2}{R^2}\frac{\partial v_R}{\partial \theta}\right)$$

$$+ \frac{1}{\rho}\left(\frac{\mu}{3} + \kappa\right)\left(\frac{1}{R}\frac{\partial^2 v_R}{\partial \theta\partial R} + \frac{1}{R^2}\frac{\partial v_R}{\partial \theta} + \frac{1}{R^2}\frac{\partial^2 v_\theta}{\partial \theta^2} + \frac{1}{R}\frac{\partial^2 v_z}{\partial \theta\partial z}\right) + b_\theta$$

$$\frac{\partial v_z}{\partial t} + v_R\frac{\partial v_z}{\partial R} + \frac{v_\theta}{R}\frac{\partial v_z}{\partial \theta} + v_z\frac{\partial v_z}{\partial z} = -\frac{1}{\rho}\frac{\partial p}{\partial z} + \frac{\mu}{\rho}\nabla^2 a_z$$

$$+ \frac{1}{\rho}\left(\frac{\mu}{3} + \kappa\right)\left(\frac{\partial^2 v_R}{\partial z\partial R} + \frac{1}{R}\frac{\partial v_R}{\partial z} + \frac{1}{R}\frac{\partial^2 v_\theta}{\partial z\partial \theta} + \frac{\partial^2 v_z}{\partial z^2}\right) + b_z$$

$$\nabla^2 = \frac{1}{R}\frac{\partial}{\partial R}\left(R\frac{\partial}{\partial R}\right) + \frac{1}{R^2}\frac{\partial^2}{\partial \theta^2} + \frac{\partial^2}{\partial z^2} = \frac{\partial^2}{\partial R^2} + \frac{1}{R}\frac{\partial}{\partial R} + \frac{1}{R^2}\frac{\partial^2}{\partial \theta^2} + \frac{\partial^2}{\partial z^2}$$

Example 7.14 Pipe Flow

A linearly viscous fluid with viscosity μ flows through a vertical straight pipe with internal diameter d. The body force is the gravitational force per unit mass g. We aim to determine the pressure p and the velocity distribution in the pipe.

We assume that the fluid sticks to the pipe wall and that the flow is steady. We may then assume the following velocity profile expressed in cylindrical coordinates (R, θ, z), with the z-axis along the axis of the pipe and positive upwards:

$$v_z(R) = v(R), \quad v(d/2) = 0, \quad v_R = v_\theta = 0 \qquad (7.9.20)$$

The body force is given by $b_z = g$, $b_R = b_\theta = 0$. The flow is isochoric, i.e. $\nabla \cdot \mathbf{v} = 0$, and the Navier-Stokes equations from Example 7.9 are in this case reduced to:

$$\frac{\partial \mathbf{v}}{\partial t} + (\mathbf{v} \cdot \nabla)\mathbf{v} = -\frac{1}{\rho}\nabla p + \frac{\mu}{\rho}\nabla^2 \mathbf{v} + \mathbf{b} \quad \Rightarrow$$

$$0 = -\frac{1}{\rho}\frac{\partial p}{\partial R}, \quad 0 = -\frac{1}{\rho R}\frac{\partial p}{\partial \theta}$$

$$0 = -\frac{1}{\rho}\frac{\partial p}{\partial z} + \frac{\mu}{\rho}\nabla^2 v_z + g = -\frac{1}{\rho}\frac{\partial p}{\partial z} + \frac{\mu}{\rho}\frac{1}{R}\frac{d}{dR}\left(R\frac{dv}{dR}\right) + g$$

It follows from these equations that the pressure p is only a function of the coordinate z, i.e. $p = p(z)$, and that the pressure gradient dp/dz is a constant c. We thus have the result with four constants of integration: C_1, C_2, C_3, and C_4:

$$p = p(z) = cz + C_1$$
$$\frac{1}{R}\frac{d}{dR}\left(R\frac{dv}{dR}\right) = \frac{1}{\mu}(c - \rho g) \Rightarrow \frac{d}{dR}\left(R\frac{dv}{dR}\right) = \frac{1}{\mu}(c - \rho g)R \Rightarrow R\frac{\partial v}{\partial R} = \frac{1}{\mu}(c - \rho g)\frac{R^2}{2} + C_2$$
$$\Rightarrow \frac{\partial v}{\partial R} = \frac{1}{\mu}(c - \rho g)\frac{R}{2} + \frac{C_2}{R} \Rightarrow v = \frac{1}{\mu}(c - \rho g)\frac{R^2}{4} + C_2 \ln R + C_3$$

With the boundary conditions: $v(d/2) = 0$, $v(0) \neq \infty$, and $p(0) = p_0$, we obtain the solution to problem:

$$v_R = v(R) = v_0\left[1 - \left(\frac{2R}{d}\right)^2\right], \quad v_0 = \frac{(c - \rho g)d^2}{16\mu}, \quad p = p(z) = cz + p_0$$

The velocity profile $v_R = v(R)$ is shown in Fig. 7.7. In order to obtain this velocity profile the fluid must have a rather high viscosity for the flow to be laminar. Otherwise the flow becomes turbulent and the Navier-Stokes equations do not apply for solution of the present flow problem.

Example 7.15 Navier-Stokes Equations in Spherical Coordinates
In spherical coordinates: $h_1 = 1$, $h_2 = r$, $h_3 = r \sin\theta$, and $h = r^2 \sin\theta$. We apply the formulas (7.9.10), (7.9.16), and (7.9.17), and obtain:

$$\frac{\partial \mathbf{v}}{\partial t} + (\mathbf{v} \cdot \nabla)\mathbf{v} = -\frac{1}{\rho}\nabla p + \frac{\mu}{\rho}\nabla^2 \mathbf{v} + \frac{1}{\rho}\left(\frac{\mu}{3} + \kappa\right)\nabla(\nabla \cdot \mathbf{v}) + \mathbf{b} \quad \Rightarrow$$

Fig. 7.7 Pipe flow. Velocity profile

Example 7.14 Pipe Flow

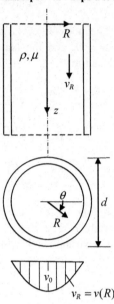

$v_R = v(R)$

$$\frac{\partial v_r}{\partial t} + v_r \frac{\partial v_r}{\partial r} + \frac{v_\theta}{r} \frac{\partial v_r}{\partial \theta} + \frac{v_\phi}{r \sin \theta} \frac{\partial v_r}{\partial \phi} - \frac{v_\theta^2 + v_\phi^2}{r}$$

$$= -\frac{1}{\rho} \frac{\partial p}{\partial r} + b_r + \frac{\mu}{\rho} \left(\nabla^2 v_r - \frac{2}{r^2} v_r - \frac{2}{r^2} \frac{\partial v_\theta}{\partial \theta} - \frac{2 \cot \theta}{r^2} v_\theta - \frac{2}{r^2 \sin \theta} \frac{\partial v_\phi}{\partial \phi} \right)$$

$$+ \frac{1}{\rho} \left(\frac{\mu}{3} + \kappa \right) \left(\frac{\partial^2 a_r}{\partial r^2} + \frac{2}{r} \frac{\partial a_r}{\partial r} - \frac{2}{r^2} a_r + \frac{1}{r} \frac{\partial^2 a_\theta}{\partial r \partial \theta} + \frac{\cot \theta}{r} \frac{\partial a_\theta}{\partial r} - \frac{1}{r^2} \frac{\partial a_\theta}{\partial \theta} \right.$$

$$\left. - \frac{\cot \theta}{r^2} a_\theta + \frac{1}{r \sin \theta} \frac{\partial^2 a_\phi}{\partial r \partial \phi} - \frac{1}{r^2 \sin \theta} \frac{\partial a_\phi}{\partial \phi} \right)$$

$$\frac{\partial v_\theta}{\partial t} + v_r \frac{\partial v_\theta}{\partial r} + \frac{v_\theta}{r} \frac{\partial v_\theta}{\partial \theta} + \frac{v_\phi}{r \sin \theta} \frac{\partial v_\theta}{\partial \phi} + \frac{v_r v_\theta}{r} - \frac{\cot \theta}{r} v_\phi^2$$

$$= -\frac{1}{\rho r} \frac{\partial p}{\partial \theta} + b_\theta + \frac{\mu}{\rho} \left(\nabla^2 v_\theta - \frac{1}{r^2 \sin^2 \theta} v_\theta + \frac{2}{r^2} \frac{\partial v_r}{\partial \theta} - \frac{\cos \theta}{r^2 \sin^2 \theta} \frac{\partial v_\phi}{\partial \phi} \right)$$

$$+ \frac{1}{\rho} \left(\frac{\mu}{3} + \kappa \right) \left(\frac{1}{r^2} \frac{\partial^2 a_\theta}{\partial \theta^2} + \frac{\cot \theta}{r^2} \frac{\partial a_\theta}{\partial \theta} - \frac{1}{r^2 \sin^2 \theta} a_\theta + \frac{1}{r} \frac{\partial^2 a_r}{\partial \theta \partial r} + \frac{2}{r^2} \frac{\partial a_r}{\partial \theta} \right.$$

$$\left. - \frac{\cos \theta}{r^2 \sin^2 \theta} \frac{\partial a_\phi}{\partial \phi} + \frac{1}{r^2 \sin \theta} \frac{\partial^2 a_\phi}{\partial \theta \partial \phi} \right)$$

$$\frac{\partial v_\phi}{\partial t} + v_r \frac{\partial v_\phi}{\partial r} + \frac{v_\theta}{r}\frac{\partial v_\phi}{\partial \theta} + \frac{v_\phi}{r \sin\theta}\frac{\partial v_\phi}{\partial \phi} + \frac{v_r v_\phi}{r} + \frac{v_\theta v_\phi}{r}\cot\theta$$

$$= -\frac{1}{\rho r \sin\theta}\frac{\partial p}{\partial \phi} + b_\phi$$

$$+ \frac{\mu}{\rho}\left(\nabla^2 v_\phi - \frac{1}{r^2 \sin^2\theta}v_\phi + \frac{2}{r^2 \sin^2\theta}\frac{\partial v_r}{\partial \phi} + \frac{2}{r^2 \sin^2\theta}\cos\theta\frac{\partial v_\theta}{\partial \phi}\right)$$

$$+ \frac{1}{\rho}\left(\frac{\mu}{3} + \kappa\right)\left(\frac{1}{r^2 \sin^2\theta}\frac{\partial^2 v_\phi}{\partial \phi^2} + \frac{2}{r^2 \sin^2\theta}\frac{\partial v_r}{\partial \phi} + \frac{1}{r^2 \sin\theta}\frac{\partial^2 v_r}{\partial \theta \partial \phi}\right.$$

$$\left. + \frac{1}{r^2 \sin^2\theta}\frac{\partial^2 v_r}{\partial \phi \partial r} + \frac{1}{r^2 \sin^2\theta}\cos\theta\frac{\partial v_\theta}{\partial \phi}\right)$$

$$\nabla^2 = \frac{\partial^2}{\partial r^2} + \frac{2}{r}\frac{\partial}{\partial r} + \frac{1}{r^2}\frac{\partial^2}{\partial \theta^2} + \frac{1}{r^2}\cot\theta\frac{\partial}{\partial \theta} + \frac{1}{r^2 \sin^2\theta}\frac{\partial^2}{\partial \phi^2}$$

Problems 7.1–7.9 with solutions see Appendix

References

1. Irgens F (2014) Rheology and non-Newtonian fluids. Springer, London
2. Oldroyd JG (1950) On the formulation of rheological equations of state. Proc Roy Soc London A 200:523–541
3. Green AE, Zerna W (1968) Theoretical elasticity, 2nd edn. Oxford University Press, London
4. Truesdell C (1953) Physical components of vectors and tensors. Z angew Math Mech 33:345–356

Chapter 8
Surface Geometry. Tensors in Riemannian Space R_2

8.1 Surface Coordinates. Base Vectors. Fundamental Parameters

Points or places in three-dimensional Euclidean space E_3 are given by a place vector $\mathbf{r}$ as a function of Cartesian coordinates x_i as shown in Fig. 8.1, or by general coordinates y^i. A *surface* imbedded in E_3 is defined by the points given by the place vector $\mathbf{r}$, the Cartesian coordinates x_i, and the general coordinates y^i as functions of two independent parameters u^1 and u^2 called *surface coordinates*:

$$\mathbf{r} = \mathbf{r}(u^1, u^2) \equiv \mathbf{r}(u), \quad x_i = x_i(u^1, u^2) \equiv x_i(u), \quad y^i = y^i(u^1, u^2) \equiv y^i(u) \quad (8.1.1)$$

In Fig. 8.1 the surface is described by coordinate lines, u^1-lines and u^2-lines. The surface defined by the formulas (8.1.1) is said to represent a *two-dimensional Riemannian space* R_2 imbedded in the three-dimensional Euclidean space E_3

Figure 8.2 shows the tangent plane in point P to the surface in Fig. 8.1. The *base vectors* $\mathbf{a}_\alpha$ for the u-system as shown in Fig. 8.2, are tangent vectors to the coordinate lines and defined by:

$$\mathbf{a}_\alpha = \frac{\partial \mathbf{r}}{\partial u^\alpha} \equiv \mathbf{r}_{,\alpha} = \frac{\partial \mathbf{r}}{\partial y^i}\frac{\partial y^i}{\partial u^\alpha} = \mathbf{g}_i \frac{\partial y^i}{\partial u^\alpha} \quad (8.1.2)$$

It follows that:

$$\frac{\partial y^i}{\partial u^\alpha} = \mathbf{g}^i \cdot \mathbf{a}_\alpha \quad (8.1.3)$$

The unit normal vector $\mathbf{a}_3$ to the surface is determined by:

$$\mathbf{a}_3 = \frac{\mathbf{a}_1 \times \mathbf{a}_2}{|\mathbf{a}_1 \times \mathbf{a}_2|} \Rightarrow \mathbf{a}_3 \cdot \mathbf{a}_\alpha = 0 \quad (8.1.4)$$

© Springer Nature Switzerland AG 2019
F. Irgens, *Tensor Analysis*,
https://doi.org/10.1007/978-3-030-03412-2_8

Fig. 8.1 Two-dimensional surface imbedded in three-dimensional space E_3. Coordinate lines: u^1-lines and u^2-lines

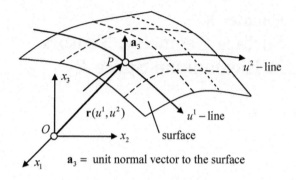

Fig. 8.2 Tangent plane to a surface in point P. Base vectors $\mathbf{a}_1$ and $\mathbf{a}_2$. Reciprocal base vectors $\mathbf{a}^1$ and $\mathbf{a}^2$

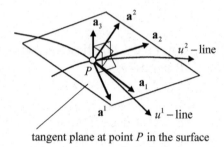

tangent plane at point P in the surface

The unit normal vector $\mathbf{a}_3$ points out from what will be defined as the *positive side of the surface*. The three vectors $\mathbf{a}_i$ represent a *right-handed system* of three vectors. *Reciprocal base vectors* $\mathbf{a}^\alpha$ for the u-system are defined by the relations:

$$\mathbf{a}^\alpha \cdot \mathbf{a}_i = \delta_i^\alpha \tag{8.1.5}$$

These vectors, shown in Fig. 8.2, lie in the tangent plane to the surface and are normal vectors to the coordinate lines. Because the surface base vectors $\mathbf{a}_\alpha$ and $\mathbf{a}^\alpha$ are all vectors in the same plane, we may write:

$$\mathbf{a}_\alpha = a_{\alpha\beta}\,\mathbf{a}^\beta \quad \text{and} \quad \mathbf{a}^\alpha = a^{\alpha\beta}\,\mathbf{a}_\beta \tag{8.1.6}$$

The components $a_{\alpha\beta}$ are called the *fundamental parameters of the first order* for the u-system. The components $a^{\alpha\beta}$ are called the *reciprocal fundamental parameters of the first order* for the u-system. Scalar products of the vectors in the formulas (8.1.6) by the base vectors $\mathbf{a}_\beta$ and $\mathbf{a}^\beta$, followed by applications of the formulas (8.1.2), (8.1.5), and (8.1.6) provide the results:

$$\mathbf{a}_\alpha \cdot \mathbf{a}_\beta = a_{\alpha\beta} = a_{\beta\alpha} = g_{ij}\frac{\partial y^i}{\partial u^\alpha}\frac{\partial y^j}{\partial u^\beta},\ \mathbf{a}^\alpha \cdot \mathbf{a}^\beta = a^{\alpha\beta} = a^{\beta\alpha},\ \mathbf{a}_\alpha \cdot \mathbf{a}^\beta = \mathbf{a}^\beta \cdot \mathbf{a}_\alpha = \delta_\alpha^\beta$$
$$\mathbf{a}_\alpha \cdot \mathbf{a}^\beta = \left(a_{\alpha\gamma}\,\mathbf{a}^\gamma\right) \cdot \left(a^{\beta\lambda}\,\mathbf{a}_\lambda\right) = a_{\alpha\gamma}a^{\beta\lambda}\mathbf{a}^\gamma \cdot \mathbf{a}_\lambda = a_{\alpha\gamma}a^{\beta\lambda}\delta_\lambda^\gamma = a_{\alpha\gamma}a^{\beta\gamma} \Rightarrow a_{\alpha\gamma}a^{\beta\lambda} = \delta_\alpha^\beta \tag{8.1.7}$$

It is convenient to define a parameter a for the u-system:

$$\alpha = \det\left(a_{\alpha\beta}\right) = a_{11}\, a_{22} - \left(a_{12}\right)^2 \tag{8.1.8}$$

Solution of Eq. $(8.1.7)_4$: $a_{\alpha\gamma}\, a^{\beta\gamma} = \delta_\alpha^\beta$, with respect to the components $a^{\beta\gamma}$ yields:

$$a^{11} = \frac{a_{22}}{\alpha}, \quad a^{22} = \frac{a_{11}}{\alpha}, \quad a^{12} = -\frac{a_{12}}{\alpha} \tag{8.1.9}$$

The angle (u^1, u^2) between two coordinate lines, e.g. a u^1-line and a u^2-line, may be computed from the equation:

$$\cos\left(u^1, u^2\right) = \frac{\mathbf{a}_1 \cdot \mathbf{a}_2}{|\mathbf{a}_1||\mathbf{a}_2|} = \frac{a_{12}}{\sqrt{a_{11}a_{22}}} \tag{8.1.10}$$

With the permutation symbol $e_{\alpha\beta}$ defined by the formula (1.1.16) we introduce the following two sets of *permutation symbols*: $\varepsilon_{\alpha\beta}$ and $\varepsilon^{\alpha\beta}$:

$$\varepsilon_{\alpha\beta} = e_{\alpha\beta}\sqrt{\alpha}, \quad \varepsilon^{\alpha\beta} = e_{\alpha\beta}/\sqrt{\alpha}, \quad e_{\alpha\beta} = \begin{pmatrix} 0 & 1 \\ -1 & 0 \end{pmatrix} \tag{8.1.11}$$

The following relationships may be derived, see Problem 8.1:

$$\varepsilon_{\alpha\beta}\, \varepsilon^{\gamma\rho} = \delta_\alpha^\gamma\, \delta_\beta^\rho - \delta_\beta^\gamma\, \delta_\alpha^\rho \;\Rightarrow\; \varepsilon_{\alpha\beta}\, \varepsilon^{\gamma\beta} = \delta_\alpha^\gamma \tag{8.1.12}$$

$$|\mathbf{a}_1 \times \mathbf{a}_2| = \sqrt{\alpha}, \quad \mathbf{a}_1 \times \mathbf{a}_2 = \sqrt{\alpha}\, \mathbf{a}_3$$
$$\mathbf{a}_\alpha \times \mathbf{a}_\beta = \varepsilon_{\alpha\beta}\, \mathbf{a}_3, \quad \mathbf{a}^\alpha \times \mathbf{a}^\beta = \varepsilon^{\alpha\beta}\, \mathbf{a}_3, \tag{8.1.13}$$
$$\mathbf{a}_3 \times \mathbf{a}_\alpha = \varepsilon_{\alpha\beta}\, \mathbf{a}^\beta, \quad \mathbf{a}_3 \times \mathbf{a}^\alpha = \varepsilon^{\alpha\beta}\, \mathbf{a}_\beta$$

The components of the unit normal vector $\mathbf{a}_3$ in a general curvilinear coordinate system y in three-dimensional space E_3 are denoted by $\underset{i}{a}$, i.e. $\mathbf{a}_3 = \underset{i}{a}\, \mathbf{g}^i$, and may be found to be, see Problem 8.2:

$$\underset{i}{a} = \frac{1}{2}\varepsilon_{ijk}\frac{\partial y^j}{\partial u^\alpha}\frac{\partial y^k}{\partial u^\beta}\varepsilon^{\alpha\beta} \tag{8.1.14}$$

Christoffel symbols of the second kind $\Gamma_{\alpha\beta}^\gamma$ and *fundamental parameters of second order* $B_{\alpha\beta}$ *for the u-system on the surface* are defined through the relationships:

$$\mathbf{a}_{\alpha,\beta} = \Gamma_{\alpha\beta}^\gamma\mathbf{a}_\gamma + B_{\alpha\beta}\mathbf{a}_3 \tag{8.1.15}$$

It follows that:

$$\Gamma^{\gamma}_{\alpha\beta} = \Gamma^{\gamma}_{\beta\alpha} = \mathbf{a}_{\alpha,\beta} \cdot \mathbf{a}^{\gamma}, \quad B_{\alpha\beta} = B_{\beta\alpha} = \mathbf{a}_{\alpha,\beta} \cdot \mathbf{a}_3 \qquad (8.1.16)$$

In a plane surface the vectors $\mathbf{a}_{\alpha,\beta}$ lie in the surface and the formulas $(8.1.16)_2$ show that the elements $B_{\alpha\beta} = 0$. It will be shown in Sect. 8.7 that the elements $B_{\alpha\beta}$ are components in the u-system of the *curvature tensor for the surface*, which represents the surface curvatures. *Christoffel symbols of the first kind for the u-system on the surface* are defined by:

$$\Gamma_{\alpha\beta\gamma} = \mathbf{a}_{\alpha,\beta} \cdot \mathbf{a}_{\gamma} = \Gamma^{\rho}_{\alpha\beta} a_{\rho\gamma} \Leftrightarrow \Gamma^{\gamma}_{\alpha\beta} = \Gamma_{\alpha\beta\rho} a^{\rho\gamma} \qquad (8.1.17)$$

The following formulas may be derived, see Problem 8.3:

$$\mathbf{a}_{3,\alpha} = -B_{\alpha\beta}\,\mathbf{a}^{\beta}, \quad \mathbf{a}^{\alpha}{}_{,\beta} = -\Gamma^{\alpha}_{\gamma\beta}\,\mathbf{a}^{\gamma} + B_{\gamma\beta}\,a^{\gamma\alpha}\,\mathbf{a}_3 \qquad (8.1.18)$$

$$a_{\alpha\beta,\gamma} = \Gamma_{\alpha\gamma\beta} + \Gamma_{\beta\gamma\alpha}, \quad \Gamma_{\alpha\beta\gamma} = \frac{1}{2}\left(a_{\alpha\gamma,\beta} + a_{\beta\gamma,\alpha} - a_{\alpha\beta,\gamma}\right) \qquad (8.1.19)$$

$$\Gamma^{\gamma}_{\gamma\beta} = \frac{1}{2\alpha}\alpha_{,\beta} = \frac{1}{\sqrt{\alpha}}\left(\sqrt{\alpha}\right)_{,\beta} \qquad (8.1.20)$$

Example 8.1 Circular Cylindrical Surface

Figure 8.3 shows a circular cylindrical surface of radius R. With respect to cylindrical coordinates (R, θ, z) the surface is described by the surface coordinates: $u^1 = \theta$ and $u^2 = z$, with the z-axis equal to the x_3-axis in a Cartesian coordinate system Ox. A point on the surface is given by the place vector:

$$\mathbf{r}(u^1, u^2) = R\,\mathbf{e}_R(u^1) + u^2\,\mathbf{e}_z = R\mathbf{e}_R(\theta) + z\,\mathbf{e}_z \qquad (8.1.21)$$

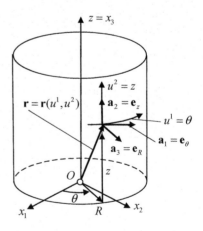

Fig. 8.3 Circular cylindrical surface of radius R: $\mathbf{r} = \mathbf{e}_R(\theta) + z\mathbf{e}_z$. Surface coordinates: $u^1 = \theta$, $u^2 = z = x_3$. Base vectors $\mathbf{a}_1 = R\mathbf{e}_\theta$ and $\mathbf{a}_2 = \mathbf{e}_z$. Surface normal $\mathbf{a}_3 = \mathbf{e}_R$.

In Cartesian coordinates the surface is described by:

$$x_1 = R\cos\theta, \quad x_2 = R\sin\theta, \quad x_3 = z \tag{8.1.22}$$

The unit normal vector to the surface is $\mathbf{a}_3 = \mathbf{e}_R$. Using the result $d\mathbf{e}_R/du_1 \equiv d\mathbf{e}_R/d\theta = \mathbf{e}_\theta$, obtained from the formulas (6.2.36), we find the base vectors as:

$$\mathbf{a}_\alpha = \frac{\partial \mathbf{r}}{\partial u^\alpha} \quad \Rightarrow \quad \mathbf{a}_1 = \frac{\partial \mathbf{r}}{\partial u^1} = R\,\mathbf{e}_\theta, \quad \mathbf{a}_2 = \frac{\partial \mathbf{r}}{\partial u^2} = \mathbf{e}_z \tag{8.1.23}$$

By the formulas (8.1.5) and (8.1.24) we find the reciprocal base vectors:

$$\mathbf{a}^1 = \mathbf{e}_\theta/R, \quad \mathbf{a}^2 = \mathbf{e}_z \tag{8.1.24}$$

The fundamental parameters of first order become:

$$\begin{aligned}
\left(a_{\alpha\beta}\right) &= \left(\mathbf{a}_\alpha \cdot \mathbf{a}_\beta\right) = \begin{pmatrix} R^2 & 0 \\ 0 & 1 \end{pmatrix} \\
\left(a^{\alpha\beta}\right) &= \left(\mathbf{a}^\alpha \cdot \mathbf{a}^\beta\right) = \begin{pmatrix} 1/R^2 & 0 \\ 0 & 1 \end{pmatrix} \\
\alpha &= \det\left(a_{\alpha\beta}\right) = R^2
\end{aligned} \tag{8.1.25}$$

In order to obtain the Christoffel symbols and the fundamental parameters of second order we need the expressions: $d\mathbf{e}_\theta/d\theta = -\mathbf{e}_R$ and $d\mathbf{e}_R/d\theta = \mathbf{e}_\theta$, obtained from the formulas (6.2.36). Then we find from the general formulas (8.1.16):

$$\Gamma^\gamma_{\alpha\beta} = \Gamma_{\alpha\beta\gamma} = 0, \quad B = \left(B_{\alpha\beta}\right) = \begin{pmatrix} -R & 0 \\ 0 & 0 \end{pmatrix} \tag{8.1.26}$$

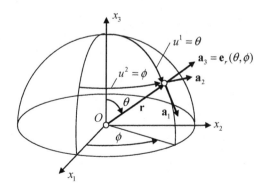

Fig. 8.4 Spherical surface of radius r: $\mathbf{r} = \mathbf{r}(u^1, u^2) = r\mathbf{e}_r(\theta, \phi)$. Surface coordinates: $u^1 = \theta$, $u^2 = \phi$. Base vectors: $\mathbf{a}_1 = r\mathbf{e}_\theta, \mathbf{a}_2 = r\sin\theta\mathbf{e}_\phi$. Surface normal $\mathbf{a}_3 = \mathbf{e}_r(\theta, \phi)$.

Example 8.2 Spherical Surface

Figure 8.4 illustrates a spherical surface of radius r with the centre of the sphere at the origin of the Cartesian coordinate system Ox. A point on the surface is given by the place vector:

$$\mathbf{r} = r\mathbf{e}_r(\theta, \phi) \tag{8.1.27}$$

The two spherical coordinates θ and ϕ in the spherical coordinate system (r, θ, ϕ) are selected as surface coordinates such that: $u^1 = \theta$, $u^2 = \phi$. A point on the surface is given by the place vector:

$$\mathbf{r} = \mathbf{r}(u^1, u^2) = r\mathbf{e}_r(u^1, u^2) = r\,\mathbf{e}_r(\theta, \phi) \tag{8.1.28}$$

The unit normal vector to the surface is $\mathbf{a}_3 = \mathbf{e}_r$. In order to find the base vectors and the reciprocal base vectors, we need the formulas (6.2.41):

$$\frac{\partial \mathbf{e}_r}{\partial \theta} = \mathbf{e}_\theta, \quad \frac{\partial \mathbf{e}_r}{\partial \phi} = \sin\theta\,\mathbf{e}_\phi, \quad \frac{\partial \mathbf{e}_\theta}{\partial \theta} = -\mathbf{e}_r, \quad \frac{\partial \mathbf{e}_\theta}{\partial \phi} = \cos\theta\,\mathbf{e}_\phi$$

$$\frac{d\mathbf{e}_\phi}{d\theta} = \mathbf{0}, \quad \frac{d\mathbf{e}_\phi}{d\phi} = -\sin\theta\,\mathbf{e}_r - \cos\theta\,\mathbf{e}_\theta \tag{8.1.29}$$

Then:

$$\mathbf{a}_1 = \frac{\partial \mathbf{r}}{\partial u^1} = r\frac{\partial \mathbf{e}_r}{\partial \theta} = r\,\mathbf{e}_\theta, \quad \mathbf{a}_2 = \frac{\partial \mathbf{r}}{\partial u^2} = r\frac{\partial \mathbf{e}_r}{\partial \phi} = r\sin\theta\,\mathbf{e}_\phi,$$

$$\mathbf{a}^1 = \frac{1}{r}\mathbf{e}_\theta, \quad \mathbf{a}^2 = \frac{1}{r\sin\theta}\mathbf{e}_\phi \tag{8.1.30}$$

The fundamental parameters of the first order become:

$$(a_{\alpha\beta}) = (\mathbf{a}_\alpha \cdot \mathbf{a}_\beta) = \begin{pmatrix} r^2 & 0 \\ 0 & r^2\sin^2\theta \end{pmatrix}, \quad \alpha = \det(a_{\alpha\beta}) = r^4\sin^2\theta$$

$$(a^{\alpha\beta}) = (\mathbf{a}^\alpha \cdot \mathbf{a}^\beta) = \begin{pmatrix} 1/r^2 & 0 \\ 0 & 1/r^2\sin^2\theta \end{pmatrix} \tag{8.1.31}$$

The non-zero Christoffel symbols and the fundamental parameters of second order are found from the formulas (8.1.16):

$$\Gamma^2_{12} = \cot\theta, \quad \Gamma^1_{22} = -\sin\theta\cos\theta,$$

$$\Gamma_{122} = r^2\sin\theta\cos\theta, \quad \Gamma_{221} = -r^2\sin\theta\cos\theta \tag{8.1.32}$$

$$B = (B_{\alpha\beta}) = \begin{pmatrix} -r & 0 \\ 0 & -r\sin^2\theta \end{pmatrix} \tag{8.1.33}$$

8.2 Surface Vectors

Let $\mathbf{r}(u^1, u^2) \equiv \mathbf{r}(u)$ and $y^i(u^1, u^2) \equiv y^i(u)$ define a surface imbedded in the space E_3. On the surface a curve may be defined by the coordinate functions $u^\alpha(s)$ with the parameter s being the arc length along the curve. The *tangent vector* $\mathbf{t}$ to the curve is the unit vector defined by formula (1.4.8):

$$\mathbf{t} = \frac{d\mathbf{r}}{ds} = \frac{\partial \mathbf{r}}{\partial y^i} \frac{dy^i}{ds} = \frac{dy^i}{ds} \mathbf{g}_i = \frac{\partial \mathbf{r}}{\partial u^\alpha} \frac{du^\alpha}{ds} = \frac{du^\alpha}{ds} \mathbf{a}_\alpha \tag{8.2.1}$$

The tangent vectors to all surface curves on the surface $y^i(u)$ and through a place $\mathbf{r}(u)$ lie in the *tangent plane* to the surface at that place. For this reason the tangent vector to a surface curve is called a *surface vector*. Vectors in the tangent plane to the place $\mathbf{r}(u)$ on the surface $y^i(u)$ may be represented by *surface vector fields*. A steady surface vector field $\mathbf{c}(u)$ has *contravariant components* c^α and *covariant components* c_α in the u-system:

$$\mathbf{c}(u) = c^\alpha \mathbf{a}_\alpha = c_\alpha \mathbf{a}^\alpha \quad \Rightarrow \quad c^\alpha = a^{\alpha\beta} c_\beta, \quad c_\alpha = a_{\alpha\beta} c^\beta \tag{8.2.2}$$

Note that throughout this chapter we shall only consider steady fields, in contrast to what was assumed in Chap. 6. In order to distinguish between the components with indices 1 and 2 in the space coordinate system y and the components c^α and c_α in the surface coordinate system u, we introduce the symbols c^k and c_k for the components of the surface vector in the space coordinate system y. We apply the formula (8.1.3) and obtain:

$$\mathbf{c} = c^k \mathbf{g}_k = c^\alpha \mathbf{a}_\alpha \Rightarrow \mathbf{c} \cdot \mathbf{g}^i = c^k \mathbf{g}_k \cdot \mathbf{g}^i = c^\alpha \mathbf{a}_\alpha \cdot \mathbf{g}^i$$

$$\Rightarrow c^k \delta_k^i = c^\alpha \frac{\partial y^i}{\partial u^\alpha} \Rightarrow c^i = c^\alpha \frac{\partial y^i}{\partial u^\alpha} \tag{8.2.3}$$

A space vector field $\mathbf{b}(y)$ may be connected to places $\mathbf{r}(u)$ on the surface $y^i(u)$ and decomposed into a vector in the direction of the surface normal $\mathbf{a}_3$, and a surface vector field $\mathbf{c}(u)$ equal to the projection of $\mathbf{b}$ onto the tangent plane to the surface $y^i(u)$:

$$\mathbf{b} = (\mathbf{b} \cdot \mathbf{a}_3)\mathbf{a}_3 + \mathbf{c}, \quad \mathbf{b}(y) = b_i \mathbf{g}^i, \quad \mathbf{c}(u) = c_\beta \mathbf{a}^\beta$$

Because $\mathbf{a}_3 \cdot \mathbf{a}_\alpha = 0$ and $\mathbf{a}^\beta \cdot \mathbf{a}_\alpha = \delta_\alpha^\beta$ we obtain, using the formula (8.1.3):

$$\mathbf{b} \cdot \mathbf{a}_\alpha = \mathbf{c} \cdot \mathbf{a}_\alpha \Rightarrow b_i \mathbf{g}^i \cdot \mathbf{a}_\alpha = c_\beta \mathbf{a}^\beta \cdot \mathbf{a}_\alpha \quad \Rightarrow \quad b_i \frac{\partial y^i}{\partial u^\alpha} = c_\beta \delta_\alpha^\beta = c_\alpha \quad \Rightarrow$$

$$c_\alpha = b_i \frac{\partial y^i}{\partial u^\alpha} \tag{8.2.4}$$

We may check the results (8.2.3) and (8.2.4) by considering $\underset{\sim}{c^k}$ as components of a space vector. Then by the formulas (8.2.4), (8.2.3), and (8.1.7)$_1$ we get:

$$c_\alpha = c_i \frac{\partial y^i}{\partial u^\alpha} = \left(g_{ik} c^k \right) \frac{\partial y^i}{\partial u^\alpha} = g_{ik} \left(c^\beta \frac{\partial y^k}{\partial u^\beta} \right) \frac{\partial y^i}{\partial u^\alpha} = \left(g_{ik} \frac{\partial y^i}{\partial u^\alpha} \frac{\partial y^k}{\partial u^\beta} \right) c^\beta = a_{\alpha\beta} c^\beta = c_\alpha$$

ok

When we apply the results (8.2.3) and (8.2.4) to the tangent unit tangent vector **t** presented by the formulas (8.2.1) we find:

$$\mathbf{t} = \frac{d\mathbf{r}}{ds} = t^i \mathbf{g}_i = t^\alpha \mathbf{a}_\alpha, \quad t^i = \frac{dy^i}{ds}, \quad t^\alpha = \frac{du^\alpha}{ds}, \quad t^i = t^\alpha \frac{\partial y^i}{\partial u^\alpha}, \quad t_\alpha = \underset{\sim}{t}_i \frac{\partial y^i}{\partial u^\alpha}$$
$$\tag{8.2.5}$$

Example 8.3 Surface Vector c as a Projection of a Space Vector b onto a Surface

A space vector field **b** with components b_i in a Cartesian coordinate system Ox, i.e.$\mathbf{b} = b_i \mathbf{e}_i$, is projected as surface vector **c** onto the circular cylindrical surface presented in Example 8.1. Figure 8.5 indicates the surface and shows the components b_1 and b_2 of the space vector. The base vectors and the reciprocal base vectors for the surface coordinate system: $u^1 = \theta$ and $u^2 = z = x_3$, are obtained from the formulas (8.1.23) and (8.1.24): $\mathbf{a}_1 = R\mathbf{e}_\theta$, $\mathbf{a}^1 = \mathbf{e}_\theta/R$, $\mathbf{a}^1 = \mathbf{e}_\theta/R$, $\mathbf{a}_2 = \mathbf{a}^2 = \mathbf{e}_z$. The surface vector field **c** as a projection of the space vector **b** is now given by:

$$\mathbf{c} = c_1 \mathbf{a}^1 + c_2 \mathbf{a}^2 = c^1 \mathbf{a}_1 + c^2 \mathbf{a}_2 = c_\theta \mathbf{e}_\theta + c_z \mathbf{e}_z \Rightarrow c_\theta = c_1/R = Rc^1, c_z = c_2 = c^2$$

Fig. 8.5 Projection $\mathbf{c} = c_\alpha \mathbf{a}^\alpha$ of a space vector $\mathbf{b} = b_i \mathbf{e}_i$ onto a circular cylindrical surface of radius R. Space vector components $\underset{\sim}{c}_i$ of **c** in a Cartesian coordinate system Ox

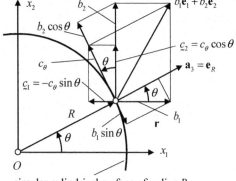

circular cylindrical surface of radius R

The physical components c_θ and c_z will be computed below, Fig. 8.5 shows the component c_θ. From the coordinate transformation formula (8.1.22) we obtain the transformation matrix:

$$\left(\frac{\partial x_i}{\partial u^\alpha}\right) = \begin{pmatrix} -R\sin\theta & 0 \\ R\cos\theta & 0 \\ 0 & 1 \end{pmatrix}$$

The formulas (8.2.4) yield:

$$c_\alpha = b_i \frac{\partial x_i}{\partial u^\alpha} \quad \Rightarrow \quad c_1 = b_1 \frac{\partial x_1}{\partial u^1} + b_2 \frac{\partial x_2}{\partial u^1} + b_3 \frac{\partial x_3}{\partial u^1} = -b_1 R\sin\theta + b_2 R\cos\theta$$

$$\Rightarrow \quad c_\theta = b_2\cos\theta - b_1\sin\theta$$

$$c_2 = c_z = b_1 \frac{\partial x_1}{\partial u^2} + b_2 \frac{\partial x_2}{\partial u^2} + b_3 \frac{\partial x_3}{\partial u^2} = b_3$$

The result for c_θ is illustrated in Fig. 8.5.

Next we shall compute the space vector components $\c_i$ of a surface vector $\mathbf{c}$ in the Cartesian coordinate system Ox: $\mathbf{c} = \c_i\mathbf{e}_i$. First we find for the components c^α in the surface coordinate system u:

$$\mathbf{c} = c^\alpha\mathbf{a}_\alpha = c^1\mathbf{a}_1 + c^2\mathbf{a}_2 = c^1 R\mathbf{e}_\theta + c^2\mathbf{e}_z = c_\theta\mathbf{e}_\theta + c_z\mathbf{e}_z$$

$$\Rightarrow c^1 = c_\theta/R, c^2 = c_z$$

The formulas (8.2.3) yield:

$$c^i \equiv c_i = c^\alpha \frac{\partial x_i}{\partial u^\alpha} \Rightarrow c_1 = c^1 \frac{\partial x_1}{\partial u^1} + c^2 \frac{\partial x_1}{\partial u^2} = \frac{c_\theta}{R}(-R\sin\theta) = -c_\theta\sin\theta,$$

$$c_2 = c^1 \frac{\partial x_2}{\partial u^1} + c^2 \frac{\partial x_2}{\partial u^2} = \frac{c_\theta}{R}(R\cos\theta) = c_\theta\cos\theta, \quad c_3 = c^1 \frac{\partial x_3}{\partial u^1} + c^2 \frac{\partial x_3}{\partial u^2} = c^2 = c_z = b_3$$

The results for the Cartesian components $\c_1$ and $\c_2$ are illustrated in Fig. 8.5.

8.2.1 Scalar Product and Vector Product of Surface Vectors

The scalar product of two surface vectors $\mathbf{b}$ and $\mathbf{c}$ is expressed by:

$$\mathbf{b} \cdot \mathbf{c} = b^\alpha c_\alpha = b_\alpha c^\alpha \tag{8.2.6}$$

The vector product of two surface vectors $\mathbf{b}$ and $\mathbf{c}$ results in a vector normal to the surface. Using the formulas (8.1.13), we obtain the result:

$$\mathbf{b} \times \mathbf{c} = \varepsilon_{\alpha\beta} b^\alpha c^\beta \mathbf{a}_3 = \varepsilon^{\alpha\beta} b_\alpha c_\beta \mathbf{a}_3 \tag{8.2.7}$$

We introduce the concept of the *rotation* (**b**, **c**) *of two surface vectors* **b** *and* **c**:

The *rotation* (**b**, **c**) *of two surface vectors* **b** *and* **c** is positive/negative if the vector product $\mathbf{b} \times \mathbf{c}$ is directed in the direction of/the opposite direction of the surface normal vector $\mathbf{a}_3$. In other words:

$$\begin{aligned} & \text{The rotation } (\mathbf{b}, \mathbf{c}) \text{ is positive if: } (\mathbf{b} \times \mathbf{c}) \cdot \mathbf{a}_3 = \varepsilon_{\alpha\beta} b^\alpha c^\beta > 0. \\ & \text{The rotation } (\mathbf{b}, \mathbf{c}) \text{ is negative if: } (\mathbf{b} \times \mathbf{c}) \cdot \mathbf{a}_3 = \varepsilon_{\alpha\beta} b^\alpha c^\beta < 0. \end{aligned} \tag{8.2.8}$$

8.3   Coordinate Transformations

A new $\bar{u}$-system of surfaces coordinates on the surface is now introduced, such that there is a one-to-one correspondence in the functional relationship between the sets of coordinates u and $\bar{u}$:

$$\bar{u}^\alpha = \bar{u}^\alpha(u^1, u^2) \Leftrightarrow u^\beta = u^\beta(\bar{u}^1, \bar{u}^2) \tag{8.3.1}$$

The *Jacobian* J^α_β to the mapping (8.3.1) must be non-zero:

$$J^\alpha_\beta \equiv \det\left(\frac{\partial \bar{u}^\alpha}{\partial u^\beta}\right) \neq 0 \tag{8.3.2}$$

Confer the condition (6.2.3) in the case of transformation of general coordinates in E_3.

The base vectors $\bar{\mathbf{a}}_\alpha$ and the reciprocal base vectors $\bar{\mathbf{a}}^\alpha$ in $\bar{u}$-system are defined by

$$\bar{\mathbf{a}}_\alpha = \frac{\partial \mathbf{r}}{\partial \bar{u}^\alpha} = \mathbf{g}_i \frac{\partial y^i}{\partial \bar{u}^\alpha}, \quad \bar{\mathbf{a}}^\alpha \cdot \bar{\mathbf{a}}_i = \delta^\alpha_i, \quad \bar{\mathbf{a}}_3 \equiv \mathbf{a}_3 \tag{8.3.3}$$

The fundamental parameters and the reciprocal fundamental parameters of the first order in the $\bar{u}$-system are:

$$\bar{a}_{\alpha\beta} = \bar{a}_{\beta\alpha} = \bar{\mathbf{a}}_\alpha \cdot \bar{\mathbf{a}}_\beta = g_{ij} \frac{\partial y^i}{\partial \bar{u}^\alpha} \frac{\partial y^j}{\partial \bar{u}^\beta}, \quad \bar{a}^{\alpha\beta} = \bar{a}^{\beta\alpha} = \bar{\mathbf{a}}^\alpha \cdot \bar{\mathbf{a}}^\beta \tag{8.3.4}$$

From the definitions (8.3.3) and (8.1.2) of the base vectors $\bar{\mathbf{a}}_\alpha$ and $\mathbf{a}_\beta$ we obtain the relations:

$$\bar{\mathbf{a}}_\alpha = \frac{\partial \mathbf{r}}{\partial \bar{u}^\alpha} = \frac{\partial \mathbf{r}}{\partial u^\beta}\frac{\partial u^\beta}{\partial \bar{u}^\alpha} = \mathbf{a}_\beta \frac{\partial u^\beta}{\partial \bar{u}^\alpha} \quad \Rightarrow$$

$$\bar{\mathbf{a}}_\alpha = \frac{\partial u^\beta}{\partial \bar{u}^\alpha}\mathbf{a}_\beta \Leftrightarrow \mathbf{a}_\beta = \frac{\partial \bar{u}^\alpha}{\partial u^\beta}\bar{\mathbf{a}}_\alpha \tag{8.3.5}$$

The transformation (8.3.5) of base vectors is called a *covariant transformation* and the base vectors are alternatively called *covariant base vectors*.

A surface vector $\mathbf{c}$ has *contravariant components* $\bar{c}^\alpha$ in the $\bar{u}$-system and c^β in the u-system: $\mathbf{c} = \bar{c}^\alpha \bar{\mathbf{a}}_\alpha = c^\beta \mathbf{a}_\beta$. We find that:

$$\mathbf{c} = \bar{c}^\alpha \bar{\mathbf{a}}_\alpha = c^\beta \mathbf{a}_\beta = c^\beta \left(\frac{\partial \bar{u}^\alpha}{\partial u^\beta}\bar{\mathbf{a}}_\alpha\right) = \frac{\partial \bar{u}^\alpha}{\partial u^\beta}c^\beta \bar{\mathbf{a}}_\alpha \Rightarrow$$

$$\bar{c}^\alpha = \frac{\partial \bar{u}^\alpha}{\partial u^\beta}c^\beta, \quad c^\beta = \frac{\partial u^\beta}{\partial \bar{u}^\alpha}\mathbf{a}_\alpha \tag{8.3.6}$$

The transformation (8.3.6) of contravariant vector components is called a *contravariant transformation*, compare with the formulas (6.3.3) for the three-dimensional case.

For any surface vector $\mathbf{c}$ we may state the identity: $\mathbf{c} = (\mathbf{c} \cdot \mathbf{a}_\beta)\mathbf{a}^\beta$. We use this identity, the result (8.3.5), and the definition (8.3.3)$_2$ of the reciprocal base vectors $\bar{\mathbf{a}}^\alpha$ to write:

$$\bar{\mathbf{a}}^\alpha = (\bar{\mathbf{a}}^\alpha \cdot \mathbf{a}_\beta)\mathbf{a}^\beta = \left(\bar{\mathbf{a}}^\alpha \cdot \bar{\mathbf{a}}_\gamma \frac{\partial \bar{u}^\gamma}{\partial u^\beta}\right)\mathbf{a}^\beta = \delta_\gamma^\alpha \frac{\partial \bar{u}^\gamma}{\partial u^\beta}\mathbf{a}^\beta = \frac{\partial \bar{u}^\alpha}{\partial u^\beta}\mathbf{a}^\beta \Rightarrow$$

$$\bar{\mathbf{a}}^\alpha = \frac{\partial \bar{u}^\alpha}{\partial u^\beta}\mathbf{a}^\beta \Leftrightarrow \mathbf{a}^\beta = \frac{\partial u^\beta}{\partial \bar{u}^\alpha}\bar{\mathbf{a}}^\alpha \tag{8.3.7}$$

Due to the transformation rule (8.3.7) the reciprocal base vectors are also called *contravariant base vectors*

The fundamental parameter of the first order and the reciprocal fundamental parameter of the first order in the two coordinate systems are related through the formulas:

$$\bar{a}_{\alpha\beta} = \bar{a}_{\beta\alpha} = \bar{\mathbf{a}}_\alpha \cdot \bar{\mathbf{a}}_\beta = \frac{\partial u^\gamma}{\partial \bar{u}^\alpha}\frac{\partial u^\rho}{\partial \bar{u}^\beta}a_{\gamma\rho}, \quad \bar{a}^{\alpha\beta} = \bar{a}^{\beta\alpha} = \bar{\mathbf{a}}^\alpha \cdot \bar{\mathbf{a}}^\beta = \frac{\partial \bar{u}^\alpha}{\partial u^\gamma}\frac{\partial \bar{u}^\beta}{\partial u^\rho}a^{\gamma\rho} \tag{8.3.8}$$

We say that the fundamental parameters of the first order follow a *covariant transformation rule* while the reciprocal fundamental parameters of the first order follow a *contravariant transformation rule*.

Christoffel symbols of the first kind and the second kind and fundamental parameters of the second order in $\bar{u}$-system are:

$$\bar{\Gamma}_{\alpha\beta\gamma} = \bar{\mathbf{a}}_{\alpha,\beta} \cdot \bar{\mathbf{a}}_\gamma, \quad \bar{\Gamma}^\gamma_{\alpha\beta} = \bar{\Gamma}^\gamma_{\beta\alpha} = \bar{\mathbf{a}}_{\alpha,\beta} \cdot \bar{\mathbf{a}}^\gamma, \quad \bar{B}_{\alpha\beta} = \bar{B}_{\beta\alpha} = \bar{\mathbf{a}}_{\alpha,\beta} \cdot \mathbf{a}_3 \qquad (8.3.9)$$

In order to develop the relations between Christoffel symbols $\bar{\Gamma}^\gamma_{\alpha\beta} = \bar{\mathbf{a}}_{\alpha,\,beta} \cdot \bar{\mathbf{a}}^\gamma$ for the $\bar{u}$-system and Christoffel symbols $\Gamma^\theta_{\lambda\rho} = \mathbf{a}_{\lambda,\rho} \cdot \mathbf{a}^\theta$ for the u-system, we first obtain from the formulas $(8.3.5)_1$:

$$\begin{aligned}
\bar{\mathbf{a}}_{\alpha,\beta} &\equiv \frac{\partial \bar{\mathbf{a}}_\alpha}{\partial \bar{u}^\beta} = \frac{\partial}{\partial \bar{u}^\beta}\left(\frac{\partial u^\lambda}{\partial \bar{u}^\alpha}\mathbf{a}_\lambda\right) = \frac{\partial^2 u^\lambda}{\partial \bar{u}^\beta \partial \bar{u}^\alpha}\mathbf{a}_\lambda \\
&+ \frac{\partial u^\lambda}{\partial \bar{u}^\alpha}\frac{\partial \mathbf{a}_\lambda}{\partial u_\rho}\frac{\partial u^\rho}{\partial \bar{u}^\beta} = \frac{\partial^2 u^\lambda}{\partial \bar{u}^\beta \partial \bar{u}^\alpha}\mathbf{a}_\lambda + \frac{\partial u^\lambda}{\partial \bar{u}^\alpha}\frac{\partial u^\rho}{\partial \bar{u}^\beta}\mathbf{a}_{\lambda,\rho} \Rightarrow \\
\bar{\mathbf{a}}_{\alpha,\beta} &= \frac{\partial^2 u^\lambda}{\partial \bar{u}^\beta \partial \bar{u}^\alpha}\mathbf{a}_\lambda + \frac{\partial u^\lambda}{\partial \bar{u}^\alpha}\frac{\partial u^\rho}{\partial \bar{u}^\beta}\mathbf{a}_{\lambda,\rho}
\end{aligned} \qquad (8.3.10)$$

From the formula $(8.3.9)_2$ we obtain, using the formulas $(8.3.10)$, $(8.3.7)_1$, and $(8.1.16)_1$ the result:

$$\begin{aligned}
\bar{\Gamma}^\gamma_{\alpha\beta} = \bar{\mathbf{a}}_{\alpha,\beta} \cdot \bar{\mathbf{a}}^\gamma &= \frac{\partial^2 u^\lambda}{\partial \bar{u}^\beta \partial \bar{u}^\alpha}\mathbf{a}_\lambda \cdot \left(\frac{\partial \bar{u}^\gamma}{\partial u^\theta}\mathbf{a}^\theta\right) + \frac{\partial u^\lambda}{\partial \bar{u}^\alpha}\frac{\partial u^\rho}{\partial \bar{u}^\beta}\mathbf{a}_{\lambda,\rho} \cdot \mathbf{a}^\theta \frac{\partial \bar{u}^\gamma}{\partial u^\theta} \\
&= \frac{\partial^2 u^\lambda}{\partial \bar{u}^\beta \partial \bar{u}^\alpha}\frac{\partial \bar{u}^\gamma}{\partial u^\lambda} + \frac{\partial u^\lambda}{\partial \bar{u}^\alpha}\frac{\partial u^\rho}{\partial \bar{u}^\beta}\frac{\partial \bar{u}^\gamma}{\partial u^\theta}\Gamma^\theta_{\lambda\rho} \Rightarrow \\
\bar{\Gamma}^\gamma_{\alpha\beta} &= \frac{\partial^2 u^\lambda}{\partial \bar{u}^\alpha \partial \bar{u}^\beta}\frac{\partial \bar{u}^\gamma}{\partial u^\lambda} + \frac{\partial u^\lambda}{\partial \bar{u}^\alpha}\frac{\partial u^\rho}{\partial \bar{u}^\beta}\frac{\partial \bar{u}^\gamma}{\partial u^\theta}\Gamma^\theta_{\lambda\rho}
\end{aligned} \qquad (8.3.11)$$

8.3.1 Geodesic Coordinate System on a Surface

It is not possible in general to find a surface coordinate system for which the Christoffel symbols are zero everywhere. However, as we now shall demonstrate, it is always possible to construct a surface coordinate system such that the Christoffel symbols are all zero in one, arbitrarily selected surface point. Such a coordinate system is called a *geodesic coordinate system* for the point, and the special selected point is called the *pole* for the geodesic coordinate system.

Let the u-system be defined such that $u^\alpha = 0$ in an arbitrarily selected point P in the surface. In order for the $\bar{u}$-system to be geodesic for the point P, the following condition must be fulfilled:

$$\bar{\Gamma}^\gamma_{\alpha\beta} = 0 \text{ in } P \qquad (8.3.12)$$

The following coordinate system $\bar{u}$ satisfies this condition:

$$\bar{u}^{\alpha} = u^{\alpha} + \frac{1}{2}\left(\Gamma^{\alpha}_{\sigma\tau}\right)\big|_{P} u^{\sigma} u^{\tau} \tag{8.3.13}$$

It follows that:

$$\frac{\partial \bar{u}^{\gamma}}{\partial \bar{u}^{\alpha}} = \delta^{\gamma}_{\alpha}, \quad \frac{\partial u^{\gamma}}{\partial \bar{u}^{\alpha}} = \frac{\partial \bar{u}^{\gamma}}{\partial \bar{u}^{\alpha}} - \frac{1}{2}\left(\Gamma^{\gamma}_{\sigma\tau}\right)\big|_{P}\left[\frac{\partial u^{\sigma}}{\partial \bar{u}^{\alpha}} u^{\tau} + u^{\sigma} \frac{\partial u^{\tau}}{\partial \bar{u}^{\alpha}}\right] \Rightarrow$$

$$\frac{\partial^{2} u^{\gamma}}{\partial \bar{u}^{\beta} \partial \bar{u}^{\alpha}} = -\frac{1}{2}\left(\Gamma^{\gamma}_{\sigma\tau}\right)\big|_{P}\left[\frac{\partial^{2} u^{\sigma}}{\partial \bar{u}^{\beta} \partial \bar{u}^{\alpha}} u^{\tau} + \frac{\partial u^{\sigma}}{\partial \bar{u}^{\alpha}} \frac{\partial u^{\tau}}{\partial \bar{u}^{\beta}} + \frac{\partial u^{\sigma}}{\partial \bar{u}^{\beta}} \frac{\partial u^{\tau}}{\partial \bar{u}^{\alpha}} + u^{\sigma} \frac{\partial^{2} u^{\tau}}{\partial \bar{u}^{\beta} \partial \bar{u}^{\alpha}}\right]$$

Because $u^{\alpha} = 0$ in the pole P and $\Gamma^{\gamma}_{\beta\alpha} = \Gamma^{\gamma}_{\alpha\beta}$, we obtain from these equations the results:

$$\frac{\partial u^{\gamma}}{\partial \bar{u}^{\alpha}} = \frac{\partial \bar{u}^{\gamma}}{\partial u^{\alpha}} = \delta^{\gamma}_{\alpha} \text{ in } P, \quad \frac{\partial^{2} u^{\gamma}}{\partial \bar{u}^{\beta} \partial \bar{u}^{\alpha}} = -\left(\Gamma^{\gamma}_{\alpha\beta}\right)\big|_{P} \text{ in } P \tag{8.3.14}$$

When the results (8.3.14) are substituted into the formula (8.3.11), we obtain the result (8.3.12) for the coordinate system (8.3.13). The coordinate system defined by the formulas (8.3.13) is not unique to obtain the condition (8.3.12). Note that the results (8.3.14)$_1$ imply that the base vectors for the $\bar{u}$-system and the u-system, related by the formulas (8.3.5) and (8.3.7), coincide in the point P.

Because the Christoffel symbols are zero for Cartesian coordinate systems in E_3, we say that a Cartesian coordinate system is geodesic in every point in the three-dimensional Euclidean space E_3.

Let the $\bar{u}$-system be geodesic on a surface for a point P, and the u-system uniquely defined by a coordinate transformation:

$$\bar{u}^{\alpha}(u) \Leftrightarrow u^{\alpha}(\bar{u}) \tag{8.3.15}$$

It then follows from the formulas (8.3.11) that the Christoffel symbols in the point P for the u-system may be derived from the formula:

$$\Gamma^{\gamma}_{\alpha\beta} = \frac{\partial^{2} \bar{u}^{\lambda}}{\partial u^{\alpha} \partial u^{\beta}} \frac{\partial u^{\gamma}}{\partial \bar{u}^{\lambda}} \quad \text{in } P \tag{8.3.16}$$

This formula is analogous to the formula (6.5.2)$_2$ in three-dimensional space E_3.

8.4 Surface Tensors

A *surface tensor of order n* is defined as a coordinate invariant quantity that represents a multilinear scalar-valued function of n surface vectors. Let for example **D** be a surface tensor of second order and $\mathbf{a}_{\alpha}$ and $\mathbf{a}^{\alpha}$ the base vectors for a coordinate system u on a surface in R_2. Then for any two surface vectors **b** and **c**:

$$\mathbf{b} = b^\alpha \mathbf{a}_\alpha = b_\alpha \mathbf{a}^\alpha, \quad \mathbf{c} = c^\alpha \mathbf{a}_\alpha = c_\alpha \mathbf{a}^\alpha$$

the tensor $\mathbf{D}$ represents a scalar given by the bilinear function:

$$\alpha = \mathbf{D}[\mathbf{b}, \mathbf{c}] = \mathbf{D}\left[b^\alpha \mathbf{a}_\alpha, c^\beta \mathbf{a}_\beta\right] = \mathbf{D}\left[\mathbf{a}_\alpha, \mathbf{a}_\beta\right] b^\alpha c^\beta = \mathbf{D}\left[b_\alpha \mathbf{a}^\alpha, c_\beta \mathbf{a}^\beta\right] = \mathbf{D}\left[\mathbf{a}^\alpha, \mathbf{a}^\beta\right] b_\alpha c_\beta$$
$$\alpha = \mathbf{D}[\mathbf{b}, \mathbf{c}] = \mathbf{D}\left[b^\alpha \mathbf{a}_\alpha, c_\beta \mathbf{a}^\beta\right] = \mathbf{D}\left[\mathbf{a}_\alpha, \mathbf{a}^\beta\right] b^\alpha c_\beta = \mathbf{D}\left[b_\alpha \mathbf{a}^\alpha, c^\beta \mathbf{a}_\beta\right] = \mathbf{D}\left[\mathbf{a}^\alpha, \mathbf{a}_\beta\right] b_\alpha c^\beta$$
$$\tag{8.4.1}$$

Thus the tensor is represented by any of the following four sets of components in the coordinate system u on the surface:

$$\begin{aligned}
D_{\alpha\beta} &= \mathbf{D}\left[\mathbf{a}_\alpha, \mathbf{a}_\beta\right] \quad \text{covariant components} \\
D^{\alpha\beta} &= \mathbf{D}\left[\mathbf{a}^\alpha, \mathbf{a}^\beta\right] \quad \text{contravariant components} \\
D_\alpha^\beta &= \mathbf{D}\left[\mathbf{a}_\alpha, \mathbf{a}^\beta\right], \quad D_\beta^\alpha = \mathbf{D}\left[\mathbf{a}^\alpha, \mathbf{a}_\beta\right] \quad \text{mixed components}
\end{aligned} \tag{8.4.2}$$

The relations (8.4.1) may now be expressed as:

$$\alpha = \mathbf{D}[\mathbf{b}, \mathbf{c}] = D_{\alpha\beta} b^\alpha c^\beta = D^{\alpha\beta} b_\alpha c_\beta = D_\alpha^\beta b^\alpha c_\beta = D_\beta^\alpha b_\alpha c^\beta \tag{8.4.3}$$

Using the relations (8.1.6) for the base vectors, we get the relationships:

$$D_\alpha^\beta = \mathbf{D}\left[\mathbf{a}_\alpha, \mathbf{a}^\beta\right] = \mathbf{D}\left[\mathbf{a}_\alpha, a^{\beta\gamma} \mathbf{a}_\gamma\right] = a^{\beta\gamma} \mathbf{D}\left[\mathbf{a}_\alpha, \mathbf{a}_\gamma\right] = a^{\beta\gamma} D_{\alpha\gamma} \quad \text{etc.} \tag{8.4.4}$$

The algebra for surface tensors is analogous to the algebra for space tensors. The transformation rules for the components of the tensor $\mathbf{D}$ when transforming from a surface coordinate system u to another surface coordinate system $\bar{u}$, are found as follows: We apply the formulas (8.3.5) and (8.3.7) and obtain:

$$\bar{D}_\alpha^\beta = \mathbf{D}\left[\bar{\mathbf{a}}_\alpha, \bar{\mathbf{a}}^\beta\right] = \mathbf{D}\left[\frac{\partial u^\gamma}{\partial \bar{u}^\alpha} \mathbf{a}_\gamma, \frac{\partial \bar{u}^\beta}{\partial u^\rho} \mathbf{a}^\rho\right] = \frac{\partial u^\gamma}{\partial \bar{u}^\alpha} \frac{\partial \bar{u}^\beta}{\partial u^\rho} D_\gamma^\rho \quad \text{etc.} \tag{8.4.5}$$

It may be shown, Problem 8.4, that the permutation symbols $\varepsilon_{\alpha\beta}$ and $\varepsilon^{\alpha\beta}$ defined by the formulas (8.1.11) represent components of a surface tensor: the *permutation tensor for surfaces*.

8.4.1   Symmetric Surface Tensors of Second Order

A *symmetric second order surface tensor* $\mathbf{S}$ satisfies the requirement:

$$\mathbf{S}[\mathbf{b}, \mathbf{c}] = \mathbf{S}[\mathbf{c}, \mathbf{b}] \quad \Rightarrow \quad \mathbf{S}\left[\mathbf{a}_\alpha, \mathbf{a}_\beta\right] = \mathbf{S}\left[\mathbf{a}_\beta, \mathbf{a}_\alpha\right] \quad \Rightarrow \quad S_{\alpha\beta} = S_{\beta\alpha} \quad \text{etc.} \tag{8.4.6}$$

Fig. 8.6 u^1-line and u^2-line
on a surface in E_3. Tangent
plane on the surface. Vector **s**
for the direction **b**, normal
component σ for the direction
b, and shear component τ for
the directions **b** and **c**

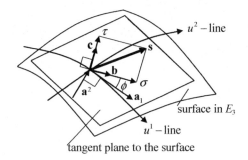

Let **b** and **c** be two orthogonal unit surface vectors at a point P on a surface $y^i(u)$.
For a second order symmetric tensor **S** we define a *vector* **s** *for the direction* **b**, a
normal component σ for the direction **b**, and a *shear component* τ for the directions
b and **c** (Fig. 8.6):

$$\mathbf{s} = \mathbf{Sb} \equiv \mathbf{S} \cdot \mathbf{b} \Leftrightarrow s^\alpha = S_\beta^\alpha b^\beta$$
$$\sigma = \mathbf{b} \cdot \mathbf{s} = \mathbf{b} \cdot \mathbf{S} \cdot \mathbf{b} = b_\alpha S^{\alpha\beta} b_\beta \qquad (8.4.7)$$
$$\tau = \mathbf{c} \cdot \mathbf{s} = \mathbf{c} \cdot \mathbf{S} \cdot \mathbf{b} = c_\alpha S^{\alpha\beta} b_\beta$$

The principal values $\sigma = \sigma_1$ and σ_2, and the principal directions $\mathbf{b} = \mathbf{b}_1$ and $\mathbf{b}_2$,
are defined by the by the equations:

$$\sigma\mathbf{b} = \mathbf{S} \cdot \mathbf{b} \Rightarrow \sigma b_\beta = S_\beta^\alpha b_\alpha \Rightarrow$$

$$\left(\sigma\delta_\beta^\alpha - S_\beta^\alpha\right) b_\alpha = 0 \qquad (8.4.8)$$

The solution of this set of equations requires that:

$$\det\left(\sigma\delta_\beta^\alpha - S_\beta^\alpha\right) = 0 \quad \Rightarrow$$

$$\sigma^2 - S_\alpha^\alpha \sigma + \det S = 0 \Leftrightarrow \sigma^2 - (\operatorname{tr}\mathbf{S})\sigma + \det\mathbf{S} = 0 \qquad (8.4.9)$$

The *trace* $\operatorname{tr}\mathbf{S}$ and the *determinant* $\det\mathbf{S}$ of the surface tensor **S** are the two
principal invariants of the surface tensor **S**:

$$\text{trace of } \mathbf{S} : \operatorname{tr}\mathbf{S} = \operatorname{tr} S = \operatorname{tr}\left(S_\beta^\alpha\right) = S_\alpha^\alpha = S_1^1 + S_2^2 \qquad (8.4.10)$$

$$\text{determinant of } \mathbf{S} : \quad \det\mathbf{S} = \det S = \det\left(S_\beta^\alpha\right) = S_1^1 S_2^2 - S_2^1 S_1^2 \qquad (8.4.11)$$

Equation (8.4.9) is the *characteristic equation* of the surface tensor **S** and determines the *principal values* σ_1 *and* σ_2 *of the surface tensor* **S**:

$$\begin{matrix} \sigma_1 \\ \sigma_2 \end{matrix} = \frac{1}{2}\operatorname{tr}\mathbf{S} \pm \sqrt{\frac{1}{4}(\operatorname{tr}\mathbf{S})^2 - \det\mathbf{S}} = \frac{1}{2}(S_1^1 + S_2^2) \pm \sqrt{\frac{1}{4}(S_1^1 - S_2^2)^2 + S_2^1 S_1^2}$$

$$(8.4.12)$$

Due to symmetry $S_1^2 = S_2^1$, which implies that the principal values are real. The directions $\mathbf{b}_1$ and $\mathbf{b}_2$ corresponding to the principal values σ_1 and σ_2 are called the *principal directions of the surface tensor* $\mathbf{S}$ and are determined from the linear equation (8.4.8). The development from the formula (8.4.8) to the formula (8.4.12) is analogous to the development from the formula $(2.3.6)_1$ to the formula (2.3.9) in Sect. 2.3.1. Based on the discussion in Sect. 2.3.1 we may conclude that if $\sigma_1 \neq \sigma_2$, the principal directions are orthogonal, and if $\sigma_1 = \sigma_2$, any direction on the surface is a principal direction for the surface tensor $\mathbf{S}$.

The angle ϕ between a principal direction $\mathbf{b}$, corresponding to the principal value σ, and the base vector $\mathbf{a}_1$ is given by the formula, see Problem 8.5:

$$\tan\phi = \frac{\sigma - S_1^1}{S_1^2}\frac{a_{11}}{\sqrt{\alpha}} - \frac{a_{12}}{\sqrt{\alpha}} \qquad (8.4.13)$$

8.4.2   Space-Surface Tensors

Let $\mathbf{E}$ be a space tensor of second order. The covariant components of the tensor $\mathbf{E}$ in a general space coordinate system y are:

$$E_{ij} = \mathbf{E}[\mathbf{g}_i, \mathbf{g}_j] \qquad (8.4.14)$$

The components $D_{\alpha\beta} = \mathbf{E}(\mathbf{a}_\alpha, \mathbf{a}_\beta)$ represent a surface tensor $\mathbf{D}$ of second order. The relationship between the space components E_{ij} and the surface components $D_{\alpha\beta}$ are found by application of the formulas (8.1.2):

$$D_{\alpha\beta} = \mathbf{E}[\mathbf{a}_\alpha, \mathbf{a}_\beta] = \mathbf{E}\left[\mathbf{g}_i\frac{\partial y^i}{\partial u^\alpha}, \mathbf{g}_j\frac{\partial y^j}{\partial u^\beta}\right] = \frac{\partial y^i}{\partial u^\alpha}\frac{\partial y^j}{\partial u^\beta}E_{ij} \qquad (8.4.15)$$

The fundamental parameters $a_{\alpha\beta}$ and $a^{\alpha\beta}$ represent components in the u-system of the *unit tensor of second order* $\mathbf{1}$. The following relations apply:

$$\mathbf{1}[\mathbf{a}_\alpha, \mathbf{a}_\beta] = \frac{\partial y^i}{\partial u^\alpha}\frac{\partial y^j}{\partial u^\beta}\mathbf{1}[\mathbf{g}_i, \mathbf{g}_j] = a_{\alpha\beta}, \quad \mathbf{1}[\mathbf{a}^\alpha, \mathbf{a}^\beta] = a^{\alpha\gamma}a^{\beta\lambda}\mathbf{1}[\mathbf{a}_\gamma, \mathbf{a}_\lambda] = a^{\alpha\beta}$$

$$\mathbf{1}[\mathbf{a}^\alpha, \mathbf{a}_\beta] = \mathbf{1}[\mathbf{a}_\beta, \mathbf{a}^\alpha] = a^{\alpha\gamma}\mathbf{1}[\mathbf{a}_\gamma, \mathbf{a}_\beta] = a^{\alpha\gamma}a_{\gamma\beta} = \delta_\beta^\alpha$$

$$(8.4.16)$$

Note that the formula $(8.4.16)_1$ is in agreement with the formula $(8.1.7)_1$

A *space-surface tensor* of order n,m is an invariant quantity representing a multilinear scalar-valued function of n space vectors and m surface vectors. As an example let $\mathbf{F}$ be space-surface tensor of order 1,1. The tensor has then the following component sets with respect to the coordinate systems y in E_3 and u in R_2 :

$$F_{i\alpha} = \mathbf{F}[\mathbf{g}_i, \mathbf{a}_\alpha] = \mathbf{g}_i \cdot \mathbf{F} \cdot \mathbf{a}_\alpha, \quad F_i{}^\alpha = \mathbf{F}[\mathbf{g}_i, \mathbf{a}^\alpha] = a^{\alpha\beta} F_{i\beta}$$
$$F^i_\alpha = \mathbf{F}\left[\mathbf{g}^i, \mathbf{a}_\alpha\right] = g^{ij} F_{j\alpha}, \quad F^{i\alpha} = \mathbf{F}\left[\mathbf{g}^i, \mathbf{a}^\alpha\right] = g^{ij} a^{\alpha\beta} F_{j\beta} \tag{8.4.17}$$

Let $\mathbf{c}$ be a surface vector and $\mathbf{d}$ a space vector. Then:

$$\mathbf{h} = \mathbf{F} \cdot \mathbf{c} \text{ with components } h_i = F_{i\alpha} c^\alpha \text{ is a space vector}$$
$$\mathbf{k} = \mathbf{b} \cdot \mathbf{F} \text{ with components } k_\alpha = b^i F_{i\alpha} \text{ is a surface vector} \tag{8.4.18}$$

The components $\partial y^i / \partial u^\alpha$ represent a space-surface tensor of order 1,1 with the property of projecting a space vector $\mathbf{b}$ onto the tangent plane to a surface, as shown by the relationships (8.2.4).

8.5 Differentiation of Surface Tensors

Let $\mathbf{c}(u)$ be a surface vector field on the surface $y^i(u)$:

$$\mathbf{c}(u) = c^\alpha(u)\mathbf{a}_\alpha(u) = c_\alpha(u)\mathbf{a}^\alpha(u) \tag{8.5.1}$$

We compute the vectors $\mathbf{c}_{,\beta}$ and use the formulas (8.1.15) and (8.1.18)$_2$:

$$\mathbf{c}_{,\beta} = c^\alpha{}_{,\beta}\,\mathbf{a}_\alpha + c^\alpha \mathbf{a}_{\alpha,\beta} = c_{\alpha,\beta}\,\mathbf{a}^\alpha + c_\alpha \mathbf{a}^\alpha{}_{,\beta}$$
$$= c^\alpha{}_{,\beta}\,\mathbf{a}_\alpha + c^\alpha \left(\Gamma^\gamma_{\alpha\beta}\mathbf{a}_\gamma + B_{\alpha\beta}\mathbf{a}_3 \right) = c_{\alpha,\beta}\,\mathbf{a}^\alpha + c_\alpha \left(-\Gamma^\alpha_{\gamma\beta}\,\mathbf{a}^\gamma + B_{\gamma\beta}\,a^{\gamma\alpha}\,\mathbf{a}_3 \right) \Rightarrow$$

$$\mathbf{c}_{,\beta} = \left(c^\alpha{}_{,\beta} + c^\gamma\Gamma^\alpha_{\gamma\beta} \right)\mathbf{a}_\alpha + c^\alpha B_{\alpha\beta}\mathbf{a}_3 = \left(c_{\alpha,\beta} - c_\gamma\Gamma^\gamma_{\alpha\beta} \right)\mathbf{a}^\alpha + c_\alpha B^\alpha_\beta\mathbf{a}_3 \tag{8.5.2}$$

Here we define the *covariant derivatives of the vector components*:

$$c^\alpha|_\beta = c^\alpha{}_{,\beta} + c^\gamma\Gamma^\alpha_{\gamma\beta}, \quad c_\alpha|_\beta = c_{\alpha,\beta} - c_\gamma\Gamma^\gamma_{\alpha\beta} \tag{8.5.3}$$

From the results (8.5.2) and the definitions (8.5.3) we obtain:

$$\mathbf{c}_{,\beta} = c^\alpha|_\beta\mathbf{a}_\alpha + c^\alpha B_{\alpha\beta}\mathbf{a}_3 = c_\alpha|_\beta\mathbf{a}^\alpha + c_\alpha B^\alpha_\beta\mathbf{a}_3 \tag{8.5.4}$$

The tangent vector $\mathbf{t}$ to the curve $u^\alpha(s)$ on the surface $y^i(u)$ is a surface vector determined according to formula the formulas (8.2.5) by the components t^α in the u-system and by the components $\underset{\sim}{t}{}^{\,k}$ in the y-system. We now compute:

$$\frac{d\mathbf{c}}{ds} = \mathbf{c}_{,\beta}\frac{du^{\beta}}{ds} = \mathbf{c}_{,\beta}\, t^{\beta} = c^{\alpha}|_{\beta}t^{\beta}\mathbf{a}_{\alpha} + c^{\alpha}B_{\alpha\beta}t^{\beta}\mathbf{a}_3 = c_{\alpha}|_{\beta}t^{\beta}\mathbf{a}^{\alpha} + c_{\alpha}B^{\alpha}_{\beta}t^{\beta}\mathbf{a}_3 \qquad (8.5.5)$$

We define the *gradient of the surface vector* $\mathbf{c}(u)$, denoted by grad $\mathbf{c}$, as the second order tensor:

$$\operatorname{grad}\mathbf{c} = c^{\alpha}|_{\beta}\mathbf{a}_{\alpha} \otimes \mathbf{a}^{\beta} = c_{\alpha}|_{\beta}\mathbf{a}^{\alpha} \otimes \mathbf{a}_{\beta} \;\Leftrightarrow$$
$$\operatorname{grad}\mathbf{c}\big[\mathbf{a}_{\alpha}, \mathbf{a}^{\beta}\big] = c^{\alpha}|_{\beta}, \quad \operatorname{grad}\mathbf{c}\big[\mathbf{a}_{\alpha}, \mathbf{a}_{\beta}\big] = c_{\alpha}|_{\beta} \qquad (8.5.6)$$

We also define the vector: the *absolute derivative* $\delta\mathbf{c}/\delta s$ *of the surface vector field* $\mathbf{c}(u)$ along the curve $u^{\alpha}(s)$ as the vector:

$$\frac{\delta\mathbf{c}}{\delta s} = \frac{\delta c^{\alpha}}{\delta s}\mathbf{a}_{\alpha} = \frac{\delta c_{\alpha}}{\delta s}\mathbf{a}^{\alpha} \;\Leftrightarrow\; \frac{\delta c^{\alpha}}{\delta s} = c^{\alpha}|_{\beta}t^{\beta}, \quad \frac{\delta c_{\alpha}}{\delta s} = c_{\alpha}|_{\beta}t^{\beta} \qquad (8.5.7)$$

The results (8.5.5) and (8.5.7) are now combined to give:

$$\frac{d\mathbf{c}}{ds} = \frac{\delta\mathbf{c}}{\delta s} + (\mathbf{c}\cdot\mathbf{B}\cdot\mathbf{t})\mathbf{a}_3, \quad \frac{\delta\mathbf{c}}{\delta s} = (\operatorname{grad}\mathbf{c})\cdot\mathbf{t} \qquad (8.5.8)$$

For plane surfaces the fundamental parameters of second order $B_{\alpha\beta}$ are all zero, which implies that:

$$\frac{d\mathbf{c}}{ds} = \frac{\delta\mathbf{c}}{\delta s} \quad \text{for plane surfaces} \qquad (8.5.9)$$

From the formulas (8.5.7) and (8.5.3) we get the results:

$$\frac{\delta c^{\alpha}}{\delta s} = \frac{dc^{\alpha}}{ds} + c^{\gamma}\Gamma^{\alpha}_{\gamma\beta}t^{\beta}, \quad \frac{\delta c_{\alpha}}{\delta s} = \frac{dc_{\alpha}}{ds} - c_{\lambda}\Gamma^{\gamma}_{\alpha\beta}t^{\beta} \qquad (8.5.10)$$

Let $\mathbf{b}(y)$ be a steady space vector field and $u^{\alpha}(s)$ a curve on the surface $y^i(u)$, where s is the arc length parameter along the curve. Along the curve we compute the vector $d\mathbf{b}/ds$ from the gradient of the vector field $\mathbf{b}(y)$, as presented in Sect. 6.5.2 by the formulas (6.5.33), (6.5.30), and (6.5.31)$_1$:

$$\frac{d\mathbf{b}}{ds} = (\operatorname{grad}\mathbf{b})\cdot\mathbf{t} \;\Leftrightarrow\; \frac{\delta b^i}{\delta s} = b^i|_j\, t^{\,j}\ ,\quad b^i|_j = \operatorname{grad}\mathbf{b}\big[\mathbf{g}^i, \mathbf{g}_j\big] = b^i_{,j} + b^k\Gamma^i_{kj}$$

$$(8.5.11)$$

We introduce the *tensor-derivatives* $b^i|_{\alpha}$ of the components of the space vector $\mathbf{b}$ on the surface $y^i(u)$:

$$b^i|_\alpha = \operatorname{grad}\left[\mathbf{g}^i, \mathbf{a}_\alpha\right] = \operatorname{grad}\left[\mathbf{g}^i, \frac{\partial y^j}{\partial u^\alpha}\mathbf{g}_j\right] = \operatorname{grad}\left[\mathbf{g}^i, \mathbf{g}_j\right]\frac{\partial y^j}{\partial u^\alpha} = b^i|_j\frac{\partial y^j}{\partial u^\alpha} \quad \Rightarrow$$

$$b^i|_\alpha = b^i|_j\frac{\partial y^j}{\partial u^\alpha}, \quad b^i|_\alpha = \left(b^i{}_{,j} + b^k\Gamma^i_{kj}\right)\frac{\partial y^j}{\partial u^\alpha} = b^i{}_{,\alpha} + b^k\Gamma^i_{kj}\frac{\partial y^j}{\partial u^\alpha} \tag{8.5.12}$$

We let the quantity grad $\mathbf{b}$ be the space tensor field with components $b^i|_j$ i.e. the covariant derivatives of the vector components b^i, but also a space-surface tensor called the *tensor-derivative of* $\mathbf{b}$ *on the surface* $y_i(u)$ with components $b^i|_\alpha$:

$$b^i|_j = \operatorname{grad}\mathbf{b}[\mathbf{g}^i, \mathbf{g}_j] = b^i{}_{,j} + b^k\Gamma^i_{kj} \tag{8.5.13}$$

$$b^i|_\alpha = \operatorname{grad}\mathbf{b}[\mathbf{g}^i, \mathbf{a}_\alpha] = b^i|_j\frac{\partial y^j}{\partial u^\alpha} = b^i{}_{,\alpha} + b^k\Gamma^i_{kj}\frac{\partial y^j}{\partial u^\alpha} \tag{8.5.14}$$

Thus the symbol grad $\mathbf{b}$ represents both a space tensor field and a space-surface tensor field.

From the formulas (8.5.11), (8.5.12), and (8.2.5) we obtain the following alternative expressions for the component form of the vector $d\mathbf{b}/ds$:

$$\frac{\delta b^i}{\delta s} = b^i|_j\, t^{\,j} = b^i|_\alpha t^\alpha \tag{8.5.15}$$

Let $\mathbf{A}$ be a space-surface tensor of order 1,1 that connects a space vector $\mathbf{b}$ and a surface vector $\mathbf{c}$:

$$\mathbf{b} = \mathbf{A}\cdot\mathbf{c} \Rightarrow b^i = A^i_\alpha c^\alpha \tag{8.5.16}$$

The following results will now be derived:

$$b^i|_\beta = A^i_\alpha|_\beta\, c^\alpha + A^i_\alpha c^\alpha|_\beta \quad \text{where} \quad A^i_\alpha|_\beta = A^i_{\alpha,\beta} - A^i_\gamma\Gamma^\gamma_{\alpha\beta} + A^j_\alpha\Gamma^i_{kj}\frac{\partial y_j}{\partial u_\beta} \tag{8.5.17}$$

We use the formulas (8.5.12) and (8.5.3)$_1$ and obtain:

$$\left.\begin{array}{l} (8.5.12) \Rightarrow: b^i|_\beta = b^i{}_{,\beta} + b^k\Gamma^i_{kj}\dfrac{\partial y^j}{\partial u^\beta} \\[2mm] b^i = A^i_\alpha c^\alpha \Rightarrow b^i{}_{,\beta} = A^i_{\alpha,\beta}\, c^\alpha + A^i_\alpha c^\alpha{}_{,\beta} \\[2mm] (8.5.3)_1 \Rightarrow: c^\alpha{}_{,\beta} = c^\alpha|_\beta - c^\gamma\Gamma^\alpha_{\gamma\beta} \end{array}\right\}$$

$$\Rightarrow b^i|_\beta = \left[A^i_{\alpha,\beta} - A^i_\gamma\Gamma^\gamma_{\alpha\beta} + A^k_\alpha\Gamma^i_{kj}\frac{\partial y^j}{\partial u^\beta}\right]c^\alpha + A^i_\alpha c^\alpha|_\beta \quad \Rightarrow$$

$$b^i|_\beta = A^i_\alpha|_\beta\, c^\alpha + A^i_\alpha c^\alpha|_\beta \text{ where } A^i_\alpha|_\beta = A^i_{\alpha,\beta} - A^i_\gamma\Gamma^\gamma_{\alpha\beta} + A^i_\alpha\Gamma^i_{kj}\frac{\partial y^j}{\partial u^\beta} \Rightarrow (8.5.17)$$

The components $A_\alpha^i|_\beta$ represent a space-surface tensor of order 1,2 called the *tensor-derivative* of $\mathbf{A}$ on the surface $y_i(u)$ and may be presented as the gradient of the a space-surface tensor $\mathbf{A}$ in the following sense:

$$\text{grad } \mathbf{A}[\mathbf{g}^i; \mathbf{a}_\alpha, \mathbf{a}_\beta] = A_\alpha^i|_\beta \qquad (8.5.18)$$

The *absolute-derivative* of $\mathbf{A}$ along the curve $u_\alpha(s)$ is a tensor of order 1,1 defined by:

$$\frac{\delta \mathbf{A}}{\delta s} = (\text{grad } \mathbf{A}) \cdot \mathbf{t}, \quad \frac{\delta A_\alpha^i}{\delta s} = A_\alpha^i|_\beta t^\beta \qquad (8.5.19)$$

By combining the formulas (8.5.15), (8.5.19), and (8.5.7) we may present the formula (8.5.17)$_1$ as:

$$\frac{d\mathbf{b}}{ds} = \frac{\delta \mathbf{A}}{\delta s} \cdot \mathbf{c} + \mathbf{A} \cdot \frac{\delta \mathbf{c}}{\delta s} \Leftrightarrow \frac{db^i}{ds} = \frac{\delta A_\alpha^i}{\delta s} c^\alpha + A_\alpha^i \frac{\delta c^\alpha}{\delta s} \qquad (8.5.20)$$

From the formulas (8.5.17) and (8.5.19) it follows that when the space coordinates are Cartesian and the surface coordinates are geodesic in a point P, the tensor-derivative components are reduced to a partial derivatives, and the absolute derivative components are reduced to an ordinary derivatives in the point P:

$$A_\alpha^i|_\beta = A_{\alpha,\beta}^i \equiv \frac{\partial A_\alpha^i}{\partial u_\beta}, \quad \frac{\delta A_\alpha^i}{\delta s} = A_\alpha^i|_\beta t^\beta = \frac{\partial A_\alpha^i}{\partial u^\beta} \frac{du^\beta}{ds} = \frac{dA_\alpha^i}{ds} \quad \text{in the point } P \quad (8.5.21)$$

This fact may be used in a general definition of the tensor-derivative grad $\mathbf{A}$ and the absolute-derivative $\delta \mathbf{A}/\delta s$ of a tensor space-surface tensor $\mathbf{A}$ of order m,n. At a point P on a surface imbedded in a three-dimensional space we introduce a Cartesian coordinate system x and surface coordinate system $\bar{u}$ that is geodesic with respect to the point P. The tensor-derivative grad $\mathbf{A}$ of the tensor $\mathbf{A}$ is a tensor of order $m,n + 1$ which in point P is represented by the partial derivative of the components $A_{..j..\beta}^{i..\alpha}$ of with respect to $\bar{u}_\gamma$, i.e. $A_{..j..\beta,\gamma}^{i..\alpha} = A_{i..j\alpha..\beta,\gamma}$. In a general space coordinates system y and a general system u of surface coordinates, the formula (8.5.17)$_2$ indicates how the components $A_{..j..\beta}^{i..\alpha}|_\gamma$ of grad $\mathbf{A}$ are presented.

The absolute-derivative of $\mathbf{A}$ of order m,n along a curve $u(s)$ on the surface is a space-surface tensor of order m,n and defined by:

$$\frac{\delta \mathbf{A}}{\delta s} = (\text{grad } \mathbf{A}) \cdot \mathbf{t} \Leftrightarrow \frac{\delta A_\alpha^i}{\delta s} = A_{..j..\beta}^{i..\alpha}|_\gamma t^\gamma \qquad (8.5.22)$$

In a Cartesian coordinate system x and a geodesic surface coordinate system $\bar{u}$:

$$\frac{\delta \mathbf{A}}{\delta s} = \frac{d \mathbf{A}}{ds} \Leftrightarrow \frac{\delta A^i \cdot\cdot^\alpha_{\cdot j} \cdot\cdot_\beta}{\delta s} = \frac{d A^i \cdot\cdot^\alpha_{\cdot j} \cdot\cdot_\beta}{ds} \qquad (8.5.23)$$

From the definitions above we can conclude that tensor-differentiation and absolute-differentiation follow ordinary rules of differentiation.

The tensor-derivative of a surface tensor is called the *surface gradient* of the tensor and are represented by the covariant-derivative components on the surface. It follows that for a space tensor A:

$$\frac{d\mathbf{A}}{ds} = \frac{\delta \mathbf{A}}{\delta s} \qquad (8.5.24)$$

The gradient of a surface tensor $\mathbf{A}$ on a surface is represented by covariant derivative components in space or by tensor-derivative components on the surface, as indicated by the formulas (8.5.14). A general rule is:

$$()|_\alpha = ()|_i \frac{\partial y^i}{\partial u^\alpha} \qquad (8.5.25)$$

For convenience we present the tensor-derivatives of the base vectors $\mathbf{g}_i$, $\mathbf{g}^i$, $\mathbf{a}_\alpha$ and $\mathbf{a}^\alpha$:

$$\mathbf{g}_i|_\alpha = \mathbf{g}_i|_j \frac{\partial y^i}{\partial u^\alpha} = \mathbf{0}, \quad \mathbf{g}^i|_\alpha = \mathbf{g}^i|_j \frac{\partial y^j}{\partial u^\alpha} = \mathbf{0} \qquad (8.5.26)$$

$$\mathbf{a}_\alpha|_\beta = \mathbf{a}_{\alpha,\beta} - \mathbf{a}_\gamma \Gamma^\gamma_{\alpha\beta}, \quad \mathbf{a}^\alpha|_\beta = \mathbf{a}^\alpha_{,\beta} - \mathbf{a}^\gamma \Gamma^\alpha_{\gamma\beta} \qquad (8.5.27)$$

It follows that $\mathbf{a}_\alpha|_\beta$ and $\mathbf{a}^\alpha|_\beta$ are vectors that, when a change is made from a surface coordinate system u to another $\bar{u}$-system transform as surface tensors of second order. By the formula (8.1.15) the formulas (8.5.27) may presented as:

$$\mathbf{a}_\alpha|_\beta = B_{\alpha\beta}\mathbf{a}_3, \quad \mathbf{a}^\alpha|_\beta = B^\alpha_\beta \mathbf{a}_3 \qquad (8.5.28)$$

For a geodesic coordinate system u for a point P on the surface $y^i(u)$ we can show, see Problem 8.6, that:

$$a_{\alpha\beta,\gamma} = a^{\alpha\beta}_{,\gamma} = \alpha_{,\alpha} = \varepsilon_{\alpha\beta,\gamma} = \varepsilon^{\alpha\beta}_{,\gamma} = 0 \text{ in } P \qquad (8.5.29)$$

These results are used to argue that for any coordinate system u on the surface $y^i(u)$, the following holds true, see Problem 8.7:

$$a_{\alpha\beta}|_\gamma = a^{\alpha\beta}|_\gamma = \varepsilon_{\alpha\beta}|_\gamma = \varepsilon^{\alpha\beta}|_\gamma = 0, \quad \frac{\delta a_{\alpha\beta}}{\delta s} = \frac{\delta a^{\alpha\beta}}{\delta s} = \frac{\delta \varepsilon_{\alpha\beta}}{\delta s} = \frac{\delta \varepsilon^{\alpha\beta}}{\delta s} = 0 \quad (8.5.30)$$

The following results can be shown for covariant derivatives of the components of a second order surface tensor **D**, see Problem 8.8:

$$D_{\alpha\beta}|_\gamma = D_{\alpha\beta,\gamma} - D_{\lambda\beta}\Gamma^\lambda_{\alpha\gamma} - D_{\alpha\lambda}\Gamma^\lambda_{\beta\gamma}, \quad D^\alpha_\beta|_\gamma = D^\alpha_{\beta,\gamma} + D^\lambda_\beta\Gamma^\alpha_{\lambda\gamma} - D^\alpha_\lambda\Gamma^\lambda_{\beta\gamma} \quad (8.5.31)$$

These formulas provide a pattern for constructing the covariant-derivatives of the components of surface tensors.

For the tensor-derivatives of the fundamental parameters and the components of the permutation tensor P we will find on the surface $y^i(u)$, see Problem 8.9:

$$g_{ij}|_\alpha = g^{ij}|_\alpha = 0, \quad \varepsilon_{ijk}|_\alpha = \varepsilon^{ijk}|_\alpha = 0 \quad (8.5.32)$$

For the components of the curvature tensor **B** we shall find, see Problem 8.10:

$$B_{\alpha\beta} = \frac{\partial y^i}{\partial u^\alpha}\Big|_\beta \underset{\sim}{a}_i, \quad \frac{\partial y^i}{\partial u^\alpha}\Big|_\beta = B_{\alpha\beta} \underset{\sim}{a}^i \quad (8.5.33)$$

The covariant-derivative and tensor-derivatives of second order of the components of surface tensors shall now be developed. For a surface vector **c** we use the formulas (8.5.3) and (8.5.31):

$$c_\alpha|_{\beta\gamma} = \left(c_{\alpha,\beta} - c_\sigma\Gamma^\sigma_{\alpha\beta}\right)_{,\gamma} - \left(c_{\lambda,\beta} - c_\sigma\Gamma^\sigma_{\lambda\beta}\right)\Gamma^\lambda_{\alpha\gamma} - \left(c_{\alpha,\lambda} - c_\sigma\Gamma^\sigma_{\alpha\lambda}\right)\Gamma^\lambda_{\beta\gamma}$$

$$= \left(c_{\alpha,\beta\gamma} - c_{\sigma,\gamma}\Gamma^\sigma_{\alpha\beta} - c_\sigma\Gamma^\sigma_{\alpha\beta,\gamma}\right) - \left(c_{\lambda,\beta}\Gamma^\lambda_{\alpha\gamma} - c_\sigma\Gamma^\sigma_{\lambda\beta}\Gamma^\lambda_{\alpha\gamma}\right) - \left(c_{\alpha,\lambda}\Gamma^\lambda_{\beta\gamma} - c_\sigma\Gamma^\sigma_{\alpha\lambda}\Gamma^\lambda_{\beta\gamma}\right) \Rightarrow$$

$$c_\alpha|_{\beta\gamma} = c_{\alpha,\beta\gamma} - c_\sigma\left(\Gamma^\sigma_{\alpha\beta,\gamma} - \Gamma^\sigma_{\lambda\beta}\Gamma^\lambda_{\alpha\gamma} - \Gamma^\sigma_{\alpha\lambda}\Gamma^\lambda_{\beta\gamma}\right) - c_{\sigma,\gamma}\Gamma^\gamma_{\alpha\beta} - c_{\lambda,\beta}\Gamma^\lambda_{\alpha\gamma} - c_{\alpha,\lambda}\Gamma^\lambda_{\beta\gamma}$$

$$(8.5.34)$$

We introduce the components:

$$R^\sigma_{\alpha\beta\gamma} = \Gamma^\sigma_{\alpha\gamma,\beta} - \Gamma^\sigma_{\alpha\beta,\gamma} + \Gamma^\sigma_{\lambda\beta}\Gamma^\lambda_{\alpha\gamma} - \Gamma^\sigma_{\lambda\gamma}\Gamma^\lambda_{\alpha\beta} \quad (8.5.35)$$

From the formulas (8.5.34) and (8.5.35) we derive the result:

$$c_\alpha|_{\beta\gamma} - c_\alpha|_{\gamma\beta} = c_\sigma R^\sigma_{\alpha\beta\gamma} \quad (8.5.36)$$

It follows from this result that the components $R^\sigma_{\alpha\beta\gamma}$ represent the components of a fourth order surface tensor **R**. This tensor is called the *Riemann-Christoffel tensor* for the surface.

The result (8.5.36) shows that the order of covariant differentiation is not immaterial, i.e. $c_\alpha|_{\beta\gamma} \neq c_\alpha|_{\gamma\beta}$, unless the Riemann-Christoffel tensor is a *zero tensor*. We shall discuss this situation in Sect. 8.7.

For a scalar field $\phi(y)$ the surface gradient on a surface $y^i(u)$ is defined by: grad $\phi = (\partial\phi/\partial u^\alpha)\,\mathbf{a}^\alpha$. Because grad ϕ is a vector it is convenient to introduce the symbol $\phi|_\alpha \equiv \phi_{,\alpha}$, such that:

$$\text{grad } \phi \equiv \frac{\partial\phi}{\partial u^\alpha}\mathbf{a}^\alpha \equiv \phi_{,\alpha}\,\mathbf{a}^\alpha \equiv \phi|_\alpha\mathbf{a}^\alpha \tag{8.5.37}$$

It follows, see Problem 8.11, that:

$$\phi|_{\alpha\beta} = \phi|_{\beta\alpha} \tag{8.5.38}$$

In Problem 8.12 we are asked to show that for the second covariant derivative of the components $D_{\alpha\beta}$ of a second order surface tensor $\mathbf{D}$ we shall find:

$$D_{\alpha\beta}|_{\gamma\delta} - D_{\alpha\beta}|_{\delta\gamma} = D_{\sigma\beta}R^\sigma_{\alpha\gamma\delta} + D_{\alpha\sigma}R^\sigma_{\beta\gamma\delta}, \quad D_{\alpha\beta}|_{\gamma\delta} = D_{\alpha\beta}|_{\delta\gamma}\text{only when } \mathbf{R} = \mathbf{0} \tag{8.5.39}$$

Similar results may be shown for surface tensors of higher order and for space-surface tensors.

From the formula (8.5.35) we can derive the formula:

$$R_{\delta\alpha\beta\gamma} = a_{\delta\sigma}R^\sigma_{\alpha\beta\gamma} = \Gamma_{\alpha\gamma\delta,\beta} - \Gamma_{\alpha\beta\delta,\gamma} + \Gamma_{\alpha\beta\sigma}\Gamma^\sigma_{\delta\gamma} - \Gamma_{\alpha\lambda\sigma}\Gamma^\sigma_{\delta\beta} \Rightarrow$$
$$\text{Symmetry properties: } R_{\delta\alpha\beta\gamma} = -R_{\delta\alpha\gamma\beta} = R_{\alpha\delta\beta\gamma} = R_{\beta\gamma\delta\alpha} \tag{8.5.40}$$

The symmetry properties for the components $R_{\delta\alpha\beta\gamma}$ imply that the tensor $\mathbf{R}$ only has one distinct component γ different from zero in any specified coordinate system u, and the properties are satisfied by setting:

$$R_{\delta\alpha\beta\gamma} = \gamma\varepsilon_{\delta\alpha}\varepsilon_{\beta\gamma} \Rightarrow R_{1212} = \gamma\alpha \tag{8.5.41}$$

By application of the formulas $(8.1.12)_2$ and (8.5.41) it follows that, see Problem 8.13:

$$\gamma = \frac{1}{4}\varepsilon^{\delta\sigma}\varepsilon^{\beta\gamma}R_{\delta\alpha\beta\gamma} \tag{8.5.42}$$

The result shows that the component γ is an invariant in the surface, i.e. γ is independent of the coordinate system u on the surface. The component γ is called the *total curvature* or the *Gauss curvature* for the surface. The geometrical interpretation of γ will be demonstrated in Sect. 8.7.

The divergence of a surface vector field $\mathbf{c}$ in a surface is defined as the scalar div $\mathbf{c} = c^\alpha|_\alpha$. We use the formula (8.1.20) to obtain the result:

$$\text{div}\mathbf{c} = c^{\alpha}\big|_{\alpha} = c^{\alpha}{}_{,\alpha} + c^{\beta}\Gamma^{\alpha}_{\beta\alpha} = c^{\alpha}{}_{,\alpha} + c^{\beta}\frac{1}{\sqrt{\alpha}}\left(\sqrt{\alpha}\right)_{,\beta} = \frac{1}{\sqrt{\alpha}}\left(c^{\beta}\sqrt{\alpha}\right)_{,\beta} \Rightarrow$$

$$\text{div}\mathbf{c} = \frac{1}{\sqrt{\alpha}}\left(c^{\beta}\sqrt{\alpha}\right)_{,\beta} \equiv \frac{1}{\sqrt{\alpha}}\frac{\partial}{\partial u^{\beta}}\left(c^{\beta}\sqrt{\alpha}\right) \tag{8.5.43}$$

The *rotation of a surface vector field* $\mathbf{c}$ in a surface is a vector field defined by the formula:

$$\text{rot}\,\mathbf{c} = \varepsilon^{\alpha\beta}c_{\beta}\big|_{\alpha}\mathbf{a}_3 = \varepsilon^{\alpha\beta}c_{\beta,\alpha}\,\mathbf{a}_3 \tag{8.5.44}$$

Weingarten [1836–1910] has developed the following formula for the tensor-derivative of the space components $\underset{\sim}{a}^i$ of the surface normal $\mathbf{a}_3 = \underset{\sim}{a}^i\mathbf{g}_i$ to a surface $u^{\alpha}(y)$,, see Problem 8.14:

$$\underset{\sim}{a}^i\big|_{\alpha} = -B^{\beta}_{\alpha}\frac{\partial y^i}{\partial u^{\beta}}\ \textit{Weingarten's formula} \tag{8.5.45}$$

For a space vector $\mathbf{b} = b_{\beta}\,\mathbf{a}^{\beta} + b_3\mathbf{a}_3$, we shall find, see Problem 8.15:

$$\mathbf{b}_{,\alpha} = \left(b_{\beta}\big|_{\alpha} - b_3 B_{\beta\alpha}\right)\mathbf{a}^{\beta} + \left(b_{3,\alpha} + b_{\beta}B^{\beta}_{\alpha}\right)\mathbf{a}_3 \tag{8.5.46}$$

8.6 Intrinsic Surface Geometry

8.6.1 The Metric of a Surface Imbedded in the Euclidean Space $\mathbf{E}_3$

Let a curve on a surface $y^i(u)$ be described by the place vector $\mathbf{r}(u)$ with the surface coordinates given by the functions $u^{\alpha}(s)$ and where the curve parameter s is the length of the curve between two points P and Q on the curve. The arc length formula (1.4.5) is now presented as:

$$s = \int_{P}^{Q}\sqrt{\frac{d\mathbf{r}}{d\bar{s}}\cdot\frac{d\mathbf{r}}{d\bar{s}}}\,d\bar{s} \tag{8.6.1}$$

The curve differential $d\mathbf{r}$ and its length ds are given by:

$$d\mathbf{r} = \frac{\partial\mathbf{r}}{\partial u^{\alpha}}du^{\alpha} = \mathbf{a}_{\alpha}du^{\alpha},\quad |d\mathbf{r}| = ds$$

$$(ds)^2 = d\mathbf{r}\cdot d\mathbf{r} = \left(\mathbf{a}_{\alpha}du^{\alpha}\right)\cdot\left(\mathbf{a}_{\beta}du^{\beta}\right) = a_{\alpha\beta}\,du^{\alpha}du^{\beta} \tag{8.6.2}$$

The quadratic form $a_{\alpha\beta}\,du^\alpha du^\beta$, which represents the scalar ds, is called the *first fundamental form of the surface* and the *metric of the surface*.

If it is possible to choose the coordinate systems on two different surfaces such that the surfaces have the same metric, the surfaces are said to be *isometric surfaces*. Surfaces that are isometric with a plane are called *developable surfaces*, e.g. cylindrical and conical surfaces. For developable surfaces it is possible to introduce a coordinate system for which the fundamental parameters of first order and the metric take the forms:

$$a_{\alpha\beta} = \delta_{\alpha\beta} \Rightarrow a_{\alpha\beta}du^\alpha du^\beta = du^\alpha du^\alpha \qquad (8.6.3)$$

On a plane this coordinate system u is Cartesian and the Pythagoras formula, Pythagoras [ca. 580 – 495 BCE], may be used to calculate lengths. A surface that is isometric with a plane is also called a *Euclidian surface*, and the surface represents a *two-dimensional Euclidian space* E_2.

A general surface imbedded in the three-dimensional Euclidean space E_3 is said to represent a *two-dimensional Riemannian space* R_2. A general theorem for n-dimensional spaces [1] may be stated that:

$R_2 = E_2 \Leftrightarrow \mathbf{R} = \mathbf{0}$ i.e. the *Riemann – Christoffel tensor* is a zero – tensor

$\Leftrightarrow$ for a Euclidian surface the *Gauss curvature* γ is zero.

$$(8.6.4)$$

Note that for Euclidian surfaces the order of covariant-differentiation and tensor-differentiation is immaterial, e.g. $A^i_\alpha|_{\beta\lambda} = A^i_\alpha|_{\lambda\beta}$.

Surface geometry related only to coordinate systems on the surface and the corresponding metric is called *intrinsic surface geometry*. Absolute-differentiation and covariant-differentiation of components of surface tensors are intrinsic operations. Isometric surfaces have identical intrinsic surface geometry. This implies that intrinsic surface properties are common for isometric surfaces, and they have the same *Riemann-Christoffel tensor* $\mathbf{R}$ and the same *Gaussian curvature* γ. Isometric surfaces may have different forms in E_3 and this form is determined by the curvature tensor $\mathbf{B}$, which will properly be presented in Sect. 8.7.

8.6.2 Surface Curves

The unit *tangent vector* $\mathbf{t}$ to a curve $u_\alpha(s)$ on the surface $y_i(u)$ is the surface vector defined by the formulas (8.2.1):

$$\mathbf{t} = \frac{d\mathbf{r}}{ds} = \frac{\partial \mathbf{r}}{\partial y^i}\frac{dy^i}{ds} = \frac{dy^i}{ds}\mathbf{g}_i = \frac{\partial \mathbf{r}}{\partial u^\alpha}\frac{du^\alpha}{ds} = \frac{du^\alpha}{ds}\mathbf{a}_\alpha \qquad (8.6.5)$$

The derivative of **t** with respect to the arc length parameter s is according to the formulas (1.4.10) and (8.5.8):

$$\frac{d\mathbf{t}}{ds} = \kappa\mathbf{n} = \frac{\delta\mathbf{t}}{\delta s} + (\mathbf{t} \cdot \mathbf{B} \cdot \mathbf{t})\mathbf{a}_3, \quad \kappa = \left|\frac{d\mathbf{t}}{ds}\right| \tag{8.6.6}$$

The vector **n** is the *principal normal* to the curve and κ is *curvature* of the curve. The contribution $\delta\mathbf{t}/\delta s$ is a surface vector and because **t** is a unit vector this surface vector is a normal vector to the curve: $\mathbf{t} \cdot \mathbf{t} = 1 \Rightarrow (\delta\mathbf{t}/\delta s) \cdot \mathbf{t} = 0$. We introduce the unit normal vector **m** parallel to the vector $\delta\mathbf{t}/\delta s$, defined such that the rotation (**t**, **m**) according to the definition (8.2.8) is positive. Because **t** and **m** are unit vectors, the rotation (**t**, **m**) is positive when:

$$\varepsilon_{\alpha\beta} t^\alpha m^\beta = +1 \tag{8.6.7}$$

Figure 8.7, which will be presented later, shows the normal vector **m**, the tangent vector **t**, the unit normal vector $\mathbf{a}_3$ to the surface, and the principal normal vector **n** to the curve $u^\alpha(s)$. From the relation (8.6.7) and the definition of the permutation symbols $\varepsilon_{\alpha\beta}$ it follows that:

$$t^\alpha = \varepsilon^{\alpha\beta} m_\beta, \quad m^\alpha = \varepsilon^{\beta\alpha} t_\beta \tag{8.6.8}$$

The *geodesic curvature* σ for the curve is defined by the expressions:

$$\sigma = \frac{\delta t^\alpha}{\delta s} m_\alpha \Leftrightarrow \frac{\delta t^\alpha}{\delta s} = \sigma m^\alpha \tag{8.6.9}$$

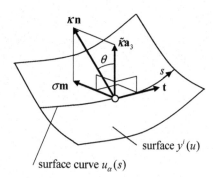

Fig. 8.7 Surface $y^i(u)$ and surface curve $u^\alpha(s)$. **t** = tangent vector to the curve $u^\alpha(s)$. $\mathbf{a}_3$ = unit normal vector to the surface $y^i(u)$. **n** = principal normal vector to the curve $u^\alpha(s)$. **m** = unit normal vector to the curve $u^\alpha(s)$ in the surface $y^i(u)$. κ = curvature of the curve $u^\alpha(s)$. $\tilde{\kappa}$ = curvature of surface for the direction **t**. σ = geodesic curvature of the curve $u^\alpha(s)$ in the surface $y^i(u)$

From the result $\delta\varepsilon^{\beta\alpha}/\delta s = 0$ in the formulas (8.5.30) and the formulas (8.6.8) and (8.6.9) it follows that:

$$\frac{\delta m^\alpha}{\delta s} = \varepsilon^{\beta\alpha}\frac{\delta t_\beta}{\delta s} = \varepsilon^{\beta\alpha}\sigma m_\beta = -\sigma t^\alpha \tag{8.6.10}$$

The following formulas are called the *Frenet-Serret formulas for a surface curve*:

$$\frac{\delta t^\alpha}{\delta s} = \sigma m_\alpha, \quad \frac{\delta m^\alpha}{\delta s} = -\sigma t^\alpha \tag{8.6.11}$$

Confer with the corresponding formulas (1.4.13) for space curves.

We shall now develop equations that may be used to determine the shortest surface curve $u^\alpha(s)$ between two points P and Q on the given surface $y^i(u)$. A neighbouring curve to the curve $u^\alpha(s)$ is defined by the coordinates:

$$v^\alpha(s,\varepsilon) = u^\alpha(s) + \varepsilon w^\alpha(s), \quad w^\alpha(s_P) = w^\alpha(s_Q) = 0 \tag{8.6.12}$$

ε is a small number and $w^\alpha(s)$ are two arbitrary functions of s. The length $L = L(\varepsilon)$ of the curve $v^\alpha(s,\varepsilon)$ between the points P and Q is given by an integral obtained from the arc length formula (8.6.1):

$$L(\varepsilon) = \int_P^Q \sqrt{a_{\alpha\beta}\frac{dv^\alpha}{ds}\frac{dv^\beta}{ds}}\,ds \tag{8.6.13}$$

Because the fundamental parameters $a_{\alpha\beta}$ are functions of the coordinates $u^\alpha(s)$ the integrand in the integral can be regarded as a function of $v^\alpha(s,\varepsilon)$ and of $dv^\alpha(s,\varepsilon)/ds$:

$$f\left(v,\frac{dv}{ds}\right) = \sqrt{a_{\alpha\beta}\frac{dv^\alpha}{ds}\frac{dv^\beta}{ds}} \tag{8.6.14}$$

The requirement for the functions $w^\alpha(s)$ in the formulas (8.6.12) is that the length $L(\varepsilon)$ shall have a minimum for $\varepsilon = 0$, which implies that:

$$\frac{dL(\varepsilon)}{d\varepsilon}\bigg|_{\varepsilon=0} = \left[\int_P^Q \left\{\frac{\partial f}{\partial v^\alpha}\frac{\partial v^\alpha}{\partial \varepsilon}\right\}ds\right]_{\varepsilon=0} + \left[\int_P^Q \left\{\frac{\partial f}{\partial(dv^\alpha/ds)}\frac{\partial}{\partial\varepsilon}\left(\frac{dv^\alpha}{ds}\right)\right\}ds\right]_{\varepsilon=0} = 0 \tag{8.6.15}$$

From the formulas (8.6.12) it follows that:

$$\frac{\partial v^\alpha}{\partial \varepsilon} = w^\alpha, \quad \frac{\partial}{\partial \varepsilon}\left(\frac{dv^\alpha}{ds}\right) = \frac{dw^\alpha}{ds} \tag{8.6.16}$$

The first integral on the right-hand side of Eq. (8.6.15) is now reduced to:

$$\left[\int_P^Q \left\{\frac{\partial f}{\partial v^\alpha}\frac{\partial v^\alpha}{\partial \varepsilon}\right\}ds\right]_{\varepsilon=0} = \int_P^Q \left\{\frac{\partial f}{\partial u^\alpha}w^\alpha\right\}ds \tag{8.6.17}$$

The second integral on the right-hand side of Eq. (8.6.15) is developed by a partial integration:

$$\left[\int_P^Q \left\{\frac{\partial f}{\partial(dv^\alpha/ds)}\frac{\partial}{\partial\varepsilon}\left(\frac{dv^\alpha}{ds}\right)\right\}ds\right]_{\varepsilon=0} = \int_P^Q \left\{\frac{\partial f}{\partial(du^\alpha/ds)}\frac{dw^\alpha}{ds}\right\}ds$$

$$= \left\{\frac{\partial f}{\partial(du^\alpha/ds)}w^\alpha\right\}_P^Q - \int_P^Q \frac{d}{ds}\left\{\frac{\partial f}{\partial(du^\alpha/ds)}\right\}w^\alpha ds$$

$$= -\int_P^Q \frac{d}{ds}\left\{\frac{\partial f}{\partial(du^\alpha/ds)}\right\}w^\alpha ds \Rightarrow \left[\int_P^Q \left\{\frac{\partial f}{\partial(dv^\alpha/ds)}\frac{\partial}{\partial\varepsilon}\left(\frac{dv^\alpha}{ds}\right)\right\}ds\right]_{\varepsilon=0} \tag{8.6.18}$$

$$= -\int_P^Q \frac{d}{ds}\left\{\frac{\partial f}{\partial(du^\alpha/ds)}\right\}w^\alpha ds$$

When the results (8.6.17) and (8.6.18) are substituted into Eq. (8.6.15), we obtain:

$$\left.\frac{dL(\varepsilon)}{d\varepsilon}\right|_{\varepsilon=0} = \int_P^Q \left\{\frac{\partial f}{\partial u^\alpha} - \frac{d}{ds}\frac{\partial f}{\partial(du^\alpha/ds)}\right\}w^\alpha ds = 0 \tag{8.6.19}$$

For this integral to be zero for any choice of the functions $w^\alpha(s)$ the integrand must be zero, see Theorem 9.2. Hence we have found that the functions describing the shortest curve $u^\alpha(s)$ between the points P and Q on the surface $y^i(u)$ must satisfy the equations:

$$\frac{\partial f}{\partial u^\alpha} - \frac{d}{ds}\frac{\partial f}{\partial(du^\alpha/ds)} = 0 \tag{8.6.20}$$

In standard *calculus of variation* Eq. (8.6.20) are called the *Euler equations* to following problem: Find the functions $u^\alpha(s)$ representing a curve between the points P and Q on the given surface $y^i(u)$ that makes the integral:

$$L = \int_P^Q f\left(u, \frac{du}{ds}\right) ds = \text{a minimum between the points } P \text{ and } Q \qquad (8.6.21)$$

when compared with value of the integral for any choice of neighbour curves $v^\alpha(s, \varepsilon)$ as given by the formula (8.6.12). The solutions $u^\alpha(s)$ of the Eq. (8.6.20) are called the *extremes of the functional L*.

In the special case where the integrand f is given by formula (8.6.14), the functional L represents the length of curves between two points P and Q on a surface $y^i(u)$, and the Euler equations determine the curve $u^\alpha(s)$ on the surface with the shortest length between the two points. Such curves are called *geodesic curves* or *geodesics*. We shall now find the differential equations that determine geodesic curves. We compute:

$$\left.\begin{array}{l}
\frac{\partial f}{\partial u^\alpha} = \frac{1}{2f} a_{\gamma\beta,\alpha} \frac{du^\gamma}{ds} \frac{du^\beta}{ds}, \frac{\partial f}{\partial(du^\alpha/ds)} = \frac{1}{2f} 2 a_{\alpha\beta} \frac{du^\alpha}{ds} = \frac{1}{f} a_{\alpha\beta} \frac{du^\alpha}{ds} \\[2mm]
\qquad\qquad f = 1 \text{ along the curve } u_\alpha(s) \Rightarrow \\[2mm]
\frac{d}{ds} \frac{\partial f}{\partial(du^\alpha/ds)} = \frac{d}{ds}\left(a_{\alpha\beta} \frac{du^\beta}{ds}\right) = a_{\alpha\beta,\gamma} \frac{du^\gamma}{ds} \frac{du^\beta}{ds} + a_{\alpha\beta} \frac{d^2 u^\beta}{ds^2} \\[2mm]
\frac{\partial f}{\partial u^\alpha} = \frac{1}{2} a_{\gamma\beta,\alpha} \frac{du^\gamma}{ds} \frac{du^\beta}{ds}, \frac{d}{ds} \frac{\partial f}{\partial(du^\alpha/ds)} = a_{\alpha\beta,\gamma} \frac{du^\gamma}{ds} \frac{du^\beta}{ds} + a_{\alpha\beta} \frac{d^2 u^\beta}{ds^2}
\end{array}\right\} \Rightarrow \qquad (8.6.22)$$

The results (8.6.22) are substituted into Eq. (8.6.20) and the result is:

$$a_{\alpha\beta} \frac{d^2 u^\beta}{ds^2} - \left(\frac{1}{2} a_{\gamma\beta,\alpha} - a_{\alpha\beta,\gamma}\right) \frac{du^\gamma}{ds} \frac{du^\beta}{ds} = 0$$

Here we use the formula $(8.1.19)_1$ and the symmetry properties of the Christoffel symbols to obtain the result:

$$a_{\alpha\beta} \frac{d^2 u^\beta}{ds^2} - \left(\Gamma_{\alpha\gamma\beta} + \Gamma_{\beta\gamma\alpha} - \frac{1}{2}\Gamma_{\gamma\alpha\beta} - \frac{1}{2}\Gamma_{\beta\alpha\gamma}\right) \frac{du^\gamma}{ds} \frac{du^\beta}{ds} = 0$$

$$\Rightarrow a_{\alpha\beta} \frac{d^2 u^\beta}{ds^2} + \Gamma_{\beta\gamma\alpha} \frac{du^\gamma}{ds} \frac{du^\beta}{ds} = 0$$

Finally this result is multiplied by $a^{\alpha\rho}$, summation is performed over the index α, the index ρ is changed to the index α, and the formula $(8.1.17)_2$ is applied. We then obtained the following set of *differential equations for the geodesic curve* $u^\alpha(s)$ on the surface $y^i(u)$:

$$\frac{d^2 u^\alpha}{ds^2} + \Gamma^\alpha_{\beta\gamma} \frac{du^\beta}{ds} \frac{du^\gamma}{ds} = 0 \qquad (8.6.23)$$

The formulas (8.6.5) show that du^α/ds are the contravariant components of the unit tangent vector **t** to the curve. When Eq. (8.6.23) are compared with Eq. $(8.5.10)_1$, we see that for a geodesic curve:

$$\frac{\delta t^\alpha}{\delta s} = 0 \qquad (8.6.24)$$

From Eq. $(8.6.9)_1$ it then follows that *the geodesic curvature σ is zero for a geodesic curve.*

With respect to the x_3-plane in a Cartesian coordinate system Ox Eq. (8.6.24) for the geodesics on the plane are:

$$\frac{d^2 x_\alpha}{ds^2} = 0$$

$\Rightarrow x_\alpha = a_\alpha s + b_\alpha$, where a_α and b_α are constant surface vector in the plane.

Thus, as expected: the geodesics on a plane are straight lines.

On developable surfaces like cylindrical and conical surfaces we shall find that the geodesics are helices, see Example 8.5. Geodesic curves on a spherical surface are great circles.

Let **a** (u) be a surface vector field defined on the surface $y^i(u)$ and $u^\alpha(s)$ a curve on the surface. The vector field **a** (u) is called a *parallel vector field* with respect to the surface and the curve if:

$$\frac{\delta \mathbf{a}}{\delta s} = \mathbf{0} \Leftrightarrow \frac{\delta a^\alpha}{\delta s} = \frac{da^\alpha}{ds} + a^\gamma \Gamma^\alpha_{\gamma\beta} \frac{du^\beta}{ds} = 0 \quad \text{along the curve } u^\alpha(s) \qquad (8.6.25)$$

The component equations are two first order ordinary simultaneous equations with a unique solutions for the vector field $\mathbf{a}(u)$ when the vector **a** is specified in a point on the curve $u^\alpha(s)$. From Eqs. (8.6.25) and (8.5.8) it follows that for a parallel vector field:

$$\frac{d\mathbf{a}}{ds} = (\mathbf{a} \cdot \mathbf{B} \cdot \mathbf{t})\mathbf{a}_3 \qquad (8.6.26)$$

This result shows that the vector $d\mathbf{a}/ds$ is perpendicular to the surface vector **a**, which implies that the magnitude $|\mathbf{a}|$ of the vector **a** is constant. It may be shown, see Problem 8.14, that along a geodesic curve the unit tangent vectors **t** represents a parallel vector field related to the geodesic curve.

Let P and Q be two points on the surface $y^i(u)$, and $\mathbf{a}(u)$ and $\mathbf{b}(u)$ two parallel vector fields related to two different curves between P and Q on the surface, and such that the vectors $\mathbf{a}$ and $\mathbf{b}$ are parallel in P. It may be shown, see Problem 8.15, that the angle $(\mathbf{a},\mathbf{b})$ between two parallel vector fields $\mathbf{a}(u)$ and $\mathbf{b}(u)$ related to the same curve $u^\alpha(s)$ on the surface $y^i(u)$, is constant.

8.7   Curvatures on a Surface

The curvature κ and the principal normal vector $\mathbf{n}$ of a space curve are defined in Sect. 1.4 by the formulas (1.4.9) and (1.4.10):

$$\kappa = \left|\frac{d\mathbf{t}}{ds}\right| = \left|\frac{d^2\mathbf{r}}{ds^2}\right|, \quad \mathbf{n} = \frac{1}{\kappa}\frac{d\mathbf{t}}{ds} = \frac{1}{\kappa}\frac{d^2\mathbf{r}}{ds^2} \tag{8.7.1}$$

The formula (8.6.6) is now be presented as:

$$\kappa\mathbf{n} = \sigma\mathbf{m} + (\mathbf{t}\cdot\mathbf{B}\cdot\mathbf{t})\mathbf{a}_3 \tag{8.7.2}$$

The three normal vectors $\mathbf{a}_3$, $\mathbf{n}$, and $\mathbf{m}$ are all in a plane perpendicular to the tangent vector $\mathbf{t}$, as shown in Fig. 8.7. The scalar invariant $\mathbf{t}\cdot\mathbf{B}\cdot\mathbf{t}$ is called the *curvature $\tilde{\kappa}$ of the surface for the direction $\mathbf{t}$*, and is also seen to be the normal component to the tensor $\mathbf{B}$ for the direction $\mathbf{t}$:

$$\tilde{\kappa} = \mathbf{t}\cdot\mathbf{B}\cdot\mathbf{t} \tag{8.7.3}$$

From Fig. 8.7 we derive the results:

$$\tilde{\kappa} = \kappa\cos\theta = \mathbf{t}\cdot\mathbf{B}\cdot\mathbf{t}, \quad \sigma = \kappa\sin\theta \tag{8.7.4}$$

The three curvatures σ, κ, and $\tilde{\kappa}$ are related through the formula:

$$\kappa\mathbf{n} = \sigma\mathbf{m} + \tilde{\kappa}\mathbf{a}_3 \tag{8.7.5}$$

The second order tensor field $\mathbf{B}$ is now called the *curvature tensor for the surface* $y^i(u)$.

The curvatures of the surface in the direction of the coordinate lines are found to be:

$$_\alpha\tilde{\kappa} = \left(\frac{\mathbf{a}_\alpha}{\sqrt{a_{\alpha\alpha}}}\right)\cdot\mathbf{B}\cdot\left(\frac{\mathbf{a}_\alpha}{\sqrt{a_{\alpha\alpha}}}\right) = \frac{1}{a_{\alpha\alpha}}\mathbf{a}_\alpha\cdot\mathbf{B}\cdot\mathbf{a}_\alpha \Rightarrow$$

$$_1\tilde{\kappa} = \frac{B_{11}}{a_{11}}, \quad _2\tilde{\kappa} = \frac{B_{22}}{a_{22}} \tag{8.7.6}$$

The *principal curvatures* $\tilde{\kappa} = \tilde{\kappa}_1$ or $\tilde{\kappa}_2$ and the *principal curvature directions* $\mathbf{e} = \mathbf{e}_1$ or $\mathbf{e}_2$ are found from the equations:

$$\left(\tilde{\kappa}\delta_\alpha^\beta - B_\alpha^\beta\right)e_\beta = 0, \quad \text{non-trivial solution} \Rightarrow \det\left(\tilde{\kappa}\delta_\alpha^\beta - B_\alpha^\beta\right) = 0 \qquad (8.7.7)$$

The last equation leads to the characteristic equation for the curvature tensor $\mathbf{B}$:

$$\tilde{\kappa}^2 - \text{tr}\mathbf{B}\tilde{\kappa} + \det\mathbf{B} = 0, \quad \text{the characteristic equation for the curvature tensor } \mathbf{B}$$

$$\Rightarrow \tilde{\kappa}^2 - 2\mu\tilde{\kappa} + \gamma = 0 \begin{cases} \mu = \frac{1}{2}B_\alpha^\alpha = \frac{1}{2}(\tilde{\kappa}_1 + \tilde{\kappa}_2) & \text{the } mean\,curvature \\ \gamma = \det\mathbf{B} = \det(B_\alpha^\beta) = \tilde{\kappa}_1\tilde{\kappa}_2 & \text{the } Gauss\,curvature \end{cases} \qquad (8.7.8)$$

A surface curve, for which the tangent vector in every point is in the principal curvature direction, is called a *curvature line*.

Through every point on a surface pass two curvature lines, which implies that the curvature lines on a surface may be used as coordinate lines. Because the principal directions $\mathbf{e}_1$ and $\mathbf{e}_2$ are orthogonal, it follows that when the coordinate lines are curvature lines:

$$a_{12} = B_{12} = 0 \qquad (8.7.9)$$

Let θ be the angle between direction $\mathbf{t}$ on the surface $y^i(u)$ and the principal curvature direction $\mathbf{e}_1$:

$$\mathbf{t} = \cos\theta\,\mathbf{e}_1 + \sin\theta\,\mathbf{e}_2 \qquad (8.7.10)$$

It follows from the formulas (8.7.3), (8.7.9), and (8.7.10) that:

$$\tilde{\kappa}_1 = \mathbf{e}_1 \cdot \mathbf{B} \cdot \mathbf{e}_1, \quad \tilde{\kappa}_2 = \mathbf{e}_2 \cdot \mathbf{B} \cdot \mathbf{e}_2, \quad \mathbf{e}_1 \cdot \mathbf{B} \cdot \mathbf{e}_2 = 0$$

The curvature $\tilde{\kappa}$ for the direction $\mathbf{t}$ on the surface $y^i(u)$ is then found to be:

$$\tilde{\kappa} = \mathbf{t} \cdot \mathbf{B} \cdot \mathbf{t} = (\cos\theta\mathbf{e}_1 + \sin\theta\mathbf{e}_2) \cdot \mathbf{B} \cdot (\cos\theta\mathbf{e}_1 + \sin\theta\mathbf{e}_2) = \tilde{\kappa}$$

$$= \mathbf{e}_1 \cdot \mathbf{B} \cdot \mathbf{e}_1 \cos^2\theta + \mathbf{e}_2 \cdot \mathbf{B} \cdot \mathbf{e}_2 \sin^2\theta \Rightarrow$$

$$\tilde{\kappa} = \tilde{\kappa}_1 \cos^2\theta + \tilde{\kappa}_2 \sin^2\theta \qquad (8.7.11)$$

Example 8.4 Curvatures of Circular Cylindrical Surface

A circular cylindrical surface of radius R is shown in Fig. 8.3 and is defined by the place vector:

$$\mathbf{r}(u^1, u^2) = R\,\mathbf{e}_R(u^1) + u^2\,\mathbf{e}_z, \quad u^1 = \theta, \quad u^2 = z,$$

The unit normal vector to the surface is $\mathbf{a}_3 = \mathbf{e}_R$. The base vectors and the fundamental parameters are:

$$\mathbf{a}_1 = R\,\mathbf{e}_\theta, \mathbf{a}_2 = \mathbf{e}_z, \mathbf{a}^1 = \mathbf{e}_\theta/R, \mathbf{a}^2 = \mathbf{e}_z, \quad (a_{\alpha\beta}) = \begin{pmatrix} R^2 & 0 \\ 0 & 1 \end{pmatrix}, \quad (a^{\alpha\beta})$$

$$= \begin{pmatrix} 1/R^2 & 0 \\ 0 & 1 \end{pmatrix}$$

The covariant components $B_{\alpha\beta}$ of the curvature tensor $\mathbf{B}$ are given in the formula (8.1.26), and the mixed components B_α^β are computed from the formulas $B_\alpha^\beta = a^{\beta\gamma} B_{\alpha\gamma}$:

$$(B_{\alpha\beta}) = \begin{pmatrix} -R & 0 \\ 0 & 0 \end{pmatrix}, \quad (B_\alpha^\beta) = \begin{pmatrix} -\frac{1}{R} & 0 \\ 0 & 0 \end{pmatrix}$$

The coordinate lines are lines of curvature and we find for the principal curvatures, the mean curvature μ, and the Gauss curvature γ:

$$\tilde{\kappa}_1 = -\frac{1}{R}, \quad \tilde{\kappa}_2 = 0, \quad \mu = -\frac{1}{2R}, \quad \gamma = 0$$

Example 8.5 Helix on a Circular Cylindrical Surface

We shall consider the helix presented in Example 1.1 as a curve $u^\alpha(s)$ on the cylindrical surface. In cylindrical coordinates (R, θ, z) and with surface coordinates: $u^1 = \theta$ and $u^2 = z$, the cylinder is defined by $R = $ constant, and the helix is defined by the place vector:

$$\mathbf{r} = \mathbf{r}(u^1, u^2) = R\mathbf{e}_R(u^1) + u^2 \mathbf{e}_z \Rightarrow \mathbf{r} = \mathbf{r}(u^1, u^2) = R\mathbf{e}_R(\theta) + (b\theta + z_0)\mathbf{e}_z$$

The tangent vector $\mathbf{t}$ for the curve is:

$$\mathbf{t} = \frac{d\mathbf{r}}{ds} = \frac{d\mathbf{r}}{d\theta}\frac{d\theta}{ds} = \left(R\frac{d\mathbf{e}_R}{d\theta} + b\mathbf{e}_z \right)\frac{d\theta}{ds} = (R\mathbf{e}_\theta(\theta) + b\mathbf{e}_z)\frac{1}{\sqrt{R^2 + b^2}} \Rightarrow$$

$$\mathbf{t} = \frac{1}{\sqrt{R^2 + b^2}}\mathbf{a}_1 + \frac{b}{\sqrt{R^2 + b^2}}\mathbf{a}_2 \Rightarrow \mathbf{t} = \frac{d\mathbf{r}}{ds} = t^\alpha \mathbf{a}_\alpha \Rightarrow$$

$$t^1 = \frac{1}{\sqrt{R^2 + b^2}}, \quad t^2 = \frac{b}{\sqrt{R^2 + b^2}} \Rightarrow$$

$$\mathbf{t} = \frac{1}{\sqrt{R^2 + b^2}}(R\mathbf{e}_\theta + b\mathbf{e}_z)$$

The principal normal vector $\mathbf{n}$ and the curvature κ for the curve are found from:

$$\kappa\mathbf{n} = \frac{d^2\mathbf{r}}{ds^2} = \left(R\frac{d\mathbf{e}_\theta}{d\theta}\frac{d\theta}{ds} \right)\frac{d\theta}{ds} = \frac{-R}{R^2 + b^2}\mathbf{e}_R \Rightarrow \mathbf{n} = -\mathbf{e}_R, \quad \kappa = \frac{R}{R^2 + b^2}$$

Because the components t^α are constants and $\Gamma_{\gamma\beta}^\alpha = 0$ according to the formula (8.1.27), it follows from the formulas (8.5.7) and (8.5.3) that:

$$\frac{\delta t^\alpha}{\delta s} = t^\alpha|_\beta t^\beta = \left(t^\alpha,_\beta + t^\gamma \Gamma^\alpha_{\gamma\beta}\right) t^\beta = t^\alpha,_\beta t^\beta = 0 \Rightarrow$$

$$\frac{\delta t^\alpha}{\delta s} = 0 \qquad\qquad (8.7.12)$$

The result shows that the helix is a geodesic curve on the circular cylindrical surface, which agrees with the fact that the helix is a straight line when the cylindrical surface is opened into a plane.

From the formulas $(8.6.8)_2$, $(8.7.12)$ and $(8.6.9)_1$ we determine the unit normal vector $\mathbf{m}$ and the torsion σ:

$$m^\alpha = \varepsilon^{\beta\alpha} t_\beta = \varepsilon^{\beta\alpha} a_{\beta\gamma} t^\gamma \Rightarrow \begin{cases} m^1 = \frac{1}{\sqrt{a}} e^{21} a_{22} t^2 = \frac{1}{R}(-1)\cdot 1 \frac{b}{\sqrt{R^2+b^2}} = \frac{-b}{R\sqrt{R^2+b^2}} \\ m^2 = \frac{1}{\sqrt{a}} e^{12} a_{11} t^1 = \frac{1}{R} 1 \cdot R^2 \frac{1}{\sqrt{R^2+b^2}} = \frac{R}{\sqrt{R^2+b^2}} \end{cases}$$

$$\mathbf{m} = m^1 \mathbf{a}_1 + m^2 \mathbf{a}_2 = \frac{-b}{\sqrt{R^2+b^2}} \mathbf{e}_\theta + \frac{R}{\sqrt{R^2+b^2}} \mathbf{e}_z, \quad \sigma = 0$$

The result that the geodesic curvature σ is zero agrees with the fact that the helix is a geodesic curve on the circular cylindrical surface.

Control of the result for the unit normal vector $\mathbf{m}$:

$$\mathbf{m}\cdot\mathbf{m} = \left(\frac{-b}{\sqrt{R^2+b^2}}\right)^2 + \left(\frac{R}{\sqrt{R^2+b^2}}\right)^2 = 1$$

$$\mathbf{m}\cdot\mathbf{t} = \left(\frac{-b}{\sqrt{R^2+b^2}}\right)\left(\frac{R}{\sqrt{R^2+b^2}}\right) + \left(\frac{R}{\sqrt{R^2+b^2}}\right)\left(\frac{b}{\sqrt{R^2+b^2}}\right) = 0$$

$$\mathbf{t}\times\mathbf{m} = \left(\frac{R}{\sqrt{R^2+b^2}}\mathbf{e}_\theta + \frac{b}{\sqrt{R^2+b^2}}\mathbf{e}_z\right) \times \left(\frac{-b}{\sqrt{R^2+b^2}}\mathbf{e}_\theta + \frac{R}{\sqrt{R^2+b^2}}\mathbf{e}_z\right)$$

$$= \frac{R}{\sqrt{R^2+b^2}}\frac{R}{\sqrt{R^2+b^2}}\mathbf{e}_\theta\times\mathbf{e}_z + \left(\frac{b}{\sqrt{R^2+b^2}}\right)\left(\frac{-b}{\sqrt{R^2+b^2}}\right)\mathbf{e}_z\times\mathbf{e}_\theta = \mathbf{e}_\theta\times\mathbf{e}_z = \mathbf{e}_R = \mathbf{a}_3$$

Example 8.6 Curve with Geodesic Curvature on a Circular Cylindrical Surface

On the cylindrical surface: $\mathbf{r} = \mathbf{r}(u^1, u^2) = R\mathbf{e}_R(u^1) + u^2\mathbf{e}_{z,,}$ with surface coordinates: $u_1 = \theta$ and $u_2 = z$, we shall investigate the curve $u^\alpha(s)$ defined by the place vector:

$$\mathbf{r} = \mathbf{r}(u_1, u_2) = R\mathbf{e}_R(u_1) + u_2\mathbf{e}_z, \quad u_1 = \theta, \ u_2 = z = R\theta^2/2 \Rightarrow$$
$$\mathbf{r} = \mathbf{r}(u_1) = \mathbf{r}(\theta) = \mathbf{r}(u_1) = R\mathbf{e}_R(\theta) + (R\theta^2/2)\mathbf{e}_z$$

The arc length formula (1.4.5) gives with $d\mathbf{e}_R/d\theta = \mathbf{e}_\theta$:

$$s(\theta) = \int\limits_0^\theta \sqrt{\frac{d\mathbf{r}}{d\theta} \cdot \frac{d\mathbf{r}}{d\theta}}\, d\theta \Rightarrow \frac{ds}{d\theta} = \sqrt{\frac{d\mathbf{r}}{d\theta} \cdot \frac{d\mathbf{r}}{d\theta}} = \sqrt{R^2 + R^2\theta^2} \Leftrightarrow \frac{d\theta}{ds} = \frac{1}{R\sqrt{1+\theta^2}}$$

The tangent vector $\mathbf{t}$ for the curve is:

$$\mathbf{t} = \frac{d\mathbf{r}}{ds} = \frac{d\mathbf{r}}{d\theta}\frac{d\theta}{ds} = \left(R\frac{d\mathbf{e}_R}{d\theta} + R\theta\mathbf{e}_z\right)\frac{d\theta}{ds}$$

$$= (R\mathbf{e}_\theta + R\theta\mathbf{e}_z)\frac{1}{R\sqrt{1+\theta^2}} = \frac{1}{R\sqrt{1+\theta^2}}\mathbf{a}_1 + \frac{\theta}{\sqrt{1+\theta^2}}\mathbf{a}_2 \Rightarrow$$

$$\mathbf{t} = \frac{d\mathbf{r}}{ds} = t^\alpha \mathbf{a}_\alpha = \frac{1}{\sqrt{1+\theta^2}}(\mathbf{e}_\theta + \theta\mathbf{e}_z) \Rightarrow t^1 = \frac{1}{R\sqrt{1+\theta^2}},\ t^2 = \frac{\theta}{\sqrt{1+\theta^2}}$$

The principal normal vector $\mathbf{n}$ and the curvature κ for the curve are, with $d\mathbf{e}_\theta/d\theta = -\mathbf{e}_R$ found from:

$$\kappa\mathbf{n} = \frac{d^2\mathbf{r}}{ds^2} = \left(R\frac{d\mathbf{e}_\theta}{d\theta} + R\mathbf{e}_z\right)\frac{d\theta}{ds}\frac{d\theta}{ds} = \frac{-1}{R(1+\theta^2)}\mathbf{e}_R + \frac{1}{R(1+\theta^2)}\mathbf{e}_z \Rightarrow$$

$$\mathbf{n} = \frac{-\sqrt{2}}{2}\mathbf{e}_R + \frac{\sqrt{2}}{2}\mathbf{e}_z,\ \kappa = \frac{\sqrt{2}}{R(1+\theta^2)}$$

From the formulas (8.5.7), (8.5.3), and (8.1.27) we obtain:

$$\frac{\delta t^\alpha}{\delta s} = t^\alpha|_\beta t^\beta = \left(t^\alpha{}_{,\beta} + t^\gamma \Gamma^\alpha_{\gamma\beta}\right)t^\beta = t^\alpha{}_{,\beta}\,t^\beta \equiv \frac{\partial t^\alpha}{\partial u^\beta}t^\beta \Rightarrow$$

$$\frac{\delta t^1}{\delta s} = t^1{}_{,1}\,t^1 = \frac{-1\cdot 2\theta}{2R\left(\sqrt{1+\theta^2}\right)^3}\frac{1}{R\sqrt{1+\theta^2}} = \frac{-\theta}{R^2(1+\theta^2)^2}$$

$$\frac{\delta t^2}{\delta s} = t^2{}_{,1}\,t^1 = \left(\frac{1}{\sqrt{1+\theta^2}} + \frac{-\theta\cdot 2\theta}{2\left(\sqrt{1+\theta^2}\right)^3}\right)\frac{1}{R\sqrt{1+\theta^2}} = \frac{1}{R(1+\theta^2)^2}$$

From the formulas (8.6.9)$_2$ and (8.6.10)$_1$ we get the results:

$$m^{\alpha} = \varepsilon^{\beta\alpha} t_{\beta} = \varepsilon^{\beta\alpha} a_{\beta\gamma} t^{\gamma} \Rightarrow m^1 = \frac{1}{\sqrt{\alpha}} e^{21} a_{22} t^2 = \frac{1}{R}(-1)1\frac{\theta}{\sqrt{1+\theta^2}} = \frac{-\theta}{R\sqrt{1+\theta^2}}$$

$$m^2 = \frac{1}{\sqrt{\alpha}} e^{12} a_{11} t^1 = \frac{1}{R} 1 \cdot R^2 \frac{1}{R\sqrt{1+\theta^2}} = \frac{1}{\sqrt{1+\theta^2}}$$

$$\mathbf{m} = m^1 \mathbf{a}_1 + m^2 \mathbf{a}_2 = \frac{-\theta}{\sqrt{1+\theta^2}} \mathbf{e}_{\theta} + \frac{R}{\sqrt{1+\theta^2}} \mathbf{e}_z$$

$$\sigma = \frac{\delta t^{\alpha}}{\delta s} m_{\alpha} = \frac{\delta t^{\alpha}}{\delta s} a_{\alpha\beta} m^{\beta} = \frac{\delta t^1}{\delta s} a_{11} m^1 + \frac{\delta t^2}{\delta s} a_{22} m^2$$

$$= \frac{-\theta}{R^2(1+\theta^2)^2} R^2 \frac{-\theta}{R\sqrt{1+\theta^2}} + \frac{1}{R(1+\theta^2)^2} 1 \frac{1}{\sqrt{1+\theta^2}} = \frac{1-\theta^2}{R(1+\theta^2)^{3/2}} \Rightarrow$$

$$\sigma = \frac{1}{R(1+\theta^2)^{3/2}} \quad \text{geodesic curvature}$$

Control of the result for the unit normal vector $\mathbf{m}$:

$$\mathbf{m} \cdot \mathbf{m} = \left(\frac{-\theta}{\sqrt{1+\theta^2}}\right)^2 + \left(\frac{1}{\sqrt{1+\theta^2}}\right)^2 = 1$$

$$\mathbf{m} \cdot \mathbf{t} = \left(\frac{-\theta}{\sqrt{1+\theta^2}}\right)\left(\frac{1}{\sqrt{1+\theta^2}}\right) + \left(\frac{1}{\sqrt{1+\theta^2}}\right)\left(\frac{\theta}{\sqrt{1+\theta^2}}\right) = 0$$

$$\mathbf{t} \times \mathbf{m} = \left(\frac{1}{\sqrt{1+\theta^2}} \mathbf{e}_{\theta} + \frac{\theta}{\sqrt{1+\theta^2}} \mathbf{e}_z\right) \times \left(\frac{-\theta}{\sqrt{1+\theta^2}} \mathbf{e}_{\theta} + \frac{1}{\sqrt{1+\theta^2}} \mathbf{e}_z\right) =$$

$$= \frac{1}{\sqrt{1+\theta^2}} \frac{1}{\sqrt{1+\theta^2}} \mathbf{e}_{\theta} \times \mathbf{e}_z + \left(\frac{\theta}{\sqrt{1+\theta^2}}\right)\left(\frac{-\theta}{\sqrt{1+\theta^2}}\right) \mathbf{e}_z \times \mathbf{e}_{\theta} = \mathbf{e}_{\theta} \times \mathbf{e}_z = \mathbf{e}_R = \mathbf{a}_3$$

Example 8.7 Curvatures of Spherical Surface

In Example 8.2 we have presented the fundamental parameters of second order $B_{\alpha\beta}$ for a spherical surface of radius r with the centre of the sphere at the origin of the Cartesian coordinate system Ox. These quantities are also components of the curvature tensor $\mathbf{B}$ for the spherical surface and will be recorded here together with the mixed components $B_{\alpha}^{\beta} = a^{\beta\gamma} B_{\alpha\gamma}$ of $\mathbf{B}$:

$$(B_{\alpha\beta}) = \begin{pmatrix} -r & 0 \\ 0 & -r\sin^2\theta \end{pmatrix}, \quad (B_{\alpha}^{\beta}) = \begin{pmatrix} -1/r & 0 \\ 0 & -1/r \end{pmatrix}$$

The coordinate lines are lines of curvature and we find for the principal curvatures, the mean curvature and the Gauss curvature:

$$\tilde{\kappa}_1 = \tilde{\kappa}_2 = -\frac{1}{r}, \quad \mu = -\frac{1}{r}, \quad \gamma = \frac{1}{r^2}$$

8.7.1 The Codazzi Equations and the Gauss Equation

From the formulas (8.5.28) and (8.1.18)$_1$ we compute:

$$\mathbf{a}_\alpha|_{\beta\gamma} = B_{\alpha\beta}|_\gamma \, \mathbf{a}_3 + B_{\alpha\beta}\mathbf{a}_3|_\gamma = B_{\alpha\beta}|_\gamma \, \mathbf{a}_3 - B_{\alpha\beta}B_{\gamma\rho}\mathbf{a}^\rho$$

and then by applying the formulas (8.5.36), we obtain the result:

$$\mathbf{a}_\alpha|_{\beta\gamma} - \mathbf{a}_\alpha|_{\gamma\beta} = R^\sigma_{\alpha\beta\gamma}\mathbf{a}_\sigma \Rightarrow \left(B_{\alpha\beta}|_\gamma - B_{\alpha\gamma}|_\beta\right)\mathbf{a}_3 - \left(B_{\alpha\beta}B_{\gamma\rho} - B_{\alpha\gamma}B_{\beta\rho}\right)\mathbf{a}^\rho = R_{\rho\alpha\beta\gamma}\,\mathbf{a}^\rho$$

Because $\mathbf{a}_3$ and $\mathbf{a}^\beta$ are independent vectors, we conclude that:

$$B_{\alpha\beta}|_\gamma - B_{\alpha\gamma}|_\beta = 0, \quad B_{\alpha\gamma}B_{\beta\rho} - B_{\alpha\beta}B_{\gamma\rho} = R_{\rho\alpha\beta\gamma}$$

Due to the symmetry properties of the curvature tensor $\mathbf{B}$ and the Riemann-Christoffel tensor $\mathbf{R}$, these two equations reduce to:

$$B_{\alpha 1}|_2 - B_{\alpha 2}|_1 = 0 \quad \text{the } \textit{Codazzi equations} \quad (\text{D.Codazzi } [1824 - 1875])$$
$$(8.7.13)$$

$$B_{11}B_{22} - B_{12}B_{12} = R_{1212} \quad \text{the } \textit{Gauss equation} \qquad (8.7.14)$$

From the Gauss equation we derive the result, confer the formula (8.5.42):

$$\gamma = \det\left(B^\alpha_\beta\right) = \frac{\det\left(B_{\alpha\beta}\right)}{\det\left(a_{\alpha\beta}\right)} = \frac{R_{1212}}{\alpha} \qquad (8.7.15)$$

It may be shown that the form of a surface in space, apart from its position in space, is uniquely determined if the following is specified in relation to a surface coordinate system u: The *fundamental parameters of first order* $a_{\alpha\beta}(u)$ and the *fundamental parameters of second order* $B_{\alpha\beta}(u)$ are given, the metric $a_{\alpha\beta}\,du^\alpha du^\beta$ is positive definite, and $B_{\alpha\beta}(u)$ satisfy the Godazzi equations and the Gauss equation.

The *fundamental parameters of third order* for a coordinate system u are defined by the components:

$$C_{\alpha\beta} = \underset{\sim}{a}{}^i|_\alpha a_i|_\beta \qquad (8.7.16)$$

The components $\underset{\sim}{a}{}^i|_\alpha$ and $a_i|_\beta$ are tensor-derivatives of the surface normal $\mathbf{a}_3$. The components $C_{\alpha\beta}$ represent a symmetric 2 second order surface tensor $\mathbf{C}$. It follows from Weingarten's formula (8.5.45) that:

$$\mathbf{C} = \mathbf{B}^2 \Leftrightarrow C_{\alpha\beta} = B_\alpha^\gamma B_{\gamma\beta} \tag{8.7.17}$$

It may be shown that, see Problem 8.16:

$$C_{\alpha\beta} - 2\mu B_{\alpha\beta} + \gamma a_{\alpha\beta} = 0 \tag{8.7.18}$$

Problem 8.1–8.16 with solutions see Appendix.

Reference

1. Sokolnikoff IS (1951) Tensor analysis. Wiley, New York

Chapter 9
Integral Theorems

9.1 Integration Along a Space Curve

In this section we shall use a Cartesian coordinate system Ox in the three-dimensional Euclidean space. A function in space is given by the field $f(\mathbf{r}) = f(x_1, x_2, x_3)$. A curve C in space is given by the place vector $\mathbf{r}(s)$, where s is the arc length parameter. The integral of the field $f(\mathbf{r})$ along the curve C from a place $\mathbf{r}_1 = \mathbf{r}(s_1)$ to a place $\mathbf{r}_2 = \mathbf{r}(s_2)$ is defined by:

$$\int_C f(\mathbf{r})\, ds \equiv \int_{s_1}^{s_2} f(\mathbf{r}(s))\, ds \qquad (9.1.1)$$

For the integration of a field $f(\mathbf{r})$ along a *closed curve* C in the x_1x_2-plane we introduce as positive direction of integration the *clockwise direction* when looking in the positive x_3-direction, i.e. counter-clockwise for the curve C shown in Fig. 9.1. We use the following symbol for integration along a closed space curve C:

$$\oint_C f(\mathbf{r})\, ds = \text{integration of field } f(\mathbf{r}) \text{ along a closed space curve } C \qquad (9.1.2)$$

Theorem 9.1. Integration Independent of the Integration Path *Let* $\mathbf{a}(\mathbf{r})$ *be a vector field and* $\mathbf{r}_1$ *and* $\mathbf{r}_2$ *two places in space. Then:*

(1) *If the vector field may be expressed as the gradient of a scalar field* $\alpha(\mathbf{r})$:

$$\mathbf{a} = \operatorname{grad} \alpha \Leftrightarrow a_i = \alpha_{,i} \qquad (9.1.3)$$

© Springer Nature Switzerland AG 2019
F. Irgens, *Tensor Analysis*,
https://doi.org/10.1007/978-3-030-03412-2_9

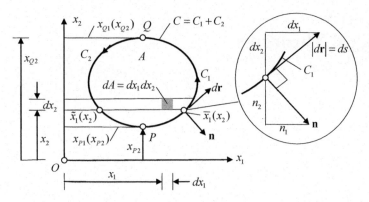

Fig. 9.1 Surface A in the $x_1 x_2$-plane bordered by the curve $C = C_1$ (from P to Q) + C_2 (from Q to P). Unit normal $\mathbf{n} = [n_1, n_2]$ to the curve. Line element $d\mathbf{r} = [dx_1, dx_2], |d\mathbf{r}| = ds$

then the integral

$$\int_{\mathbf{r}_1}^{\mathbf{r}_2} \mathbf{a} \cdot d\mathbf{r} \equiv \int_{\mathbf{r}_1}^{\mathbf{r}_2} a_i \, dx_i \text{ is independent of the integration path between } \mathbf{r}_1 \text{ and } \mathbf{r}_2$$

$$(9.1.4)$$

(2) If the proposition (9.1.4) is true, then the vector field $\mathbf{a}(\mathbf{r})$ may be expressed as the gradient of a scalar field $\alpha(\mathbf{r})$, such that the result (9.1.3) is true.

Proof of (1) The assumption (9.1.3) $\Rightarrow$ the result (9.1.4. Formula (9.1.3)$\Rightarrow \mathbf{a} \cdot d\mathbf{r} = \alpha_{,i} \, dx_i = d\alpha$. The integral from $\mathbf{r}_1$ to $\mathbf{r}_2$ of the vector field $\mathbf{a}$ ($\mathbf{r}$) becomes:

$$\int_{\mathbf{r}_1}^{\mathbf{r}_2} \mathbf{a} \cdot d\mathbf{r} = \int_{\mathbf{r}_1}^{\mathbf{r}_2} a_i \, dx_i = \int_{\mathbf{r}_1}^{\mathbf{r}_2} \alpha_{,i} \, dx_i = \int_{\mathbf{r}_1}^{\mathbf{r}_2} d\alpha(\mathbf{r}) = [\alpha(\mathbf{r})]_{\mathbf{r}_1}^{\mathbf{r}_2} = \alpha(\mathbf{r}_2) - \alpha(\mathbf{r}_1)$$

which is independent of the integration path between the places $\mathbf{r}_1$ and $\mathbf{r}_2$. Thus the assumption (9.1.3) implies the result (9.1.4).

Proof of (2) The assumption (9.1.4) $\Rightarrow$ the result (9.1.3). Let $\mathbf{r}_1$ be a fixed place and $\mathbf{r}_2 = \mathbf{r} = x_i \mathbf{e}_i (x_1 + h, x_2, x_3)$. The integral in assumption (9.1.4) may be considered to be a scalar field $\alpha(\mathbf{r})$:

$$\alpha(\mathbf{r}) = \int_{\mathbf{r}_1}^{\mathbf{r}} \mathbf{a} \cdot d\mathbf{r}$$

Let $\mathbf{r}_3 = \mathbf{r} + h\mathbf{e}_1 = (x_1 + h)\mathbf{e}_1 + x_2\mathbf{e}_2 + x_3\mathbf{e}_3$, where h is any scalar variable. Then the assumption (9.1.4) implies that:

$$\overset{\mathbf{r}_3}{\underset{\mathbf{r}_1}{\int}} \mathbf{a} \cdot d\mathbf{r} \equiv \overset{\mathbf{r}}{\underset{\mathbf{r}_1}{\int}} a_i \, dx_i + \overset{\mathbf{r}+h\mathbf{e}_1}{\underset{\mathbf{r}}{\int}} a_1 \, dx_1 = \alpha(x_1 + h, x_2, x_3) = \alpha(x_1, x_2, x_3) + \overset{x_1+h}{\underset{x_1}{\int}} a_1 \, dx_1 \Rightarrow$$

$$\tfrac{\partial}{\partial h}\left(\alpha(x_1 + h, x_2, x_3)\right)\big|_{h=0} = \tfrac{\partial}{\partial h}\left(\overset{x_1+h}{\underset{x_1}{\int}} a_i \, dx_i\right)\Bigg| \Rightarrow \tfrac{\partial \alpha(x_1, x_2, x_3)}{\partial x_1} = \tfrac{\partial \alpha(\mathbf{r})}{\partial x_1} = a_1(x_1, x_2, x_3)$$

The result is generalized to: $a_i = \partial \alpha / \partial x_i \equiv \alpha_{,i} \Rightarrow \mathbf{a} = \mathrm{grad}\,\alpha \Rightarrow$ (9.1.3). Thus the assumption (9.1.4) implies the result (9.1.3). This completes the proof of theorem 9.1.

Theorem 9.2 *Let $f(x)$ and $g(x)$ be continuous functions of the variable x and I the integral:*

$$I = \int_{x_1}^{x_2} f(x)\,g(x)\,dx$$

If the integral I vanishes for any function $g(x)$ satisfying the conditions $g(x_1) = g(x_2) = 0$, then:

$$f(x) = 0 \quad \text{for } x_1 \le x \le x_2$$

Proof Assume that: $f(\bar{x}) > 0$ for some value of $\bar{x}$ in the interval: $x_1 \le \bar{x} \le x_2$. Continuity of the function $f(x)$ implies then that there exists a value $\varepsilon > 0$ such that:

$$f(x) > 0 \quad \text{for } \bar{x} - \varepsilon \le x \le \bar{x} + \varepsilon$$

Select the following function:

$$g(x) = 0 \quad \text{for } x_1 \le x < \bar{x} - \varepsilon \text{ and } \bar{x} + \varepsilon < x \le x_2$$
$$g(x) > 0 \quad \text{for } \bar{x} - \varepsilon \le x \le \bar{x} + \varepsilon$$

Under these conditions the value of the integral I is positive, which contradicts the assumption $f(\bar{x}) > 0$. Hence, if the integral I vanishes for any function $g(x)$ satisfying the condition $g(x_1) = g(x_2) = 0$, then the function $f(x) = 0$ for all values of x in the interval $x_1 \le x \le x_2$. This completes the proof of Theorem 9.2.

9.2 Integral Theorems in a Plane

Figure 9.1 shows a plane surface A bounded by the curve C in a two-dimensional Cartesian coordinate system Ox. The area A of the surface may be calculated from the double integral:

$$A = \int_A dA = \iint_A dx_1 dx_2 \equiv \int_A dx_1 dx_2 \qquad (9.2.1)$$

For simplicity we use the symbol A for the surface as well as for the region that the surface occupies in the $x_1 x_2$-plane, and the area A of the surface. The symbol $dA = dx_1\, dx_2$ is called the *differential element of area in the Cartesian coordinate system Ox.*

The double integral in formula (9.2.1) is further expanded to:

$$A = \int_A dA = \int_A dx_1 dx_2 = \int_{x_{P2}}^{x_{Q2}} \left[\int_{\tilde{x}_1(x_2)}^{\bar{x}_1(x_2)} dx_1 \right] dx_2 \Rightarrow$$

$$A = \int_{x_{P2}}^{x_{Q2}} [\bar{x}_1(x_2) - \tilde{x}_1(x_2)] \, dx_2 = \int_{x_{P2}}^{x_{Q2}} \bar{x}_1(x_2)\, dx_2 - \int_{x_{P2}}^{x_{Q2}} \tilde{x}_1(x_2)\, dx_2 \qquad (9.2.2)$$

The first integral on the right-hand side of Eq. (9.2.2) represents the area within the boundary described by the curve C_1 from the point P to the point Q, the x_2-axis, and the horizontal lines for $x_2 = x_{2P}$ and $x_2 = x_{2Q}$. The second integral on the right-hand side of Eq. (9.2.2) represents the area within the boundary described by the curve C_2 from the point P to the point Q, the x_2-axis, and the horizontal lines for $x_2 = x_{2P}$ and $x_2 = x_{2Q}$. With the parameter s as the arc length parameter along the curves C, C_1, and C_2, and $\mathbf{n}$ as the unit normal vector pointing out from the curve $C = C_1 + C_2$, we find from Fig. 9.1 that along the curves C, C_1, and C_2 :

$$dx_1 = -n_2\, ds, \quad dx_2 = n_1\, ds \Leftrightarrow dx_\alpha = -e_{\alpha\beta}\, n_\beta\, ds \qquad (9.2.3)$$

The integrals on the right-hand side of Eq. (9.2.2) may now be presented as:

$$A_1 = \int_{x_{P2}}^{x_{Q2}} \bar{x}_1(x_2)\, dx_2 = \int_{C_1} \bar{x}_1(x_2(s))\, n_1 ds,$$

$$A_2 = \int_{x_{P2}}^{x_{Q2}} \tilde{x}_1(x_2)\, dx_2 = - \int_{C_2} \tilde{x}_1(x_2(s))\, n_1 ds$$

The area A of the plane surface A bounded by the curve C then becomes:

$$A = A_1 - A_2 = \int_{C_1} \bar{x}_1(x_2(s))\, n_1 ds + \int_{C_2} \tilde{x}_1(x_2(s))\, n_1 ds = \oint_C x_1(x_2(s))\, n_1 ds \Rightarrow$$

$$A = \int_A dx_1\, dx_2 = \oint_C x_1(x_2)\, dx_2 = \oint_C x_1(x_2(s))\, n_1\, ds \qquad (9.2.4)$$

For a surface integral of a function $f(\mathbf{r}) = f(x_1, x_2)$ over a surface A in the $x_1 x_2$-plane we use the symbols:

$$\int_A f(\mathbf{r})\, dA = \int_A f(\mathbf{r})\, dx_1 dx_2 \tag{9.2.5}$$

Theorem 9.3. Gauss' Integral Theorem in a Plane *Let A be a surface in the $x_1 x_2$-plane and bordered by the curve C, Fig. 9.1. The unit normal $\mathbf{n}$ to C lies in the $x_1 x_2$-plane and is pointing out from the curve C. Then for any field function $f(\mathbf{r}) = f(x_1, x_2)$ the following result holds true:*

$$\oint_C f\,\mathbf{n}\, ds = \int_A \operatorname{grad} f\, dA \Leftrightarrow \oint_C f\, n_\alpha\, ds = \int_A f_{,\alpha}\, dA \tag{9.2.6}$$

Proof The proof will be given for the index value $\alpha = 1$. First we consider the situation in Fig. 9.1, in which a straight line parallel to the x_1-axis only intersects the curve C in two points $\tilde{x}_1(x_2)$ and $\bar{x}_1(x_2)$. Then:

$$\int_A f_{,1}\, dA = \int_{x_{P2}}^{x_{Q2}} \left[\int_{\tilde{x}_1}^{\bar{x}_1} f_{,1}\, dx_1 \right] dx_2 = \int_{x_{P2}}^{x_{Q2}} \left[f(\bar{x}_1(x_2), x_2) - f(\tilde{x}_1(x_2), x_2) \right] dx_2 \Rightarrow$$

$$\int_A f_{,1}\, dA = \int_{x_{P2}}^{x_{Q2}} f(\bar{x}_1(x_2), x_2)\, dx_2 - \int_{x_{P2}}^{x_{Q2}} f(\tilde{x}_1(x_2), x_2)\, dx_2$$

$$\tag{9.2.7}$$

Following the procedures that led from the Eq. (9.2.1) to the Eq. (9.2.4), we may present the result (9.2.7) as:

$$\int_A f_{,1}\, dA = \int_A f_{,1}\, dx_1\, dx_2 = \oint_C f(x_1, x_2)\, n_1 \tag{9.2.8}$$

Thus the formula $(9.2.6)_2$ has been proved for the index value $\alpha = 1$ when a straight line parallel to the x_1-axis intersects the curve C in only two points $\tilde{x}_1(x_2)$ and $\bar{x}_1(x_2)$. For the special case: $f = x_1$, Eq. (9.2.6) provides the result (9.2.4).

If a straight line parallel to the x_1-axis intersects the curve C in more than two points, as indicated in Fig. 9.2, the surface A may be divided into parts as shown, each of which satisfies the condition of only two points of intersection between a line parallel to the x_1-axis and the bordering curve.

The result (9.2.6) is now applied to each part of the area A and the results are added. The contributions to the integrals along the lines, $\bar{C}$ in Fig. 9.2, dividing the

Fig. 9.2 Surface A in the x_1x_2-plane bordered by the curve $C = C_1 + C_2 + C_3...$ A straight line parallel with the x_1-axis intersects the curve C in more than two points

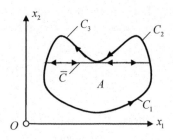

area A, add up to zero. Thus the Gauss's theorem (9.2.6) has been proved in general for the index value $\alpha = 1$, and this result is generalized as a general proof of the formula (9.2.6).

Theorem 9.4. The Divergence Theorem in a Plane *Let A be a surface in the x_1x_2-plane and bordered by the curve C, Figs. 9.1 and 9.2. The unit normal $\mathbf{n}$ to C lies in the x_1x_2-plane and is pointing out from the curve C. Then for any vector field $\mathbf{a}(\mathbf{r}) = \mathbf{a}(x_1, x_2)$ in the x_1x_2-plane the following result holds true:*

$$\int_A \text{div}\,\mathbf{a}\,dA = \oint_C \mathbf{a}\cdot\mathbf{n}\,ds \Leftrightarrow \int_A a_{\alpha,\alpha}dA = \oint_C a_\alpha n_\alpha\,ds \qquad (9.2.9)$$

Proof: Theorem 9.2 is applied to each components $a_\alpha(x_1, x_2)$ of the vector $\mathbf{a}$ and the results are added to give formula (9.2.9).

Theorem 9.5 Stokes' Theorem in a Plane *Let A be a surface in the x_1x_2-plane and bordered by the curve C, Figs. 9.1 and 9.2. The unit normal vector to the plane area A is the base vector $\mathbf{e}_3$ of the Cartesian coordinate system Ox. The unit tangent vector to the curve is $\mathbf{t} = d\mathbf{r}/ds$. The Stokes' theorem in a plane states that for any vector field $\mathbf{a}(\mathbf{r}) = \mathbf{a}(x_1, x_2)$ the following result holds true:*

$$\int_A (\nabla \times \mathbf{a})\cdot\mathbf{e}_3\,dA = \oint_C \mathbf{a}\cdot\mathbf{t}\,ds \qquad (9.2.10)$$

Proof Applying the formulas (9.2.3), with $\mathbf{e}_i$ as the base vectors of the system Ox, we obtain for the tangent vector $\mathbf{t}$:

$$\mathbf{t} = \frac{d\mathbf{r}}{ds} = \frac{\partial\mathbf{r}}{\partial x_\alpha}\frac{dx_\alpha}{ds} = \mathbf{e}_\alpha\frac{dx_\alpha}{ds} = -\mathbf{e}_1 n_2 + \mathbf{e}_2 n_1 \qquad (9.2.11)$$

The integral on the right-hand side of the Eq. (9.2.10) is rewritten to:

$$\oint_C \mathbf{a}\cdot\mathbf{t}\,ds = \oint_C a_\alpha t_\alpha\,ds = \oint_C [a_2 n_1 - a_1 n_2]\,ds$$

Application of Theorem 9.3 to the integral on the left-hand side of the Eq. (9.2.10) gives:

$$\int_A (\nabla \times \mathbf{a}) \cdot \mathbf{e}_3 \; dA = \int_A \left(e_{ijk} a_{k,j} \, \mathbf{e}_i \right) \cdot \mathbf{e}_3 \; dA = \int_A [a_{2,1} - a_{1,2}] \, dA$$

$$= \oint_C [a_2 n_1 - a_1 n_2] \, ds$$

Hence, the integrals on both sides of the Eq. (9.2.10) are shown to be equal. This result proves Stokes' theorem in a plane.

Stokes' theorem is also called Green's theorem, George Green [1793–1841].

9.2.1   Integration Over a Plane Region in Curvilinear Coordinates

In the $x_1 x_2$-plane of a Cartesian coordinate system Ox we introduce *surface coordinates* u^α, Fig. 9.3a, such that there exits a one - to - one correspondence between the point (x_1, x_2) and the coordinate pair (u^1, u^2):

$$x_\alpha = x_\alpha(u^1, u^2) \Leftrightarrow u^\alpha = u^\alpha(x_1, x_2) \tag{9.2.12}$$

We say that the formulas (9.2.12) represent a one-to-one *mapping between the coordinates* (x_1, x_2) *and the coordinates* (u^1, u^2). The functions $x_\alpha(u^1, u^2)$ and $u^\alpha = u^\alpha(x_1, x_2)$ must satisfy the following requirements:

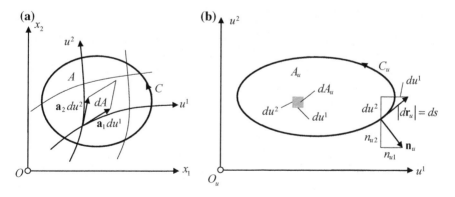

Fig. 9.3 a Surface A in the $x_1 x_2$-plane. Surrounding curve C. Curvilinear coordinate u_α. Surface element $dA = J du^1 du^2$. **b** "Cartesian" coordinate system $O_u u$. Surface A_u. Surrounding curve C_u. Surface element $dA_u = du^1 du^2$

(a) $x_\alpha = x_\alpha(u^1, u^2)$ are continuous functions of u^α with continuous partial derivative of first order, and $u^\alpha = u^\alpha(x_1, x_2)$ are continuous functions of x_α with continuous partial derivative of first order.

(b) The *Jacobi determinants* $J \equiv J_u^x$ and J_x^u of the mapping must be non-zero:

$$J \equiv J_u^x = \det\left(\frac{\partial x_\alpha}{\partial u^\beta}\right) \equiv \det\begin{pmatrix} \frac{\partial x_1}{\partial u^1} & \frac{\partial x_1}{\partial u^2} \\ \frac{\partial x_2}{\partial u^1} & \frac{\partial x_2}{\partial u^2} \end{pmatrix} = \frac{\partial x_1}{\partial u^1}\frac{\partial x_1}{\partial u^2} - \frac{\partial x_1}{\partial u^2}\frac{\partial x_2}{\partial u^1} \neq 0$$

$$J_x^u = \det\left(\frac{\partial u^\alpha}{\partial x_\beta}\right) \equiv \det\begin{pmatrix} \frac{\partial u^1}{\partial x_1} & \frac{\partial u^1}{\partial x_2} \\ \frac{\partial u^2}{\partial x_1} & \frac{\partial u^2}{\partial x_2} \end{pmatrix} = \frac{\partial u^1}{\partial x_1}\frac{\partial u^2}{\partial x_2} - \frac{\partial u^1}{\partial x_2}\frac{\partial u^2}{\partial x_1} \neq 0$$

$$(9.2.13)$$

When one u-coordinate is kept constant, e.g. $u^2 = u^{2c}$, the functions $x_\alpha = x_\alpha(u^1, u^{2c})$ describe a *coordinate line*, a u^1-line, in the $x_1 x_2$-plane. Coordinate lines are shown in Fig. 9.3a. A surface A in the $x_1 x_2$-plane is bordered by the closed curve C. Figure 9.3b shows images A_u and C_u in a "Cartesian" coordinate system $O_u u$ of the area A and the curve C. The formulas (9.2.12) represent a one-to-one mapping of points (x_1, x_2) in the region A in the $x_1 x_2$-plane to points (u^1, u^2) in the region A_u in the "Cartesian" coordinate system $O_u u$.

It will now be shown that the area A of the surface is given by the formula:

$$A = \int_A dA = \int_A dx_1\, dx_2 = \int_{A_u} J\, du^1\, du^2 \Rightarrow A = \int_{A_u} \sqrt{\alpha}\, du^1\, du^2 \qquad (9.2.14)$$

du^1 and du^2 are differentials of the *curvilinear coordinates* u^α, as illustrated in Fig. 9.3a.

From the formula (9.2.4) we obtain for the area A:

$$A = \int_A dx_1\, dx_2 = \oint_C x_1\, dx_2 \qquad (9.2.15)$$

We introduce the arc length parameter σ for the curve C_u and apply the results (9.2.3), where $\mathbf{n}_u$ is now the unit normal vector to the curve C_u in the $u^1 u^2$-plane in Fig. 9.3b. Then we may write:

$$dx_2 = \frac{\partial x_2}{\partial u^\alpha}\frac{du^\alpha}{d\sigma}\, d\sigma = \frac{\partial x_2}{\partial u^1}\frac{du^1}{d\sigma}\, d\sigma + \frac{\partial x_2}{\partial u^2}\frac{du^2}{d\sigma}\, d\sigma = -\frac{\partial x_2}{\partial u^1} n_{u2}\, d\sigma + \frac{\partial x_2}{\partial u^2} n_{u1}\, d\sigma$$

$$(9.2.16)$$

From the formula (9.2.4) and the result (9.2.16) we get:

$$A = \int_A dx_1\, dx_2 = \oint_C x_1\, dx_2 = \oint_{C_u} \left[-x_1 \frac{\partial x_2}{\partial u^1} n_{u2} + x_1 \frac{\partial x_2}{\partial u^2} n_{u1} \right] d\sigma \qquad (9.2.17)$$

By Gauss' integral theorem in a plane (9.2.6) applied to the area A_u and the surrounding curve C_u, we obtain:

$$A = \int_A dx_1\, dx_2 = \oint_{C_u} \left[-x_1 \frac{\partial x_2}{\partial u^1} n_{u2} + x_1 \frac{\partial x_2}{\partial u^2} n_{u1} \right] d\sigma$$

$$= \int_{A_u} \left[-\frac{\partial}{\partial u^2} \left(x_1 \frac{\partial x_2}{\partial u^1} \right) + \frac{\partial}{\partial u^1} \left(x_1 \frac{\partial x_2}{\partial u^2} \right) \right] du^1\, du^2 \Rightarrow$$

$$A = \int_{A_u} \left[-\frac{\partial x_1}{\partial u^2} \frac{\partial x_2}{\partial u^1} - x_1 \frac{\partial^2 x_2}{\partial u^2 \partial u^1} + \frac{\partial x_1}{\partial u^1} \frac{\partial x_2}{\partial u^2} + x_1 \frac{\partial^2 x_2}{\partial u^1 \partial u^2} \right] du^1\, du^2$$

$$= \int_{A_u} \left[-\frac{\partial x_1}{\partial u^2} \frac{\partial x_2}{\partial u^1} + \frac{\partial x_1}{\partial u^1} \frac{\partial x_2}{\partial u^2} \right] du^1\, du^2 \Rightarrow A = \int_A dx_1\, dx_2 == \int_{A_u} J\, du^1\, du^2 \Rightarrow (9.2.14)_1$$

The result $(9.2.14)_2$ is found as follows. First we introduce the base vectors $\mathbf{a}_\alpha$ and the *fundamental parameters* $a_{\alpha\beta}$ for the u-system referring to the formulas (8.1.2), (8.1.7), (8.1.8) and (8.1.13):

$$\mathbf{a}_\alpha = \frac{\partial \mathbf{r}}{\partial u_\alpha}, \qquad a_{\alpha\beta} = \mathbf{a}_\alpha \cdot \mathbf{a}_\beta, \qquad \alpha = \det(a_{\alpha\beta}), \qquad \mathbf{a}_1 \times \mathbf{a}_2 = \sqrt{\alpha}\, \mathbf{a}_3 \qquad (9.2.18)$$

$\mathbf{a}_3$ is a unit normal vector to the surface, and will in this chapter be denoted by $\mathbf{n}$, i.e. $\mathbf{n} = \mathbf{a}_3$:

$$\mathbf{n} = \frac{\mathbf{a}_1 \times \mathbf{a}_2}{|\mathbf{a}_1 \times \mathbf{a}_2|} \Rightarrow \mathbf{n} \cdot \mathbf{a}_\alpha = 0 \qquad (9.2.19)$$

The three vectors $\mathbf{a}_\alpha$ and $\mathbf{n}$ represent a *right-handed system*, and the unit normal vector $\mathbf{n}$ points out from what will be defined as the *positive side A^+ of the surface*. Because the Cartesian coordinate x_3 is independent of the coordinates u^α, we obtain from the formulas (8.1.7) and (8.1.8) that:

$$a_{\alpha\beta} = \frac{\partial x_\gamma}{\partial u^\alpha} \frac{\partial x_\gamma}{\partial u^\beta} \Rightarrow \alpha = \det(a_{\alpha\beta}) = \det\left(\frac{\partial x_\gamma}{\partial u^\alpha} \frac{\partial x_\gamma}{\partial u^\beta} \right) = \left[\det\left(\frac{\partial x_\alpha}{\partial u^\beta} \right) \right]^2 = J^2 \Rightarrow$$
$$J = \sqrt{\alpha} = \sqrt{\det(a_{\alpha\beta})} \qquad (9.2.20)$$

Now the result $(9.2.14)_2$ follows from $(9.2.14)_1$

The differential line elements along the coordinate lines are defined by the vectors:

$$ds_\alpha = \mathbf{a}_\alpha \, du^\alpha \tag{9.2.21}$$

The *element of area* is represented by scalar dA, shown in Fig. 9.3a, and by the vector $d\mathbf{A}$:

$$d\mathbf{A} = d\mathbf{s}_1 \times d\mathbf{s}_2 = \mathbf{n} \, dA \tag{9.2.22}$$

From the formulas (9.2.22, 9.2.21, and 9.2.18) we obtain:

$$d\mathbf{A} = d\mathbf{s}_1 \times d\mathbf{s}_2 = \left(\mathbf{a}_1 du^1\right) \times \left(\mathbf{a}_2 du^2\right) = \mathbf{a}_1 \times \mathbf{a}_2 du^1 du^2 = \sqrt{\alpha} \, du^1 du^2 \mathbf{n} \Rightarrow$$
$$d\mathbf{A} = \sqrt{\alpha} \, du^1 du^2 \, \mathbf{n}, \quad dA = \sqrt{\alpha} \, du^1 du^2$$

$$\tag{9.2.23}$$

$$A = \int_A dA = \int_A dx_1 \, dx_2 = \int_{A_u} \sqrt{\alpha} \, du^1 \, du^2 \tag{9.2.24}$$

Example 9.1 Area of a Quarter Circle

Figure 9.4 shows a quarter of a circle of radius r. Polar coordinates are chosen as curvilinear coordinates: $u^1 = R, u^2 = \theta$. The mapping (9.2.12) is given by:

$$x_\alpha = x_\alpha(u^1, u^2) \Rightarrow x_1 = R \cos \theta = u^1 \cos u^2, \quad x_2 = R \sin \theta = u^1 \sin u^2$$

The Jacobian J for the mapping becomes, according to the formulas (9.2.13):

$$J = \det \begin{pmatrix} \frac{\partial x_1}{\partial u^1} & \frac{\partial x_1}{\partial u^2} \\ \frac{\partial x_2}{\partial u^1} & \frac{\partial x_2}{\partial u^2} \end{pmatrix} = \det \begin{pmatrix} \cos u_2 & -u_1 \sin u_2 \\ \sin u_2 & u_1 \cos u_2 \end{pmatrix} = \sqrt{\alpha} = u^1 = R$$

The coordinate differentials are: $du^1 = dR$ and $du^2 = d\theta$. The area of the quarter circle is found from:

Fig. 9.4 Area A of a quarter circle. Cartesian coordinate system Ox. Polar coordinates $R\theta$. Curvilinear coordinates $u^1 = R, u^2 = \theta$. Differential element of area $dA = J du^1 du^2$

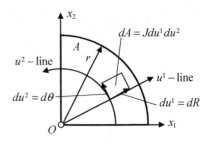

$$A = \int_A dA = \int_{A_u} J\, du^1\, du^2 = \int_{A_u} R\, dR\, d\theta = \int_0^{\pi/2} \left(\int_0^r R\, dR \right) d\theta = \int_0^{\pi/2} \frac{r^2}{2}\, d\theta = \frac{\pi r^2}{4}$$

We now turn to a general definition of double integral of a function $f(x_1, x_2)$ over a region A with area A in the $x_1 x_2$-plane in the Cartesian coordinate system Ox:

$$I = \int_A f(x_1, x_2)\, dx_1\, dx_2 \tag{9.2.25}$$

In the $x_1 x_2$-plane we introduce curvilinear coordinate system u through a mapping (9.2.12) with the Jacobian J from the formulas (9.2.13). Let A_u be the region in the $u^1 u^2$-plane corresponding to the region A in the $x_1 x_2$-plane Fig. 9.3b. It will then be shown that:

$$I = \int_A f\, dA \equiv \int_A f(x_1, x_2)\, dx_1\, dx_2 = \int_{A_u} f(x_1(u), x_2(u)) J\, du^1\, du^2 \tag{9.2.26}$$

The region A is divided into n small regions A_n of areas A_n, and with $x_{\alpha n}$ as an arbitrarily chosen point in A_n. Let δ be the maximum diameter (extension) of all the regions A_n. The general definition of the double integral of the function $f(x_1, x_2)$ over the region A may then be presented as:

$$I = \int_A f(x_1, x_2)\, dx_1\, dx_2 = \lim_{\delta \to 0} \sum_n f(x_{1n}, x_{2n}) A_n \tag{9.2.27}$$

Let $f_{n,\min}$ be the minimum value and $f_{n,\max}$ the maximum value of the function $f(x_{1n}, x_{2n})$ in the region A_n. such that: $f_{n,\min} \le f(x_{1n}, x_{2n}) \le f_{n,\max}$ Then it follows that:

Formula $(9.2.14) \Rightarrow A_n = \int_{A_{nu}} J\, du^1\, du^2$, and

$$f_{n,\min} A_n = f_{n,\min} \int_{A_{nu}} J\, du^1\, du^2 \le \int_{A_{nu}} f(x_1(u), x_2(u)) J\, du^1\, du^2 \le f_{n,\max} \int_{A_{nu}} J\, du^1\, du^2 = f_{n,\max} A_n \Rightarrow$$

$$L_n \equiv \sum_n f_{n,\min} A_n \le \sum_n \int_{A_{nu}} f(x_1(u), x_2(u)) J\, du^1\, du^2 \le \sum_n f_{n,\max} A_n \equiv U_n$$

$$L_n \equiv \sum_n f_{n,\min} A_n \le \sum_n f(x_{1n}, x_{2n}) A_n \le \sum_n f_{n,\max} A_n \equiv U_n$$

$$\tag{9.2.28}$$

The bounds:L_n(lower bound) and U_n(upper bound) approach each other when we let: $n \to \infty$ and $\delta \to 0$, we conclude from the result (9.2.28) and formula (9.2.27) that:

$$I = \int_A f \, dA \equiv \int_A f(x_1, x_2) dx_1 \, dx_2 = \int_{A_u} f(x_1(u), x_2(u)) J \, du^1 \, du^2$$
$$= \lim_{\delta \to 0} \sum_n f(x_{1n}, x_{2n}) A_n$$

This result proves the formula (9.2.26).

9.3   Integral Theorems in Space

As presented in Sect. 8.1 a two-dimensional surface imbedded in a three-dimensional *Euclidean space* E_3, Fig. 8.1 and Fig. 9.5, is defined by the place vector $\mathbf{r}$ as a function of two parameters u^1 and u^2 called *surface coordinates*:

$$\mathbf{r} = \mathbf{r}(u) \equiv \mathbf{r}(u^1, u^2) \Leftrightarrow x_i = x_i(u) \equiv x_i(u^1, u^2), \quad y^i = y^i(u) \equiv y^i(u^1, u^2)$$
$$(9.3.1)$$

The base vectors and the fundamental parameters for the u − system are as given by the formulas (9.2.18). The unit normal vector $\mathbf{n}$ for the surface is shown in Figs. 9.5 and 9.6, and defined by the formulas (9.2.19).

The differential line elements ds_α along the coordinate lines and the differential element of area $d\mathbf{A}$ are defined by the vectors in the formulas (9.2.21–9.2.23), and shown in Fig. 9.6. A Cartesian coordinate system Ox is shown in Fig. 9.5. The parameters y^i are general curvilinear coordinates in E_3.

Fig. 9.5 Two-dimensional surface imbedded in three-dimensional space E_3. Cartesian coordinate system Ox. Surface coordinates u^1 and u^2. Coordinate lines. Base vectors $\mathbf{a}_1$ and $\mathbf{a}_2$. Unit normal vector $\mathbf{n}$

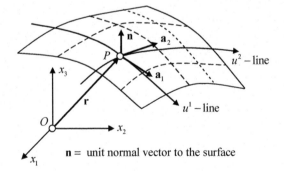

$\mathbf{n}$ = unit normal vector to the surface

Fig. 9.6 Line elements ds_1 and ds_2. Differential element of area $d\mathbf{A} = \mathbf{n}\, dA$

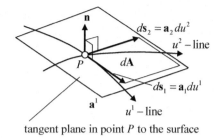

tangent plane in point P to the surface

9.3.1   Integration Over a Surface Imbedded in the Euclidean Space E_3

Based on the development in Sect. 9.2.1 it seems reasonable to define the area of a curved surface by the integral:

$$A = \int_A dA = \int_{A_u} \sqrt{\alpha}\, du^1\, du^2 \qquad (9.3.2)$$

In addition to comply with the concept of the area of a plane surface in the $x_1 x_2$-plane, this definition of area satisfies, as we shall show, the natural requirement that the integral expressing the area is coordinate invariant with respect to both space coordinates and surface coordinates.

Example 9.2 Area of the Surface of a Sphere

Figure 9.7 shows an eighth of a sphere of radius r. Spherical coordinates are chosen as curvilinear surface coordinates: $u^1 = \theta$, $u^2 = \phi$. The mapping (9.2.12) is expressed by:

$$x_\alpha = x_\alpha(u^1, u^2) \Rightarrow x_1 = r \sin\theta \cos\phi = r \sin u^1 \cos u^2,$$
$$x_2 = r \sin\theta \sin\phi = r \sin u^1 \sin u^2$$

Fig. 9.7 An eighth of a sphere. Surface coordinates: $u^1 = \theta, u^2 = \phi$. Element of area: $dA = r^2 \sin\theta d\theta d\phi$

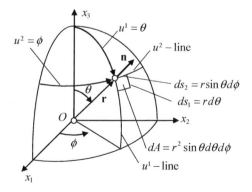

From the formulas (8.1.30–8.1.31) we have:

$$\mathbf{a}_1 = r\,\mathbf{e}_\theta, \quad \mathbf{a}_2 = r\sin\theta\,\mathbf{e}_\phi, \quad (a_{\alpha\beta}) = \begin{pmatrix} r^2 & 0 \\ 0 & r^2\sin^2\theta \end{pmatrix}, \quad \alpha = \det(a_{\alpha\beta}) = r^4\sin^2\theta$$

$$d\mathbf{s}_\alpha = \mathbf{a}_\alpha\,du^\alpha \Rightarrow d\mathbf{s}_1 = r\,d\theta\,\mathbf{e}_\theta, \quad d\mathbf{s}_2 = r\sin\theta\,d\phi\,\mathbf{e}_\phi, \quad dA = \sqrt{\alpha}\,du^1\,du^2 = r^2\sin\theta\,d\theta\,d\phi$$

The area A of the surface of the sphere becomes:

$$A = 8\int_{A/8} dA = 8\int_{A_u/8}\sqrt{\alpha}\,du^1\,du^2 = 8\int_{A_u/8} r^2\sin\theta\,d\theta\,d\phi = 8\int_0^{\pi/2}\left(\int_0^{\pi/2} r^2\sin\theta\,d\theta\right)d\phi$$

$$= 8\int_0^{\pi/2} r^2\,d\phi = 8r^2\frac{\pi}{2} \Rightarrow \quad A = 4\pi r^2$$

9.3.2   Integration Over a Volume in the Euclidean Space $\mathbf{E}_3$

Let V represent a region in E_3. The volume V of the region is defined by the triple integral:

$$V = \int_V dV = \iiint_V dx_1 dx_2 dx_3 \equiv \int_V dx_1\,dx_2\,dx_3 \tag{9.3.3}$$

The *volume element* $dV = dx_1\,dx_2\,dx_3$ is shown in Fig. 9.8. An alternative form for the volume element integral dV will be presented below.

The integral of a field function $f(\mathbf{r})$ over the region V is presented as the triple integral:

$$I = \int_V f(\mathbf{r})dV = \int_V f(x_1, x_2, x_3)\,dx_1\,dx_2\,dx_3 \tag{9.3.4}$$

Theorem 9.6 Gauss' Integral Theorem in Space *Let V represent a volume in three-dimensional space bounded by the surface A with an outward unit normal vector $\mathbf{n}$. Then for any field $f(\mathbf{r})$:*

$$\int_V \frac{\partial f}{\partial x_i}\,dV = \int_A f\,n_i\,dA \tag{9.3.5}$$

Proof for $i = 3$: Consider first the case, Fig. 9.9, where a straight line parallel to the x_3-axis intersects the surface A in only two points: $\tilde{x}_3(x_1, x_2)$ and $\bar{x}_3(x_1, x_2)$.

Fig. 9.8 Volume element dV in a Cartesian a coordinate system Ox. $dV = dx_1 dx_2 dx_3$

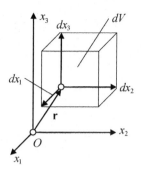

Fig. 9.9 Volume V with surface $A = \tilde{A} + \bar{A}$. Lines parallel with the x_3-axis intersect the surface A in only two points: $\tilde{x}_3(x_1, x_2)$ and $\tilde{x}_3(x_1, x_2)$. Elements of area $d\tilde{A}$ and $d\bar{A}$. Outward unit normals $\mathbf{n}$ to the surfaces $\tilde{A}$ and $\bar{A}$

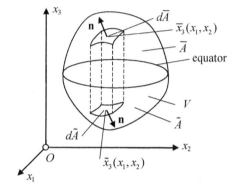

The sets of points $\tilde{x}_3(x_1, x_2)$ and $\bar{x}_3(x_1, x_2)$ describe respectively the surfaces $\tilde{A}$ below the equator and $\bar{A}$ above the equator, such that $\tilde{A} + \bar{A} = A$. Then:

$$\int_V \frac{\partial f}{\partial x_3} dV = \int_V \frac{\partial f}{\partial x_3} dx_1 dx_2 dx_3 = \int_A \left(\int_{\tilde{x}}^{\bar{x}} \frac{\partial f}{\partial x_3} dx_3 \right) dx_1 \, dx_2$$

$$= \int_{\bar{A}} f \, dx_1 \, dx_2 - \int_{\tilde{A}} f \, dx_1 \, dx_2$$

From the formula $(2.2.24)_2$ we can deduce that $dx_1 \, dx_2 = n_3 \, dA$ on $\bar{A}$ and $dx_1 \, dx_2 = -n_3 \, dA$ on $\tilde{A}$. Hence, we have found that:

$$\int_V \frac{\partial f}{\partial x_3} dV = \int_V \frac{\partial f}{\partial x_3} dx_1 dx_2 dx_3 = \int_A \left(\int_{\tilde{x}}^{\bar{x}} \frac{\partial f}{\partial x_3} dx_3 \right) dx_1 \, dx_2 = \int_{\bar{A}} f \, n_3 \, dA + \int_{\tilde{A}} f \, n_3 \, dA = \int_A f \, n_3 \, dA$$

$$\Rightarrow \int_V \frac{\partial f}{\partial x_3} dV = \int_A f \, n_3 \, dA$$

This result is generalized to:

$$\int_V \frac{\partial f}{\partial x_i}\,dV = \int_A f\,n_i\,dA \quad \text{for the situation in Fig. 9.9} \tag{9.3.6}$$

If lines parallel with the x_3-axis intersect the surface in more than two points, Fig. 9.10, the volume may be divided into parts for which each satisfies the condition about only two points of intersection with lines parallel with the x_3-axis. The formula (9.3.6) is applied to the volume parts and the results are summed. The sum of the volume integrals add up to the volume integral over the volume V. The sum of the surface integrals provides the surface integral over the surface area A plus the contributions from the interfaces separating the volume parts. However, these latter contributions will cancel each other, since they appear with contributions of opposite signs. By these arguments and the result (9.3.6) the formula (9.3.5) is true and Theorem 9.6 is proved.

Let $\mathbf{B}(\mathbf{r})$ be a tensor field of order n with components $B_{i\cdot\cdot j}$ in a Cartesian coordinate system Ox. Then we may apply the result (9.3.5) to write:

$$\int_V \frac{\partial B_{i\cdot\cdot j}}{\partial x_k}\,dV = \int_A B_{i\cdot\cdot j}\,n_k\,dA \tag{9.3.7}$$

This may be called the Cartesian representation of the *gradient theorem*:

$$\int_V \operatorname{grad}\mathbf{B}\,dV = \int_A \mathbf{B}\otimes\mathbf{n}\,dA \tag{9.3.8}$$

The integrands in the integrals on both sides of the formula (9.3.8) are tensor fields. In Cartesian coordinates the tensor property of both sides of the integral relation (9.3.8) is preserved. However, although the integrands in the relation (9.3.8) have components representation in a general coordinate system y, the integral relation will in general be meaningless in component presentation. This is due to the fact

Fig. 9.10 Volume V with surface $A = \tilde{A} + \bar{A}$. Lines parallel with the x_3-axis intersect the surface A in more than two points. Normals $\mathbf{n}$ and $-\mathbf{n}$ to the interface

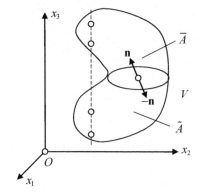

that when transforming from one general coordinate system y to another system $\bar{y}$ the transformation elements are functions of place.

Theorem 9.7 The Divergence Theorem in Space *Let V represent a volume in three-dimensional space bounded by the surface A with an outward unit normal vector $\mathbf{n}$. Then for any vector field $\mathbf{a}(\mathbf{r})$:*

$$\int_V \operatorname{div} \mathbf{a}\, dV = \int_A \mathbf{a} \cdot \mathbf{n}\, dA \Leftrightarrow \begin{cases} \int_V a_{i,i}\, dV = \int_A a_i\, n_i\, dA & \text{in Cartesian coordinates} \\ \int_V a^i\,|_i\, dV = \int_A a^i\, n_i\, dA & \text{in general coordinates} \end{cases}$$

$$(9.3.9)$$

Proof In Cartesian coordinates the theorem follows directly from Theorem 9.6 by replacing the scalar field $f(\mathbf{r})$ in the formula (9.3.5) by the vector components a_i and summing the results over the index (i). Because the expressions $\operatorname{div} \mathbf{a} = a_{i,i}$ and $\mathbf{a} \cdot \mathbf{n} = a_i n_i$ are scalar fields the Cartesian version of the theorem may be directly transformed to general coordinates.

The result (9.3.9) may be generalized to a general divergence theorem for a tensor field $\mathbf{B}(\mathbf{r})$ of order n with components $B_{i..j}$ in a Cartesian coordinate system Ox:

$$\int_V \operatorname{div} \mathbf{B}\, dV = \int_A \mathbf{B} \cdot \mathbf{n}\, dA$$

$$\Leftrightarrow \int_V B_{i..k,k}\, dV = \int_A B_{i..k}\, n_k\, dA \quad \text{in a Cartesian coordinates}$$

$$(9.3.10)$$

Theorem 9.8 Stokes' Theorem for a Curved Surface *Let $\mathbf{b}(\mathbf{r})$ be a vector field and A a surface in space with an outward unit normal vector $\mathbf{n}$. The surface is bordered by the curve C. Then:*

$$\int_A (\operatorname{rot} \mathbf{b}) \cdot d\mathbf{A} \equiv \int_A (\operatorname{rot} \mathbf{b}) \cdot \mathbf{n}\, dA \equiv \int_A (\nabla \times \mathbf{b}) \cdot \mathbf{n}\, dA = \oint_C \mathbf{b} \cdot d\mathbf{r} \Leftrightarrow$$

$$\int_A e_{ijk} b_{k,j}\, n_i\, dA \equiv \oint_C b_k\, dx_k \quad \text{in Cartesian coordinates}$$

$$\int_A \varepsilon^{ijk} b_{k,j}\, n_i\, dA \equiv \oint_C b_k\, dy^k \quad \text{in general coordinates}$$

$$(9.3.11)$$

The positive direction of integration along the curve C is determined such that the vector $\mathbf{n} \times d\mathbf{r}$ points to the side of C connected to the surface A.

Proof With respect to a Cartesian coordinate system Ox, with base vectors $\mathbf{e}_i$, and a surface coordinate system u, with base vectors $\mathbf{a}_\alpha$, we obtain for the vectors $\operatorname{rot} \mathbf{b}$ and $\mathbf{n} dA$:

$$\text{rot}\,\mathbf{b} = e_{ijk}b_{k\cdot j}\,\mathbf{e}_i$$

$$\mathbf{n}dA = \mathbf{a}_1 \times \mathbf{a}_2\, du^1\, du^2 = \left(\mathbf{e}_r \tfrac{\partial x_r}{\partial u^1}\right) \times \left(\mathbf{e}_s \tfrac{\partial x_s}{\partial u^2}\right) du^1\, du^2 = e_{rst}\mathbf{e}_t \tfrac{\partial x_r}{\partial u^1}\tfrac{\partial x_s}{\partial u^2}\, du^1\, du^2$$

Applying the identity (1.1.19) we get:

$$(\text{rot}\,\mathbf{b}) \cdot \mathbf{n}dA = \left(e_{ijk}b_{k\cdot j}\,\mathbf{e}_i\right) \cdot \left(e_{rst}\mathbf{e}_t \tfrac{\partial x_r}{\partial u^1}\tfrac{\partial x_s}{\partial u^2}\, du^1\, du^2\right) = e_{ijk}e_{irs}b_{k\cdot j}\tfrac{\partial x_r}{\partial u^1}\tfrac{\partial x_s}{\partial u^2}\, du^1\, du^2$$

$$= \left(\delta_{jr}\delta_{ks} - \delta_{js}\delta_{kr}\right)b_{k\cdot j}\tfrac{\partial x_r}{\partial u^1}\tfrac{\partial x_s}{\partial u^2}\, du^1\, du^2 = (b_{s,r} - b_{r,s})\tfrac{\partial x_r}{\partial u^1}\tfrac{\partial x_s}{\partial u^2}\, du^1\, du^2$$

$$= \left(\tfrac{\partial}{\partial u_1}\left(b_s \tfrac{\partial x_s}{\partial u^2}\right) - \tfrac{\partial}{\partial u_2}\left(b_r \tfrac{\partial x_r}{\partial u^1}\right)\right)du^1\, du^2$$

The integral on the left-hand side of Eq. (9.3.11) now becomes:

$$\int_A (\text{rot}\,\mathbf{b}) \cdot \mathbf{n}\,dA = \int_{A_u} \left[\frac{\partial}{\partial u^1}\left(b_s \frac{\partial x_s}{\partial u^2}\right) - \frac{\partial}{\partial u^2}\left(b_r \frac{\partial x_r}{\partial u^1}\right)\right] du^1\, du^2 \qquad (9.3.12)$$

Gauss' integration theorem in the $u^1 u^2$-plane is now applied to transform the surface integral on the right-hand side of Eq. (9.3.12) to an integral along the curve C_u in the $u^1 u^2$-plane which is in turn transformed to an integral along the bordering curve C of the original surface A:

$$\int_A (\text{rot}\,\mathbf{b}) \cdot \mathbf{n}\,dA = \int_{A_u} \left[\tfrac{\partial}{\partial u^1}\left(b_s \tfrac{\partial x_s}{\partial u^2}\right) - \tfrac{\partial}{\partial u^2}\left(b_r \tfrac{\partial x_r}{\partial u^1}\right)\right] du^1\, du^2$$

$$= \oint_{C_u} \left[(b_s \tfrac{\partial x_s}{\partial u^2})n_{u1} - (b_r \tfrac{\partial x_r}{\partial u^1})n_{u2}\right] d\sigma = \oint_{C_u} \left[(b_s \tfrac{\partial x_s}{\partial u^2})\tfrac{du^2}{d\sigma} - (b_r \tfrac{\partial x_r}{\partial u^1})\left(-\tfrac{du^1}{d\sigma}\right)\right] d\sigma$$

$$= \oint_{C_u} \left[b_s \tfrac{\partial x_s}{\partial u^2}\, du^2 + b_r \tfrac{\partial x_r}{\partial u^1}\, du^1\right] = \oint_C b_s dx_s = \oint_C \mathbf{b} \cdot d\mathbf{r} \Rightarrow \int_A (\text{rot}\,\mathbf{b}) \cdot \mathbf{n}\,dA = \oint_C \mathbf{b} \cdot d\mathbf{r}$$

This is a proof of Theorem 9.8 Stokes' theorem, formula (9.3.8), for a curved surface. The theorem includes Theorem 9.5 Stokes' theorem in a plane, formula (9.2.10).

Theorem 9.9 The Mean Value Theorem *Let $f(\mathbf{r})$ and $g(\mathbf{r})$ be two continuous field functions and V a volume in space. Then at least one point $\bar{\mathbf{r}}$ in V exists such that:*

$$\int_V f(\mathbf{r})\,g(\mathbf{r})dV = f(\bar{\mathbf{r}}) \int_V g(\mathbf{r})dV \qquad (9.3.13)$$

For the special case $g(\mathbf{r}) = 1$ in V the value $f(\bar{\mathbf{r}})$ is called the mean value of the function $f(\mathbf{r})$ in the volume V and is given by:

$$f(\bar{\mathbf{r}}) = \frac{1}{V} \int_V f(\mathbf{r})dV \qquad (9.3.14)$$

The theorem has analogous formulations and proofs for line and surface integrals.

Proof Let f_{min} be the maximum value and f_{max} the maximum value of the function $f(\mathbf{r})$ in the volume V. Then because $f(\mathbf{r})$ is a continuous function in V, at least one point $\bar{\mathbf{r}}$ exists in V such that: $f_{min} \leq f(\bar{\mathbf{r}}) \leq f_{max}$, and:

$$f_{min} \int\limits_V g(\mathbf{r})\, dV \leq \int\limits_V f(\mathbf{r}) g(\mathbf{r})\, dV = f(\bar{\mathbf{r}}) \int\limits_V g(\mathbf{r})\, dV \leq f_{max} \int\limits_V g(\mathbf{r})\, dV$$

This result proves theorem 9.9.

Theorem 9.10 Let $f(\mathbf{r})$ *be a continuous field function in a volume* V_1 *in the space* E_3. *Then the following holds true:*

If: $\quad \int\limits_V f(\mathbf{r})\, dV = 0$ for any choice of volume V in V_1, then: $\quad f(\mathbf{r}) = 0$ in V_1.

$$(9.3.15)$$

Proof If it is assumed that $f(\bar{\mathbf{r}}) > 0$ in a point $\bar{\mathbf{r}}$ in V_1, then according to the condition of continuity for the function $f(\mathbf{r})$, the function must be positive, i.e. $f(\mathbf{r}) > 0$, in a small volume ΔV surrounding the point $\bar{\mathbf{r}}$, which implies that:

$$\int\limits_{\Delta V} f(\mathbf{r})\, dV > 0$$

However, this result contradicts the proposition that the integral is zero for any choice of volume V in V_1. Hence the assumption $f(\bar{\mathbf{r}}) > 0$ is impossible. Hence, $f(\mathbf{r}) = 0$ throughout the volume V_1. This proves the Theorem 9.10.

By similar arguments we obtain the extension of the theorem:

If: $\quad \int\limits_A f(\mathbf{r})\, dA = 0$ for any choice of surface A in V, then: $\quad f(\mathbf{r}) = 0$ in V

$$(9.3.16)$$

Theorem 9.11 Change of Variables in a Volume Integral *Let* y *represent a curvilinear coordinate system in space. The relationship between the curvilinear coordinates* y^i *and the coordinates* x_i *in a Cartesian coordinate system* Ox *of a place* $\mathbf{r}$ *in space is given by a one - to - one mapping:*

$$y^i = y^i(x) \Leftrightarrow x_i = x_i(y) \qquad (9.3.17)$$

It is assumed that the coordinates y^i are ordered such that the Jacobian of the mapping is positive:

$$J = \det\left(\frac{\partial x_i}{\partial y^j}\right) > 0 \qquad (9.3.18)$$

Let $f(\mathbf{r})$ be a field function, V a volume in space, and V_y the mapping of V onto an orthogonal "Cartesian" coordinate system y. *Then*:

$$\int_V f(\mathbf{r})\,dV = \int_V f(\mathbf{r})\,dx_1 dx_2 dx_3 = \int_{V_y} f(\mathbf{r})\,J\,dV_y = \int_{V_y} f(\mathbf{r})\,J dy^1 dy^2 dy^3 \quad (9.3.19)$$

Proof In order to simplify the expressions in the proof we introduce a special notation:

$$x_{ik} \equiv \frac{\partial x_i(y^1, y^2, y^3)}{\partial y^k} \qquad (9.3.20)$$

The condition $J > 0$ implies that at least one of the elements x_{3k} is different from zero. Assume that $x_{33} \equiv \partial x_3/\partial y^3 \neq 0$ in the region V. If in a part of the region V, we may exclude this part of V and then performed the proof for this part by selecting another $x_{3k} \equiv \partial x_3/\partial y^k \neq 0.$,

We solve for the coordinate y^3 from the equation $x_3 = x_3(y) = x_3(y^1, y^2, y^3)$ and set $y^3 = \phi(y^1, y^2, x_3)$. Now we write, without specifying proper boundaries for the last integral:

$$\int_V f(\mathbf{r})\,dV = \int_V f(\mathbf{r})\,dx_1 dx_2 dx_3 = \int \left[\int f(\mathbf{r})dx_1 dx_2\right] dx_3 \qquad (9.3.21)$$

In the double integral we introduce the function:

$$f(\mathbf{r}) = f(x_1, x_2, x_3) = f(x_1(y^1, y^2, \phi), x_2(y^1, y^2, \phi), x_3) \qquad (9.3.22)$$

For the double integral we now use the integral formula (9.2.26) and write:

$$\int f(\mathbf{r})dx_1 dx_2 = \int f(x_1, x_2, x_3)dx_1 dx_2$$
$$= \int f(x_1(y^1, y^2, \phi), x_2(y^1, y^2, \phi), x_3)J_1\, dy^1\, dy^2 \qquad (9.3.23)$$

J_1 is the Jacobian to the mapping $\bar{x}_\alpha(y^1, y^2, x_3) = x_\alpha(y^1, y^2, \phi(y^1, y^2, x_3))$:

$$J_1 = \det\left(\frac{\partial x_\alpha}{\partial y^\beta}\right), \quad \frac{\partial x_\alpha}{\partial y^\beta} = x_{\alpha\beta} + \frac{\partial x_\alpha}{\partial y^3}\frac{\partial \phi}{\partial y^\beta} \qquad (9.3.24)$$

Because x_3 is constant in the double integral we find:

$$\frac{\partial x_3(y^1, y^2, \phi)}{\partial y^\beta} = 0 \Rightarrow x_{3\beta} + x_{33}\frac{\partial \phi}{\partial y^\beta} = 0 \Rightarrow \frac{\partial \phi}{\partial y^\beta} = -\frac{x_{3\beta}}{x_{33}} \tag{9.3.25}$$

From the formula (9.3.24–9.3.25) we obtain for the Jacobian J_1:

$$J_1 = \det\left(\frac{\partial x_\alpha}{\partial y^\beta}\right) = \det\left(x_{\alpha\beta} + \frac{\partial x_\alpha}{\partial y^3}\frac{\partial \phi}{\partial y^\beta}\right) = \det\left(x_{\alpha\beta} - \frac{\partial x_\alpha}{\partial y^3}\frac{x_{3\beta}}{x_{33}}\right)$$

$$= [x_{11}x_{22}x_{33} - x_{11}x_{23}x_{32} \quad + x_{12}x_{23}x_{31} - x_{12}x_{21}x_{33} + x_{13}x_{21}x_{32}$$

$$- x_{13}x_{22}x_{31} - x_{13}x_{23}x_{31}x_{32}/x_{33} + x_{13}x_{23}x_{32}x_{31}/x_{33}]/x_{33} \Rightarrow$$

$$J_1 = J/x_{33} \tag{9.3.26}$$

The integrand in the integral with respect to the variable x_3 in the formula (9.3.21) is only dependent on the variable x_3 (or y^3). We introduce y^3 as a new variable and obtain $dx_3 = x_{33}dy^3$. From the Eqs. (9.3.21, 9.3.23 and 9.3.26) we get:

$$\int_V f(\mathbf{r})\, dV = \int_V f(\mathbf{r})\, dx_1 dx_2 dx_3 = \int_{V_y} f(\mathbf{r}) J dy^1 dy^2 dy^3 \Rightarrow (9.3.19)$$

We may consider the volume element to be alternatively an orthogonal parallelepiped $dV = dx_1\, dx_2\, dx_3$ in Cartesian coordinates x, or in general curvilinear coordinates y, and shown in Fig. 9.11, to be a parallelepiped with sides given by the line elements along the coordinate lines:

$$d\mathbf{s}_i = \frac{\partial \mathbf{r}}{\partial y^i}dy^i = \mathbf{g}_i\, dy^i = \sum_k \frac{\partial x_k}{\partial y^i}dy^i\, \mathbf{e}_k \tag{9.3.27}$$

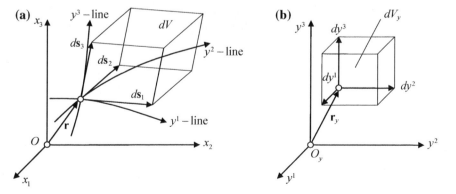

Fig. 9.11   a Volume element in curvilinar coordinates: $dV = [d\mathbf{s}_1 d\mathbf{s}_2 d\mathbf{s}_3]dy^1 dy^2 dy^3 = J dy^1 dy^2 dy^3$
b Volume element in "Cartesian" coordinates: $dV_y = dy^1 dy^2 dy^3$

The volume element dV in the general coordinate system y becomes:

$$dV = [ds_1 \, ds_2 \, ds_3] = J \, dy^1 \, dy^2 \, dy^3 = J dV_y \tag{9.3.28}$$

Example 9.3 Volume of a Cone with Circular Base
A straight cone has the height h, a circular base with radius r, a symmetry axis along the x_3-axis of a Cartesian coordinate system Ox, and its apex at the origin O. We shall find the volume of the cone using cylindrical coordinates $(R, \theta, z) \equiv (y^1, y^2, y^3)$. The mapping is:

$$x_1 = R \cos \theta \equiv y^1 \cos y^2, \quad x_2 = R \sin \theta \equiv y^1 \sin y^2, \quad x_2 = z \equiv y^3$$

The Jacobian J when transforming from the Cartesian coordinates x_i to the cylindrical coordinates becomes:

$$J = \det\left(\frac{\partial x_i}{\partial y_j}\right) = \det\begin{pmatrix} \cos \theta & -R \sin \theta & 0 \\ \sin \theta & R \cos \theta & 0 \\ 0 & 0 & 1 \end{pmatrix} = R$$

The volume of the cylinder is:

$$V = \int_V J \, dy^1 \, dy^2 \, dy^3 = \int_V R \, dR \, d\theta \, dz$$

$$= \int_o^h \left(\int_0^{2\pi} \left(\int_0^{rz/h} R dR \right) d\theta \right) dz = \int_o^h \left(\int_0^{2\pi} \left(\frac{r^2 z^2}{2h^2}\right) d\theta \right) dz = \int_o^h \left(2\pi \frac{r^2 z^2}{2h^2}\right) dz = \frac{\pi r^2 h}{3}$$

Theorem 9.12 From Volume Integration to Surface Integration *A body of a continuum has at the time t the volume V and the surface A with the outward unit normal vector* **n**. *Let* $f(\mathbf{r}, t)$ *be an intensive quantity defined per unit volume at the place* **r** *in V and at the time t, and let* $g(\mathbf{r}, t, \mathbf{n})$ *be an intensive quantity at the place* **r** *and at the time t on A defined per unit area of the surface A.*
  If the following integral equation is true for any volume V with the surface A:

$$\int_V f(\mathbf{r}, t) \, dV = \int_A g(\mathbf{r}, t, \mathbf{n}) \, dA \tag{9.3.29}$$

then the fields $f(\mathbf{r}, t)$ and $g(\mathbf{r}, t, \mathbf{n})$ are related through a vector field $\mathbf{a}(\mathbf{r}, t)$ such that:

$$g(\mathbf{r}, t, \mathbf{n}) = \mathbf{a} \cdot \mathbf{n} \quad \text{and} \quad f(\mathbf{r}, t) = \text{div } \mathbf{a}$$
$$a_i = g(\mathbf{r}, t, \mathbf{e}_i) \text{ in a Cartesian coodinate system with base vectors } \mathbf{e}_i \tag{9.3.30}$$

The integral Eq. (9.3.29) may then be rewritten to:

$$\int_V \operatorname{div} \mathbf{a} \, dV = \int_A \mathbf{a} \cdot \mathbf{n} \, dA \tag{9.3.31}$$

The result confirms the divergence theorem in space, formula (9.3.9).

Proof Let the volume V be subdivided into the volumes V_1 and V_2 by the interface surface A', Fig. 9.12. Equation (9.3.29) is now applied to the three volumes V, V_1, and V_2 and the integral contributions from the volumes V_1 and V_2 are subtracted from the contribution from the volume V. The result is that:

$$\int_{A'} [g(\mathbf{r}, t, \mathbf{n}) + g(\mathbf{r}, t, -\mathbf{n})] \, dA = 0 \tag{9.3.32}$$

Because the volume V and thus the surface A' may be chosen arbitrarily, the integrand in the result (9.3.32) must be zero, which implies that:

$$g(\mathbf{r}, t, \mathbf{n}) = -g(\mathbf{r}, t, -\mathbf{n}) \tag{9.3.33}$$

Next we choose for the volume the tetrahedron shown in Fig. 9.13. This is similar to the Cauchy tetrahedron shown in Fig. 2.11. From Sect. 2.2 we cite the relations (2.2.24):

$$\bar{V} = \bar{A}h/3, \quad \bar{A}_i = \bar{A}n_i \tag{9.3.34}$$

The integral Eq. (9.3.29) applied to the tetrahedron yields:

$$\int_{\bar{V}} f(\mathbf{r}, t) \, dV = \int_{\bar{A}} g(\mathbf{r}, t, \mathbf{n}) \, dA + \sum_i \int_{\bar{A}_i} g(\mathbf{r}, t, -\mathbf{e}_i) \, dA$$

Let the functions $f(\mathbf{r}, t)$, $g(\mathbf{r}, t, \mathbf{n})$, and $g(\mathbf{r}, t, -\mathbf{e}_i)$ represent mean values over the volume V and the surfaces $\bar{A}$ and $\bar{A}_i$ respectively. Then the integral equation above may be presented as:

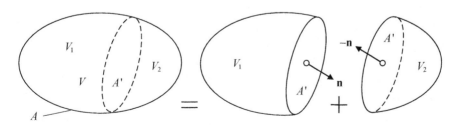

Fig. 9.12 Volume V subdivided into V_1 and V_2. Interface surface A'

Fig. 9.13 Cauchy
tetrahedron. Body of volume
$\bar{V}$. Surface consisting of four
triangles: three orthogonal
triangles in coordinate planes
through the particle P and
with areas $\bar{A}_i$. The fourth
triangle with area $\bar{A}$ and unit
normal **n**

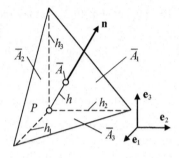

$$f(\mathbf{r}, t)\,\bar{V} = g(\mathbf{r}, t, \mathbf{n})\bar{A} + g(\mathbf{r}, t, -\mathbf{e}_i)\bar{A}_i$$

Now we use the relations (9.3.34), to obtain:

$$f(\mathbf{r}, t)\,\bar{A}h/3 = g(\mathbf{r}, t, \mathbf{n})\bar{A} + g(\mathbf{r}, t, -\mathbf{e}_i)\bar{A}n_i$$

We divide this equation by $\bar{A}$ and let $h \to 0$. The result is:

$$g(\mathbf{r}, t, \mathbf{n}) + g(\mathbf{r}, t, -\mathbf{e}_i)n_i = 0$$

Now we use the formula (9.3.33) and write:

$$g(\mathbf{r}, t, \mathbf{n}) = g_i(\mathbf{r}, t)n_i, \quad \text{where } g_i(\mathbf{r}, t) = g(\mathbf{r}, t, \mathbf{e}_i) \tag{9.3.35}$$

We now return to the integral Eq. (9.3.29) for the volume V with the surface A.
We apply the result (9.3.35) and the Gauss' theorem, formula (9.3.5), and obtain:

$$\int_V f(\mathbf{r}, t)\, dV = \int_A g(\mathbf{r}, t, \mathbf{n})\, dA = \int_A g_i(\mathbf{r}, t)\, n_i\, dA = \int_V g_{i,i}\, dV$$

$$\Rightarrow \int_V [f(\mathbf{r}, t) - g_{i,i}]\, dV = 0$$

The last integral equation is true for any choice of volume V. Hence by
Theorem 9.10: $f(\mathbf{r}, t) - g_{i,i} = 0$ or:

$$f(\mathbf{r}, t) = g_{i,i} \Rightarrow f(\mathbf{r}, t) = \operatorname{div} \mathbf{g} \tag{9.3.36}$$

The results (9.3.35–9.3.36) prove theorem 9.12.

The results of theorem 9.12 may be generalized by replacing the function $f(\mathbf{r}, t)$
with a vector or a tensor. The function $g(\mathbf{r}, t, \mathbf{n})$ will then represent a tensor of one
order higher than $f(\mathbf{r}, t)$. For example:

$$\int_V \text{div}\,\mathbf{F}(\mathbf{r},t)\,dV = \int_A \mathbf{F}(\mathbf{r},t,\mathbf{n})\cdot\mathbf{n}\,dA \Leftrightarrow$$
$$\int_V F(\mathbf{r},t)_{ij\cdot\cdot k,k}\,dV = \int_A F(\mathbf{r},t,\mathbf{n})_{ij\cdot\cdot k}n_k\,dA \quad \text{in Cartesian coordinate systems}$$

$$(9.3.37)$$

Theorem 9.13 Necessary and Sufficient Condition for a Vector Field to be Irrotational *A vector field* $\mathbf{v}(\mathbf{r},t)$ *is irrotational if* rot $\mathbf{v}(\mathbf{r},t) \equiv \nabla \times \mathbf{v}(\mathbf{r},t) = \mathbf{0}$. *A necessary and sufficient condition for* $\mathbf{v}(\mathbf{r},t)$ *to be irrotational is that the vector field may be expressed as the gradient of a scalar field:*

$$\text{rot}\,\mathbf{v}(\mathbf{r},t) \equiv \nabla \times \mathbf{v}(\mathbf{r},t) = \mathbf{0} \Leftrightarrow \mathbf{v}(\mathbf{r},t) = \text{grad}\,\phi(\mathbf{r},t) \equiv \nabla\phi(\mathbf{r},t) \quad (9.3.38)$$

Proof of the implication: $\mathbf{v}(\mathbf{r},t) = \text{grad}\,\phi(\mathbf{r},t) \Rightarrow \text{rot}\,\mathbf{v}(\mathbf{r},t) = \mathbf{0}$. For the sake of simplicity we use Cartesian coordinates. Because $\phi_{,jk} = \phi_{,kj}$ and $e_{ijk} = -e_{ikj}$ for the permutation symbols :

$$\text{rot}\,\mathbf{v}(\mathbf{r},t) = e_{ijk}v_{k,j}\mathbf{e}_i = e_{ijk}\phi_{,kj}\mathbf{e}_i = \mathbf{0}$$

Proof of the implication: rot $\mathbf{v}(\mathbf{r},t) = \mathbf{0} \Rightarrow \mathbf{v}(\mathbf{r},t) = \text{grad}\,\phi(\mathbf{r},t)$. Let A be any surface bordered by the curve C. Then according to Theorem 9.8 Stokes theorem for a curved surface:

$$\int_C \mathbf{v}\cdot d\mathbf{r} = \int_A (\text{rot}\,\mathbf{v})\cdot\mathbf{n}\,dA = 0$$

By Theorem 9.1 Integration independent of the integration path, this result implies that scalar field $\phi(\mathbf{r},t)$ exists such that $\mathbf{v}(\mathbf{r},t) = \text{grad}\,\phi(\mathbf{r},t)$.

Theorem 9.14 Material Derivative of an Extensive Quantity *Let* $f(\mathbf{r},t)$ *be a place function representing an intensive physical quantity defined per unit mass and with* $F(t)$ *as the corresponding extensive quantity for a material body of volume* $V(t)$, *such that:*

$$F(t) = \int_{V(t)} f\rho\,dV \quad (9.3.39)$$

$\rho = \rho(\mathbf{r},t)$ *is the mass density, i.e. mass per unit volume, of the material body. The material derivative of the extensive quantity* $F(t)$ *may then be given by:*

$$\dot{F}(t) = \frac{dF}{dt} = \int_{V(t)} \dot{f}\rho\,dV \quad (9.3.40)$$

Proof of formula (9.3.40): Sect. 2.1.3 presents a physical argument as a proof for the theorem. Here we shall use Theorem 9.11 Change of variables in the volume integral, to present a somewhat more stringent mathematical proof of Theorem 9.14.

The motion of the body may be interpreted as a one-to-one mapping between the place coordinates X_i of the particles in the reference configuration K_0 at the reference time t_0, and the place coordinates x_i of the particles in the present configuration K at the time t:

$$x_i = x_i(X, t) \Leftrightarrow X_i = X_i(x, t) \tag{9.3.41}$$

The volume of the body in the reference configuration K_0 is $V_0 = V(t_0)$. The element of volume is $dV = dx_1 \, dx_2 \, dx_3$ in K and $dV_0 = dX_1 \, dX_2 \, dX_3$ in K_0. The integral in the formula (9.3.39) is now by Theorem 9.11 transformed to:

$$F(t) = \int_{V(t)} f\rho \, dV = \int_{V_0} f\rho J \, dV_0 \tag{9.3.42}$$

J is the Jacobian to the mapping (9.3.41) and matrix of the deformation gradient tensor $\mathbf{F}$:

$$J = \det \mathbf{F} = \det(F_{ij}) = \det\left(\frac{\partial x_i}{\partial X_j}\right)$$

If we as a special case choose the place function $f(\mathbf{r}, t) = 1$ in the volume $V(t)$, then $F(t)$ by the formula (9.3.39) represents the constant mass m of the body:

$$m = \int_{V(t)} \rho \, dV = \int_{V_0} \rho J \, dV_0 \tag{9.3.43}$$

The density ρ is mass per unit volume in K and the combination ρJ is mass per unit volume in K_0. Due to mass conservation the combination is independent of time, which implies that: $d(\rho J)/dt = 0$. Because the volume V_0 is independent of time, the material derivative of the integral on the right-hand side in the formula (9.3.42) may be obtained by differentiation of the integrand. Hence:

$$\dot{F}(t) = \int_{V_0} \left[\dot{f}\rho J + f\frac{d(\rho J)}{dt}\right] dV_0 = \int_{V_0} \dot{f}\rho J \, dV_0 = \int_{V(t)} \dot{f}\rho \, dV_0 \Rightarrow (9.3.40)$$

This completes the proof of Theorem 9.14.

Theorem 9.15 Reynolds' Transport Theorem *Let, at the present time t, B(t) be an extensive quantity for a body of a continuous medium with the volume V(t) and* *surface $A(t)$, and let $\beta(t)$ be the intensive quantity related to $B(t)$ and expressing quantity per unit volume. $\beta(t)$ is called the density of the quantity:*

Extensive quantity $B(t)$ expressed by the intensive quantity $\beta(\mathbf{r}, t)$:

$$B(t) = \int_{V(t)} \beta(\mathbf{r}, t) dV \qquad (9.3.44)$$

A *control volume V* is a region in space coinciding with the volume of the body at the present time t, i.e. $V = V(t)$. The surface A of the control volume V is called a *control surface* and coincides with the surface of the body at the present time t, i.e. $A = A(t)$. The control volume and control surface are fixed relative to the reference Rf chosen to describe the motion of the continuum. The particle velocity is denoted by $\mathbf{v} = \mathbf{v}(\mathbf{r}, t)$ and the unit outward unit normal vector to the control surface is given as $\mathbf{n} = \mathbf{n}(\mathbf{r})$.

Then *Reynolds' transport theorem*, Osborne Reynolds [1842–1912], states that:

$$\dot{B} = \int_V \frac{\partial \beta}{\partial t} dV + \int_A \beta(\mathbf{v} \cdot \mathbf{n}) dA \qquad (9.3.45)$$

Proof The formula (9.3.19) in Theorem 9.11 is first used to transform the integral in Eq. (9.3.44):

$$B(t) = \int_{V(t)} \beta \, dV = \int_{V_o} \beta J \, dV_o \qquad (9.3.46)$$

V_o is the volume of the body in the reference configuration K_o, and J is the Jacobian to the deformation gradient:

$$J = \det \mathbf{F} = \det F_{ij} \equiv \det \left(\frac{\partial x_i}{\partial X_j} \right) \qquad (9.3.47)$$

Because the volume V_o is independent of the time t, the material derivative of the second integral in Eq. (9.3.38) may be obtained by performing the differentiation under the integral sign:

$$\dot{B} = \int_{V_o} \left(\dot{\beta} J + \beta \dot{J} \right) dV_o \qquad (9.3.48)$$

Now we apply the formula (2.1.17) for the material derivative of an intensive quantity:

$$\dot{\beta} = \frac{\partial \beta}{\partial t} + (\mathbf{v} \cdot \nabla) \beta,$$

the formula (4.5.33): $\dot{J} = J \operatorname{div} \mathbf{v}$, and finally the formula:

$$v_i \beta_{,i} + \beta v_{i,i} = (\beta v_i)_{,i} \Leftrightarrow (\mathbf{v} \cdot \nabla)\beta + \beta \operatorname{div} \mathbf{v} = \operatorname{div}(\beta \mathbf{v})$$

Then we obtain from the integral Eq. (9.3.40) the result:

$$\dot{B} = \int_{V_o} \left(\dot{\beta} J + \beta \dot{J} \right) dV_o = \int_{V_o} \left(\dot{\beta} + \beta \operatorname{div} \mathbf{v} \right) J \, dV_o = \int_{V_o} \left(\frac{\partial \beta}{\partial t} + (\mathbf{v} \cdot \nabla)\beta + \beta \operatorname{div} \mathbf{v} \right) J \, dV_o \Rightarrow$$

$$\dot{B} = \int_{V_o} \left(\frac{\partial \beta}{\partial t} + \operatorname{div}(\beta \mathbf{v}) \right) J \, dV_o$$

By applying the formula (9.3.19) in Theorem 9.11 we transform this result to:

$$\dot{B} = \int_{V(t)} \frac{\partial \beta}{\partial t} \, dV + \int_{V(t)} \operatorname{div}(\beta \mathbf{v}) \, dV$$

By Theorem 9.6 Gauss' Integral Theorem in Space the last volume integral above is transformed into a surface integral and we obtain the following formulafor the material derivative of the extensive quantity $B(t)$:

$$\dot{B} = \int_{V} \frac{\partial \beta}{\partial t} \, dV + \int_{A} \beta(\mathbf{v} \cdot \mathbf{n}) \, dA \Rightarrow (9.3.45)$$

This completes the proof of Theorem9.15 Reynolds' Transport Theorem.

An illustration of Reynolds' transport theorem will be presented by Fig. 9.14 which shows a material body that at the time t has the volume $V(t)$ and the surface $A(t)$. At a time $t + \Delta t$, where Δt is a small time increment, the volume of the body is $V(t + \Delta t)$ and the surface is $A(t + \Delta t)$. For the extensive quantity $B(t)$ expressed by the intensive quantity $\beta(\mathbf{r}, t)$ we write with respect to the times t and $t + \Delta t$:

$$B(t) = \int_{V(t)} \beta(\mathbf{r}, t) \, dV$$

$$B(t + \Delta t) = \int_{V(t + \Delta t)} \beta(\mathbf{r}, t + \Delta t) \, dV = \int_{V(t)} \beta(\mathbf{r}, t + \Delta t) \, dV + \int_{\Delta V(t, \Delta t)} \beta(\mathbf{r}, t + \Delta t) \, d(\Delta V)$$

$$(9.3.49)$$

As illustrated in Fig. 9.14 the volume $\Delta V(t, \Delta t)$ and the differential volume element $d(\Delta V)$ may be presented as:

$$\Delta V(t, \Delta t) = V(t + \Delta t) - V(t), \quad d(\Delta V) = dA \cdot [(\mathbf{v} \cdot \mathbf{n})\Delta t]$$

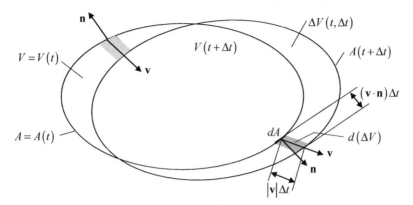

Fig. 9.14 Body of volume $V(t)$ and surface $A(t)$ at time t, and volume $V(t+\Delta t)$ and surface $A(t+\Delta t)$ at time $t+\Delta t$. Particle velocity $\mathbf{v}$. Unit surface normal $\mathbf{n}$. Control volume $V = V(t)$. Control surface $A = A(t)$. Differential volume element: $d(\Delta V) = dA[(v \cdot n)\Delta t]$

We may now write:

$$\int\limits_{\Delta V(t,\Delta t)} \beta(\mathbf{r}, t+\Delta t)\, d(\Delta V) = \int\limits_{A(t)} \beta(\mathbf{r}, t+\Delta t)\,[(\mathbf{v}\cdot\mathbf{n})\Delta t]dA \qquad (9.3.50)$$

From the Eqs. (9.3.49) and (9.3.50) we obtain:

$$\frac{B(t+\Delta t) - B(t)}{\Delta t} = \int\limits_{V(t)} \frac{\beta(\mathbf{r}, t+\Delta t) - \beta(\mathbf{r}, t)}{\Delta t}\, dV + \int\limits_{A(t)} \beta(\mathbf{r}, t+\Delta t)\,[(\mathbf{v}\cdot\mathbf{n})]dA$$

The material derivative of the extensive quantity $B(t)$ for the body is now found from:

$$\dot{B} = \lim_{\Delta t \to 0} \frac{B(t+\Delta t) - B(t)}{\Delta t} = \int\limits_{V} \frac{\partial\beta}{\partial t}\, dV + \int\limits_{A} \beta(\mathbf{v}\cdot\mathbf{n})dA \Rightarrow \qquad (9.3.45)$$

Example 9.4 Resultant Axial Force on the Bolts of a Nozzle
Fluid of density ρ flows through a circular pipe with internal diameter d. A nozzle with maximum internal diameter d and minimum internal diameter $d/2$ is attached to the pipe with bolts, Fig. 9.15a. The volume flow through the system is Q. It is assumed that the velocity and pressure over the cross-sections of the system are uniform. The atmospheric pressure is p_a. The pressure p in the pipe is found to be:

$$p = p_a + \frac{120\rho Q^2}{\pi^2 d^4}$$

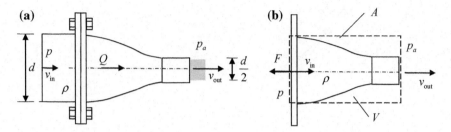

(a) **(b)**

Fig. 9.15 a Pipe flow through a nozzle. Fluid density ρ.Volume flow Q. Atmospheric pressure p_a.
b Control volume V with control surface A

We shall find an expression for the axial force F transmitted through the bolts due
to the flow of fluid.

Figure 9.15b shows a control volume V with a control surface A. The uniform
fluid velocity into the nozzle is v_{in} and the uniform fluid velocity out of the nozzle
v_{out}. The force F is transmitted through the bolts from the pipe. The continuity
equation applied to the control volume yields for the velocities v_{in} and v_{out}:

$$Q = Q_{in} = Q_{out} \Rightarrow Q = v_{in}\frac{\pi d^2}{4} = v_{out}\frac{\pi(d/2)^2}{4} \Rightarrow v_{in} = \frac{4Q}{\pi d^2}, \quad v_{out} = \frac{16Q}{\pi d^2}$$

The law of balance of linear momentum for the control volume results in:

$$\int_A \mathbf{v}\rho(\mathbf{v}\cdot\mathbf{n})\,dA = \mathbf{F} \Rightarrow v_{in}\,\rho(-v_{in})\tfrac{\pi d^2}{4} + v_{out}\,\rho\,v_{out}\tfrac{\pi(d/2)^2}{4} = -F + p\cdot\tfrac{\pi d^2}{4} - p_a\cdot\tfrac{\pi d^2}{4} \Rightarrow$$

$$F = (p - p_a)\tfrac{\pi d^2}{4} + \rho v_{in}^2\tfrac{\pi d^2}{4} - \rho v_{out}^2\tfrac{\pi(d/2)^2}{4}$$

With the given pressure p and the velocities v_{in} and v_{out} we obtain for the force F:

$$F = \frac{120\rho Q^2}{\pi^2 d^4}\frac{\pi d^2}{4} + \rho\left(\frac{4Q}{\pi d^2}\right)^2\frac{\pi d^2}{4} - \rho\left(\frac{16Q}{\pi d^2}\right)^2\frac{\pi(d/2)^2}{4} = \frac{\rho Q^2}{\pi^2 d^2}[30 + 4 - 16] \Rightarrow F$$
$$= \frac{18\rho Q^2}{\pi^2 d^2}a$$

Appendix
Problems with Solutions

Chapter 1
Problem 1.1

(a) Validate the identity (1.1.19) through some test examples.

(b) Compute: $\delta_{ii}, \delta_{ij}\,\delta_{ij}, e_{ijk}\,e_{rjk}$ {use formula 1.1.19)}, $e_{ijk}\,e_{ijk}$

(c) Use formula $(1.1.22)_2$ to show that for a 3×3 matrix A: $\det A = e_{ijk}e_{rst}A_{ir}A_{js}A_{kt}/6$

Solution

(a) Validation of the identity (1.1.19): $e_{ijk}e_{rsk} = \delta_{ir}\delta_{js} - \delta_{is}\delta_{jr}$

$$\text{Test (1) } i = j = 1 \Rightarrow e_{11k}e_{rsk} = \delta_{1r}\delta_{1s} - \delta_{1s}\delta_{1r} = 0$$

$$\text{Test (2) } i = 1, j = 2 \Rightarrow \left.\begin{array}{c} e_{12k}e_{rsk} = e_{123}e_{rs3} = e_{rs3} \\ = \delta_{1r}\delta_{2s} - \delta_{1s}\delta_{2r} = \end{array}\right\}$$

$$= \begin{cases} = +1 \text{ for } r = 1, s = 2 \\ = -1 \text{ for } r = 2, s = 1 \\ = 0 \text{ otherwise} \end{cases}$$

with similar results for other choices of the indices.

(b) $\delta_{ii} = \delta_{11} + \delta_{22} + \delta_{33} = 1 + 1 + 1 = 3.$ $\delta_{ij}\,\delta_{ij} = \delta_{ii} = 3.$ $e_{ijk}e_{rsk} = \delta_{ir}\delta_{js} - \delta_{is}\delta_{jr} \Rightarrow$
$e_{ijk}e_{rjk} = \delta_{ir}\delta_{jj} - \delta_{ij}\delta_{jr} = \delta_{ir}3 - \delta_{ir} = 2\delta_{ir}.e_{ijk}e_{ijk} = 2\delta_{ii} = 2 \cdot 3 = 6.$

(c) Formula $(1.1.22)_2$. $\Rightarrow e_{ijk}A_{ir}A_{js}A_{kt} = (\det A)e_{rst} \Rightarrow$
$e_{ijk}A_{ir}A_{js}A_{kt}e_{rst} = (\det A)e_{rst}e_{rst} = (\det A)6 \Rightarrow \det A = e_{ijk}e_{rst}A_{ir}A_{js}A_{kt}/6.$

© Springer Nature Switzerland AG 2019
F. Irgens, *Tensor Analysis*,
https://doi.org/10.1007/978-3-030-03412-2

Problem 1.2 Determine the inverse matrix U^{-1} of the matrix: $U =$
$\begin{pmatrix} 1 & 1 & 0 \\ 1 & 3 & 0 \\ 0 & 0 & \sqrt{2} \end{pmatrix} \frac{1}{\sqrt{2}}$

Solution Formula (1.1.29) implies:

$$U^{-1} \det U = \text{Co } U^T \Rightarrow U^{-1} = \frac{1}{\det U} \text{Co } U^T$$

$$\det U = \left[\left(1 \cdot 3 \cdot \sqrt{2} - 1 \cdot 0 \cdot 0 \right) + \left(1 \cdot 0 \cdot 0 - 1 \cdot 1 \cdot \sqrt{2} \right) + (0 \cdot 1 \cdot 0 - 0 \cdot 3 \cdot 0) \right]$$
$$\left(1/\sqrt{2} \right)^3 = 1$$

The cofactor Co U is obtained from the formula (1.1.26):

$$\det U = \sum_{i=1}^{3} U_{ij} \text{Co } U_{ij} \quad \text{for } j = 1, 2, \text{ or } 3 \Rightarrow$$

$$\text{Co } U = \begin{pmatrix} 3\sqrt{2} & -\sqrt{2} & 0 \\ -\sqrt{2} & \sqrt{2} & 0 \\ 0 & 0 & 2 \end{pmatrix} \left(\frac{1}{\sqrt{2}} \right)^2 = \text{Co } U^T$$

Hence:

$$U^{-1} = \frac{1}{\det U} \text{Co } U^T = \frac{1}{1} \begin{pmatrix} 3\sqrt{2} & -\sqrt{2} & 0 \\ -\sqrt{2} & \sqrt{2} & 0 \\ 0 & 0 & 2 \end{pmatrix} \left(\frac{1}{\sqrt{2}} \right)^2 = \begin{pmatrix} 3 & -1 & 0 \\ -1 & 1 & 0 \\ 0 & 0 & \sqrt{2} \end{pmatrix} \frac{1}{\sqrt{2}}$$

Control: $\Rightarrow UU^{-1} = 1$ OK.

Problem 1.3 Referred to the Cartesian coordinate system Ox, the Cartesian coordinate system $\bar{O}\bar{x}$ has the base vectors: $\bar{e}_1 = [1, 1, 1](1/\sqrt{3})$, $\bar{e}_2 = [1, 0, -1](1/\sqrt{2})$.

Determine the base vector $\bar{e}_3$ for the system $\bar{O}\bar{x}$ and the transformation matrix $Q = [\bar{e}_i \cdot e_j]$.

Solution The base vector $\bar{e}_3 = \bar{e}_1 \times \bar{e}_2 = \det \begin{pmatrix} \mathbf{e}_1 & \mathbf{e}_2 & \mathbf{e}_3 \\ 1 & 1 & 1 \\ 1 & 0 & -1 \end{pmatrix} \left(\frac{1}{\sqrt{2}} \frac{1}{\sqrt{3}} \right) \Rightarrow$

$\bar{e}_3 = [-1, 2, -1](1/\sqrt{6})$

The transformation matrix: $Q = (\bar{e}_i \cdot e_j) = \begin{pmatrix} \sqrt{2} & \sqrt{2} & \sqrt{2} \\ \sqrt{3} & 0 & -\sqrt{3} \\ -1 & 2 & -1 \end{pmatrix} \frac{1}{\sqrt{6}}$. Control:

$\det Q = +1$ OK

Problem 1.4 Three Cartesian coordinate systems are denoted Ox, $\bar{O}\bar{x}$, and $\tilde{O}\tilde{x}$. The transformation matrices relating the base vectors of the three systems are:

$$Q_{ij} = \bar{\mathbf{e}}_i \cdot \mathbf{e}_j, \quad \tilde{Q}_{ij} = \tilde{\mathbf{e}}_i \cdot \mathbf{e}_j, \quad \bar{Q}_{ij} = \bar{\mathbf{e}}_i \cdot \tilde{\mathbf{e}}_j$$

(a) Show that $Q = \bar{Q}\tilde{Q}$. (b) Compute Q when: $\bar{Q} = \begin{pmatrix} 2 & 0 & 0 \\ 0 & \sqrt{3} & 1 \\ 0 & -1 & \sqrt{3} \end{pmatrix} \frac{1}{2}$,

$\tilde{Q} = \begin{pmatrix} 0 & \sqrt{3} & 1 \\ 2 & 0 & 0 \\ 0 & 1 & -\sqrt{3} \end{pmatrix} \frac{1}{2}$

Check that $Q, \bar{Q}$, and $\tilde{Q}$ are orthogonal matrices.

Solution

(a) Proof that $Q = \bar{Q}\tilde{Q}$: $\bar{\mathbf{e}}_i = \bar{Q}_{ik}\tilde{\mathbf{e}}_k \Rightarrow Q_{ij} = \bar{\mathbf{e}}_i \cdot \mathbf{e}_j = (\bar{Q}_{ik}\tilde{\mathbf{e}}_k) \cdot \mathbf{e}_j = \bar{Q}_{ik}\tilde{Q}_{kj} \Rightarrow$
$Q = \bar{Q}\tilde{Q}$ QED.

(b) Computation of Q: $Q = \bar{Q}\tilde{Q} \Rightarrow (Q_{ij}) = (\bar{Q}_{ik}\tilde{Q}_{kj}) = \begin{pmatrix} 0 & 2\sqrt{3} & 2 \\ 2\sqrt{3} & 1 & -\sqrt{3} \\ -2 & \sqrt{3} & -3 \end{pmatrix} \frac{1}{4}$

Control of orthogonality: $\bar{Q}\bar{Q}^T = 1, \tilde{Q}\tilde{Q}^T = 1, QQ^T = 1, \det Q = \det \bar{Q} = \det$
$\tilde{Q} = +1$ OK.

Problem 1.5 The rotation rot $\mathbf{a}$ of a vector field $\mathbf{a}$ may be defined as the vector represented by the Cartesian components defined in formula (1.6.15). Show that rot $\mathbf{a}$ is a proper vector, i.e. the components obey the transformation rule (1.3.9).

Solution Let Ox and $\bar{O}\bar{x}$ be two Cartesian coordinate systems with base vectors $\mathbf{e}_i$ and $\bar{\mathbf{e}}_r$. The transformation matrix for the transformation from Ox to $\bar{O}\bar{x}$ is Q_{ri}. With $\mathbf{a}$ and $\mathbf{b} = \text{rot}\,\mathbf{a}$ as two vector fields the definition (1.6.15) gives:

$$\mathbf{b} = \text{rot}\,\mathbf{a} = e_{ijk}a_{k,j}\,\mathbf{e}_i \equiv e_{ijk}\frac{\partial a_k}{\partial x_j}\mathbf{e}_i = b_i\mathbf{e}_i \quad \text{in } Ox.$$

$$\mathbf{b} = \text{rot}\,\mathbf{a} = e_{rst}\bar{a}_{t,s}\,\bar{\mathbf{e}}_r \equiv e_{rst}\frac{\partial \bar{a}_t}{\partial \bar{x}_s}\bar{\mathbf{e}}_r = \bar{b}_r\bar{\mathbf{e}}_r \quad \text{in } \bar{O}\bar{x}$$

For $\mathbf{b} = \text{rot}\,\mathbf{a}$ to be a proper vector the components b_i and $\bar{b}_r$ must define the same vector which implies that the vector defined by $e_{rst}\bar{a}_{t,s}\,\bar{\mathbf{e}}_r$ shall be identical to the vector defined by $e_{ijk}a_{k,j}\mathbf{e}_i$. The formulas (1.3.2), (1.3.7) and (1.3.9) and (1.1.22) yield:

$$\bar{\mathbf{e}}_r = Q_{ri}\mathbf{e}_i, \quad \frac{\partial x_j}{\partial \bar{x}_s} = Q_{sj}, \bar{a}_t = Q_{tk}a_k, \quad e_{rst}Q_{ri}Q_{sj}Q_{tk} = e_{ijk}\det Q = e_{ijk}$$

Now we obtain:

$$\bar{b}_r\bar{\mathbf{e}}_r = e_{rst}\frac{\partial \bar{a}_t}{\partial \bar{x}_s}\bar{\mathbf{e}}_r = e_{rst}\frac{\partial(Q_{tk}a_k)}{\partial x_j}\frac{\partial x_j}{\partial \bar{x}_s}(Q_{ri}\mathbf{e}_i) = e_{ijk}Q_{ri}Q_{sj}Q_{tk}\frac{\partial a_k}{\partial x_j}\mathbf{e}_i = e_{ijk}\frac{\partial a_k}{\partial x_j}\mathbf{e}_i \Rightarrow$$
$$e_{rst}\frac{\partial \bar{a}_t}{\partial \bar{x}_s}\bar{\mathbf{e}}_r \equiv e_{ijk}\frac{\partial a_k}{\partial x_j}\mathbf{e}_i$$

The result proves that the vector $\mathbf{b} = \mathrm{rot}\,\mathbf{a}$ defined by the formula (1.6.15) is a proper vector.

Problem 1.6 Use identity (1.1.19) to prove the identities

(a) $\mathbf{a} \times (\mathbf{b} \times \mathbf{c}) = (\mathbf{a}\cdot\mathbf{c})\mathbf{b} - (\mathbf{a}\cdot\mathbf{b})\cdot\mathbf{c}$
(b) $\nabla^2\mathbf{a} = \nabla(\nabla\cdot\mathbf{a}) - \nabla \times (\nabla \times \mathbf{a})$
c) $(\mathbf{a}\cdot\nabla)\mathbf{a} = (\nabla \times \mathbf{a}) \times \mathbf{a} + \nabla(\mathbf{a}\cdot\mathbf{a}/2)$
(d) $\nabla \times (\alpha\mathbf{a}) = (\nabla\alpha) \times \mathbf{a} + \alpha(\nabla \times \mathbf{a})$

Solution

(a) $\mathbf{a} \times (\mathbf{b} \times \mathbf{c}) = (\mathbf{a}\cdot\mathbf{c})\mathbf{b} - (\mathbf{a}\cdot\mathbf{b})\cdot\mathbf{c}$. Proof: The formula (1.2.33) and the identity (1.1.19) imply:

$$\mathbf{b} \times \mathbf{c} = e_{rsk}b_rc_s\mathbf{e}_k, \quad \mathbf{a} \times (\mathbf{b} \times \mathbf{c}) = e_{ijk}a_j(e_{rsk}b_rc_s)\mathbf{e}_i = e_{ijk}e_{rsk}a_jb_rc_s\mathbf{e}_i$$
$$= (\delta_{ir}\delta_{js} - \delta_{is}\delta_{jr})a_jb_rc_s\mathbf{e}_i = (a_sc_s)(b_i\mathbf{e}_i) - (a_rb_r)(c_i\mathbf{e}_i) \Rightarrow$$
$$\mathbf{a} \times (\mathbf{b} \times \mathbf{c}) = (\mathbf{a}\cdot\mathbf{c})\mathbf{b} - (\mathbf{a}\cdot\mathbf{b})\mathbf{c} \quad \text{QED}$$

(b) $\nabla^2\mathbf{a} = \nabla(\nabla\cdot\mathbf{a}) - \nabla \times (\nabla \times \mathbf{a})$. Proof: The formulas (1.6.14), (1.6.10), (1.6.15), (1.2.33), and (1.1.19) imply:

$$\nabla^2\mathbf{a} = a_{r,ss}\mathbf{e}_r, \quad \nabla(\nabla\cdot\mathbf{a}) = a_{s,sr}\mathbf{e}_r, \quad \nabla \times \mathbf{a} = e_{ijk}a_{k,j}\mathbf{e}_r,$$
$$\nabla \times (\nabla \times \mathbf{a}) = e_{rsi}(e_{ijk}a_{k,j}),_s\mathbf{e}_r = e_{rsi}e_{jki}a_{k,js}\mathbf{e}_r = (\delta_{rj}\delta_{sk} - \delta_{rk}\delta_{sj})a_{k,js}\mathbf{e}_r$$
$$= a_{s,sr}\mathbf{e}_r - a_{r,ss}\mathbf{e}_r = \nabla(\nabla\cdot\mathbf{a}) - \nabla^2\mathbf{a} \Rightarrow \nabla^2\mathbf{a}$$
$$= \nabla(\nabla\cdot\mathbf{a}) - \nabla \times (\nabla \times \mathbf{a}) \quad \text{QED}$$

(c) $(\mathbf{a}\cdot\nabla)\mathbf{a} = (\nabla \times \mathbf{a}) \times \mathbf{a} + \nabla(\mathbf{a}\cdot\mathbf{a}/2)$. Proof: The formulas (1.6.10), (1.6.15), (1.2.33), and (1.1.19) imply:

$$\left.\begin{array}{l}(\mathbf{a}\cdot\nabla)\mathbf{a} = \left(a_k\frac{\partial}{\partial x_k}\right)(a_i\mathbf{e}_i) = a_ka_{i,k}\mathbf{e}_i, \quad \nabla(\mathbf{a}\cdot\mathbf{a}/2) = \left(\mathbf{e}_i\frac{\partial}{\partial x_i}\right)(a_ka_k/2) = a_ka_{k,i}\mathbf{e}_i \\ (\nabla \times \mathbf{a}) \times \mathbf{a} = e_{ijk}(e_{jrs}a_{s,r})a_k\mathbf{e}_i = e_{kij}e_{rsj}a_{s,r}a_k\mathbf{e}_i = (\delta_{kr}\delta_{is} - \delta_{ks}\delta_{ir})a_{s,r}a_k\mathbf{e}_i = (a_{i,k}a_k - a_{k,i}a_k)\mathbf{e}_i\end{array}\right\} \Rightarrow$$
$$(\nabla \times \mathbf{a}) \times \mathbf{a} = (\mathbf{a}\cdot\nabla)\mathbf{a} - \nabla(\mathbf{a}\cdot\mathbf{a}/2) \Rightarrow (\mathbf{a}\cdot\nabla)\mathbf{a} = (\nabla \times \mathbf{a}) \times \mathbf{a} + \nabla(\mathbf{a}\cdot\mathbf{a}/2) \quad \text{QED}$$

(d) $\nabla \times (\alpha\mathbf{a}) = (\nabla\alpha) \times \mathbf{a} + \alpha(\nabla \times \mathbf{a})$. Proof: The formula (1.6.15) implies:
$$\nabla \times (\alpha\mathbf{a}) = e_{ijk}(\alpha a_k),_j\mathbf{e}_i = e_{ijk}(\alpha,_ja_k)\mathbf{e}_i + e_{ijk}(\alpha a_k,_j)\mathbf{e}_i = (\nabla\alpha) \times \mathbf{a} + \alpha\nabla \times \mathbf{a} \quad \text{QED}$$

Chapter 2

Problem 2.1 A velocity field for a continuous material in motion is given as:

$$v_1 = \frac{\alpha\, x_1}{t - t_0}, \qquad v_2 = -\frac{\alpha\, x_2}{t - t_0}, \qquad v_3 = 0, \qquad \alpha \text{ and } t_0 \text{ are constants}$$

(a) Show that the flow is *isochoric* = volume preserving, i.e. div $\mathbf{v} = 0$.
(b) Determine the local acceleration, the convective acceleration, and the particle acceleration $\dot{\mathbf{v}}$.
(c) Show that the flow is irrotational, i.e. rot $\mathbf{v} = \mathbf{0}$, and determine the velocity potential ϕ from the formula $\mathbf{v} = \nabla\phi$.

Solution

(a) The divergence of the velocity field $\mathbf{v}$:

$$\text{div } \mathbf{v} = v_{i,i} = \frac{\partial v_1}{\partial x_1} + \frac{\partial v_2}{\partial x_2} + \frac{\partial v_3}{\partial x_3} = \frac{\alpha}{t - t_0} + \frac{-\alpha}{t - t_0} + 0 \Rightarrow \text{div } \mathbf{v} = 0$$

$$\Rightarrow \text{ isochoric flow}$$

(b) The local acceleration $\partial_t \mathbf{v} : \partial_t v_1 = \frac{\alpha x_1}{(t-t_0)^2}(-1), \quad \partial_t v_2 = -\frac{\alpha x_2}{(t-t_0)^2}(-1),$

$\partial_t v_3 = 0 \Rightarrow \partial_t \mathbf{v} = [-\alpha x_1, \alpha x_2, 0]\frac{1}{(t-t_0)^2}$

The convective acceleration $(\mathbf{v} \cdot \nabla)\mathbf{v} = (v_k \partial_k)(v_i \mathbf{e}_i) = v_k v_{i,k}\, \mathbf{e}_i$:

$$v_k v_{1,k} = v_1 v_{1,1} + v_2 v_{1,2} + v_3 v_{1,3} = v_1 v_{1,1} = \frac{\alpha^2 x_1}{(t - t_0)^2}, \qquad v_k v_{2,k} = \frac{\alpha^2 x_2}{(t - t_0)^2}, \qquad v_k v_{3,k} = 0$$

$$\Rightarrow \quad (\mathbf{v} \cdot \nabla)\mathbf{v} = [\alpha^2 x_1, \alpha^2 x_2, 0]\frac{1}{(t - t_0)^2}$$

The particle acceleration $\dot{\mathbf{v}} = \partial_t \mathbf{v} + (\mathbf{v} \cdot \nabla)\mathbf{v} = [(\alpha^2 - \alpha) x_1, (\alpha^2 + \alpha) x_2, 0]$
$\frac{1}{(t-t_0)^2}$

(c) rot $\mathbf{v} = e_{ijk} v_{k,j}\, \mathbf{e}_i = [v_{3,2} - v_{2,3}, v_{1,3} - v_{3,1}, v_{2,1} - v_{1,2}] = \mathbf{0} \Leftrightarrow$ irrotational flow

Velocity potential: $\phi \Rightarrow \mathbf{v} = \nabla\phi \Rightarrow v_i = \frac{\partial \phi}{\partial x_i} \Rightarrow$ by partial integration:

$$\phi = \frac{\alpha x_1^2}{2(t - t_0)} + f_1(x_2, x_3, t) = -\frac{\alpha x_2^2}{2(t - t_0)} + f_2(x_3, x_1, t) = f_3(x_1, x_2, t)$$

$$= \frac{\alpha\left(x_1^2 - x_2^2\right)}{2(t - t_0)} + f(t)$$

Problem 2.2 The state of stress in a particle is represented by the following stress matrix with respect to the Cartesian coordinate system Ox:

$$T = \begin{pmatrix} 90 & -30 & 0 \\ -30 & 120 & -30 \\ 0 & -30 & 90 \end{pmatrix} \text{ MPa}$$

(a) Determine the stress vector, the normal stress and the shear stress on a surface with unit normal: $\mathbf{n} = [1, 1, 0]/\sqrt{2}$.

(b) Determine the stress matrix $\bar{T}$ with respect to a Cartesian coordinate system $\bar{O}\bar{x}$ when the transformation matrix is:

$$Q = [\bar{\mathbf{e}}_i \cdot \mathbf{e}_k] = \begin{pmatrix} 1 & 0 & 0 \\ 0 & \sqrt{3}/2 & 1/2 \\ 0 & -1/2 & \sqrt{3}/2 \end{pmatrix}$$

Solution

(a) By definition: Stress vector: a) $\mathbf{t} = \mathbf{T} \cdot \mathbf{n} = t_i \mathbf{e}_i \Rightarrow t_i = T_{ik} n_k \Rightarrow$

normal stress: $\sigma = \mathbf{n} \cdot \mathbf{t} = n_i T_{ik} n_k$, shear stress: $\tau = \sqrt{\mathbf{t} \cdot \mathbf{t} - \sigma^2} = \sqrt{t_i t_i - \sigma^2}$

Computations:

$$t_1 = T_{1k} n_k = 90 \cdot \frac{1}{\sqrt{2}} + (-30) \cdot \frac{1}{\sqrt{2}}, t_2 = T_{2k} n_k = (-30) \cdot \frac{1}{\sqrt{2}}$$

$$+ 120 \cdot \frac{1}{\sqrt{2}}, t_3 = T_{3k} n_k = (-30) \cdot \frac{1}{\sqrt{2}} \Rightarrow$$

$$\mathbf{t} = [60, 90, -30] \frac{1}{\sqrt{2}} \text{MPa}, \quad \sigma = \mathbf{n} \cdot \mathbf{t} = \frac{1}{\sqrt{2}} 60 \frac{1}{\sqrt{2}} + \frac{1}{\sqrt{2}} 90 \frac{1}{\sqrt{2}} = 75 \text{ MPa}$$

$$\tau = \sqrt{\mathbf{t} \cdot \mathbf{t} - \sigma^2} = \sqrt{\left(\frac{60}{\sqrt{2}}\right)^2 + \left(\frac{90}{\sqrt{2}}\right)^2 + \left(\frac{-30}{\sqrt{2}}\right)^2 - 75^2} = 26 \text{ MPa}$$

(b) The stress matrix $\bar{T}$ in the $\bar{O}\bar{x}-system$: $\bar{T} = QTQ^T \Rightarrow \bar{T}_{ij} = Q_{ik} T_{kl} Q_{jl} \Rightarrow$

$$\bar{T}_{11} = 1 \cdot T_{11} \cdot 1 = 90, \quad \bar{T}_{12} = 1 \cdot T_{12} \cdot \left(\sqrt{3}/2\right) = -26,$$

$$\bar{T}_{22} = \left(\sqrt{3}/2\right) \cdot T_{22} \cdot \left(\sqrt{3}/2\right) + \left(\sqrt{3}/2\right) \cdot T_{23} \cdot (1/2) + (1/2) T_{32} \cdot \left(\sqrt{3}/2\right)$$

$$+ (1/2) \cdot T_{33} \cdot (1/2)$$

$$= 87, \quad \text{etc.}$$

$$\Rightarrow \quad \bar{T} = \begin{pmatrix} 90 & -26 & 15 \\ -26 & 87 & -28 \\ 15 & -28 & 123 \end{pmatrix} \text{ MPa.} \quad \text{A control: } \bar{T}_{kk} = T_{kk} = 300 \text{ MPa.}$$

Problem 2.3 A thin-walled circular tube has middle radius $r = 150$ mm and wall thickness $h = 7$ mm. The tube is closed in both ends, and is subjected to an internal pressure $p = 8$ MPa, a torque $m_t = 60$ kNm, and an axial force $N = 180$ kN. The state of stress in the wall of the tube may be found by superposition of the states of stress given in the Examples 2.1, 2.2, and 2.5.

(a) Determine the coordinate stresses with respect to the local Cartesian coordinate system presented in the Figs. 2.9 and 2.18.

(b) Determine the principal stresses and the principal stress directions in the tube wall.

(c) Determine the maximum shear stress in the tube wall.

Solution

(a) Non-zero coordinate stresses and the stress matrix T:

$$T_{11} = \frac{N}{2\pi rh} + \frac{r}{2h}p = 113.0 \text{ MPa}, \quad T_{22} = \frac{r}{h}p = 171.4 \text{ MPa}, \quad T_{12} = \frac{m_t}{2\pi r^2 h} = 60.6 \text{ MPa}$$

$$T = \begin{pmatrix} T_{11} & T_{12} & T_{13} \\ T_{21} & T_{22} & T_{23} \\ T_{31} & T_{32} & T_{33} \end{pmatrix} = \begin{pmatrix} 113.0 & 60.6 & 0 \\ 60.6 & 171.4 & 0 \\ 0 & 0 & 0 \end{pmatrix} \text{ MPa}$$

b) Principal stresses and principal stress directions in the tube wall. The formula (2.3.32) for the state of plane stress gives:

$$\frac{\sigma_1}{\sigma_2} = \frac{1}{2}(T_{11} + T_{22}) \pm \sqrt{\left(\frac{T_{11} - T_{22}}{2}\right)^2 + (T_{12})^2} \Rightarrow \frac{\sigma_1}{\sigma_2} = \frac{210 \text{ MPa}}{75 \text{ MPa}}, \quad \sigma_3 = 0$$

Formula (2.3.33) gives: $\phi_1 = \arctan\frac{\sigma_1 - T_{11}}{T_{12}} = 58°$, angle between the x_1-direction and the principal direction of σ_1

c) Maximum shear stress. Formula (2.3.31) gives:

$$\tau_{max} = \frac{1}{2}(\sigma_{max} - \sigma_{min}) = \frac{1}{2}(\sigma_1 - \sigma_3) = \frac{1}{2}\sigma_1 = 105 \text{ MPa}$$

Problem 2.4 The state of stress in a particle is represented by the stress matrix:

$$T = \begin{pmatrix} 200 & -100 & -100 \\ -100 & 100 & 0 \\ -100 & 0 & 100 \end{pmatrix} \text{ MPa}$$

(a) Determine the matrices for the stress isotrop and the stress deviator.

(b) Determine the principal stresses and the principal stress directions.

(c) Compute the maximum shear stress, and determine the unit normal to one of the planes of maximum shear stress.

(d) Compute the normal stress and the shear stress on the plane with unit normal:
$\mathbf{n} = [1, 1, 0]/\sqrt{2}$

Solution

(a) Matrices for the stress isotrop $\mathbf{T}^0$ and for the stress deviator $\mathbf{T}'$:

$$T^0 = \frac{1}{3}(\operatorname{tr} T)1, \quad \operatorname{tr} T = T_{kk} = T_{11} + T_{22} + T_{33} = 400 \text{ MPa}, \quad T' = T - T^0 \Rightarrow$$

$$T^0 = \begin{pmatrix} 1 & 0 & 0 \\ 0 & 1 & 0 \\ 0 & 0 & 1 \end{pmatrix} \frac{400}{3} \text{ MPa}, \quad T' = \begin{pmatrix} 2 & -3 & -3 \\ -3 & -1 & 0 \\ -3 & 0 & -1 \end{pmatrix} \frac{100}{3} \text{ MPa}$$

(b) Principal stresses σ_i and principal stress directions $\mathbf{n}_i$.
The stress tensor $\mathbf{T}$ and the stress deviator $\mathbf{T}'$ are coaxial tensors and have the same principal directions $\mathbf{n}_i$. We first search a principal stress σ' for the stress deviator $\mathbf{T}'$ and the corresponding principal direction $\mathbf{n}$. The principal invariants for the stress deviator $\mathbf{T}'$ are:

$$I' = T'_{ii} = 0, \quad II' = -\frac{1}{2}T'_{ik}T'_{ik} = -\frac{70{,}000}{3} \text{ MPa},$$

$$III' = \det T' = \frac{20}{27} \cdot 10^6 \text{ MPa}$$

The characteristic equation (2.3.21) for $\mathbf{T}'$ becomes: $(\sigma')^3 + II'\sigma' - III' = 0$. The solution of this equation, following a standard procedure, is found by introducing the angle θ from the formula:

$$\cos\theta = \frac{III'}{2\sqrt{-(II'/3)^3}}, \quad 0 \le \theta \le 180° \Rightarrow \theta = 57.32°$$

Then the principal stresses σ'_i for the stress deviator $\mathbf{T}'$ are determined from the formulas:

$$\sigma'_1 = 2\sqrt{-\frac{II'}{3}}\cos\frac{\theta}{3}, \quad \sigma'_2 = -2\sqrt{-\frac{II'}{3}}\cos\left(\frac{\theta}{3} + 60°\right) \le \sigma'_1,$$

$$\sigma'_3 = -2\sqrt{-\frac{II'}{3}}\cos\left(\frac{\theta}{3} - 60°\right) \le \sigma'_2$$

The principal stresses σ_i for the stress tensor $\mathbf{T}$ are determined from the formulas:

$$\sigma_i = \sigma_i' + \text{tr}\, T/3 \Rightarrow \sigma_1 = 300\,\text{MPa}, \quad \sigma_2 = 100\,\text{MPa}, \quad \sigma_3 = 0$$

The principal stress directions n_i are found from the Eq. (2.3.2) as follows.

$$(\sigma_1 \delta_{ik} - T_{ik})n_{1k} = 0 \Rightarrow \begin{pmatrix} 100n_{11} + 100n_{12} + 100n_{13} = 0 \\ 100n_{11} + 200n_{12} + 000n_{13} = 0 \\ 100n_{11} + 000n_{12} + 200n_{13} = 0 \end{pmatrix} \Rightarrow \mathbf{n}_1 = [2, -1, -1]1/\sqrt{6}$$

$$(\sigma_2 \delta_{ik} - T_{ik})n_{2k} = 0 \Rightarrow \begin{pmatrix} -100n_{21} + 100n_{22} + 100n_{23} = 0 \\ 100n_{11} + 000n_{12} + 000n_{13} = 0 \\ 100n_{11} + 000n_{12} + 000n_{13} = 0 \end{pmatrix} \Rightarrow \mathbf{n}_2 = [0, 1, -1]1/\sqrt{2}$$

$$(\sigma_3 \delta_{ik} - T_{ik})n_{3k} = 0 \Rightarrow \begin{pmatrix} -200n_{21} + 100n_{22} + 100n_{23} = 0 \\ 100n_{11} - 100n_{12} + 000n_{13} = 0 \\ 100n_{11} + 000n_{12} - 100n_{13} = 0 \end{pmatrix} \Rightarrow \mathbf{n}_3 = [1, 1, 1]1/\sqrt{3}$$

(c) The maximum shear stress and unit normal $\mathbf{n}$ to one of the planes of maximum shear stress.

$$\text{Formula (2.3.31)} \Rightarrow \tau_{\max} = (\sigma_{\max} - \sigma_{\min})/2 = (\sigma_1 - \sigma_3)/2 = 150\,\text{MPa}$$

$$\mathbf{n} = (\mathbf{n}_1 + \mathbf{n}_3)/\sqrt{2} = \left[2 + \sqrt{2}, \sqrt{2} - 1, \sqrt{2} - 1\right]/\left(2\sqrt{3}\right)$$

(d) Normal stress σ and shear stress τ on the plane with unit normal: $\mathbf{n} = [1, 1, 0]/\sqrt{2}$:

$$\sigma = n_i T_{ik} n_k = 50\,\text{MPa. Stress vector } \mathbf{t} : t_i = T_{ik} n_k \Rightarrow \mathbf{t} = [100, 0, 100]/\sqrt{2},$$
$$\tau = \sqrt{t_i t_i - \sigma^2} = 87\,\text{MPa}$$

Chapter 3

Problem 3.1 Show that a completely antisymmetric tensor of third order $\mathbf{A}$ only has one distinct component different from zero, and that the tensor is represented by the product of a scalar α and the permutation tensor $\mathbf{P}$, i.e. $\mathbf{A} = \alpha\mathbf{P}$.

Solution Antisymmetry implies:

$$A_{ijk} = -A_{kji} = A_{kij} = -A_{ikj} \Rightarrow A_{123} = -A_{312} = A_{231} = -A_{321} = -A_{123} = -A_{213} = -A_{132}$$

With no summation with respect to k: $A_{kjk} = -A_{jkk} = 0$, $A_{ikk} = -A_{ikk} = 0$,
$A_{kkj} = -A_{kkj} = 0$

The tensor matrix $A_{ijk} = \alpha e_{ijk}$ satisfies all these equations. Hence: $\mathbf{A} = \alpha\mathbf{P}$.

Problem 3.2 Show that an isotropic tensor of second order $\mathbf{I}_2$ always is a product of a scalar and the unit tensor as given by formula (3.2.8): $\mathbf{I}_2 = \alpha\mathbf{1}$.

Solution Let **B** be an isotropic tensor of second order which implies that the tensor matrix is the same in any Cartesian coordinate system. Three Cartesian coordinate systems are now introduced: Ox with base vectors $\mathbf{e}_i$, $O\bar{x}$ with base vectors $\bar{\mathbf{e}}_i$, and $O\tilde{x}$ with base vectors $\tilde{\mathbf{e}}_i$. The system $O\bar{x}$ is obtained by a $90°$-rotation of the system Ox about the x_3-axis, and the system $O\tilde{x}$ is obtained by a $90°$-rotation of the system Ox about the x_1-axis. The transformation matrices relating the three described coordinate systems Ox, $O\bar{x}$, and $O\tilde{x}$ are:

$$\bar{Q} = (\bar{\mathbf{e}}_i \cdot \mathbf{e}_k) = \begin{pmatrix} 0 & 1 & 0 \\ -1 & 0 & 0 \\ 0 & 0 & 1 \end{pmatrix}, \quad \tilde{Q} = (\tilde{\mathbf{e}}_i \cdot \mathbf{e}_k) = \begin{pmatrix} 1 & 0 & 0 \\ 0 & 0 & 1 \\ 0 & -1 & 0 \end{pmatrix}$$

The tensor matrices in the three coordinate systems are $B = \bar{B} = \tilde{B}$ and must satisfy the transformation formulas:

$$\bar{B} = \bar{Q}B\bar{Q}^T = B \Rightarrow \begin{pmatrix} B_{22} & -B_{21} & B_{23} \\ -B_{12} & B_{11} & -B_{13} \\ B_{32} & -B_{31} & B_{33} \end{pmatrix} = \begin{pmatrix} B_{11} & B_{12} & B_{13} \\ B_{21} & B_{22} & B_{23} \\ B_{31} & B_{32} & B_{33} \end{pmatrix} \Rightarrow$$

$$B_{22} = B_{11} = \beta, \text{ a scalar}, \quad B_{23} = B_{13} = -B_{13} = 0, \quad B_{32} = B_{31} = -B_{31} = 0 \quad (1)$$

$$\tilde{B} = \tilde{Q}B\tilde{Q}^T = B \Rightarrow \begin{pmatrix} B_{11} & B_{13} & -B_{12} \\ B_{31} & B_{33} & -B_{32} \\ -B_{21} & -B_{23} & B_{22} \end{pmatrix} = \begin{pmatrix} B_{11} & B_{12} & B_{13} \\ B_{21} & B_{22} & B_{23} \\ B_{31} & B_{32} & B_{33} \end{pmatrix} \Rightarrow$$

$$B_{22} = B_{33} = \beta, \quad B_{13} = B_{12} = -B_{12} = 0, \quad B_{31} = B_{21} = -B_{21} = 0 \quad (2)$$

From the results (1) and (2) we conclude that:

$$\begin{pmatrix} B_{11} & B_{12} & B_{13} \\ B_{21} & B_{22} & B_{23} \\ B_{31} & B_{32} & B_{33} \end{pmatrix} = \begin{pmatrix} \beta & 0 & 0 \\ 0 & \beta & 0 \\ 0 & 0 & \beta \end{pmatrix} = \beta 1 \Rightarrow \mathbf{B} = \beta \mathbf{1} \quad \text{QED}$$

Problem 3.3 Show by induction that the result (3.3.32) follows from the formula (3.3.30).

$$S_{ij} = \sum_k \sigma_k a_{ki} a_{kj} \Leftrightarrow \mathbf{S} = \sum_k \sigma_k \mathbf{a}_k \otimes \mathbf{a}_k \quad (3.3.30)$$

$$\mathbf{S}^n = \sum_k (\sigma_k)^n \mathbf{a}_k \otimes \mathbf{a}_k, \quad n = 1, 2, 3, \ldots \quad (3.3.32)$$

Solution In a Cartesian coordinate system Ox with base vectors $\mathbf{e}_k \equiv \mathbf{a}_k = \delta_{ki}\mathbf{e}_i$ the tensor **S** is in accordance with the formula (3.3.30) represented by the components:

$$S_{ij} = \sum_k \sigma_k a_{ki} a_{kj} = \sum_k \sigma_k \delta_{ki} \delta_{kj} = \sigma_i \delta_{ij} \quad (1)$$

If we assume that the formula is correct for a natural number n, the tensors $\mathbf{S}^n$ and $\mathbf{S}^{n+1} = \mathbf{S}^n \mathbf{S}$ has in the system Ox the components:

$$S_{ij}^n = \sum_k (\sigma_k)^n a_{ki} a_{kj} = (\sigma_i)^n \delta_{ij} \quad (2),$$

$$S_{ij}^{n+1} = S_{ir}^n S_{rj} = \left(\sum_r (\sigma_i)^n \delta_{ir} \right) (\sigma_r \delta_{rj}) = (\sigma_i)^{n+1} \delta_{ij} \quad (3)$$

Because the formula (2) is correct for $n = 1$, according to formula (1), the formula (3) shows that the formula (2) is correct for $n = 1 + 1 = 2$. By induction the formula (2) is correct for any natural number $n = 1, 2, 3, \ldots$.

Problem 3.4 Show that the definition (3.3.33) of a positive definite and symmetric second order tensor $\mathbf{S}$ implies that all the principal values and the principal invariants of the tensor are positive.

Solution Let $\mathbf{S}$ be a positive and symmetric second order tensor. Then by definition:

$$\mathbf{c} \cdot \mathbf{S} \cdot \mathbf{c} > 0 \quad \text{for all vectors } \mathbf{c} \neq 0 \quad (3.3.33)$$

Let the principal values and the principal directions of $\mathbf{S}$ be σ_k and $\mathbf{a}_k$. Then by definition of σ_k and $\mathbf{a}_k$: $\mathbf{S} \cdot \mathbf{a}_k = \sigma_k \mathbf{a}_k$. For a vector $\mathbf{c} = \mathbf{a}_k$ the definition (3.3.33) implies:

$$\text{for } k = 1, 2, \text{ or } 3 : \ \mathbf{a}_k \cdot \mathbf{S} \cdot \mathbf{a}_k = \mathbf{a}_k \cdot (\mathbf{S} \cdot \mathbf{a}_k) = \mathbf{a}_k \cdot (\sigma_k \mathbf{a}_k) = \sigma_k \mathbf{a}_k \cdot \mathbf{a}_k = \sigma_k > 0$$

Hence the principal values σ_k of the tensor $\mathbf{S}$ are all positive. The formulas (3.3.25) now imply that all principal invariants of the tensor $\mathbf{S}$ are positive.

Problem 3.5 The general form of the linear, symmetric isotropic tensor-valued function $\mathbf{B}[\mathbf{A}]$ of a symmetric second order tensor $\mathbf{A}$ is given by the formula:

$$\mathbf{B}[\mathbf{A}] = (\gamma + \lambda \text{tr} \mathbf{A}) \mathbf{1} + 2\mu \mathbf{A} \quad (3.9.16)$$

Derive the relations:

$$\mathbf{B}'[\mathbf{A}] = 2\mu \mathbf{A}', \quad \mathbf{B}^o[\mathbf{A}] = 3\kappa \mathbf{A}^o + \gamma \mathbf{1} \quad (3.9.19)$$

Solution The tensors $\mathbf{A}$ and $\mathbf{B}$ are decomposed into deviators and isotrops:

$$\mathbf{B} = \mathbf{B}' + \mathbf{B}^o, \quad \mathbf{B}' = \mathbf{B} - \mathbf{B}^o, \quad \mathbf{B}^o = \frac{1}{3}(\text{tr}\mathbf{B})\mathbf{1}, \quad \mathbf{A} = \mathbf{A}' + \mathbf{A}^o, \quad \mathbf{A}' = \mathbf{A} - \mathbf{A}^o,$$

$$\mathbf{A}^o = \frac{1}{3}(\text{tr}\mathbf{A})\mathbf{1}$$

Formula (3.9.16) implies that: $\mathrm{tr}\mathbf{B} = (\gamma + \lambda\mathrm{tr}\mathbf{A})\mathrm{tr}\mathbf{1} + 2\mu\,\mathrm{tr}\mathbf{A} = 3\gamma + (3\lambda + 2\mu)\mathrm{tr}\mathbf{A}$. We then get:

$$\mathbf{B}^o = \frac{1}{3}[3\gamma + (3\lambda + 2\mu)\mathrm{tr}\mathbf{A}]\mathbf{1} = (3\gamma + 2\mu)\mathbf{A}^o + \gamma\mathbf{1} = 3\kappa\mathbf{A}^o + \gamma\mathbf{1}, \quad \kappa = \lambda + \frac{2}{3}\mu$$

$$\mathbf{B}' = \mathbf{B} - \mathbf{B}^o = \left[(\gamma + \lambda\mathrm{tr}\mathbf{A}^o)\mathbf{1} + 2\mu\left(\left(\left(\frac{1}{3}\mathrm{tr}\mathbf{A}^o\right)\mathbf{1}\right) + \mathbf{A}'\right)\right]$$

$$- \left[(3\lambda + 2\mu)\left(\left(\frac{1}{3}\mathrm{tr}\mathbf{A}^o\right)\mathbf{1}\right) + \gamma\mathbf{1}\right] = 2\mu\mathbf{A}'$$

$$\Rightarrow \mathbf{B}' = 2\mu\mathbf{A}', \quad \mathbf{B}^o = 3\kappa\mathbf{A}^o + \gamma\mathbf{1}, \quad \kappa = \lambda + \frac{2}{3}\mu \Rightarrow (3.9.19)$$

Chapter 4

Problem 4.1 Show that in homogeneous deformation of a body of continuous material, formula (4.2.26), planes and straight lines in the reference configuration K_0 deform into planes and straight lines in the present configuration K.

Solution Homogeneous deformation is defined by:

$$\mathbf{r}(\mathbf{r}_0, t) = \mathbf{u}_0(t) + \mathbf{F}(t) \cdot \mathbf{r}_0 \quad (4.2.26)$$

A material plane in K_0 through any two particles $\mathbf{r}_0$ and $\mathbf{r}_{10}$ and with unit normal vector $\mathbf{a}_0$ is defined by:

$$(\mathbf{r}_{10} - \mathbf{r}_0) \cdot \mathbf{a}_0 = 0 \quad (1)$$

In the present configuration K the particle $\mathbf{r}_0$ has moved to a new place $\mathbf{r}$ and the particle $\mathbf{r}_{10}$ has moved to the place $\mathbf{r}_1$. The formula (4.2.26) and Eq. (1) imply that:

$$\mathbf{r}_1 - \mathbf{r} = \mathbf{F} \cdot (\mathbf{r}_{10} - \mathbf{r}_0) = \mathbf{0} \Rightarrow \mathbf{r}_{10} - \mathbf{r}_0 = \mathbf{F}^{-1} \cdot (\mathbf{r}_1 - \mathbf{r}) = \mathbf{0} \Rightarrow$$
$$(\mathbf{r}_{10} - \mathbf{r}_0) \cdot \mathbf{a}_0 = \left[\mathbf{F}^{-1} \cdot (\mathbf{r}_1 - \mathbf{r})\right] \cdot \mathbf{a}_0 = 0 \Rightarrow$$
$$(\mathbf{r}_1 - \mathbf{r}) \cdot \mathbf{F}^{-T} \cdot \mathbf{a}_0 = 0 \quad (2)$$

Introducing the vector $\mathbf{a} = \mathbf{F}^{-T} \cdot \mathbf{a}_0$ in Eq. (2) implies that:

$$(\mathbf{r}_1 - \mathbf{r}) \cdot \mathbf{a} = 0 \quad (3)$$

By comparing the Eqs. (1) and (3) we may conclude that the material plane (1) in K_0 is deformed into the material plane (3) in K. Material straight lines represent intersections of material planes. Hence we have proved that: In homogeneous deformation of a body of continuous material planes and straight lines in the reference configuration K_0 deform into planes and straight lines in the present configuration K.

Problem 4.2 Develop the relations

$$\dot{\mathbf{F}} = \mathbf{LF} \Leftrightarrow \mathbf{L} = \dot{\mathbf{F}}\mathbf{F}^{-1} \tag{4.5.28}$$

$$\mathbf{D} = \tfrac{1}{2}\mathbf{R}\left(\dot{\mathbf{U}}\mathbf{U}^{-T} + \mathbf{U}^{-1}\dot{\mathbf{U}}\right)\mathbf{R}^T \tag{4.5.29}$$

$$\mathbf{W} = \dot{\mathbf{R}}\mathbf{R}^T + \tfrac{1}{2}\mathbf{R}\left(\dot{\mathbf{U}}\mathbf{U}^{-T} - \mathbf{U}^{-1}\dot{\mathbf{U}}\right)\mathbf{R}^T \tag{4.5.30}$$

$$\dot{\mathbf{E}} = \tfrac{1}{2}\mathbf{F}^T\left(\mathbf{L} + \mathbf{L}^T\right)\mathbf{F} = \mathbf{F}^T\mathbf{DF} \tag{4.5.31}$$

Solution By definition:

$$\mathbf{F} = \frac{\partial \mathbf{r}(\mathbf{r}_o, t)}{\partial \mathbf{r}_o} \Rightarrow F_{ik} = \frac{\partial x_i}{\partial X_k}, \quad \mathbf{v} = \dot{\mathbf{r}} = \frac{\partial \mathbf{r}(\mathbf{r}_o, t)}{\partial t} \Rightarrow v_i = \frac{\partial x_i}{\partial t},$$

$$\mathbf{L} = \frac{\partial \mathbf{v}}{\partial \mathbf{r}} \Leftrightarrow L_{ik} = \frac{\partial v_i}{\partial x_k}$$

We obtain:

$$\dot{F}_{ik} = \frac{\partial}{\partial t}\left(\frac{\partial x_i}{\partial X_k}\right) = \frac{\partial}{\partial X_k}\frac{\partial x_i}{\partial t} = \frac{\partial v_i}{\partial X_k} = \frac{\partial v_i}{\partial x_j}\frac{\partial x_j}{\partial X_k} = L_{ij}F_{jk} \Rightarrow \dot{\mathbf{F}} = \mathbf{LF} \Leftrightarrow (4.5.28)$$

By polar decomposition: $\mathbf{F} = \mathbf{RU} \Rightarrow \dot{\mathbf{F}} = \dot{\mathbf{R}}\mathbf{U} + \mathbf{R}\dot{\mathbf{U}}, \quad \mathbf{F}^{-1} = \mathbf{U}^{-1}\mathbf{R}^{-1} = \mathbf{U}^{-1}\mathbf{R}^T$ (1)
From the Eqs. (4.5.28) and (1) we get:

$$\mathbf{L} = \dot{\mathbf{F}}\mathbf{F}^{-1} = \left(\dot{\mathbf{R}}\mathbf{U} + \mathbf{R}\dot{\mathbf{U}}\right)\left(\mathbf{U}^{-1}\mathbf{R}^T\right) = \dot{\mathbf{R}}\mathbf{R}^T + \mathbf{R}\left(\dot{\mathbf{U}}\mathbf{U}^{-1}\right)\mathbf{R}^T \tag{2}$$

$$\mathbf{R}\mathbf{R}^T = 1 \Rightarrow \dot{\mathbf{R}}\mathbf{R}^T + \mathbf{R}\dot{\mathbf{R}}^T = 0 \Rightarrow \left(\dot{\mathbf{R}}\mathbf{R}^T\right)^T = -\dot{\mathbf{R}}\mathbf{R}^T \tag{3}$$

From the Eqs. (4.4.4), (2), (3), and (4.4.5) we get:

$$\mathbf{D} = \frac{1}{2}\left(\mathbf{L} + \mathbf{L}^T\right) = \frac{1}{2}\left(\dot{\mathbf{F}}\mathbf{F}^{-1} + \mathbf{F}^{-T}\dot{\mathbf{F}}^T\right) \Rightarrow \mathbf{D} = \frac{1}{2}\mathbf{R}\left(\dot{\mathbf{U}}\mathbf{U}^{-1} + \mathbf{U}^{-1}\dot{\mathbf{U}}\right)\mathbf{R}^T \Rightarrow (4.5.29)$$

$$\mathbf{W} = \frac{1}{2}\left(\mathbf{L} - \mathbf{L}^T\right) = \frac{1}{2}\left(\dot{\mathbf{F}}\mathbf{F}^{-1} - \mathbf{F}^{-T}\dot{\mathbf{F}}^T\right) \Rightarrow$$

$$\mathbf{W} = \dot{\mathbf{R}}\mathbf{R}^T + \frac{1}{2}\mathbf{R}\left(\dot{\mathbf{U}}\mathbf{U}^{-1} - \mathbf{U}^{-1}\dot{\mathbf{U}}\right)\mathbf{R}^T \Rightarrow (4.5.30)$$

The Eqs. (4.2.11), (4.2.10), (4.5.28), and (4.4.4) imply that: $\mathbf{E} = \dfrac{1}{2}(\mathbf{C} - 1) = \dfrac{1}{2}\left(\mathbf{F}^T\mathbf{F} - 1\right) \Rightarrow$

$$\dot{\mathbf{E}} = \frac{1}{2}\left(\dot{\mathbf{F}}^T\mathbf{F} + \mathbf{F}^T\dot{\mathbf{F}}\right) = \frac{1}{2}\left(\mathbf{F}^T\mathbf{L}^T\mathbf{F} + \mathbf{F}^T\mathbf{LF}\right) = \frac{1}{2}\mathbf{F}^T\left(\mathbf{L}^T + \mathbf{L}\right)\mathbf{F} = \mathbf{F}^T\mathbf{DF} \Rightarrow$$

$$\dot{\mathbf{E}} = \frac{1}{2}\mathbf{F}^T\left(\mathbf{L} + \mathbf{L}^T\right)\mathbf{F} = \mathbf{F}^T\mathbf{DF} \Leftrightarrow (4.5.31)$$

Problem 4.3 Derive the formula:

$$\dot{J} = J \operatorname{div} \mathbf{v} \quad (4.5.33).$$

Solution The formulas (1.1.24) and (1.1.27) imply:

$$\frac{\partial(\det \mathbf{F})}{\partial F_{ik}} = \operatorname{Co}F_{ik}, \quad (\operatorname{Co}F_{ik})F_{jk} = (\det \mathbf{F})\delta_{ij} \quad (1)$$

By the definition (4.5.32): $J = \det \mathbf{F}$. From the formulas (1), (4.5.28), and (4.4.3) we obtain:

$$\dot{J} = \frac{\partial(\det \mathbf{F})}{\partial F_{ik}} \dot{F}_{ik} = (\operatorname{Co}F_{ik})(L_{ij}F_{jk}) = \det \mathbf{F}\delta_{ij}L_{ij} = (\det \mathbf{F})L_{kk} = (\det \mathbf{F})v_{k,k} \Rightarrow$$
$$\dot{J} = J \operatorname{div} \mathbf{v} \Rightarrow (4.5.33)$$

Chapter 5
Problem 5.1 Develop the (5.2.7) of Hooke's law from the form (5.2.5).

$$\mathbf{E} = \frac{1+v}{\eta}\mathbf{T} - \frac{v}{\eta}(\operatorname{tr}\mathbf{T})\mathbf{1} \quad (5.2.5)$$

$$\mathbf{T} = \frac{\eta}{1+v}\left[\mathbf{E} + \frac{v}{1-2v}(\operatorname{tr}\mathbf{E})\mathbf{1}\right] \quad (5.2.7)$$

Solution The formula (5.2.5) implies:

$$\operatorname{tr}\mathbf{E} = \frac{1+v}{\eta}\operatorname{tr}\mathbf{T} - \frac{v}{\eta}(\operatorname{tr}\mathbf{T})\operatorname{tr}\mathbf{1} = \frac{1-2v}{\eta}\operatorname{tr}\mathbf{T} \Rightarrow \operatorname{tr}\mathbf{T} = \frac{\eta}{1-2v}\operatorname{tr}\mathbf{E} \quad (1)$$

The formulas (5.2.5) and (1) imply:

$$\mathbf{T} = \frac{\eta}{1+v}\left[\mathbf{E} + \frac{v}{1-2v}(\operatorname{tr}\mathbf{E})\mathbf{1}\right] \Leftrightarrow (5.2.7)$$

Problem 5.2 Develop the decomposition (5.2.14)–(5.2.16) from Hooke's law (5.2.7).

$$\mathbf{T} = \frac{\eta}{1+v}\left[\mathbf{E} + \frac{v}{1-2v}(\operatorname{tr}\mathbf{E})\mathbf{1}\right] \quad (5.2.7)$$

$$\kappa = \frac{\eta}{3(1-2v)} \quad (5.2.11)_2, \quad \mu = \frac{\eta}{2(1+v)} \quad (5.2.9)$$

$$\mathbf{T} = \mathbf{T}^o + \mathbf{T}', \quad \mathbf{E} = \mathbf{E}^o + \mathbf{E}' \tag{5.2.14}$$

$$\mathbf{T}^o = \frac{1}{3}(\text{tr}\,\mathbf{T})\mathbf{1} = \sigma^o\mathbf{1}, \quad \mathbf{E}^o = \frac{1}{3}(\text{tr}\,\mathbf{E})\mathbf{1} = \frac{1}{3}\varepsilon_v\mathbf{1} \tag{5.2.15}$$

$$\mathbf{T}^o = 3\kappa\,\mathbf{E}^o, \quad \mathbf{T}' = 2\mu\,\mathbf{E}' \tag{5.2.16}$$

Solution The formula (5.2.7) implies:

$$\text{tr}\,\mathbf{T} = \frac{\eta}{1+v}\left[\text{tr}\,\mathbf{E} + \frac{v}{1-2v}(\text{tr}\,\mathbf{E})\text{tr}\,\mathbf{1}\right] = \frac{\eta}{1+v}\left[\frac{1-2v}{1-2v} + \frac{3v}{1-2v}\right]\text{tr}\,\mathbf{E} \Rightarrow$$

$$\text{tr}\,\mathbf{T} = \frac{\eta}{1-2v}\,\text{tr}\,\mathbf{E} \quad (1)$$

The formulas (1) and $(5.2.11)_2$ imply that:

$$\mathbf{T}^o = \frac{1}{3}(\text{tr}\,\mathbf{T})\mathbf{1} = \frac{\eta}{3(1-2v)}(\text{tr}\,\mathbf{E})\mathbf{1} = 3\kappa\,\mathbf{E}^o \Leftrightarrow (5.2.16)_1$$

$$\mathbf{T}' = \mathbf{T} - \mathbf{T}^o = \frac{\eta}{1+v}\left[(\mathbf{E}^o + \mathbf{E}') + \frac{3v}{1-2v}\frac{1}{3}(\text{tr}\,\mathbf{E})\mathbf{1}\right] - \frac{\eta}{(1-2v)}\mathbf{E}^o$$

$$= \frac{\eta}{1+v}\mathbf{E}' + \frac{\eta}{1+v}\mathbf{E}^o\left[1 + \frac{3v}{1-2v} - \frac{1+v}{(1-2v)}\right] \Rightarrow \mathbf{T}' = \frac{\eta}{1+v}\mathbf{E}' = 2\mu\,\mathbf{E}' \Leftrightarrow (5.2.16)_2$$

Chapter 6
Problem 6.1 Derive the formula (6.4.21) for the determinant of a second order tensor from the definition (3.3.10).

Solution Formula (6.4.21): $\det\mathbf{B} = \det(B_{ij})/g = \det(B_i{}^j) = \det\left(B_j^i\right) = \det(B^{ij})g$

Definition (3.3.10) of the determinant of a second order tensor $\mathbf{B}$:

$$\det\mathbf{B} = \det B = \det(B_{ij})\text{in Cartesian coordinates}$$

Let Ox be a Cartesian coordinate system with base vectors $\mathbf{e}_i$, and y a general coordinate system with base vectors $\mathbf{g}_k$ and $\mathbf{g}^l$. By the formulas (6.2.11) and (6.2.19):

$$\mathbf{e}_i = \frac{\partial y^k}{\partial x_i}\mathbf{g}_k = \frac{\partial x_i}{\partial y^k}\mathbf{g}^k$$

The components of a second order tensor $\mathbf{B}$ in the two coordinate systems are by definition:

$\mathbf{B}[\mathbf{e}_i, \mathbf{e}_i]$ in the Ox-system and

$$B_{kl} = \mathbf{B}[\mathbf{g}_k, \mathbf{g}_l], \quad B_k^l = \mathbf{B}[\mathbf{g}_k, \mathbf{g}^l], \quad B_l^k = \mathbf{B}[\mathbf{g}^k, \mathbf{g}_l], \quad B^{kl} = \mathbf{B}[\mathbf{g}^k, \mathbf{g}^l] \quad \text{in the } y\text{-system}$$

It follows that:

$$\mathbf{B}[\mathbf{e}_i, \mathbf{e}_j] = \mathbf{B}\left[\frac{\partial y^k}{\partial x_i}\mathbf{g}_k, \frac{\partial y^l}{\partial x_j}\mathbf{g}_l\right] = \mathbf{B}\left[\frac{\partial y^k}{\partial x_i}\mathbf{g}_k, \frac{\partial x_j}{\partial y^l}\mathbf{g}^l\right] = \mathbf{B}\left[\frac{\partial x_i}{\partial y^k}\mathbf{g}^k, \frac{\partial y^l}{\partial x_j}\mathbf{g}_l\right] = \mathbf{B}\left[\frac{\partial x_i}{\partial y^k}\mathbf{g}^k, \frac{\partial x_j}{\partial y^l}\mathbf{g}^l\right]$$

$$= \frac{\partial y^k}{\partial x_i}\frac{\partial y^l}{\partial x_j}\mathbf{B}[\mathbf{g}_k, \mathbf{g}_l] = \frac{\partial y^k}{\partial x_i}\frac{\partial x_j}{\partial y^l}\mathbf{B}[\mathbf{g}_k, \mathbf{g}^l] = \frac{\partial x_i}{\partial y^k}\frac{\partial y^l}{\partial x_j}\mathbf{B}[\mathbf{g}^k, \mathbf{g}_l] = \frac{\partial x_i}{\partial y^k}\frac{\partial x_j}{\partial y^l}\mathbf{B}[\mathbf{g}^k, \mathbf{g}^l]$$

$$\Rightarrow \mathbf{B}[\mathbf{e}_i, \mathbf{e}_i] = \frac{\partial y^k}{\partial x_i}\frac{\partial y^l}{\partial x_j}B_{kl} = \frac{\partial y^k}{\partial x_i}\frac{\partial x_j}{\partial y^l}B_k^l = \frac{\partial x_i}{\partial y^k}\frac{\partial y^l}{\partial x_j}B_l^k = \frac{\partial x_i}{\partial y^k}\frac{\partial x_j}{\partial y^l}B^{kl}$$

Now we apply the multiplication theorem for determinants (1.1.23) and the formulas:

$$\det\left(\frac{\partial y^k}{\partial x_i}\right) = J_x^y = \frac{1}{\sqrt{g}}, \quad \det\left(\frac{\partial x_i}{\partial y^k}\right) = J_y^x = \sqrt{g} \quad \text{from the formulas (6.2.30)}$$

and obtain:

$$\det \mathbf{B} = \det(\mathbf{B}[\mathbf{e}_i, \mathbf{e}_i]) = \det\left(\frac{\partial y^k}{\partial x_i}\right)\det(B_{kl})\det\left(\frac{\partial y^l}{\partial x_j}\right) = \left(\frac{\partial y^k}{\partial x_i}\right)\det(B_k^l)\det\left(\frac{\partial x_j}{\partial y^l}\right)$$

$$= \det\left(\frac{\partial x_i}{\partial y^k}\right)\det(B_l^k)\det\left(\frac{\partial y^l}{\partial x_j}\right) = \det\left(\frac{\partial x_i}{\partial y^k}\right)\det(B^{kl})\det\left(\frac{\partial x_j}{\partial y^l}\right) \Rightarrow$$

$$\det \mathbf{B} = J_x^y \det(B_{kl})J_x^y = J_y^x \det(B_l^k)J_x^y = J_x^y \det(B_k^l)J_y^x = J_y^x \det(B^{kl})J_y^x =$$

$$\det \mathbf{B} = \det(B_{ij})/g = \det(B_i^j) = \det(B_j^i) = \det(B^{ij})g \Rightarrow (6.4.21)$$

Problem 6.2 Show that the symbols ε_{ijk} and ε^{ijk} defined by the formulas (6.2.22) and (6.2.24) are components in a general coordinate system y of the permutation tensor $\mathbf{P}$ defined by the formula:

$$\alpha = [\mathbf{abc}] \equiv [(\mathbf{a} \times \mathbf{b}) \cdot \mathbf{c}] = \mathbf{P}[\mathbf{a}, \mathbf{b}, \mathbf{c}] \quad (3.1.25)$$

Solution The formulas (3.1.25) and (6.2.25) imply that:

$$P_{ijk} = \mathbf{P}[\mathbf{g}_i, \mathbf{g}_j, \mathbf{g}_k] = \varepsilon_{ijk}, \quad P^{ijk} = \mathbf{P}[\mathbf{g}^i, \mathbf{g}^j, \mathbf{g}^k] = \varepsilon^{ijk} \quad \text{QED}$$

Problem 6.3 Prove the identity: $\varepsilon^{ijk}\varepsilon_{rsk} = \delta_r^i\delta_s^j - \delta_s^i\delta_r^j$ (6.4.23), using the identity:

$$e_{ijk}e_{rsk} = \delta_{ir}\delta_{js} - \delta_{is}\delta_{jr} \quad (1.1.19)$$

Solution By the formulas (6.2.24) and (6.2.22):

$$\varepsilon^{ijk} = J_x^y e_{ijk}, \quad \varepsilon_{rsk} = J_y^x e_{rsk}$$

The formulas (6.2.4) imply that: $J_x^y J_y^x = 1$. Applying the identity (1.1.19) we obtain:

$$\varepsilon^{ijk}\varepsilon_{rsk} = J_x^y e_{ijk} J_y^x e_{rsk} = e_{ijk} e_{rsk} = \delta_{ir}\delta_{js} - \delta_{is}\delta_{jr} \Rightarrow \varepsilon^{ijk}\varepsilon_{rsk} = \delta_r^i\delta_s^j - \delta_s^i\delta_r^j \quad \text{QED}$$

Problem 6.4 Prove the formulas (6.5.3), (6.5.4), and (6.5.5).

$$\Gamma_{ij}^k = g^{kl}\Gamma_{ijl}, \quad \Gamma_{ijk} = g_{kl}\Gamma_{ij}^l \quad (6.5.3)$$

$$g_{ij,k} = \Gamma_{ikj} + \Gamma_{jki}, \quad \Gamma_{ijk} = \frac{1}{2}\left(g_{ik,j} + g_{jk,i} - g_{ij,k}\right) \quad (6.5.4)$$

$$\mathbf{g}^k,_j = -\Gamma_{ij}^k \mathbf{g}^i \quad (6.5.5)$$

Solution Proof of the formulas (6.5.3). By the definition of the Christoffel symbols: $\mathbf{g}_{i,j} = \Gamma_{ijk}\mathbf{g}^k = \Gamma_{ij}^k \mathbf{g}_k$. From the formulas (6.5.1) it follows that:

$$\mathbf{g}_{i,j} \cdot \mathbf{g}^l = \Gamma_{ijk}\mathbf{g}^k \cdot \mathbf{g}^l = \Gamma_{ij}^k \mathbf{g}_k \cdot \mathbf{g}^l \Rightarrow \Gamma_{ijk} g^{kl} = \Gamma_{ij}^k \delta_k^l = \Gamma_{ij}^l \Rightarrow (6.5.3)_1$$
$$\mathbf{g}_{i,j} \cdot \mathbf{g}_l = \Gamma_{ijk}\mathbf{g}^k \cdot \mathbf{g}_l = \Gamma_{ij}^k \mathbf{g}_k \cdot \mathbf{g}_l \Rightarrow \Gamma_{ijk}\delta_l^k = \Gamma_{ij}^k g_{kl} = \Gamma_{ijl} \Rightarrow (6.5.3)_2$$

Proof of the formulas (6.5.4). $g_{ij,k} = \left(\mathbf{g}_i \cdot \mathbf{g}_j\right),_k = \mathbf{g}_{i,k}\cdot\mathbf{g}_j + \mathbf{g}_i \cdot \mathbf{g}_{j,k} = \left(\Gamma_{ikl}\mathbf{g}^l\right)\cdot\mathbf{g}_j + \mathbf{g}_i\cdot\left(\Gamma_{jkl}\mathbf{g}^l\right) = \Gamma_{ikj} + \Gamma_{jki} \Rightarrow (6.5.4)_1$

The formula $(6.5.4)_1$ and the symmetry: $\Gamma_{jik} = \Gamma_{ijk} \Rightarrow$

$$\frac{1}{2}\left(g_{ik,j} + g_{jk,i} - g_{ij,k}\right) = \frac{1}{2}\left(\left[\Gamma_{ijk} + \Gamma_{kji}\right] + \left[\Gamma_{jik} + \Gamma_{kij}\right] - \left[\Gamma_{ikj} + \Gamma_{jki}\right]\right)$$
$$= \frac{1}{2}\left(\Gamma_{ijk} + \Gamma_{jik} + \left[\Gamma_{kji} - \Gamma_{jki}\right] + \left[\Gamma_{kij} - \Gamma_{ikj}\right]\right) = \Gamma_{ijk} \Rightarrow (6.5.4)_2$$

Proof of the formula (6.5.5).

$$\mathbf{g}^k \cdot \mathbf{g}_i = \delta_i^k \Rightarrow \mathbf{g}^k,_j \cdot\mathbf{g}_i + \mathbf{g}^k \cdot \mathbf{g}_{i,j} = 0 \text{ and with the formula } (6.5.1) \Rightarrow$$
$$\mathbf{g}^k,_j \cdot\mathbf{g}_i = -\mathbf{g}^k \cdot \mathbf{g}_{i,j} = -\mathbf{g}^k \cdot \left(\Gamma_{ij}^l \mathbf{g}_l\right) = -\Gamma_{ij}^k \Rightarrow \mathbf{g}^k,_j = -\Gamma_{ij}^k\mathbf{g}^i \Rightarrow (6.5.5)$$

Problem 6.5 Use the formulas $(6.5.4)_1$, $(1.1.24)$, and $(6.2.31)$:

$$g_{ij,k} = \Gamma_{ikj} + \Gamma_{jki}\,(6.5.4)_1, \quad Co\,g_{jk} = \frac{\partial g}{\partial g_{jk}} \quad (1.1.24), \quad g^{jk} = \frac{Co\,g_{jk}}{g} \quad (6.2.31)$$

To prove the formulas:

$$\Gamma^k_{ik} = \frac{1}{2g}g_{,i} = \frac{1}{\sqrt{g}}\left(\sqrt{g}\right)_{,i} \quad (6.5.6)$$

Solution We use the formulas $(1.1.24)$ and $(6.2.31)$ and obtain:

$$\frac{\partial g}{\partial g_{jk}} = Co\,g_{jk} = gg^{jk} \quad (1)$$

From the formulas (1), $(6.5.4)_1$, and $(6.5.3)$ we get:

$$g_{,i} = \frac{\partial g}{\partial g_{jk}}g_{jk,i} = gg^{jk}g_{jk,i} = gg^{jk}\left[\Gamma_{jik} + \Gamma_{kij}\right] = g\left[\Gamma^j_{ji} + \Gamma^k_{ki}\right] = 2g\Gamma^k_{ik} \Rightarrow \Gamma^k_{ik}$$
$$= \frac{1}{2g}g_{,i} \Rightarrow (6.5.6)_1$$

Furthermore with the formuls $(6.5.6)_1$:

$$\left(\sqrt{g}\right)_{,i} = \frac{d\sqrt{g}}{dg}g_{,i} = \frac{1}{2\sqrt{g}}g_{,i} = \frac{1}{2\sqrt{g}}\left(2g\Gamma^k_{ik}\right) \Rightarrow \Gamma^k_{ik} = \frac{1}{\sqrt{g}}\left(\sqrt{g}\right)_{,i} \Rightarrow (6.5.6)_2$$

Problem 6.6 Derive the transformation rule $(6.5.10)$ for the Christoffel symbols of the second kind.

$$\bar{\Gamma}^k_{ij} = \frac{\partial y^r}{\partial \bar{y}^i}\frac{\partial y^s}{\partial \bar{y}^j}\frac{\partial \bar{y}^k}{\partial y^t}\Gamma^t_{rs} + \frac{\partial^2 y^r}{\partial \bar{y}^i \partial \bar{y}^j}\frac{\partial \bar{y}^k}{\partial y^r} \quad (6.5.10)$$

Solution By the definitions $(6.5.1)$:

$$\frac{\partial \bar{g}_i}{\partial \bar{y}^j} \equiv \bar{g}_{i,j} = \bar{\Gamma}^k_{ij}\bar{g}_k, \quad \frac{\partial g_r}{\partial y^s} \equiv g_{r,s} = \Gamma^t_{rs}g_t \quad (1)$$

The relations between the base vectors in two general coordinate systems y and $\bar{y}$ are:

$$\bar{g}_i = \frac{\partial y^r}{\partial \bar{y}^i}g_r, \quad g_r = \frac{\partial \bar{y}^k}{\partial y^r}\bar{g}_k$$

We compute:

$$\bar{\mathbf{g}}_{i,j} = \frac{\partial y^r}{\partial \bar{y}^i}\frac{\partial \mathbf{g}_r}{\partial y^s}\frac{\partial y^s}{\partial \bar{y}^j} + \frac{\partial^2 y^r}{\partial \bar{y}^j \partial \bar{y}^i}\mathbf{g}_r$$

$$= \frac{\partial y^r}{\partial \bar{y}^i}\Gamma^t_{rs}\mathbf{g}_t\frac{\partial y^s}{\partial \bar{y}^j} + \frac{\partial^2 y^r}{\partial \bar{y}^j \partial \bar{y}^i}\frac{\partial \bar{y}^k}{\partial y^r}\bar{\mathbf{g}}_k = \frac{\partial y^r}{\partial \bar{y}^i}\Gamma^t_{rs}\frac{\partial y^s}{\partial \bar{y}^j}\frac{\partial \bar{y}^k}{\partial y^t}\bar{\mathbf{g}}_k + \frac{\partial^2 y^r}{\partial \bar{y}^j \partial \bar{y}^i}\frac{\partial \bar{y}^k}{\partial y^r}\bar{\mathbf{g}}_k \quad \Rightarrow$$

$$\bar{\mathbf{g}}_{i,j} = \left[\frac{\partial y^r}{\partial \bar{y}^i}\frac{\partial y^s}{\partial \bar{y}^j}\frac{\partial \bar{y}^k}{\partial y^t}\Gamma^t_{rs} + \frac{\partial^2 y^r}{\partial \bar{y}^j \partial \bar{y}^i}\frac{\partial \bar{y}^k}{\partial y^r}\right]\bar{\mathbf{g}}_k \quad (2)$$

Equations (1) and (2) imply:

$$\bar{\Gamma}^k_{ij} = \frac{\partial y^r}{\partial \bar{y}^i}\frac{\partial y^s}{\partial \bar{y}^j}\frac{\partial \bar{y}^k}{\partial y^t}\Gamma^t_{rs} + \frac{\partial^2 y^r}{\partial \bar{y}^j \partial \bar{y}^i}\frac{\partial \bar{y}^k}{\partial y^r} \quad \Rightarrow \quad (6.5.10)$$

Problem 6.7 Use the formula: $a_k|_j = a_{k,j} - a_l\Gamma^l_{kj}$ from the formula $(6.5.31)_2$ to prove the last two equalities in Eq. (6.5.50).

$$\text{rot } \mathbf{a} \equiv \text{curl } \mathbf{a} \equiv \nabla \times \mathbf{a} = \varepsilon^{ijk}a_k|_j\mathbf{g}_i = \varepsilon^{ijk}a_{k,j}\mathbf{g}_i = \frac{1}{\sqrt{g}}\det\begin{pmatrix} \mathbf{g}_1 & \mathbf{g}_2 & \mathbf{g}_3 \\ \frac{\partial}{\partial y^1} & \frac{\partial}{\partial y^2} & \frac{\partial}{\partial y^3} \\ a_1 & a_2 & a_3 \end{pmatrix} \quad (6.5.50)$$

Solution Symmetry and antisymmetry imply:

$$\Gamma^l_{kj} = \Gamma^l_{jk}, \quad \varepsilon^{ijk} = -\varepsilon^{ikj} \Rightarrow \varepsilon^{ijk}\Gamma^l_{kj} = -\varepsilon^{ikj}\Gamma^l_{kj} = -\varepsilon^{ijk}\Gamma^l_{jk} = -\varepsilon^{ijk}\Gamma^l_{kj} = 0$$

We now can write: $\varepsilon^{ijk}a_k|_j\mathbf{g}_i = \varepsilon^{ijk}\left[a_{k,j} - a_l\Gamma^l_{kj}\right]\mathbf{g}_i = \varepsilon^{ijk}a_{k,j}\mathbf{g}_i$ (1)
The formulas (6.2.24) and (6.2.30) imply that: $\varepsilon^{ijk} = J^y_x e_{ijk} = e_{ijk}/\sqrt{g}$ (2)
From the results (1) and (2) we obtain:

$$\varepsilon^{ijk}a_k|_j\mathbf{g}_i = \frac{1}{\sqrt{g}}e_{ijk}a_{k,j} = \frac{1}{\sqrt{g}}\det\begin{pmatrix} \mathbf{g}_1 & \mathbf{g}_2 & \mathbf{g}_3 \\ \frac{\partial}{\partial y^1} & \frac{\partial}{\partial y^2} & \frac{\partial}{\partial y^3} \\ a_1 & a_2 & a_3 \end{pmatrix} \quad \text{QED}$$

Problem 6.8 Derive the formulas (6.5.51) for the divergence of a vector field by using the formulas (6.5.49), $(6.2.31)_1$, and (6.5.6).

$$\text{div } \mathbf{a} \equiv \nabla \cdot \mathbf{a} = \frac{1}{\sqrt{g}}\left(\sqrt{g}\,a^i\right)_{,i} \quad (6.5.51)$$

$$\text{div } \mathbf{a} \equiv \nabla \cdot \mathbf{a} = a^i|_i \quad (6.5.49),$$

$$a^i|_j = a^i_{,j} + a^k\Gamma^i_{kj} \quad (6.5.31)_1,$$

$$\Gamma^k_{ik} = \frac{1}{2g}g_{,i} = \frac{1}{\sqrt{g}}\left(\sqrt{g}\right)_{,i} \quad (6.5.6),$$

Solution We use the formulas (6.5.49), (6.2.31)$_1$, and (6.5.6) to write:

$$\text{div } \mathbf{a} = a^i|_i = a^i{}_{,i} + a^k \Gamma^i_{ki} = a^i{}_{,i} + a^k \frac{1}{\sqrt{g}} (\sqrt{g})_{,k} = \frac{1}{\sqrt{g}} \left[\sqrt{g} a^i{}_{,i} + (\sqrt{g})_{,i} a^i \right] = \frac{1}{\sqrt{g}} (\sqrt{g} a^i)_{,i} \Rightarrow$$

$$\text{div } \mathbf{a} = a^i|_i = \frac{1}{\sqrt{g}} (\sqrt{g} a^i)_{,i} \Leftrightarrow (6.5.51)$$

Problem 6.9 Derive the formula (6.5.99) using the formula (6.5.51):

$$\nabla^2 \alpha = \frac{1}{\sqrt{g}} (\sqrt{g} g^{ij} \alpha_{,j})_{,i} \quad (6.5.99), \quad \text{div } \mathbf{a} = \nabla \cdot \mathbf{a} = \frac{1}{\sqrt{g}} (\sqrt{g} a^i)_{,i} \quad (6.5.51)$$

Solution We define the vector: $\mathbf{a} = \text{grad} \alpha \equiv \nabla \alpha = g^{ij} \alpha_{,j} \, \mathbf{g}_i$. The formula (6.5.51) gives the result:

$$\nabla^2 \alpha = \nabla \cdot \nabla \alpha = \nabla \cdot \mathbf{a} = \frac{1}{\sqrt{g}} (\sqrt{g} g^{ij} \alpha_{,j})_{,i} \Rightarrow \nabla^2 \alpha = \frac{1}{\sqrt{g}} (\sqrt{g} g^{ij} \alpha_{,j})_{,i} \Leftrightarrow (6.5.99)$$

Problem 6.10 Derive the formula: $\nabla \times (\nabla \times \mathbf{a}) = \frac{1}{\sqrt{g}} \frac{\partial}{\partial y^k} \left[\sqrt{g} g^{kr} g^{is} \left(\frac{\partial a_r}{\partial y^s} - \frac{\partial a_s}{\partial y^r} \right) \right] \mathbf{g}_i$ (6.5.107)

Solution We define the vector $\mathbf{b} = \nabla \times \mathbf{a}$. The formula (6.5.50) gives:

$$\mathbf{b} = \nabla \times \mathbf{a} = \varepsilon^{ijk} a_k |_j \mathbf{g}_i = \varepsilon_{ijk} a^k |^j \mathbf{g}^i \Rightarrow b_i = \varepsilon_{ijk} a^k |^j = \varepsilon_{ijk} g^{kr} g^{js} a_r |_s$$

$$\nabla \times (\nabla \times \mathbf{a}) = \nabla \times \mathbf{b} = \varepsilon^{tpi} b_i |_p \mathbf{g}_t = \varepsilon^{tpi} b_{i,p} \, \mathbf{g}_t = \varepsilon^{tpi} \frac{\partial}{\partial y^p} \left[\varepsilon_{ijk} g^{kr} g^{js} a_r |_s \right] \mathbf{g}_t \quad (1)$$

We now use the formulas (6.2.22), (6.2.30).(6.2.24), and (1.1.19) to obtain.

$$\varepsilon_{ijk} = J^x_y \, e_{ijk} = \sqrt{g} e_{ijk}, \quad \varepsilon^{tpi} = J^y_x \, e_{tpi} = \frac{1}{\sqrt{g}} e_{tpi}, \quad e_{tpi} e_{ijk} = e_{tpi} e_{jki} = \delta_{tj} \delta_{pk} - \delta_{tk} \delta_{pj}$$

The formula (1) is then rewritten to:

$$\nabla \times (\nabla \times \mathbf{a}) = \frac{1}{\sqrt{g}} \frac{\partial}{\partial y^p} \left[\sqrt{g} g^{kr} g^{js} a_r |_s (\delta_{tj} \delta_{pk} - \delta_{tk} \delta_{pj}) \right] \mathbf{g}_t = \frac{1}{\sqrt{g}} \frac{\partial}{\partial y^p} \left[\sqrt{g} g^{pr} g^{ts} (a_r |_s - a_s |_r) \right] \mathbf{g}_t \Rightarrow$$

$$\nabla \times (\nabla \times \mathbf{a}) = \frac{1}{\sqrt{g}} \frac{\partial}{\partial y^p} \left[\sqrt{g} g^{pr} g^{ts} (a_r |_s - a_s |_r) \right] \mathbf{g}_t \quad (2)$$

We apply the formula (6.5.31)$_2$ and the symmetry of the Christoffel symbols to obtain the result:

$$a_r|_s - a_s|_r = \left(a_{r,s} - a_t\Gamma^t_{rs}\right) - \left(a_{s,r} - a_t\Gamma^t_{sr}\right) = a_{r,s} - a_{s,r} = \frac{\partial a_r}{\partial y^s} - \frac{\partial a_s}{\partial y^r} \quad (3)$$

When the result (3) is inserted into the formula (2) we obtain the formula (6.5.107).

Chapter 7
Problem 7.1 Derive the formulas (7.4.14) and (7.4.15).

$$\tilde{E}_{ij} = \frac{1}{2}\left(\mathbf{g}_i \cdot \mathbf{u}_{,j} + \mathbf{g}_j \cdot \mathbf{u}_{,i} - \mathbf{u}_{,i} \cdot \mathbf{u}_{,j}\right) = \frac{1}{2}\left(u_i|_j + u_j|_i - u_k|_i u^k|_j\right) \quad (7.4.14)$$

$$\mathbf{E} = \mathbf{F}^T \tilde{\mathbf{E}} \mathbf{F} \Leftrightarrow E_{KL} = \tilde{E}_{ij} F^i_K F^j_L \quad (7.4.15)$$

Solution Derivation of the formula (7.4.14).

First we develop the relations between the material line elements $d\mathbf{r}_0$ of length ds_0 and $d\mathbf{r}$ of length ds:

$$\mathbf{r} = \mathbf{r}_0 + \mathbf{u} \Rightarrow d\mathbf{r} = d\mathbf{r}_0 + d\mathbf{u}, \quad ds = |d\mathbf{r}|, \quad ds_0 = |d\mathbf{r}_0|, \quad (ds)^2 = d\mathbf{r} \cdot d\mathbf{r} \Rightarrow$$

$$(ds_0)^2 = d\mathbf{r}_0 \cdot d\mathbf{r}_0 = (d\mathbf{r} - d\mathbf{u}) \cdot (d\mathbf{r} - d\mathbf{u}) = (ds)^2 - 2d\mathbf{r} \cdot d\mathbf{u} + d\mathbf{u} \cdot d\mathbf{u} \Rightarrow$$

$$(ds)^2 - (ds_0)^2 = 2d\mathbf{r} \cdot d\mathbf{u} - d\mathbf{u} \cdot d\mathbf{u} \quad (1)$$

From the formulas (6.5.35) we get:

$$d\mathbf{u} = \mathbf{u}_{,j}\, dy^j = u_k|_j \mathbf{g}^k dy^j = u^k|_j \mathbf{g}_k dy^j \Rightarrow$$

$$2d\mathbf{r} \cdot d\mathbf{u} = 2\left(\mathbf{g}_i dy^i\right) \cdot \left(\mathbf{u}_{,j}\, dy^j\right) = \left(\mathbf{g}_i \cdot \mathbf{u}_{,j} + \mathbf{g}_j \cdot \mathbf{u}_{,i}\right) dy^i dy^j = \left(u_i|_j + u_j|_i\right) dy^i dy^j \quad (2)$$

$$d\mathbf{u} \cdot d\mathbf{u} = \left(\mathbf{u}_{,i}\, dy^i\right) \cdot \left(\mathbf{u}_{,j}\, dy^j\right) = u^k|_i u_k|_j dy^i dy^j \quad (3)$$

The results (2) and (3) are substituted into the formula (1) and we obtain, with reference to formula (7.4.12):

$$(ds)^2 - (ds_0)^2 = dy^i\left(\mathbf{g}_i \cdot \mathbf{u}_{,j} + \mathbf{g}_j \cdot \mathbf{u}_{,i}\right) dy^j = dy^i\left(u_i|_j + u_j|_i - u^k|_i u_k|_j\right) dy^j$$

$$= 2d\mathbf{r} \cdot \tilde{\mathbf{E}} \cdot d\mathbf{r} \quad (4) \Rightarrow$$

$$\tilde{E}_{ij} = \frac{1}{2}\left(\mathbf{g}_i \cdot \mathbf{u}_{,j} + \mathbf{g}_j \cdot \mathbf{u}_{,i} - \mathbf{u}_{,i} \cdot \mathbf{u}_{,j}\right) = \frac{1}{2}\left(u_i|_j + u_j|_i - u^k|_i u_k|_j\right) \Leftrightarrow (7.4.14)$$

Derivation of the formula (7.4.15).

The formulas (7.3.1) and (7.3.3) give:

$$d\mathbf{r} = \mathbf{F} \cdot d\mathbf{r}_0 \Leftrightarrow dy^i = F^i_K dY^K \quad (5)$$

From the formulas (4) and (5) we obtain:

$$(ds)^2 - (ds_0)^2 = 2d\mathbf{r} \cdot \tilde{\mathbf{E}} \cdot d\mathbf{r} = 2dy^i \tilde{E}_{ij} dy^j = 2\left(F_K^i dY^K\right) \tilde{E}_{ij} \left(F_L^j dY^L\right) = 2dY^K \left(F_K^i \tilde{E}_{ij} F_L^j\right) dY^L \Rightarrow$$
$$(ds)^2 - (ds_0)^2 = 2d\mathbf{r}_0 \cdot \left(\mathbf{F}^T \tilde{\mathbf{E}} \mathbf{F}\right) \cdot d\mathbf{r}_0 \quad (6)$$

Because the tensors $\mathbf{F}, \tilde{\mathbf{E}}$, and $\mathbf{E}$ are independent of $d\mathbf{r}_0$, i.e. dY^K, we conclude from the result (6) and the result formula (7.3.10): $(ds)^2 - (ds_0)^2 = 2d\mathbf{r}_0 \cdot \mathbf{E} \cdot d\mathbf{r}_0$ from the formula (7.3.10), that:

$$\mathbf{E} = \mathbf{F}^T \tilde{\mathbf{E}} \mathbf{F} \Leftrightarrow E_{KL} = \tilde{E}_{ij} F_K^i F_L^j \Leftrightarrow (7.4.15)$$

Problem 7.2 Derive the formula (7.6.3).

$$\dot{\mathbf{c}}^K = -v^K \big|\big|_L \mathbf{c}^L \quad (7.6.3)$$

Solution By definition of the base vectors $\mathbf{c}^K$ and $\mathbf{c}_L$, and by Eq. (7.6.2):

$$\mathbf{c}^K \cdot \mathbf{c}_L = \delta_L^K \quad \text{and} \quad \dot{\mathbf{c}}_L = v^N \big|\big|_L \mathbf{c}_N \Rightarrow \dot{\mathbf{c}}^K \cdot \mathbf{c}_L + \mathbf{c}^K \cdot \dot{\mathbf{c}}_L = 0 \Rightarrow \dot{\mathbf{c}}^K \cdot \mathbf{c}_L + \mathbf{c}^K \cdot \left(v^N \big|\big|_L \mathbf{c}_N\right) = 0 \Rightarrow$$
$$\dot{\mathbf{c}}^K \cdot \mathbf{c}_L = -v^K \big|\big|_L \Rightarrow \dot{\mathbf{c}}^K = -v^K \big|\big|_L \mathbf{c}^L \Leftrightarrow (7.6.3)$$

Problem 7.3 Show that Eq. (7.6.18) follows from Eq. (7.6.17).

$$\partial_c B_j^i = \dot{B}_j^i - B_j^k v^i \big|_k + B_k^i v^k \big|_j \qquad (7.6.17)$$
$$\partial_c B_j^i = \frac{\partial}{\partial t} B_j^i + B_{j,k}^i v^k - B_j^k v^i{}_{,k} + B_k^i v^k{}_{,j} \quad (7.6.18)$$

Solution The following equations are used:

$$(7.2.12) \Rightarrow \dot{B}_j^i = \frac{\partial}{\partial t} B_j^i + B_j^i \big|_k v^k, \quad (6.5.31)_1 \Rightarrow v^i \big|_k = v^i{}_{,k} + v^l \Gamma_{lk}^i$$
$$(6.5.81) \Rightarrow B_j^i \big|_k = B_{j,k}^i + B_j^l \Gamma_{lk}^i - B_l^i \Gamma_{jk}^l$$

From Eq. (7.6.17) we obtain:

$$\partial_c B_j^i = \dot{B}_j^i - B_j^k v^i \big|_k + B_k^i v^k \big|_j$$
$$= \left[\frac{\partial}{\partial t} B_j^i + \left(B_{j,k}^i + B_j^l \Gamma_{lk}^i - B_l^i \Gamma_{jk}^l \right) v^k \right] - B_j^k \left(v^i{}_{,k} + v^l \Gamma_{lk}^i \right) + B_k^i \left(v^k{}_{,j} + v^l \Gamma_{lj}^k \right) \Rightarrow$$
$$\partial_c B_j^i = \frac{\partial}{\partial t} B_j^i + B_{j,k}^i v^k - B_j^k v^i{}_{,k} + B_k^i v^k{}_{,j} \Leftrightarrow (7.6.18)$$

Problem 7.4 Derive the expression (7.6.18) for the convective derivatives of the components of an objective tensor of second order **B** by applying the Oldroyd method described in connection with Eqs. (7.6.19) and (7.6.20).

$$\partial_c B_j^i = \frac{\partial}{\partial t} B_j^i + B_{j,k}^i v^k - B_j^k v_{,k}^i + B_k^i v_{,j}^k \quad (7.6.18)$$

Solution The following equations are used:

$$(7.6.12): \partial_c B_L^K = \frac{\partial}{\partial t} B_L^K \quad (1), \quad (7.6.19): B_L^K = \frac{\partial Y^K}{\partial y^i} \frac{\partial y^j}{\partial Y^L} B_j^i \quad (2),$$

$$(7.6.20): \partial_c B_s^r = \partial_c B_L^K \frac{\partial y^r}{\partial Y^K} \frac{\partial Y^L}{\partial y^s} \quad (3), \quad (7.2.13): \dot{B}_j^i = \frac{\partial B_j^i}{\partial t} + B_{j,k}^i v^k$$

Material differentiation of Eq. (2) gives:

$$\partial_c B_L^K = \frac{\partial}{\partial t} B_L^K = \frac{\partial Y^K}{\partial y^i} \frac{\partial y^j}{\partial Y^L} \dot{B}_j^i + \frac{\partial Y^K}{\partial y^i} \frac{\partial^2 y^j}{\partial t \partial Y^L} B_j^i + \frac{\partial Y^K}{\partial y^i} \frac{\partial y^j}{\partial Y^L} \dot{B}_j^i \quad (4)$$

$(4) \rightarrow (3) \Rightarrow$

$$\partial_c B_s^r = \partial_c B_L^K \frac{\partial y^r}{\partial Y^K} \frac{\partial Y^L}{\partial y^s} = \frac{\partial Y^K}{\partial y^i} \frac{\partial y^j}{\partial Y^L} B_j^i \frac{\partial y^r}{\partial Y^K} \frac{\partial Y^L}{\partial y^s}$$

$$+ \frac{\partial Y^K}{\partial y^i} \frac{\partial^2 y^j}{\partial t \partial Y^L} B_j^i \frac{\partial y^r}{\partial Y^K} \frac{\partial Y^L}{\partial y^s} + \frac{\partial Y^K}{\partial y^i} \frac{\partial y^j}{\partial Y^L} \dot{B}_j^i \frac{\partial y^r}{\partial Y^K} \frac{\partial Y^L}{\partial y^s} \quad \Rightarrow$$

$$\partial_c B_s^r = \frac{\partial \dot{Y}^K}{\partial y^i} \frac{\partial y^r}{\partial Y^K} B_s^i + \frac{\partial v^k}{\partial y^s} B_k^r + \dot{B}_s^r \left(\frac{\partial B_s^r}{\partial t} + B_{s,k}^r v^k \right) \quad (5)$$

We need an expression for $\frac{\partial \dot{Y}^K}{\partial y^i} \frac{\partial y^r}{\partial Y^K}$ and write:

$$\frac{\partial y^r}{\partial Y^K} \frac{\partial Y^K}{\partial y^i} = \delta_i^r \Rightarrow \frac{\partial y^r}{\partial Y^K} \frac{\partial \dot{Y}^K}{\partial y^i} + \frac{\partial^2 y^r}{\partial t \partial Y^K} \frac{\partial Y^K}{\partial y^i} = 0 \Rightarrow \frac{\partial y^r}{\partial Y^K} \frac{\partial \dot{Y}^K}{\partial y^i} = -\frac{\partial v^r}{\partial Y^K} \frac{\partial Y^K}{\partial y^i}$$

$$= -v^r_{,i} \quad (6)$$

When the result (6) is substituted into Eq. (5) we obtain:

$$\partial_c B_s^r = \frac{\partial}{\partial t} B_s^r + B_{s,k}^r v^k - B_s^k v_{,k}^r + B_k^r v_{,s}^k \Leftrightarrow (7.6.18)$$

Problem 7.5 Derive the formula:

$$\partial_c B_{ij} = \dot{B}_{ij} + B_{kj} v^k |_i + B_{ik} v^k |_j \quad (7.6.23)_1$$

Solution Let **a** and **b** be any two vector fields such that according to Eq. (7.6.8):

$$\partial_c a^i = \dot{a}^i - a^k v^i|_k, \quad \partial_c b^j = \dot{b}^j - b^k v^j|_k \quad (1)$$

Let a scalar field be defined by: $\alpha = \mathbf{B}[\mathbf{a}, \mathbf{b}] = B_{ij}a^i b^j$. In a fixed y-system:

$$\dot{\alpha} = \partial_c B_{ij}a^i b^j + B_{ij}(\partial_c a^i\, b^j + a^i\, \partial_c b^j) = \dot{B}_{ij}a^i b^j + B_{ij}(\dot{a}^i\, b^j + a^i\, \dot{b}^j) \quad (2)$$

The Eqs. (1) and (2) provide the result:

$$[\partial_c B_{ij} - \dot{B}_{ij}]a^i b^j + B_{ij}[(\dot{a}^i - a^k v^i|_k - \dot{a}^i)b^j + a^i(\dot{b}^j - b^k v^j|_k - \dot{b}^j)] = 0 \Rightarrow$$
$$\left[\partial_c B_{ij} - \dot{B}_{ij} - B_{kj}v^k|_i - B_{ik}v^k|_j\right]a^i b^j = 0 \quad (3)$$

Because **a** and **b** may be any two vector fields, the term in the parenthesis [] must be zero. Hence:

$$\partial_c B_{ij} = \dot{B}_{ij} + B_{kj}v^k|_i + B_{ik}v^k|_j \Leftrightarrow (7.6.23)_1$$

Problem 7.6 Show that the two sets of physical components of stress defined respectively by Eqs. (7.7.8) and (7.7.11) are related through Eq. (7.7.12).

$$\tau_{ki} = T^{ki}\sqrt{\frac{g_{kk}}{g^{ii}}} \quad (7.7.8), \quad T(ki) = T_i^k\sqrt{\frac{g_{kk}}{g_{ii}}} \quad (7.7.11), \quad \tau_{ki} = \sum_j T(kj)g^{ij}\sqrt{\frac{g_{jj}}{g^{ii}}} \quad (7.7.12)$$

Solution The formulas (7.7.8) and (7.7.11) give:

$$\tau_{ki} = T^{ki}\sqrt{\frac{g_{kk}}{g^{ii}}} = \sum_j \left(T_j^k g^{ij}\right)\sqrt{\frac{g_{kk}}{g^{ii}}}, \quad T_j^k = T(kj)\sqrt{\frac{g_{jj}}{g_{kk}}}$$

Then:

$$\tau_{ki} = T^{ki}\sqrt{\frac{g_{kk}}{g^{ii}}} = \sum_j \left(T_j^k g^{ij}\right)\sqrt{\frac{g_{kk}}{g^{ii}}} = \sum_j \left(\left[T(kj)\sqrt{\frac{g_{jj}}{g_{kk}}}\right]g^{ij}\right)\sqrt{\frac{g_{kk}}{g^{ii}}}$$
$$= \sum_j T(kj)g^{ij}\sqrt{\frac{g_{jj}}{g^{ii}}} \Leftrightarrow (7.7.12)$$

Problem 7.7 Use the general Cauchy equations for orthogonal coordinates (7.7.18) to develop the Cauchy equations (7.7.19) in cylindrical coordinates.

$$\sum_k \left[\frac{1}{h}\frac{\partial}{\partial y^k}\left(\frac{h}{h_k}T(ik)\right) + \frac{1}{h_i h_k}\frac{\partial h_i}{\partial y^k}T(ik) - \frac{1}{h_i h_k}\frac{\partial h_k}{\partial y^i}T(kk)\right] + \rho\, b(i) = \rho\, a(i)$$
$$(7.7.18)$$

Solution In cylindrical coordinates (R, θ, z): $h_1 = 1, h_2 = R, h_3 = 1, h = R$.

$$\text{Physical coordinate stresses: } (T(ik)) = \begin{pmatrix} \sigma_R & \tau_{R\theta} & \tau_{Rz} \\ \tau_{\theta R} & \sigma_\theta & \tau_{\theta z} \\ \tau_{zR} & \tau_{z\theta} & \sigma_z \end{pmatrix}$$

Equation (7.7.18) $\Rightarrow$

$$i = 1: \frac{1}{R}\left[\frac{\partial}{\partial R}\left(\frac{R}{1}\sigma_R\right) + \frac{\partial}{\partial \theta}\left(\frac{R}{R}\tau_{R\theta}\right) + \frac{\partial}{\partial z}\left(\frac{R}{1}\tau_{Rz}\right)\right] - \frac{1}{1 \cdot R}\sigma_\theta + \rho\, b_R = \rho\, a_R$$

$$i = 2: \frac{1}{R}\left[\frac{\partial}{\partial R}\left(\frac{R}{1}\tau_{\theta R}\right) + \frac{\partial}{\partial \theta}\left(\frac{R}{R}\sigma_\theta\right) + \frac{\partial}{\partial z}\left(\frac{R}{1}\tau_{\theta z}\right)\right] + \frac{1}{1 \cdot R}\tau_{\theta R} + \rho\, b_\theta = \rho\, a_\theta$$

$$i = 3: \frac{1}{R}\left[\frac{\partial}{\partial R}\left(\frac{R}{1}\tau_{zR}\right) + \frac{\partial}{\partial \theta}\left(\frac{R}{R}\tau_{z\theta}\right) + \frac{\partial}{\partial z}\left(\frac{R}{1}\sigma_z\right)\right] + \rho\, b_z = \rho\, a_z$$

These equations are reorganized to:

$$\frac{\partial \sigma_R}{\partial R} + \frac{\sigma_R - \sigma_\theta}{R} + \frac{1}{R}\frac{\partial \tau_{R\theta}}{\partial \theta} + \frac{\partial \tau_{Rz}}{\partial z} + \rho b_R = \rho a_R$$

$$\frac{1}{R^2}\frac{\partial}{\partial R}(R^2 \tau_{\theta R}) + \frac{1}{R}\frac{\partial \sigma_\theta}{\partial \theta} + \frac{\partial \tau_{\theta z}}{\partial z} + \rho b_\theta = \rho a_\theta \quad \Leftrightarrow \quad (7.7.19)$$

$$\frac{1}{R}\frac{\partial}{\partial R}(R\tau_{zR}) + \frac{1}{R}\frac{\partial \tau_{\theta z}}{\partial \theta} + \frac{\partial \sigma_z}{\partial z} + \rho b_z = \rho a_z$$

Problem 7.8 Use the general Cauchy equations for orthogonal coordinates (7.7.18) to develop the Cauchy equations (7.7.21) in spherical coordinates.

$$\sum_k \left[\frac{1}{h}\frac{\partial}{\partial y^k}\left(\frac{h}{h_k}T(ik)\right) + \frac{1}{h_i h_k}\frac{\partial h_i}{\partial y^k}T(ik) - \frac{1}{h_i h_k}\frac{\partial h_k}{\partial y^i}T(kk)\right] + \rho\, b(i) = \rho\, a(i)$$
(7.7.18)

Solution In spherical coordinates (r, θ, ϕ) : $h_1 = h_r = 1, h_2 = h_\theta = r, h_3 = h_\phi = r\sin\theta, h = r^2 \sin\theta$.

$$\text{Physical coordinate stresses: } (T(ik)) = \begin{pmatrix} \sigma_r & \tau_{r\theta} & \tau_{r\phi} \\ \tau_{\theta r} & \sigma_\theta & \tau_{\theta\phi} \\ \tau_{\phi r} & \tau_{\phi\theta} & \sigma_\phi \end{pmatrix}$$

Equation (7.7.18) $\Rightarrow$

$$i = 1: \quad \frac{1}{r^2 \sin \theta} \left[\frac{\partial}{\partial r} \left(\frac{r^2 \sin \theta}{1} \sigma_r \right) + \frac{\partial}{\partial \theta} \left(\frac{r^2 \sin \theta}{r} \tau_{r\theta} \right) + \frac{\partial}{\partial \phi} \left(\frac{r^2 \sin \theta}{r \sin \theta} \tau_{r\phi} \right) \right]$$

$$- \frac{1}{1 \cdot r} \sigma_\theta - \frac{\sin \theta}{1 \cdot r \sin \theta} \sigma_\phi + \rho \, b_r = \rho \, a_r$$

$$i = 2: \quad \frac{1}{r^2 \sin \theta} \left[\frac{\partial}{\partial r} \left(\frac{r^2 \sin \theta}{1} \tau_{\theta r} \right) + \frac{\partial}{\partial \theta} \left(\frac{r^2 \sin \theta}{r} \sigma_\theta \right) + \frac{\partial}{\partial \phi} \left(\frac{r^2 \sin \theta}{r \sin \theta} \tau_{\theta\phi} \right) \right]$$

$$+ \frac{1}{r \cdot 1} \tau_{\theta r} - \frac{r \cos \theta}{r \cdot r \sin \theta} \sigma_\phi + \rho \, b_\theta = \rho \, a_\theta$$

$$i = 3: \quad \frac{1}{r^2 \sin \theta} \left[\frac{\partial}{\partial r} \left(\frac{r^2 \sin \theta}{1} \tau_{\phi r} \right) + \frac{\partial}{\partial \theta} \left(\frac{r^2 \sin \theta}{r} \tau_{\phi\theta} \right) + \frac{\partial}{\partial \phi} \left(\frac{r^2 \sin \theta}{r \sin \theta} \sigma_\phi \right) \right]$$

$$+ \frac{\sin \theta}{r \sin \theta \cdot 1} \tau_{\phi r} + \frac{r \cos \theta}{r \sin \theta \cdot r} \tau_{\phi\theta} + \rho \, b_\phi = \rho \, a_\phi$$

These equations are reorganized to:

$$\frac{\partial \sigma_r}{\partial r} + \frac{2\sigma_r - \sigma_\theta - \sigma_\phi}{r} + \frac{1}{r \sin \theta} \left[\frac{\partial}{\partial \theta} (\sin \theta \, \tau_{r\theta}) + \frac{\partial \tau_{r\phi}}{\partial \phi} \right] + \rho b_r = \rho a_r$$

$$\frac{1}{r^3} \frac{\partial}{\partial r} (r^3 \tau_{\theta r}) + \frac{1}{r \sin \theta} \left[\frac{\partial}{\partial \theta} (\sin \theta \, \sigma_\theta) + \frac{\partial \tau_{\theta\phi}}{\partial \phi} \right] - \frac{\cot \theta}{r} \sigma_\phi + \rho b_\theta = \rho a_\theta \Leftrightarrow (7.7.21)$$

$$\frac{1}{r^3} \frac{\partial}{\partial r} (r^3 \tau_{\phi r}) + \frac{1}{r \sin^2 \theta} \frac{\partial}{\partial \theta} (\sin^2 \theta \, \tau_{\phi\theta}) + \frac{1}{r \sin \theta} \frac{\partial \sigma_\phi}{\partial \phi} + \rho b_\phi = \rho a_\phi$$

Problem 7.9 Use the formulas (6.5.69) for the covariant derivatives of the vector components v^i of a vector field $\mathbf{v}(\mathbf{r}, t)$ to derive the formula (7.9.7) for the physical components of the particle acceleration.

$$(6.5.69) \Rightarrow v^i \big|_i = \frac{\partial}{\partial y^i} \left(\frac{v(i)}{h_i} \right) + \sum_k \frac{v(k)}{h_k} \frac{1}{h_i} \frac{\partial h_i}{\partial y^k}, \quad v^i \big|_k = \frac{1}{h_i} \frac{\partial v(i)}{\partial y^k} - \frac{v(k)}{h_i^2} \frac{\partial h_k}{\partial y^i} \quad i \neq k$$

$$a(i) = \frac{Dv(i)}{Dt} + \sum_k \frac{v(k)}{h_i h_k} \left\{ v(i) \frac{\partial h_i}{\partial y^k} - v(k) \frac{\partial h_k}{\partial y^i} \right\} \quad (7.9.7)$$

Solution The general expression for the acceleration $\mathbf{a}$ in a general coordinate system y is found in the formulas (7.9.4) as:

$$\mathbf{a} = \frac{\partial \mathbf{v}}{\partial t} + (\mathbf{v} \cdot \nabla)\mathbf{v} \Rightarrow a^i = \frac{\partial v^i}{\partial t} + v^k v^i \big|_k$$

For the index $i = 1$ we write:

$$a^1 = \frac{\partial v^1}{\partial t} + v^k v^1 \big|_k = \frac{\partial v^1}{\partial t} + v^1 v^1 \big|_1 + v^2 v^1 \big|_2 + v^3 v^1 \big|_3 \Rightarrow$$

$$a(1) = h_1 a^1 = h_1 \frac{\partial}{\partial t}\left(\frac{v(1)}{h_1}\right) + h_1 \frac{v(1)}{h_1}\left[\frac{\partial}{\partial y^1}\left(\frac{v(1)}{h_1}\right) + \sum_k \frac{v(k)}{h_k} \frac{1}{h_1} \frac{\partial h_1}{\partial y^k}\right]$$

$$+ h_1 \frac{v(2)}{h_2}\left[\frac{1}{h_1}\frac{\partial v(1)}{\partial y^2} - \frac{v(2)}{h_1^2}\frac{\partial h_2}{\partial y^1}\right] + h_1 \frac{v(3)}{h_3}\left[\frac{1}{h_1}\frac{\partial v(1)}{\partial y^3} - \frac{v(3)}{h_1^2}\frac{\partial h_3}{\partial y^1}\right] \Rightarrow$$

$$a(1) = \frac{\partial v(1)}{\partial t} + v(1)\left[\frac{1}{h_1}\frac{\partial v(1)}{\partial y^1} - \frac{v(1)}{h_1^2}\frac{\partial h_1}{\partial y^1} + \frac{v(1)}{h_1}\frac{\partial h_1}{\partial y^1} + \frac{v(2)}{h_2}\frac{\partial h_1}{\partial y^2} + \frac{v(3)}{h_3}\frac{\partial h_1}{\partial y^3}\right]$$

$$+ \frac{v(2)}{h_2}\left[\frac{\partial v(1)}{\partial y^2} - \frac{v(2)}{h_1}\frac{\partial h_2}{\partial y^1}\right] + \frac{v(3)}{h_3}\left[\frac{\partial v(1)}{\partial y^3} - \frac{v(3)}{h_1}\frac{\partial h_3}{\partial y^1}\right]$$

$$= \frac{\partial v(1)}{\partial t} + \frac{v(1)}{h_1}\frac{\partial v(1)}{\partial y^1} - \frac{v(1)}{h_1}\frac{v(1)}{h_1}\frac{\partial h_1}{\partial y^1} + \frac{v(1)}{h_1}\frac{v(1)}{h_1}\frac{\partial h_1}{\partial y^1} + \frac{v(1)}{h_1}\frac{v(2)}{h_2}\frac{\partial h_1}{\partial y^2} + \frac{v(1)}{h_1}\frac{v(3)}{h_3}\frac{\partial h_1}{\partial y^3}$$

$$+ \frac{v(2)}{h_2}\frac{\partial v(1)}{\partial y^2} - \frac{v(2)}{h_2}\frac{v(2)}{h_1}\frac{\partial h_2}{\partial y^1} + \frac{v(3)}{h_3}\frac{\partial v(1)}{\partial y^3} - \frac{v(3)}{h_3}\frac{v(3)}{h_1}\frac{\partial h_3}{\partial y^1}$$

$$= \frac{\partial v(1)}{\partial t} + \sum_k \frac{v(k)}{h_k}\frac{\partial v(1)}{\partial y^k} + \sum_k \frac{v(1)}{h_1}\frac{v(k)}{h_k}\frac{\partial h_1}{\partial y^k} - \sum_k \frac{v(k)}{h_k}\frac{v(k)}{h_1}\frac{\partial h_k}{\partial y^1} \Rightarrow$$

$$a(1) = \left[\frac{\partial v(1)}{\partial t} + \sum_k \frac{v(k)}{h_k}\frac{\partial v(1)}{\partial y^k}\right] + \left[\sum_k \frac{v(k)}{h_1 h_k}\frac{\partial h_1}{\partial y^k}\left\{\frac{\partial h_1}{\partial y^k}v(1) - \frac{\partial h_k}{\partial y^1}v(k)\right\}\right]$$

The operator (7.9.8) is utilized to define:

$$\frac{Dv(1)}{Dt} = \frac{\partial v(1)}{\partial t} + \sum_k \frac{v(k)}{h_k}\frac{\partial v(1)}{\partial y^k}$$

Hence we have the result:

$$a(1) = \frac{Dv(1)}{Dt} + \sum_k \frac{v(k)}{h_1 h_k}\frac{\partial h_1}{\partial y^k}\left\{\frac{\partial h_1}{\partial y^k}v(1) - \frac{\partial h_k}{\partial y^1}v(k)\right\}$$

The above formula is generalized to:

$$a(i) = \frac{Dv(i)}{Dt} + \sum_k \frac{v(k)}{h_i h_k}\left\{\frac{\partial h_i}{\partial y^k}v(i) - \frac{\partial h_k}{\partial y^i}v(k)\right\} \Leftrightarrow (7.9.7)$$

Chapter 8
Problem 8.1

(a) Use the formula (6.4.23) and the definitions (1.1.16) and (8.1.11) for the permutation symbols

$$e_{ijk}, \varepsilon_{ijk}, \varepsilon^{ijk}, e_{\alpha\beta}, \varepsilon_{\alpha\beta}, \quad \text{and} \quad \varepsilon^{\alpha\beta} \quad (1)$$

to prove the relationships:

$$\varepsilon_{\alpha\beta}\,\varepsilon^{\gamma\rho} = \delta^\gamma_\alpha\,\delta^\rho_\beta - \delta^\gamma_\beta\,\delta^\rho_\alpha \Rightarrow \varepsilon_{\alpha\beta}\,\varepsilon^{\gamma\beta} = \delta^\gamma_\alpha \quad (8.1.12)$$

(b) Derive the formulas (8.1.13):

$$|\mathbf{a}_1 \times \mathbf{a}_2| = \sqrt{\alpha}, \quad \mathbf{a}_1 \times \mathbf{a}_2 = \sqrt{\alpha}\,\mathbf{a}_3, \quad \mathbf{a}_\alpha \times \mathbf{a}_\beta = \varepsilon_{\alpha\beta}\mathbf{a}_3, \quad \mathbf{a}^\alpha \times \mathbf{a}^\beta = \varepsilon^{\alpha\beta}\mathbf{a}_3 \quad (8.1.13)_{1,2,3,4}$$
$$\mathbf{a}_3 \times \mathbf{a}_\alpha = \varepsilon_{\alpha\beta}\,\mathbf{a}^\beta, \quad \mathbf{a}_3 \times \mathbf{a}^\alpha = \varepsilon^{\alpha\beta}\mathbf{a}_\beta \quad (8.1.13)_{5,6}$$

Solution

(a) Proof of the formulas (8.1.12). From the definitions of the permutation symbols (1) it follows that:

$$\varepsilon_{\alpha\beta} = e_{\alpha\beta}\sqrt{\alpha} = e_{\alpha\beta3}\sqrt{\alpha}, \quad \varepsilon^{\alpha\beta} = e_{\alpha\beta}/\sqrt{\alpha} = e_{\alpha\beta3}/\sqrt{\alpha}, \quad e_{\alpha\beta k} = 0$$
for $k = 1$ or 2

From the formulas (6.4.23) and (1.1.19) it then follows that:

$$\varepsilon_{\alpha\beta k}\varepsilon^{\gamma\rho k} = \delta^\gamma_\alpha\delta^\rho_\beta - \delta^\gamma_\beta\delta^\rho_\alpha = \varepsilon_{\alpha\beta}\varepsilon^{\gamma\rho} \Rightarrow (8.1.12)_1$$
$$(8.1.12)_1 \Rightarrow \varepsilon_{\alpha\beta}\varepsilon^{\gamma\beta} = \delta^\gamma_\alpha\delta^\beta_\beta - \delta^\gamma_\beta\delta^\beta_\alpha = \delta^\gamma_\alpha 2 - \delta^\gamma_\alpha = \delta^\gamma_\alpha \Rightarrow (8.1.12)_2$$

(b) Derivation of the formulas (8.1.13).

$$|\mathbf{a}_1 \times \mathbf{a}_2|^2 = |\mathbf{a}_1|^2|\mathbf{a}_2|^2\sin^2(\mathbf{a}_1, \mathbf{a}_2) = a_{11}a_{22}\left[1 - \cos^2(\mathbf{a}_1, \mathbf{a}_2)\right]$$
$$= a_{11}a_{22}\left[1 - \frac{(a_{12})^2}{a_{11}a_{22}}\right] = a_{11}a_{22} - (a_{12})^2 = \det(a_{\alpha\beta}) = \alpha \Rightarrow$$

$$\mathbf{a}_1 \times \mathbf{a}_2 = \sqrt{\alpha}\,\mathbf{a}_3 \Rightarrow \mathbf{a}_\alpha \times \mathbf{a}_\beta = \sqrt{\alpha}\,e_{\alpha\beta}\mathbf{a}_3 = \varepsilon_{\alpha\beta}\,\mathbf{a}_3 \quad \Rightarrow \quad (8.1.13)_{1,2,3}$$

$$\mathbf{a}^1 \times \mathbf{a}^2 = (a^{1\alpha}\mathbf{a}_\alpha) \times (a^{1\beta}\mathbf{a}_\beta) = a^{1\alpha}a^{1\beta}\mathbf{a}_\alpha \times \mathbf{a}_\beta = a^{1\alpha}a^{1\beta}\varepsilon_{\alpha\beta}\,\mathbf{a}_3$$
$$= \left[a^{11}a^{22} - (a^{12})^2\right]\sqrt{\alpha}\,\mathbf{a}_3 \Rightarrow$$
$$\mathbf{a}^1 \times \mathbf{a}^2 = \det(a^{\alpha\beta})\sqrt{\alpha}\,\mathbf{a}_3 \quad (1)$$

Formula $(8.1.7)_3 \Rightarrow a_{\alpha\gamma}a^{\gamma\beta} = \delta_\alpha^\beta \Rightarrow \det(a_{\alpha\beta})\det(a^{\alpha\beta}) = 1 \Rightarrow \det(a^{\alpha\beta}) = 1/\det(a_{\alpha\beta})$
$= 1/\alpha$ (2)

The equations (1) and (2)
$\Rightarrow \mathbf{a}^1 \times \mathbf{a}^2 = \mathbf{a}_3/\sqrt{\alpha} \Rightarrow \mathbf{a}^\alpha \times \mathbf{a}^\beta = \varepsilon^{\alpha\beta}\mathbf{a}_3 \Rightarrow (8.1.13)_4$

The formulas $(8.1.11)_3$ and $(8.1.5)$, i.e. $\mathbf{a}^\alpha \cdot \mathbf{a}_i = \delta_i^\alpha$, yield:

$$\left.\begin{array}{l}(\mathbf{a}_3 \times \mathbf{a}_\alpha) \cdot \mathbf{a}_\beta = \mathbf{a}_3 \cdot (\mathbf{a}_\alpha \times \mathbf{a}_\beta) = \mathbf{a}_3 \cdot (\varepsilon_{\alpha\beta}\mathbf{a}_3) = \varepsilon_{\alpha\beta} \\ \varepsilon_{\alpha\rho}\mathbf{a}^\rho \cdot \mathbf{a}_\beta = \varepsilon_{\alpha\rho}\delta_\beta^\rho = \varepsilon_{\alpha\beta}\end{array}\right\} \Rightarrow \mathbf{a}_3 \times \mathbf{a}_\alpha = \varepsilon_{\alpha\beta}\mathbf{a}^\beta$$
$\Rightarrow (8.1.13)_5$

The formula $\mathbf{a}_3 \times \mathbf{a}^\alpha = \varepsilon^{\alpha\beta}\mathbf{a}_\beta$ $(8.1.13)_6$ is proved analogously.

Problem 8.2 The covariant components of the unit normal vector $\mathbf{a}_3$ in a general curvilinear coordinate system y are denoted by $\underset{\sim}{a}_i$ such that $\mathbf{a}_3 = \underset{\sim}{a}_i\mathbf{g}^i$. Derive the formula $(8.1.14)$:

$$\underset{\sim}{a}_i = \frac{1}{2}\varepsilon_{ijk}\frac{\partial y^j}{\partial u^\alpha}\frac{\partial y^k}{\partial u^\beta}\varepsilon^{\alpha\beta} \quad (8.1.14)$$

Solution The following formulas are applied:

$$\mathbf{a}_3\varepsilon_{\alpha\beta} = \mathbf{a}_\alpha \times \mathbf{a}_\beta = \varepsilon_{\alpha\beta}\mathbf{a}_3 \quad (8.1.13)_3, \quad \varepsilon_{\alpha\beta}\varepsilon^{\gamma\beta} = \delta_\alpha^\gamma \quad (8.1.12)$$

$$\mathbf{a}_\alpha = \mathbf{g}_i\frac{\partial y^i}{\partial u^\alpha} \quad (8.1.2), \quad \mathbf{g}_i \times \mathbf{g}_i = \varepsilon_{ijk}\mathbf{g}^k \quad (6.2.21)$$

We now find:

$$\varepsilon_{\alpha\beta}\underset{\sim}{a}_i\mathbf{g}^i = \varepsilon_{\alpha\beta}\mathbf{a}_3 = \mathbf{a}_\alpha \times \mathbf{a}_\beta = \left(\mathbf{g}_j\frac{\partial y^j}{\partial u^\alpha}\right) \times \left(\mathbf{g}_k\frac{\partial y^k}{\partial u^\beta}\right)$$

$$= \frac{\partial y^j}{\partial u^\alpha}\frac{\partial y^k}{\partial u^\beta}\varepsilon_{jki}\mathbf{g}^i \Rightarrow \varepsilon_{\alpha\beta}\underset{\sim}{a}_i = \frac{\partial y^j}{\partial u^\alpha}\frac{\partial y^k}{\partial u^\beta}\varepsilon_{jki} \Rightarrow$$

$$\varepsilon^{\alpha\beta}\varepsilon_{\alpha\beta}\underset{\sim}{a}_i = \frac{\partial y^j}{\partial u^\alpha}\frac{\partial y^k}{\partial u^\beta}\varepsilon_{jki}\varepsilon^{\alpha\beta} = \delta_\alpha^\alpha\underset{\sim}{a}_i \Rightarrow \underset{\sim}{a}_i$$

$$= \frac{1}{2}\varepsilon_{ijk}\frac{\partial y^j}{\partial u^\alpha}\frac{\partial y^k}{\partial u^\beta}\varepsilon^{\alpha\beta} \Leftrightarrow \quad (8.1.14)$$

Problem 8.3 Derive the formulas $(8.1.18)$–$(8.1.20)$.

$$\mathbf{a}_{3,\alpha} = -B_{\alpha\beta}\mathbf{a}^\beta, \quad \mathbf{a}^\alpha{}_{,\beta} = -\Gamma_{\gamma\beta}^\alpha\mathbf{a}^\gamma + B_{\gamma\beta}a^{\gamma\alpha}\mathbf{a}_3 \quad (8.1.18)$$
$$a_{\alpha\beta,\gamma} = \Gamma_{\alpha\gamma\beta} + \Gamma_{\beta\gamma\alpha}, \quad \Gamma_{\alpha\beta\gamma} = \frac{1}{2}(a_{\alpha\gamma,\beta} + a_{\beta\gamma,\alpha} - a_{\alpha\beta,\gamma}) \quad (8.1.19)$$
$$\Gamma_{\alpha\beta}^\alpha = \frac{1}{2\alpha}\alpha_{,\beta} = \frac{1}{\sqrt{\alpha}}(\sqrt{\alpha})_{,\beta} \quad (8.1.20)$$

Appendix: Problems with Solutions

Solution Derivation of the formulas (8.1.18).

We use the formula (8.1.15): $\mathbf{a}_{\alpha,\beta} = \Gamma^\gamma_{\alpha\beta}\mathbf{a}_\gamma + B_{\alpha\beta}\mathbf{a}_3$ to obtain:

$$\mathbf{a}_3 \cdot \mathbf{a}_\beta = 0 \Rightarrow \mathbf{a}_{3,\alpha} \cdot \mathbf{a}_\beta + \mathbf{a}_3 \cdot \mathbf{a}_{\beta,\alpha} = 0 \Rightarrow \mathbf{a}_{3,\alpha} \cdot \mathbf{a}_\beta$$

$$= -\mathbf{a}_3 \cdot \left[\Gamma^\gamma_{\beta\alpha}\mathbf{a}_\gamma + B_{\beta\alpha}\mathbf{a}_3\right] \Rightarrow \mathbf{a}_{3,\alpha} = B_{\alpha\beta}\mathbf{a}^\beta \Rightarrow (8.1.18)_1$$

To derive $(8.1.18)_2$ we again use the formula (8.1.15): $\mathbf{a}_{\alpha,\beta} = \Gamma^\gamma_{\alpha\beta}\mathbf{a}_\gamma + B_{\alpha\beta}\mathbf{a}_3$ to obtain:

$$\left.\begin{array}{l}\mathbf{a}^\alpha \cdot \mathbf{a}_\rho = \delta^\alpha_\rho \Rightarrow \mathbf{a}^\alpha_{,\beta} \cdot \mathbf{a}_\rho + \mathbf{a}^\alpha \cdot \mathbf{a}_{\rho,\beta} = 0 \Rightarrow \mathbf{a}^\alpha_{,\beta} \cdot \mathbf{a}_\rho = -\mathbf{a}^\alpha \cdot \left[\Gamma^\gamma_{\rho\beta}\mathbf{a}_\gamma + B_{\rho\beta}\mathbf{a}_3\right] = -\Gamma^\alpha_{\rho\beta} \\ \mathbf{a}^\alpha \cdot \mathbf{a}_3 = 0 \Rightarrow \mathbf{a}^\alpha_{,\beta} \cdot \mathbf{a}_3 + \mathbf{a}^\alpha \cdot \mathbf{a}_{3,\beta} = 0 \Rightarrow \mathbf{a}^\alpha_{,\beta} \cdot \mathbf{a}_3 = -\mathbf{a}^\alpha \cdot \left[-B_{\beta\gamma}\mathbf{a}^\gamma\right] = B_{\beta\gamma}a^{\alpha\gamma}\end{array}\right\} \Rightarrow$$

$$\mathbf{a}^\alpha = -\Gamma^\alpha_{\rho\beta}\mathbf{a}^\alpha + B_{\beta\gamma}a^{\alpha\gamma}\mathbf{a}_3 \Rightarrow (8.1.18)_2$$

Derivation of the formulas (8.1.19).

$$a_{\alpha\beta} = \mathbf{a}_\alpha \cdot \mathbf{a}_\beta \Rightarrow a_{\alpha\beta,\gamma} = \mathbf{a}_{\alpha,\gamma} \cdot \mathbf{a}_\beta + \mathbf{a}_\alpha \cdot \mathbf{a}_{\beta,\gamma} \Rightarrow$$

$$a_{\alpha\beta,\gamma} = \left(\Gamma^\rho_{\alpha\gamma}\mathbf{a}_\rho\right) \cdot \mathbf{a}_\beta + \mathbf{a}_\alpha \cdot \left(\Gamma^\rho_{\beta\gamma}\mathbf{a}_\rho\right) = \Gamma^\rho_{\alpha\gamma}a_{\rho\beta} + \Gamma^\rho_{\beta\gamma}a_{\alpha\rho} = \Gamma_{\alpha\gamma\beta} + \Gamma_{\beta\gamma\alpha} \quad \Rightarrow \quad (8.1.19)_1$$

$$a_{\alpha\gamma,\beta} + a_{\beta\gamma,\alpha} - a_{\alpha\beta,\gamma} = \left(\Gamma_{\alpha\beta\gamma} + \Gamma_{\gamma\beta\alpha}\right) + \left(\Gamma_{\beta\alpha\gamma} + \Gamma_{\gamma\alpha\beta}\right) - \left(\Gamma_{\alpha\gamma\beta} + \Gamma_{\beta\gamma\alpha}\right) = \Gamma_{\alpha\beta\gamma} + \Gamma_{\beta\alpha\gamma} = 2\Gamma_{\alpha\beta\gamma} \Rightarrow$$

$$\Gamma_{\alpha\beta\gamma} = \frac{1}{2}\left(a_{\alpha\gamma,\beta} + a_{\beta\gamma,\alpha} - a_{\alpha\beta,\gamma}\right) \quad \Rightarrow \quad (8.1.19)_2$$

Derivation of the formulas (8.1.20).

The formulas (1.1.25), (1.1.29), and (8.1.17) provide the results:

$$Co\, a_{\alpha\beta} = \frac{\partial\alpha}{\partial a_{\alpha\beta}}, \quad a^{\alpha\beta} = \frac{Co\, a_{\alpha\beta}}{\alpha}$$

Then we obtain:

$$\left.\begin{array}{l}\alpha_{,\beta} = \frac{\partial\alpha}{\partial a_{\alpha\gamma}}a_{\alpha\gamma,\beta} = \alpha a^{\alpha\gamma}a_{\alpha\gamma,\beta} = \alpha a^{\alpha\gamma}\left(\Gamma_{\alpha\beta\gamma} + \Gamma_{\gamma\beta\alpha}\right) = 2\alpha\,\Gamma^\alpha_{\alpha\beta} \\ (\sqrt\alpha)_{,\gamma} = \frac{d\sqrt\alpha}{d\alpha}\alpha_{,\gamma} = \frac{1}{2\sqrt\alpha}\alpha_{,\gamma}\end{array}\right\} \Rightarrow$$

$$\Gamma^\alpha_{\alpha\beta} = \frac{1}{2\alpha}\alpha_{,\beta} = \frac{1}{\sqrt\alpha}(\sqrt\alpha)_{,\beta} \Rightarrow (8.1.20)$$

Problem 8.4 Show that the symbols $\varepsilon_{\alpha\beta}$ and $\varepsilon^{\alpha\beta}$ defined by the formulas (8.1.11):

$$\varepsilon_{\alpha\beta} = e_{\alpha\beta}\sqrt\alpha, \quad \varepsilon^{\alpha\beta} = e_{\alpha\beta}/\sqrt\alpha, \quad (e_{\alpha\beta}) = \begin{pmatrix} 0 & 1 \\ -1 & 0 \end{pmatrix} \quad (8.1.11)$$

represent the components of a surface tensor: the permutation tensor for surfaces.

Solution Let be the components of the tensor in two different surface coordinate systems $\bar{u}$ and u be given by: the symbols $\bar{\varepsilon}_{\alpha\beta}$ and $\bar{\varepsilon}^{\alpha\beta}$ in the $\bar{u}$-system, and by $\varepsilon_{\gamma\rho}$ and $\varepsilon^{\gamma\rho}$ in the u-system. We need to show that:

$$\bar{\varepsilon}_{\alpha\beta} = \frac{\partial u^{\gamma}}{\partial \bar{u}^{\alpha}} \frac{\partial u^{\rho}}{\partial \bar{u}^{\beta}} \varepsilon_{\gamma\rho} \quad (1), \qquad \bar{\varepsilon}^{\alpha\beta} = \frac{\partial \bar{u}^{\alpha}}{\partial u^{\gamma}} \frac{\partial \bar{u}^{\beta}}{\partial u^{\rho}} \varepsilon^{\gamma\rho} \quad (2)$$

The fundamental parameters of first order in two surface coordinate systems u and $\bar{u}$ are related through the formulas:

$$\bar{a}_{\alpha\beta} = \frac{\partial u^{\gamma}}{\partial \bar{u}_{\alpha}} \frac{\partial u^{\rho}}{\partial \bar{u}_{\beta}} a_{\gamma\rho} \quad (3), \qquad \bar{a}^{\alpha\beta} = \frac{\partial \bar{u}^{\alpha}}{\partial u^{\gamma}} \frac{\partial \bar{u}^{\beta}}{\partial u^{\gamma}} a^{\gamma\rho} \quad (4)$$

Using the multiplication theorem (1.1.23) for determinants, we obtain from the formulas (3) and (4):

$$\bar{\alpha} = \det(\bar{a}_{\alpha\beta}) = \left[\det\left(\frac{\partial u^{\gamma}}{\partial \bar{u}^{\alpha}}\right)\right]^2 \det(a_{\lambda\rho}) = \left[\det\left(\frac{\partial u^{\gamma}}{\partial \bar{u}^{\alpha}}\right)\right]^2 \alpha \Rightarrow \sqrt{\bar{\alpha}} = \det\left(\frac{\partial u^{\gamma}}{\partial \bar{u}^{\alpha}}\right)\sqrt{\alpha} \quad (5)$$

$$\frac{1}{\bar{\alpha}} = \det(\bar{a}^{\alpha\beta}) = \left[\det\left(\frac{\partial \bar{u}^{\alpha}}{\partial u^{\gamma}}\right)\right]^2 \det(a^{\gamma\rho}) = \left[\det\left(\frac{\partial u^{\gamma}}{\partial \bar{u}^{\alpha}}\right)\right]^2 \frac{1}{\alpha} \Rightarrow \frac{1}{\sqrt{\bar{\alpha}}} = \det\left(\frac{\partial u^{\gamma}}{\partial \bar{u}^{\alpha}}\right)\frac{1}{\sqrt{\alpha}} \quad (6)$$

Applying the formulas (8.1.11) we obtain from the results (5) and (6):

$$\frac{\partial u^{\gamma}}{\partial \bar{u}^{\alpha}} \frac{\partial u^{\rho}}{\partial \bar{u}^{\beta}} \varepsilon_{\gamma\rho} = \frac{\partial u^{\gamma}}{\partial \bar{u}^{\alpha}} \frac{\partial u^{\rho}}{\partial \bar{u}^{\beta}} e_{\gamma\rho}\sqrt{\alpha} = \left[\frac{\partial u^1}{\partial \bar{u}^{\alpha}} \frac{\partial u^2}{\partial \bar{u}^{\beta}} - \frac{\partial u^2}{\partial \bar{u}^{\alpha}} \frac{\partial u^1}{\partial \bar{u}^{\beta}}\right]\sqrt{\alpha}$$

$$= \det\left(\frac{\partial u^{\gamma}}{\partial \bar{u}^{\rho}}\right) e_{\alpha\beta}\sqrt{\alpha} = e_{\alpha\beta}\sqrt{\bar{\alpha}} = \bar{\varepsilon}_{\alpha\beta} \Rightarrow \bar{\varepsilon}_{\alpha\beta} = \frac{\partial u^{\gamma}}{\partial \bar{u}^{\alpha}} \frac{\partial u^{\rho}}{\partial \bar{u}^{\beta}} \varepsilon_{\gamma\rho} \Rightarrow (1)$$

$$\frac{\partial \bar{u}^{\alpha}}{\partial u^{\gamma}} \frac{\partial \bar{u}^{\beta}}{\partial u^{\gamma}} \varepsilon^{\gamma\rho} = \frac{\partial \bar{u}^{\alpha}}{\partial u^{\gamma}} \frac{\partial \bar{u}^{\beta}}{\partial u^{\rho}} e_{\gamma\rho}/\sqrt{\alpha} = \left[\frac{\partial \bar{u}^{\alpha}}{\partial u^1} \frac{\partial \bar{u}^{\beta}}{\partial u^2} - \frac{\partial \bar{u}^{\alpha}}{\partial u^2} \frac{\partial \bar{u}^{\beta}}{\partial u^1}\right]\frac{1}{\sqrt{\alpha}}$$

$$= \det\left(\frac{\partial \bar{u}^{\rho}}{\partial u^{\gamma}}\right) e_{\alpha\beta}\frac{1}{\sqrt{\alpha}} = e_{\alpha\beta}\frac{1}{\sqrt{\bar{\alpha}}} = \bar{\varepsilon}^{\alpha\beta} \Rightarrow \bar{\varepsilon}^{\alpha\beta} = \frac{\partial \bar{u}^{\alpha}}{\partial u^{\gamma}} \frac{\partial \bar{u}^{\beta}}{\partial u^{\gamma}} \varepsilon^{\gamma\rho} \Rightarrow (2)$$

Problem 8.5 Let S be a symmetric surface tensor field of second order. Show that the angle ϕ between a principal direction **b** corresponding to the principal value σ, and the base vector $\mathbf{a}_1$ in a surface coordinate system u is given by the formula:

$$\tan\phi = \frac{\sigma - S_1^1}{S_1^2} \frac{a_{11}}{\sqrt{\alpha}} - \frac{a_{12}}{\sqrt{\alpha}} \quad (8.4.13)$$

Solution From Fig. 8.6 we obtain:

$$\left.\begin{array}{l} \sin\phi = \mathbf{b}\cdot\dfrac{\mathbf{a}^2}{\sqrt{a^{22}}} = (b_\alpha\mathbf{a}^\alpha)\cdot\dfrac{\mathbf{a}^2}{\sqrt{a^{22}}} = \dfrac{b_1 a^{12} + b_2 a^{22}}{\sqrt{a^{22}}} \\[2mm] \cos\phi = \mathbf{b}\cdot\dfrac{\mathbf{a}_1}{\sqrt{a_{11}}} = (b_\alpha\mathbf{a}^\alpha)\cdot\dfrac{\mathbf{a}_1}{\sqrt{a_{11}}} = \dfrac{b_1}{\sqrt{a_{11}}} \end{array}\right\} \Rightarrow \tan\phi = \dfrac{\sin\phi}{\cos\phi}$$

$$= \left(a^{12} + \frac{b_2}{b_1}a^{22}\right)\sqrt{\frac{a_{11}}{a^{22}}} \quad (1)$$

From the formulas (8.1.9) and (8.4.8) we get:

$$\sqrt{\frac{a_{11}}{a^{22}}} = \sqrt{\alpha}, \quad a^{12} = -\frac{a_{12}}{\alpha}, \quad a^{22} = \frac{a_{11}}{\alpha} \quad (2),$$

$$\sigma b_1 = S_1^1 b_1 + S_1^2 b_2 \Rightarrow \frac{b_2}{b_1} = \frac{\sigma - S_1^1}{S_1^2} \quad (3)$$

The results (1), (2), and (3) combine to: $\tan\phi = \dfrac{\sigma - S_1^1}{S_1^2}\dfrac{a_{11}}{\sqrt{\alpha}} - \dfrac{a_{12}}{\sqrt{\alpha}} \Rightarrow$ (8.4.13)

Problem 8.6 Show that for a geodesic coordinate system u for a point P on a surface $y^i(u)$ the following hold true:

$$a_{\alpha\beta,\gamma} = a^{\alpha\beta}{}_{,\gamma} = \alpha_{,\alpha} = \varepsilon_{\alpha\beta,\gamma} = \varepsilon^{\alpha\beta}{}_{,\gamma} = 0 \quad \text{in } P \quad (8.5.29)$$

Solution Because the Christoffel symbols are zero in the pole P for a geodesic coordinate system u, it follows from the formulas (8.1.9) and (8.1.20) that:

$$a_{\alpha\beta,\gamma} = 0, \quad \alpha_{,\alpha} = 0 \text{ in } P \quad (1)$$

From the formulas (8.1.7) and (8.1.11) we write:

$$a_{\alpha\gamma}a^{\beta\gamma} = \delta_\alpha^\beta, \quad \varepsilon_{\alpha\beta} = e_{\alpha\beta}\sqrt{\alpha}, \quad \varepsilon^{\alpha\beta} = e_{\alpha\beta}/\sqrt{\alpha} \quad (2)$$

The results (1) and (2) yield the formulas (8.5.29).

Problem 8.7 Use the results (8.5.29) to prove that the formulas (8.5.30).

$$a_{\alpha\beta,\gamma} = a^{\alpha\beta}{}_{,\gamma} = \alpha_{,\alpha} = \varepsilon_{\alpha\beta,\gamma} = \varepsilon^{\alpha\beta}{}_{,\gamma} = 0 \quad (8.5.29)$$

$$a_{\alpha\beta}\big|_\gamma = a^{\alpha\beta}\big|_\gamma = \varepsilon_{\alpha\beta}\big|_\gamma = \varepsilon^{\alpha\beta}\big|_\gamma = 0, \quad \frac{\delta a_{\alpha\beta}}{\delta s} = \frac{\delta a^{\alpha\beta}}{\delta s} = \frac{\delta \varepsilon_{\alpha\beta}}{\delta s} = \frac{\delta \varepsilon^{\alpha\beta}}{\delta s} = 0 \quad (8.5.30)$$

Solution In the special geodesic coordinate system u for the pole P the formulas (8.5.29) represent the tensor equations:

$$a_{\alpha\beta}\big|_\gamma = a^{\alpha\beta}\big|_\gamma = \varepsilon_{\alpha\beta}\big|_\gamma = \varepsilon^{\alpha\beta}\big|_\gamma = 0 \quad (1)$$

and must therefore be valid in any general coordinate system u on the surface $y^i(u)$. The formulas (8.5.22) imply that:

The results (1) and (2) prove the formulas 8.5.30).

Problem 8.8 Develop the formulas (8.5.31) for the covariant derivatives of the components of a second order tensor **D**:

$$D_{\alpha\beta}\big|_{\gamma} = D_{\alpha\beta,\gamma} - D_{\lambda\beta}\Gamma^{\lambda}_{\alpha\gamma} - D_{\alpha\lambda}\Gamma^{\alpha}_{\beta\gamma}, \quad D^{\alpha}_{\beta}\big|_{\gamma} = D^{\alpha}_{\beta,\gamma} + D^{\lambda}_{\beta}\Gamma^{\alpha}_{\lambda\gamma} - D^{\alpha}_{\lambda}D_{\alpha\lambda}\Gamma^{\lambda}_{\beta\gamma} \quad (8.5.31)$$

Solution Formula $(8.5.31)_1$ will be developed by the following procedure. For any two coordinate systems u and $\bar{u}$ in the surface the components $D_{\rho\lambda}$ and $\bar{D}_{\alpha\beta}$ of the tensor **D** are related through the formulas:

$$\bar{D}_{\alpha\beta} = \frac{\partial u^{\rho}}{\partial \bar{u}^{\alpha}} \frac{\partial u^{\lambda}}{\partial \bar{u}^{\beta}} D_{\rho\lambda}$$

It follows that:

$$\frac{\partial \bar{D}_{\alpha\beta}}{\partial \bar{u}^{\gamma}} = \frac{\partial u^{\rho}}{\partial \bar{u}^{\alpha}} \frac{\partial u^{\lambda}}{\partial \bar{u}^{\beta}} \frac{\partial D_{\rho\lambda}}{\partial u^{\phi}} \frac{\partial u^{\phi}}{\partial \bar{u}^{\gamma}} + \frac{\partial^2 u^{\rho}}{\partial \bar{u}^{\gamma}\partial \bar{u}^{\alpha}} \frac{\partial u^{\lambda}}{\partial \bar{u}^{\beta}} D_{\rho\lambda} + \frac{\partial u^{\rho}}{\partial \bar{u}^{\alpha}} \frac{\partial^2 u^{\lambda}}{\partial \bar{u}^{\gamma}\partial \bar{u}^{\beta}} D_{\rho\lambda} \quad (1)$$

Let the system $\bar{u}$ be a geodesic coordinate system in a point P in the surface. Then according to the formulas (8.3.14) the result (1) gives in point P:

$$\bar{D}_{\alpha\beta,\lambda} \equiv \frac{\partial \bar{D}_{\alpha\beta}}{\partial \bar{u}^{\gamma}} = \delta^{\rho}_{\alpha}\delta^{\lambda}_{\beta} \frac{\partial D_{\rho\lambda}}{\partial u^{\phi}}\delta^{\phi}_{\gamma} - \Gamma^{\rho}_{\alpha\gamma}\delta^{\lambda}_{\beta}D_{\rho\lambda} - \delta^{\rho}_{\alpha}\Gamma^{\lambda}_{\beta\gamma}D_{\rho\lambda}$$

$$= \frac{\partial D_{\alpha\beta}}{\partial u^{\gamma}} - \Gamma^{\lambda}_{\alpha\gamma}D_{\lambda\beta} - \Gamma^{\lambda}_{\beta\gamma}D_{\alpha\lambda} \quad \text{in } P \quad (2)$$

The components $\bar{D}_{\alpha\beta,\lambda}$ in Eq. (2) represent the covariant derivatives $\bar{D}_{\alpha\beta}\big|_{\lambda}$ of the tensor components $\bar{D}_{\alpha\beta}$ in P in the system $\bar{u}$. The covariant derivatives $D_{\alpha\beta}\big|_{\lambda}$ in the point P for the system u are obtained from the transformation formulas:

$$D_{\alpha\beta}\big|_{\gamma} = \frac{\partial \bar{u}^{\rho}}{\partial u^{\alpha}} \frac{\partial \bar{u}^{\lambda}}{\partial u^{\beta}} \frac{\partial \bar{u}^{\phi}}{\partial u^{\gamma}} \bar{D}_{\rho\lambda}\big|_{\phi} = \delta^{\rho}_{\alpha}\delta^{\lambda}_{\alpha}\delta^{\phi}_{\alpha}\bar{D}_{\rho\lambda,\phi} = \bar{D}_{\alpha\beta,\lambda} \quad \text{in } P \quad (3)$$

The combination of the formulas (2) and (3) gives the result $(8.5.31)_1$.

Formula $(8.5.31)_2$ will be developed by another method. Let **c** be any surface vector field and define the surface vector field $\mathbf{b} = \mathbf{D} \cdot \mathbf{c} \Rightarrow b^{\alpha} = D^{\alpha}_{\beta}c^{\beta}$. The relation $b^{\alpha} = D^{\alpha}_{\beta}c^{\beta}$ may be differentiated in two ways:

$$b^{\alpha}\big|_{\gamma} = D^{\alpha}_{\beta}\big|_{\gamma}c^{\beta} + D^{\alpha}_{\beta}c^{\beta}\big|_{\gamma}, \quad b^{\alpha}{}_{,\gamma} = D^{\alpha}_{\beta,\gamma}c^{\beta} + D^{\alpha}_{\beta}c^{\beta}{}_{,\gamma} \quad (4)$$

The formulas (8.5.3) yield:

$$b^\alpha\big|_\gamma = b^\alpha,_{\gamma} + b^\lambda\Gamma^\alpha_{\lambda\gamma}, \quad c^\beta\big|_\gamma = c^\beta,_{\gamma} + c^\lambda\Gamma^\beta_{\lambda\gamma} \quad (5)$$

From the Eqs. (4) and (5) we obtain:

$$b^\alpha\big|_\gamma = \left[D^\alpha_{\beta,\gamma}\, c^\beta + D^\alpha_\beta c^\beta,_{\gamma}\right] + b^\lambda\Gamma^\alpha_{\lambda\gamma} = D^\alpha_\beta\big|_\gamma c^\beta + D^\alpha_\beta\left[c^\beta,_{\gamma} + c^\lambda\Gamma^\beta_{\lambda\gamma}\right] \Rightarrow$$

$$D^\alpha_\beta\big|_\gamma c^\beta - D^\alpha_{\beta,\gamma}\, c^\beta - D^\alpha_\beta c^\beta \Gamma^\lambda_{\lambda\gamma} + D^\alpha_\lambda c^\beta \Gamma^\lambda_{\beta\gamma} = 0 \Rightarrow \left[D^\alpha_\beta\big|_\gamma - D^\alpha_{\beta,\gamma} - D^\alpha_\beta \Gamma^\lambda_{\lambda\gamma} + D^\alpha_\lambda \Gamma^\lambda_{\beta\gamma}\right]c^\beta = 0$$

Because the result is to be true for any choice of the vector **c**, the term in the brackets [] must be zero:

$$D^\alpha_\beta\big|_\gamma - D^\alpha_{\beta,\gamma} - D^\lambda_\beta\Gamma^\alpha_{\lambda\gamma} + D^\alpha_\lambda\Gamma^\lambda_{\beta\gamma} = 0 \Rightarrow D^\alpha_\beta\big|_\gamma = D^\alpha_{\beta,\gamma} + D^\lambda_\beta\Gamma^\alpha_{\lambda\gamma} - D^\alpha_\lambda\Gamma^\lambda_{\beta\gamma}$$
$$\Rightarrow (8.5.31)_2$$

Problem 8.9 Show that: $g_{ij}\big|_\alpha = g^{ij}\big|_\alpha = 0$, $\varepsilon_{ijk}\big|_\alpha = \varepsilon^{ijk}\big|_\alpha = 0$ (8.5.32)
Solution In a Cartesian coordinate y-system and a geodesic u-system:

$$g_{ij} = g^{ij} = \delta_{ij}, \quad \varepsilon_{ijk} = \varepsilon^{ijk} = e_{ijk} \Rightarrow$$
$$g_{ij}\big|_\alpha = g^{ij}\big|_\alpha = \frac{\partial(\delta_{ij})}{\partial u^\alpha} = 0, \quad \varepsilon_{ijk}\big|_\alpha = \varepsilon^{ijk}\big|_\alpha = \frac{\partial(e_{ijk})}{\partial u^\alpha} = 0 \quad (1)$$

Since the formulas (1) represent tensor equations the formulas also hold true in space coordinate systems y and surface coordinate systems u. This argument proves the formulas (8.5.32).

Problem   8.10 Derive   the   formulas: $B_{\alpha\beta} = \frac{\partial y^i}{\partial u^\alpha}\big|_\beta \underset{\sim}{a}_i \Leftrightarrow$
$\frac{\partial y^i}{\partial u^\alpha}\big|_\beta = B_{\alpha\beta}\, \underset{\sim}{a}^i$ (8.5.33)

Solution The following formulas will be used:

$$(8.1.2) \Rightarrow \mathbf{a}_\alpha = \mathbf{g}_i\frac{\partial y^i}{\partial u^\alpha}, \quad (8.5.26) \Rightarrow \mathbf{g}_i\big|_\alpha = \mathbf{0}, \quad (8.5.28) \Rightarrow \mathbf{a}_\alpha\big|_\beta = B_{\alpha\beta}\mathbf{a}_3,$$
$$\text{Definition}: \mathbf{a}_3 = \underset{\sim}{a}_i\mathbf{g}^i$$

We find:

$$\mathbf{a}_\alpha\big|_\beta = \mathbf{g}_i\big|_\beta\frac{\partial y^i}{\partial u^\alpha} + \mathbf{g}_i\frac{\partial y^i}{\partial u^\alpha}\big|_\beta = \mathbf{g}_i\frac{\partial y^i}{\partial u^\alpha}\big|_\beta = B_{\alpha\beta}\mathbf{a}_3 = B_{\alpha\beta}\,\underset{\sim}{a}^i\mathbf{g}_i \Rightarrow \frac{\partial y^i}{\partial u^\alpha}\big|_\beta = B_{\alpha\beta}\,\underset{\sim}{a}^i \Rightarrow (8.5.33)_2$$

$$(8.5.33)_2 \Rightarrow \frac{\partial y^i}{\partial u^\alpha}\big|_\beta\underset{\sim}{a}_i = B_{\alpha\beta}\,\underset{\sim}{a}^i\underset{\sim}{a}_i \Rightarrow B_{\alpha\beta} = \frac{\partial y^i}{\partial u^\alpha}\big|_\beta\underset{\sim}{a}_i \Rightarrow (8.5.33)_1$$

Problem 8.11 Let be $\phi(y)$ scalar field on a surface $y^i(u)$. Show that:

$$\phi|_{\alpha\beta} = \phi|_{\beta\alpha} \quad (8.5.38)$$

Solution By definition: $\phi|_{\alpha} = \phi_{,\alpha}$. Using the formula $(8.5.3)_2$ and the symmetry of the Christoffel symbols we get:

$$\phi|_{\alpha\beta} = (\phi|_{\alpha})_{,\beta} - \phi|_{\gamma}\Gamma^{\gamma}_{\alpha\beta} = \phi_{,\alpha\beta} - \phi_{,\gamma}\Gamma^{\gamma}_{\alpha\beta} = \phi|_{\beta\alpha} \Rightarrow (8.5.38)$$

Problem 8.12 Show that for the second covariant derivatives of the components $D_{\alpha\beta}$ of a second order surface tensor **D**:

$$D_{\alpha\beta}|_{\gamma\delta} - D_{\alpha\beta}|_{\delta\gamma} = D_{\sigma\beta}R^{\sigma}_{\alpha\gamma\delta} + D_{\alpha\sigma}R^{\sigma}_{\beta\gamma\delta}, \quad D_{\alpha\beta}|_{\gamma\delta} = D_{\alpha\beta}|_{\delta\gamma} \quad \text{only when } \mathbf{R} = \mathbf{0}$$
$$(8.5.39)$$

Solution Let **c** by any surface vector field and define the surface vector field $\mathbf{b} = \mathbf{D} \cdot \mathbf{c} \Rightarrow b_{\alpha} = D_{\alpha\beta}c^{\beta}$. It follows that:

$$b_{\alpha}|_{\gamma} = D_{\alpha\beta}|_{\gamma}c^{\beta} + D_{\alpha\beta}c^{\beta}|_{\gamma}, \quad b_{\alpha}|_{\gamma}|_{\delta} \equiv b_{\alpha}|_{\gamma\delta}$$

$$= D_{\alpha\beta}|_{\gamma\delta}c^{\beta} + D_{\alpha\beta}|_{\gamma}c^{\beta}|_{\delta} + D_{\alpha\beta}|_{\delta}c^{\beta}|_{\gamma} + D_{\alpha\beta}c^{\beta}|_{\gamma\delta}$$

$$\Rightarrow \left[D_{\alpha\beta}|_{\gamma\delta} - D_{\alpha\beta}|_{\delta\gamma}\right]c^{\beta}$$

$$= \left[b_{\alpha}|_{\gamma\delta} - b_{\alpha}|_{\delta\gamma}\right] - D_{\alpha\beta}\left[c^{\beta}|_{\gamma\delta} - c^{\beta}|_{\delta\gamma}\right]$$

$$5 - \left[D_{\alpha\beta}|_{\gamma}c^{\beta}|_{\delta} + D_{\alpha\beta}|_{\delta}c^{\beta}|_{\gamma} - D_{\alpha\beta}|_{\delta}c^{\beta}|_{\gamma} - D_{\alpha\beta}|_{\gamma}c^{\beta}|_{\delta}\right] \Rightarrow \quad (1)$$

Using the formula (8.5.36) and the result $R^{\sigma}_{\beta\gamma\delta} = -R\sigma_{\beta\gamma\delta}$ from the formulas (8.5.40) we can write:

$$b_{\alpha}|_{\gamma\delta} - b_{\alpha}|_{\delta\gamma} = b_{\sigma}R^{\sigma}_{\alpha\gamma\delta} = D_{\sigma\beta}c^{\beta}R^{\sigma}_{\alpha\gamma\delta} \quad (2)$$

$$D_{\alpha\beta}\left[c^{\beta}|_{\gamma\delta} - c^{\beta}|_{\delta\gamma}\right] = D_{\alpha\beta}c^{\sigma}R^{\beta}_{\sigma\gamma\delta} = -D_{\alpha\sigma}c^{\beta}R^{\sigma}_{\beta\gamma\delta} \quad (3)$$

By combining the two formulas (1), (2), and (2) we obtain:

$$\left[D_{\alpha\beta}|_{\gamma\delta} - D_{\alpha\beta}|_{\delta\gamma} - D_{\sigma\beta}R^{\sigma}_{\alpha\gamma\delta} - D_{\alpha\sigma}R^{\sigma}_{\beta\gamma\delta}\right]c^{\beta} = 0$$

Because the result is to be true for any choice of the vector **c**, the term in the brackets [] must be zero:

$$D_{\alpha\beta}|_{\gamma\delta} - D_{\alpha\beta}|_{\delta\gamma} - D_{\sigma\beta}R^{\sigma}_{\alpha\gamma\delta} - D_{\alpha\sigma}R^{\sigma}_{\beta\gamma\delta} = 0 \Rightarrow$$
$$D_{\alpha\beta}|_{\gamma\delta} - D_{\alpha\beta}|_{\delta\gamma} = D_{\sigma\beta}R^{\sigma}_{\alpha\gamma\delta} + D_{\alpha\sigma}R^{\sigma}_{\beta\gamma\delta}, D_{\alpha\beta}|_{\gamma\delta} = D_{\alpha\beta}|_{\delta\gamma} \text{ only when } \mathbf{R} = \mathbf{0} \Leftrightarrow (8.5.39)$$

Problem 8.13 Prove the formula for the Gauss curvature γ of a surface $y^i(u)$:

$$\gamma = \frac{1}{4} \varepsilon^{\delta\alpha} \varepsilon^{\beta\gamma} R_{\delta\alpha\beta\gamma} \quad (8.5.42)$$

Solution We use the formulas:

$(8.1.11) \Rightarrow \varepsilon_{\alpha\beta} = e_{\alpha\beta}\sqrt{\alpha}, \quad \varepsilon^{\alpha\beta} = e_{\alpha\beta}/\sqrt{\alpha}, \quad (8.1.12) \Rightarrow \varepsilon_{\alpha\beta}\varepsilon^{\gamma\beta} = \delta_\alpha^\gamma$

$(8.5.41) \Rightarrow R_{\delta\alpha\beta\lambda} = \gamma \varepsilon_{\delta\alpha}\varepsilon_{\beta\gamma}$

Then we obtain:

$$\left(\varepsilon^{\delta\alpha}\varepsilon^{\beta\gamma}\right)\left(R_{\delta\alpha\beta\lambda}\right) = \left(\gamma \varepsilon_{\delta\alpha}\varepsilon_{\beta\gamma}\right)\left(\varepsilon^{\delta\alpha}\varepsilon^{\beta\gamma}\right) \Rightarrow \varepsilon^{\delta\alpha}\varepsilon^{\beta\gamma}R_{\delta\alpha\beta\lambda} = \gamma\delta_\alpha^\alpha\delta_\beta^\beta = \gamma \cdot 2 \cdot 2 \Rightarrow$$
$$\gamma = \tfrac{1}{4}\varepsilon^{\delta\alpha}\varepsilon^{\beta\gamma}R_{\delta\alpha\beta\lambda} \Leftrightarrow (8.5.42)$$

Problem 8.14 Derive Weingarten's formula: $\underset{\sim}{a}^i\big|_\alpha = -B_\alpha^\beta \frac{\partial y^i}{\partial u^\beta}$ (8.5.45)

Solution From the formulas:

$$\mathbf{a}_3 = \underset{\sim}{a}^i \mathbf{g}_i, \quad (8.1.18) \Rightarrow \mathbf{a}_{3,\alpha} = -B_{\alpha\beta}\mathbf{a}^\beta = -B_\alpha^\beta \mathbf{a}_\beta, \quad (8.5.26) \Rightarrow \mathbf{g}_j\big|_\alpha = \mathbf{0},$$

$$(8.1.3) \Rightarrow \mathbf{a}_\beta \cdot \mathbf{g}^i = \frac{\partial y^i}{\partial u^\beta}$$

we obtain:

$$\mathbf{a}_{3,\alpha} = \underset{\sim}{a}^i\big|_\alpha \mathbf{g}_i + \underset{\sim}{a}^i \mathbf{g}_i\big|_\alpha = \underset{\sim}{a}^i\big|_\alpha \mathbf{g}_i = -B_\alpha^\beta \mathbf{a}_\beta \Rightarrow$$

$$\underset{\sim}{a}^i\big|_\alpha \mathbf{g}_i \cdot \mathbf{g}^j = -B_\alpha^\beta \mathbf{a}_\beta \cdot \mathbf{g}^j \quad \Rightarrow \quad \underset{\sim}{a}^i\big|_\alpha = -B_\alpha^\beta \frac{\partial y^i}{\partial u^\beta} \quad \Rightarrow \quad (8.5.45)$$

Problem 8.15 Let b be the space vector: $\mathbf{b} = b_\beta \mathbf{a}^\beta + b_3 \mathbf{a}_3$. Derive the equation:

$$\mathbf{b}_{,\alpha} = \left(b_\beta\big|_\alpha - b_3 B_{\beta\alpha}\right)\mathbf{a}^\beta + \left(b_{3,\alpha} + b_\beta B_\alpha^\beta\right)\mathbf{a}_3 \quad (8.5.46)$$

Solution We introduce a surface vector $\mathbf{c} = b_\beta \mathbf{a}^\beta \mathbf{c}$ such that $\mathbf{b} = \mathbf{c} + b_3 \mathbf{a}_3$. Then we use the result (8.5.4) to obtain:

$$\mathbf{b}_{,\alpha} = \mathbf{c}_{,\alpha} + b_{3,\alpha}\,\mathbf{a}_3 + b_3\,\mathbf{a}_{3,\alpha} = c_\beta\big|_\alpha \mathbf{a}^\beta + c_\beta \mathbf{a}^\beta\big|_\alpha + b_{3,\alpha}\,\mathbf{a}_3 + b_3\,\mathbf{a}_{3,\alpha}$$
$$= b_\beta\big|_\alpha \mathbf{a}^\beta + b_\beta \mathbf{a}^\beta\big|_\alpha + b_{3,\alpha}\,\mathbf{a}_3 + b_3\,\mathbf{a}_{3,\alpha}$$

Now we use formula (8.6.18) to write: $\mathbf{a}_{3,\alpha} = -B_{\alpha\beta}\mathbf{a}^\beta$ and obtain the result:

$$\mathbf{b}_{,\alpha} = \left(b_\beta\big|_\alpha - b_3 B_{\beta\alpha}\right)\mathbf{a}^\beta + \left(b_{3,\alpha} + b_\beta B_\alpha^\beta\right)\mathbf{a}_3 \Leftrightarrow (8.5.46)$$

Problem 8.16 Show that the fundamental parameters of third order $C_{\alpha\beta}$, of second order $B_{\alpha\beta}$, and of first order $a_{\alpha\beta}$ of a surface $y^i(u)$ satisfy the equation:

$$C_{\alpha\beta} - 2\mu B_{\alpha\beta} + \gamma a_{\alpha\beta} = 0 \quad (8.7.18)$$

μ is the mean curvature and γ is the Gauss curvature of the surface $y^i(u)$.

Solution First we obtain from the formula (8.1.2):

$$\mathbf{a}_\alpha \cdot \mathbf{a}_\beta = \left(\frac{\partial y^i}{\partial u^\alpha}\mathbf{g}_i\right) \cdot \left(\frac{\partial y^j}{\partial u^\beta}\mathbf{g}_j\right) = \frac{\partial y^i}{\partial u^\alpha}\frac{\partial y^j}{\partial u^\beta}g_{ij} = a_{\alpha\beta} \Rightarrow \frac{\partial y^i}{\partial u^\alpha}\frac{\partial y^j}{\partial u^\beta}g_{ij} = a_{\alpha\beta} \quad (1)$$

The formulas (8.7.16), (8.5.45), and (1) now yield:

$$C_{\alpha\beta} = \underset{\sim}{a}^i\big|_\alpha \underset{\sim}{a}_i\big|_\beta = \left(-B_\alpha^\rho \frac{\partial y^i}{\partial u^\rho}\right)\left(-B_\beta^\gamma \frac{\partial y^j}{\partial u^\gamma}g_{ij}\right) = B_\alpha^\rho B_\beta^\gamma a_{\rho\gamma} \Rightarrow C_\alpha^\beta = B_\alpha^\rho B_\rho^\beta \Rightarrow$$
$$\mathbf{C} = \mathbf{B}^2$$

In a surface coordinate system $\bar{u}$ with base vectors parallel to the principal directions of the tensors $\mathbf{B}$ and $\mathbf{C}$, the tensor matrices are:

$$\bar{B} = \begin{pmatrix} \tilde{\kappa}_1 & 0 \\ 0 & \tilde{\kappa}_2 \end{pmatrix}, \quad \bar{C} = \bar{B}^2 = \begin{pmatrix} \tilde{\kappa}_1^2 & 0 \\ 0 & \tilde{\kappa}_2^2 \end{pmatrix}$$

From the characteristic equation (8.7.8) of the tensor $\mathbf{B}$: $\tilde{\kappa}^2 - 2\mu\tilde{\kappa} + \gamma = 0$, we obtain the matrix equation: $C - 2\mu B + \gamma 1 = 0$. This is a matrix representation of the tensor equation: $\mathbf{C} - 2\mu\mathbf{B} + \gamma\mathbf{1} = \mathbf{0}$. In any surface coordinate system u one component representation of this tensor equation is (8.7.18).

Index

© Springer Nature Switzerland AG 2019
F. Irgens, *Tensor Analysis*,
https://doi.org/10.1007/978-3-030-03412-2

Printed by Printforce, the Netherlands